METHODS IN MOLECULAR BIOLOGY™

Series Editor
John M. Walker
School of Life Sciences
University of Hertfordshire
Hatfield, Hertfordshire, AL10 9AB, UK

For other titles published in this series, go to
www.springer.com/series/7651

DRILL HALL LIBRARY
MEDWAY

RNA Detection and Visualization

Methods and Protocols

Edited by

Jeffrey E. Gerst

Department of Molecular Genetics, Weizmann Institute of Science, Rehovot, Israel

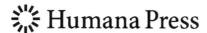

Editor
Jeffrey E. Gerst, Ph.D.
Department of Molecular Genetics
Weizmann Institute of Science
Rehovot
Israel
jeffrey.gerst@weizmann.ac.il

ISSN 1064-3745   e-ISSN 1940-6029
ISBN 978-1-61779-004-1   e-ISBN 978-1-61779-005-8
DOI 10.1007/978-1-61779-005-8
Springer New York Dordrecht Heidelberg London

Library of Congress Control Number: 2011924684

© Springer Science+Business Media, LLC 2011
All rights reserved. This work may not be translated or copied in whole or in part without the written permission of the publisher (Humana Press, c/o Springer Science+Business Media, LLC, 233 Spring Street, New York, NY 10013, USA), except for brief excerpts in connection with reviews or scholarly analysis. Use in connection with any form of information storage and retrieval, electronic adaptation, computer software, or by similar or dissimilar methodology now known or hereafter developed is forbidden.
The use in this publication of trade names, trademarks, service marks, and similar terms, even if they are not identified as such, is not to be taken as an expression of opinion as to whether or not they are subject to proprietary rights.
While the advice and information in this book are believed to be true and accurate at the date of going to press, neither the authors nor the editors nor the publisher can accept any legal responsibility for any errors or omissions that may be made. The publisher makes no warranty, express or implied, with respect to the material contained herein.

Printed on acid-free paper

Humana Press is part of Springer Science+Business Media (www.springer.com)

Preface

Of all biological molecules, RNA has perhaps the most complex and extensively regulated metabolism. Beginning with control at the transcriptional level, RNAs undergo nuclear processing, export, transport and localization, translation, storage, silencing and, ultimately, decay, all of which are mediated by cohorts of interacting proteins and additional RNA molecules. However, as most of the RNA life cycle actually occurs within the cytoplasm, it is imperative to have tools that allow for the localization of RNAs to be observed either in an *in situ* setting or, preferably, under *in vivo* conditions. In addition, there is a requisite need for tools that allow for both RNA purification from cells and the identification of bound interacting proteins and accessory RNAs. Finally, bioinformatic tools are requisite for mining the sequence and structural motifs necessary for RNAs to localize. When used in combination, these approaches can allow researchers to ask complex questions regarding where and how RNAs localize, and whether localization pertains to translational control, protein localization and complex assembly, and RNA stability.

The aim of this volume is to provide an up-to-date and in-depth description of basic methods and protocols used for detecting and visualizing mRNAs in both fixed and live cells, from bacteria to mammals. Written for novices and experts alike, this mix of classic *in situ* hybridization and advanced live imaging techniques, cell fractionation and affinity purification procedures, and bioinformatic tools is to give researchers the most complete and extensive array of research aids possible. This volume is composed of twenty eight chapters separated into six subject areas: (1) Visualizing mRNAs in situ; (2) Visualizing mRNAs *in vivo* using molecular probes or reconstituted fluorescent proteins; (3) Visualizing mRNAs *in vivo* using aptamers and intact fluorescent proteins; (4) Use of cell fractionation to demonstrate the subcellular localization of RNA; (5) Affinity purification of mRNAs and the identification of *trans*-acting factors; and (6) Use of bioinformatics to identify *cis*-acting motifs and structures in RNAs. Chapters have been contributed by some of the best and brightest investigators, established or young, that have made an impact in the study of RNA localization and its importance in biological processes. Finally, I offer my sincere gratitude to the authors who have endeavored to make this book a unique and extensive collection of useful RNA detection methods and protocols for the benefit of other researchers.

Rehovot, Israel   *Jeffrey E. Gerst*

Contents

Preface v
Contributors xi

PART I VISUALIZING mRNAs IN SITU

1 Single Molecule Imaging of RNA In Situ 3
  Mona Batish, Arjun Raj, and Sanjay Tyagi

2 FISH and Immunofluorescence Staining in *Chlamydomonas* 15
  James Uniacke, Daniel Colón-Ramos, and William Zerges

3 High Resolution Fluorescent In Situ Hybridization in *Drosophila* 31
  Eric Lécuyer

4 Localization and Anchorage of Maternal mRNAs to Cortical Structures
  of Ascidian Eggs and Embryos Using High Resolution In Situ Hybridization 49
  Alexandre Paix, Janet Chenevert, and Christian Sardet

5 Visualization of mRNA Localization in *Xenopus* Oocytes 71
  James A. Gagnon and Kimberly L. Mowry

6 Visualization of mRNA Expression in the Zebrafish Embryo 83
  Yossy Machluf and Gil Levkowitz

7 High-Resolution Fluorescence In Situ Hybridization to Detect mRNAs
  in Neuronal Compartments In Vitro and In Vivo 103
  Sharon A. Swanger, Gary J. Bassell, and Christina Gross

8 Localization of mRNA in Vertebrate Axonal Compartments
  by In Situ Hybridization 125
  *José Roberto Sotelo-Silveira, Aldo Calliari, Alejandra Kun,
  Victoria Elizondo, Lucía Canclini, and José Roberto Sotelo*

PART II VISUALIZING mRNAs IN VIVO USING MOLECULAR
         PROBES OR RECONSTITUTED FLUORESCENT PROTEINS

9 Tiny Molecular Beacons for *in vivo* mRNA Detection 141
  Diana P. Bratu, Irina E. Catrina, and Salvatore A.E. Marras

10 Delivery of Molecular Beacons for Live-Cell Imaging and Analysis of RNA 159
   Antony K. Chen, Won Jong Rhee, Gang Bao, and Andrew Tsourkas

11 Genetically-Encoded Fluorescent Probes for Imaging Endogenous
   mRNA in Living Cells 175
   Takeaki Ozawa and Yoshio Umezawa

12 Visualization of Induced RNA in Single Bacterial Cells 189
   Azra Borogovac and Natalia E. Broude

Part III Visualizing mRNAs In Vivo Using Aptamers and Intact Fluorescent Proteins

13 Visualizing mRNAs in Fixed and Living Yeast Cells..................... 203
    Franck Gallardo and Pascal Chartrand

14 In Vivo Visualization of RNA Using the U1A-Based Tagged RNA System 221
    Sunglan Chung and Peter A. Takizawa

15 Visualizing Endogenous mRNAs in Living Yeast Using m-TAG, a PCR-Based
    RNA Aptamer Integration Method, and Fluorescence Microscopy 237
    Liora Haim-Vilmovsky and Jeffrey E. Gerst

16 Imaging mRNAs in Living Mammalian Cells 249
    Sharon Yunger and Yaron Shav-Tal

17 Using the mRNA-MS2/MS2CP-FP System to Study mRNA
    Transport During *Drosophila* Oogenesis 265
    Katsiaryna Belaya and Daniel St Johnston

Part IV Use of Cell Fractionation to Demonstrate the Sub-Cellular Localization of RNA

18 Genome-Wide Analysis of RNA Extracted from Isolated Mitochondria......... 287
    Erez Eliyahu, Daniel Melamed, and Yoav Arava

19 Analyzing mRNA Localization to the Endoplasmic Reticulum via
    Cell Fractionation.. 301
    Sujatha Jagannathan, Christine Nwosu, and Christopher V. Nicchitta

20 Isolation of mRNAs Encoding Peroxisomal Proteins from Yeast
    Using a Combined Cell Fractionation and Affinity Purification Procedure....... 323
    Gadi Zipor, Cecile Brocard, and Jeffrey E. Gerst

21 Profiling Axonal mRNA Transport 335
    Dianna E. Willis and Jeffery L. Twiss

22 RNA Purification from Tumor Cell Protrusions Using Porous
    Polycarbonate Filters 353
    Jay Shankar and Ivan R. Nabi

Part V Affinity Purification of mRNAs and the Identification of Trans-Acting Factors

23 RNA-Binding Protein Immunopurification-Microarray (RIP-Chip)
    Analysis to Profile Localized RNAs................................... 369
    Alessia Galgano and André P. Gerber

24 RaPID: An Aptamer-Based mRNA Affinity Purification Technique
    for the Identification of RNA and Protein Factors Present
    in Ribonucleoprotein Complexes 387
    Boris Slobodin and Jeffrey E. Gerst

25 RIP: An mRNA Localization Technique 407
    *Sabarinath Jayaseelan, Francis Doyle, Salvatore Currenti,
    and Scott A. Tenenbaum*

26 The Dual Use of RNA Aptamer Sequences for Affinity Purification
    and Localization Studies of RNAs and RNA–Protein Complexes............ 423
    Scott C. Walker, Paul D. Good, Theresa A. Gipson, and David R. Engelke

PART VI   USE OF BIOINFORMATICS TO IDENTIFY CIS-ACTING MOTIFS
          AND STRUCTURES IN RNAS

27 Identifying and Searching for Conserved RNA Localisation Signals............ 447
    Russell S. Hamilton and Ilan Davis
28 Computational Prediction of RNA Structural Motifs Involved
    in Post-Transcriptional Regulatory Processes........................... 467
    Michal Rabani, Michael Kertesz, and Eran Segal

Index *481*

Contributors

Yoav Arava • *Technion – Israel Institute of Technology, Haifa, Israel*
Gang Bao • *Department of Biomedical Engineering, Georgia Institute of Technology and Emory University, Atlanta, GA, USA*
Gary J. Bassell • *Departments of Cell Biology and Neurology, Emory University School of Medicine, Atlanta, GA, USA*
Mona Batish • *Public Health Research Institute Center, New Jersey Medical School, University of Medicine and Dentistry of New Jersey, Newark, NJ, USA*
Katsiaryna Belaya • *The Gurdon Institute, University of Cambridge, Cambridge, UK*
Azra Borogovac • *Department of Biomedical Engineering, Center for Advanced Biotechnology, College of Engineering, Boston University, Boston, MA, USA*
Diana P. Bratu • *Biological Sciences Department, Hunter College, City University of New York, New York, NY, USA*
Cecile Brocard • *Max F. Perutz Laboratories, University of Vienna, Vienna, Austria*
Natalia E. Broude • *Department of Biomedical Engineering, Center for Advanced Biotechnology, College of Engineering, Boston University, Boston, MA, USA*
Aldo Calliari • *Department of Proteins and Nucleic Acids, Instituto de Investigaciones Biológicas Clemente Estable (IIBCE), Av. Italia 3318, CP 11600, Montevideo, Uruguay; Biophysics Area, School of Veterinary Sciences, UdelaR, Lasplaces 1550, Montevideo, Uruguay*
Lucía Canclini • *Department of Proteins and Nucleic Acids, Instituto de Investigaciones Biológicas Clemente Estable (IIBCE), Av. Italia 3318, CP 11600, Montevideo, Uruguay*
Irina E. Catrina • *Biological Sciences Department, Hunter College, City University of New York, New York, NY, USA*
Pascal Chartrand • *Département de Biochimie, Université de Montréal, Montréal, QC, Canada*
Antony K. Chen • *Department of Bioengineering, University of Pennsylvania, Philadelphia, PA, USA*
Janet Chenevert • *Observatoire Océanologique, UPMC University of Paris 06, CNRS, UMR 7009, Villefranche-sur-mer, France*
Sunglan Chung • *Department of Cell Biology, Yale University School of Medicine, New Haven, CT, USA*
Daniel Colón-Ramos • *Yale Program in Cellular Neuroscience, Neurodegeneration and Repair, Yale University School of Medicine, New Haven, CT, USA*
Salvatore Currenti • *College of Nanoscale Science and Engineering, Nanobioscience Constellation, University at Albany-SUNY, Albany, NY, USA*
Ilan Davis • *Department of Biochemistry, University of Oxford, Oxford, UK*
Francis Doyle • *College of Nanoscale Science and Engineering, Nanobioscience Constellation, University at Albany-SUNY, Albany, NY, USA*

EREZ ELIYAHU • Technion – Israel Institute of Technology, Haifa, Israel
VICTORIA ELIZONDO • Department of Proteins and Nucleic Acids, Instituto de Investigaciones Biológicas Clemente Estable (IIBCE), Av. Italia 3318, CP 11600, Montevideo, Uruguay
DAVID R. ENGELKE • Department of Biological Chemistry, University of Michigan, Ann Arbor, MI, USA
JAMES A. GAGNON • Department of Molecular Biology, Cell Biology and Biochemistry, Brown University, Providence, RI, USA
ALESSIA GALGANO • Department of Chemistry and Applied Biosciences, Institute of Pharmaceutical Sciences, ETH Zurich, Zurich, Switzerland
FRANCK GALLARDO • Département de Biochimie, Université de Montréal, Montréal, QC, Canada
ANDRÉ P. GERBER • Department of Chemistry and Applied Biosciences, Institute of Pharmaceutical Sciences, ETH Zurich, Zurich, Switzerland
JEFFREY E. GERST • Department of Molecular Genetics, Weizmann Institute of Science, Rehovot, Israel
THERESA A. GIPSON • Department of Biological Chemistry, University of Michigan, Ann Arbor, MI, USA
PAUL D. GOOD • Department of Biological Chemistry, University of Michigan, Ann Arbor, MI, USA
CHRISTINA GROSS • Department of Cell Biology, Emory University School of Medicine, Atlanta, GA, USA
LIORA HAIM-VILMOVSKY • Department of Molecular Genetics, Weizmann Institute of Science, Rehovot, Israel
RUSSELL S. HAMILTON • Department of Biochemistry, University of Oxford, Oxford, UK
SUJATHA JAGANNATHAN • Department of Cell Biology, Duke University Medical Center, Durham, NC, USA
SABARINATH JAYASEELAN • College of Nanoscale Science and Engineering, Nanobioscience Constellation, University at Albany-SUNY, Albany, NY, USA
MICHAEL KERTESZ • Department of Computer Science and Applied Mathematics, Weizmann Institute of Science, Rehoboth, Israel
ALEJANDRA KUN • Department of Proteins and Nucleic Acids, Instituto de Investigaciones Biológicas Clemente Estable (IIBCE), Av. Italia 3318, CP 11600, Montevideo, Uruguay; Department of Cell and Molecular Biology, School of Sciences, Universidad de la República(UdelaR), Igua 4225, Montevideo, Uruguay
ERIC LÉCUYER • Institut de Recherches Cliniques de Montréal (IRCM), Ouest Montréal, QC, Canada
GIL LEVKOWITZ • Department of Molecular Cell Biology, Weizmann Institute of Science, Rehovot, Israel
YOSSY MACHLUF • Department of Molecular Cell Biology, Weizmann Institute of Science, Rehovot, Israel
SALVATORE A.E. MARRAS • Department of Microbiology and Molecular Genetics, Public Health Research Institute, New Jersey Medical School, University of Medicine and Dentistry of New Jersey, Newark, NJ, USA
DANIEL MELAMED • Faculty of Biology, Technion – Israel Institute of Technology, Haifa, Israel

KIMBERLY L. MOWRY • *Department of Molecular Biology, Cell Biology and Biochemistry, Brown University, Providence, RI, USA*
IVAN R. NABI • *Department of Cellular and Physiological Sciences, Life Sciences Institute, University of British Columbia, Vancouver, BC, Canada*
CHRISTOPHER V. NICCHITTA • *Department of Cell Biology, Duke University Medical Center, Durham, NC, USA*
CHRISTINE NWOSU • *Department of Cell Biology, Duke University Medical Center, Durham, NC, USA*
TAKEAKI OZAWA • *Department of Chemistry, Graduate School of Science, The University of Tokyo, Tokyo, Japan; Japan Science and Technology Agency, Saitama, Japan*
ALEXANDRE PAIX • *Observatoire Océanologique, UPMC University of Paris 06, CNRS, UMR 7009, Villefranche-sur-mer, France; University of Nice Sophia-Antipolis, Nice, France*
MICHAL RABANI • *Department of Computer Science and Applied Mathematics, Weizmann Institute of Science, Rehoboth, Israel*
ARJUN RAJ • *Department of Bioengineering, University of Pennsylvania, Philadelphia, PA, USA*
WON JONG RHEE • *Department of Biomedical Engineering, Georgia Institute of Technology and Emory University, Atlanta, GA, USA*
CHRISTIAN SARDET • *Observatoire Océanologique, UPMC University of Paris 06, CNRS, UMR 7009 Villefranche-sur-mer, France*
ERAN SEGAL • *Department of Computer Science and Applied Mathematics, Weizmann Institute of Science, Rehoboth, Israel*
JAY SHANKAR • *Department of Cellular and Physiological Sciences, Life Sciences Institute, University of British Columbia, Vancouver, BC, Canada*
YARON SHAV-TAL • *The Mina and Everard Goodman Faculty of Life Sciences and Institute of Nanotechnology, Bar-Ilan University, Ramat Gan, Israel*
BORIS SLOBODIN • *Department of Molecular Genetics, Weizmann Institute of Science, Rehovot, Israel*
JOSÉ ROBERTO SOTELO • *Department of Proteins and Nucleic Acids, Instituto de Investigaciones Biológicas Clemente Estable (IIBCE), Av. Italia 3318, CP 11600, Montevideo, Uruguay*
JOSÉ ROBERTO SOTELO-SILVEIRA • *Department of Proteins and Nucleic Acids, Instituto de Investigaciones Biológicas Clemente Estable (IIBCE), Av. Italia 3318, CP 11600, Montevideo, Uruguay; Department of Cell and Molecular Biology, School of Sciences. Universidad de la República (UdelaR), Igua 4225, Montevideo, Uruguay*
DANIEL ST JOHNSTON • *The Gurdon Institute, University of Cambridge, Cambridge, UK*
SHARON A. SWANGER • *Department of Cell Biology, Emory University School of Medicine, Atlanta, GA, USA*
PETER A. TAKIZAWA • *Department of Cell Biology, Yale University School of Medicine, New Haven, CT, USA*
SCOTT A. TENENBAUM • *Department of Nanobioscience, University at Albany, Albany, NY, USA*

ANDREW TSOURKAS • *Department of Bioengineering, University of Pennsylvania, Philadelphia, PA, USA*

JEFFERY L. TWISS • *Department of Biology, Drexel University, Philadelphia, PA, USA*

SANJAY TYAGI • *Public Health Research Institute Center, New Jersey Medical School, University of Medicine and Dentistry of New Jersey, Newark, NJ, USA*

YOSHIO UMEZAWA • *Research Institute of Pharmaceutical Sciences, Musashino University, Tokyo, Japan*

JAMES UNIACKE • *Biology Department, Concordia University, Montreal, QC, Canada*

SCOTT C. WALKER • *Department of Biological Chemistry, University of Michigan, Ann Arbor, MI, USA*

DIANNA E. WILLIS • *Burke-Cornell Medical Research Institute, White Plains, NY, USA*

SHARON YUNGER • *The Mina and Everard Goodman Faculty of Life Sciences and Institute of Nanotechnology, Bar-Ilan University, Ramat Gan, Israel*

WILLIAM ZERGES • *Biology Department, Concordia University, Montreal, QC, Canada*

GADI ZIPOR • *Department of Molecular Genetics, Weizmann Institute of Science, Rehovot, Israel*

Part I

Visualizing mRNAs In Situ

Chapter 1

Single Molecule Imaging of RNA In Situ

Mona Batish, Arjun Raj, and Sanjay Tyagi

Abstract

This protocol describes a method to image individual mRNA molecules in situ. About 50 oligonucleotides complementary to different regions of a target mRNA species are used simultaneously. Each probe is labeled with a single fluorescent moiety. When these probes bind to their target, each mRNA molecule becomes so intensely fluorescent that it can be seen as a fine fluorescent spot. Several different mRNA species can be detected in multiplex imaging using differently colored probe sets for each species. An automated image-processing program is used to count the number of mRNA molecules of each species that are expressed in each cell, thus yielding single-cell gene expression profiles.

Key words: mRNA visualization, FISH, Single molecule detection, In situ hybridization, MATLAB image processing, Gene expression profiling

1. Introduction

In situ hybridization provides powerful insights into the mechanisms of RNA synthesis and intracellular distribution. Recently new procedures have been developed that permit detection of individual molecules of RNAs in single cells (1–3). High signal strengths required for these procedures comes from attaching the target RNA with so many fluorescent dye molecules that each RNA molecule becomes visible as a diffraction-limited fluorescent spot. Robert Singer and colleagues first introduced an effective procedure to accomplish single molecule detection of mRNAs in fixed cultured cells (1). They hybridized about five 50-nucleotide long oligonucleotides to the same RNA target. Each of these oligonucleotides was coupled to five fluorescent dyes. Since these probes were smaller than the traditional RNA probes, they could better penetrate the cell matrix and the secondary structure of RNA compared to conventional RNA probes.

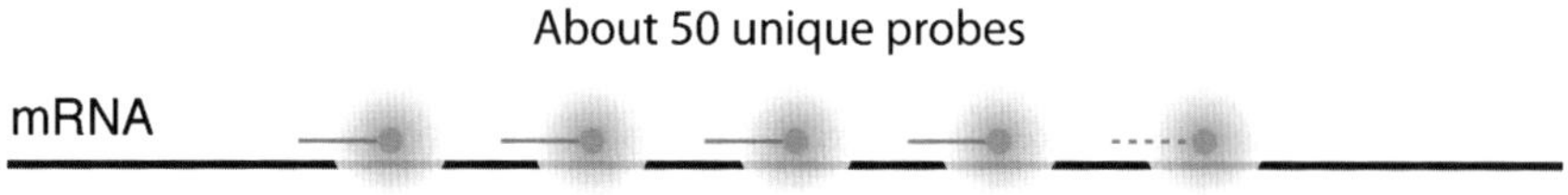

About 50 unique probes

mRNA

Fig. 1. Schematic representation of the principle of single molecule FISH method. A set of about 50 oligonucleotides, each labeled with a fluorescent dye and complementary to different region, are hybridized to the target RNA. The binding all of these probes to the same mRNA molecule makes it so intensely fluorescent that it can be seen as a fine spot in a fluorescence microscope.

Reliable labeling and purification of oligonucleotides at multiple internal sites is, however, difficult and expensive. Furthermore, when dyes are placed in close proximity they quench each other. With the availability of high-throughput DNA synthesis it is cheaper and more convenient to use many smaller singly labeled probes. We showed that about 50 singly labeled oligonucleotides yield fluorescent spots of sufficient intensity such that they can be detected and counted by automated algorithms (3) (Fig. 1). Although each oligonucleotide is synthesized separately, they can be coupled to fluorophores and purified as a single pool. A key to the success of this approach is that while all or most of the 50 probes bind to legitimate targets to produce intense spot like signals, the background generated by occasional probe binding to nonspecific sites is diffused and not particulate.

A number of experiments have been done to demonstrate the specificity and single molecule sensitivity of this approach. It was shown that the Chinese hamster ovary (CHO) cells expressing a foreign gene exhibit spot like signals while the cells not having, or expressing, the foreign gene show no spots. In the early drosophila embryos the spots are visible only in the subcellular zones in which the RNA is expected to localize. The inducible mRNAs were visible only upon induction and the level of induction measured by real-time PCR matches with the level of induction measured by counting spots. The number of molecules per cell measured by real-time PCR is similar to the number of molecules measured by spot counting. When two regions of the same mRNA are targeted by a different probe set, the spots from the two probe sets colocalize; however, when the target regions of these probe sets are segregated due to RNA processing events, the colocalization is lost (3–5). Effectiveness of this procedure in imaging single mRNA molecules of many different species in specimens ranging from bacteria, yeast, cultured mammalian cells, early embryos of fruit fly and worms, to adult worms has been demonstrated (3).

In this protocol we describe the synthesis and purification of probes, their hybridization with RNA in cultured cells, imaging and our particle detection, and counting algorithm. In addition, we describe a powerful approach towards prolonging the life of fluorophores during imaging which is important to accomplish multiplex detection of several mRNA species from the same cell.

2. Materials

2.1. Equipment

1. High-pressure liquid chromatograph (HPLC) equipped with a reverse-phase C-18 column and a dual wavelength detector.
2. Cell culture hood and 37°C CO_2 incubator.
3. A wide-field fluorescence microscope equipped with a mercury lamp, cooled CCD camera, appropriate filter sets, a high numerical aperture (>1.3) 100× objective and an image acquisition computer and software (see Note 1).
4. MATLAB software with image processing toolbox (The Math Works, Natick, MA).

2.2. Reagents

Make all buffers and reagents with RNase free DEPC treated water.

2.2.1. Labeling and Purification of Probes

1. Amino reactive dyes (see Note 2).
2. N,N-Dimethylformamide (DMF).
3. Ethanol.
4. 1 M Sodium bicarbonate, pH 8.0.
5. 3 M Sodium acetate, pH 5.2.
6. Buffer A (0.1 M triethylammonium acetate buffer pH 6).
7. Buffer B (0.1 M triethylammonium acetate in 70% v/v HPLC grade acetonitrile, pH 6).
8. 0.2 μm Filters (National Scientific, Rockwood, TN, Catalogue Number F2517-1).

2.2.2. Cell Culture of Hippocampal Neurons

1. Hippocampi dissected from E18 embryo (Brain Bits, Springfield, IL).
2. Neurobasal medium supplemented with B27 and glutamine (Invitrogen, Invitrogen, Carlsbad, CA).
3. 12-Well Falcon sterile culture dishes (Becton Dickinson, Franklin Lake, NJ).
4. Sterile glass cover slips (Fisher Scientific, Logan, UT, Catalogue Number 12-545-10018CIR-1); clean by shaking with 70% ethanol overnight, wash with 100% ethanol, and then expose to UV for 1 h in the tissue culture hood.
5. Poly-D-lysine (Sigma, St. Louis, IL Catalogue Number P6407). Make 100 μg/ml stock solution and store in aliquots at −20°C.
6. Laminin (Sigma, Catalogue Number L2020). Make 10 μg/ml stock solution and store in aliquots at −20°C.

2.2.3. Fixation and Hybridization	1. 37% Formaldehyde (Sigma, Catalogue Number, F1635). Prepare 3.7% formaldehyde in 1× PBS in fume hood.
2. 1× PBS.
3. 70% Ethanol.
4. Deionized formamide (Ambion, Austin, TX). Formamide is a teratogen that is easily absorbed through the skin and should be used in a chemical fume hood. Be sure to warm the formamide bottle to room temperature before opening.
5. Hybridization buffer: 10% Dextran sulfate (w/v) (Sigma, D8906), 1 µg/µl *Escherichia coli* tRNA (Sigma, R8759), 2 mM Vanadyl ribonucleoside complex (New England Biolab, Ipswich, MA, Catalogue Number, S142), 0.02% RNase free BSA (Ambion, AM2618), 10% formamide. Filter using 0.2 µm filters and store at −20°C in aliquots.
2.2.4. Washing and Mounting Cover Slips	1. Slides (Fisher Scientific, Catalogue Number 12-544-7).
2. Clear transparent finger nail polish.
3. RNase free water, DEPC treated.
4. Washing buffer, 10% formamide and 2× SSC.
5. Equilibration buffer, 0.4% glucose (w/v) and 2× SSC.
6. Glucose oxidase, Type VII from *Aspergillus niger* (Sigma, Catalogue Number G2133-10). Dilute to 3.7 mg/ml in water for stock and store at 4°C.
7. Catalase preparation from *A. niger* (Sigma, Catalogue Number C3515). Store at 4°C.
8. Deoxygenated mounting medium. Prepare just before use. Mix 100 µl equilibration buffer with 1 µl glucose oxidase stock solution and 1 µl catalase suspension. This solution can be used for 1 day when kept at 4°C.

3. Methods

3.1. Design of Probes

Since we use about fifty 17–22 nt oligonucleotide as probes, the length of the target RNA should ideally be about 850 nt or higher. However, when the target mRNA is smaller, it is still possible to obtain reliable signals from as few as 30 probes. When the target mRNA is longer than 850 nt, it is possible to optimize the location of probes with respect to potentially important criteria such as lack of strong hairpins and very high or very low GC content. We have developed a publicly accessible computer program (available at http://www.singlemoleculefish.com) that generates list of probe sequences from a supplied target mRNA sequence. The default parameters of this program are 17–22 nt for probe length, 2 nt gap between adjacent probe-binding sites, and GC content of each probe optimized to about 45%. The program also displays a graphical representation of where the probes will bind on the target based on the optimal binding properties. The program can be customized to get desired number, length, and GC content of the probes.

The designed oligonucleotides should be synthesized with an amino group at their 3′ ends for the subsequent coupling of fluorophore. We use Biosearch Technologies (Novato, CA, USA) for the oligonucleotide synthesis; however, any other manufacturers will be able to make oligonucleotides of acceptable quality. It is economical to order the oligonucleotides in a 96-well plate format at 25 nM (or smaller) scale. The oligonucleotides should be dissolved in water at equimolar concentrations.

3.2. Coupling of Probes with Fluorophore

It is desirable to choose fluorophores that are bright and photostable and are optimally excited and detected using the available light source and the filter set. Photostability is important because in order to obtain the optical slices, cells are repeatedly exposed with the light of high flux. Our recommendation for the primary fluorophore is tetramethylrhodamine (TMR), which, in addition to being bright and photostable, emits away from the blue and green zone of spectrum in which many cells autofluoresce. For the secondary fluorophores that can be used for multiplex imaging, we recommend Alexa 594 and Cy5. Although extremely photolabile under normal conditions, they perform as well as TMR when used with deoxygenated mounting medium described below. The emission spectra of TMR, Alexa 594, and Cy5 are so well separated that reliable imaging of three mRNA in the same image is possible. Obtain amino reactive versions (succinimidyl ester or thiocyanate) of the dye of your choice (Common sources of such dyes are Invitrogen and Amersham Bioscience, Pittsburgh, PA):

1. Pool equal volumes of all 48 oligonucleotides together. Alcohol precipitate the mixture by adding one-tenth volume of 3 M sodium acetate and 2.5 volume of alcohol, incubating at −20°C for at least 1 h, and then pelleting the DNA by centrifugation. Wash the pellet with 70% alcohol. Dissolve in 200 μl of 0.1 M sodium bicarbonate buffer.

2. Dissolve a small amount (~0.1–1.0 mg) dye in 10 μl DMF. It is not necessary to weigh the dye; just pick the approximate amount with a dry pipette tip and then dissolve it into DMF. Add 200 μl of 0.1 M sodium bicarbonate buffer to the dye and then add the mixture to the solution of oligonucleotide from the previous step mix and incubate at room temperature overnight. Keep the dyes in a desiccator at −20°C.

3. Ethanol precipitate the oligonucleotides and wash the colored pellet with 70% alcohol. This removes most of the free dye. Dissolve in 400 μl buffer and filter using 0.2 μm filter.

3.3. Purification of Coupled Probes

The pellet contains a mixture of coupled and uncoupled oligonucleotides and some free dye. Most fluorophores being hydrophobic will cause an increase in the hydrophobicity of the oligonucleotides upon coupling. Therefore on a reverse-phase column, coupled oligonucleotides will be retained longer than the uncoupled oligonucleotides. All 48 different oligonucleotides in the pool usually elute in one broad peak, which is sometimes split into multiple sub peaks, all of which should be collected:

1. Load all of the material on the HPLC column equilibrated with 10% buffer B. Initiate a program that would linearly raise the proportion of buffer B to 70% over 10 min. Monitor and record the absorption at 260 nm and at a wavelength at which the dye absorbs maximally.

2. Generally, it will result in two major peaks, separated by several minutes depending upon the fluorophore used. The first peak corresponds to uncoupled oligonucleotides, which will be detected only at 260 nm (DNA absorption). The second peak will represent pure coupled probes and will be detected by both channels. In case the second main peak resolves into sub-peaks, collect all of them, as they may represent oligonucleotides that have different retention times due to their sequence (Fig. 2).

3. Pool all fractions corresponding to the coupled oligonucleotides and alcohol precipitate as before.

4. Dissolve the pellet in water, measure concentration using a spectrophotometer, and save in aliquots at –20°C.

Pause point: The probes can be stored for years until ready to use.

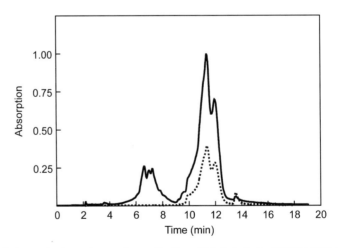

Fig. 2. A sample chromatogram for purification of labeled oligonucleotides from unlabeled ones. The continuous trace is for absorption at 260 nm and the broken trace is for absorption at the wavelength of maximum absorption of the coupled dye. The group of peaks on the right corresponds to the labeled oligonucleotides and the group on the left corresponds to unlabeled ones.

3.4. Cell Culture

This method is highly versatile and can be used for large variety of biological specimens including yeast, bacteria, worms, early drosophila embryos, cultured mammalian cells, and tissue sections. Here we will detail the procedure for culture and fixation of primary rat neurons cells:

1. Place clean sterile cover slips in wells of a 12-well plate and then add 1 ml of coating solution (1× PBS, 100 µg/ml poly-D-lysine and 10 µg/ml laminin in a ratio of 5:3:2). After 30 min, remove the coating solution and wash twice with sterile water (see Note 3).

2. Isolate hippocampus from E18 rat embryos and dissociate the tissue carefully to get the primary neurons. To dissociate neurons pipette 20 times using a sterile 1 ml pipette tip followed by five rounds of pipetting with 200 µl pipette tip. The plating density should optimally be in the range of 1,500–5,000 cells/cm^2. Culture the dissociated neurons on coated cover slips in neurobasal medium supplemented with B27 and glutamine (6).

3. Allow the cells to grow undisturbed for 5 days. Replace two-third of medium by fresh medium after 5 days and every third day afterwards. The neurons should be grown for 10–15 days in vitro before fixation. Too dense and too old neurons do not yield good quality images.

3.5. Fixation

1. Aspirate the culture medium off and wash the cells twice with 1× PBS. In this and all the subsequent steps add enough volume of the buffer (1–2 ml depending upon the well size) so that the cover slips remain submerged. Do not let the cells dry.

2. Add 4% formaldehyde made in 1× PBS to the wells containing cover slips and leave for 10 min at room temperature. This cross links the proteins and traps the RNAs in the protein matrix.

3. Remove formaldehyde, wash with 1× PBS, and add 70% ethanol to each cover slip and keep in at 4°C for 2 h. This step permeablilizes the cells so that the probes can freely permeate the cell.

Pause point: The cells can be left in 70% ethanol at 4°C overnight or for a few days.

3.6. Hybridization

Our protocol is derived from Singer lab protocol (http://singerlab.org/protocols/insitu_mammalian.htm) with an important modification; since we use smaller probes, we decrease the formamide concentration in the hybridization and washes to 10% (see Note 4). Of other two significant parameters, temperature and probe concentration, we keep the former fixed at 37°C, and the latter is varied from 0.1 to 1 ng/μl. The starting probe concentration should be 1 ng/μl, but in the case where high background is observed, the probe concentration can be decreased first to 0.3 ng/μl and then to 0.1 ng/μl:

1. Take 10 × 10 cm piece of parafilm and press it against a glass plate with the paper side up and the parafilm side in contact with the glass, rubbing with your palm so that parafilm sticks to the glass. Peel away the paper layer exposing the clean parafilm surface.

2. Add appropriate amount of probes to the hybridization solution (50 μl for each cover slip). The probe stock should be concentrated so that the hybridization solution is not diluted substantially upon the addition of probes. Mix the viscous solution well by pipetting several times. Spin the contents to remove bubbles.

3. Replace the 70% alcohol from the wells containing the cover slips with the washing buffer and allow the cover slips to equilibrate for at least 2 min.

4. Place 50 μl droplets of this solution on the parafilm. Pick the cover slips with fine forceps, wipe the underside against a tissue, and remove any excess buffer. Place the cover slips over the droplets of the hybridization mixtures with cell side facing the liquid. Do not let any bubbles form.

5. Place a flat object in a 37°C bath making a platform that rises above the water level. Place the glass plate with cover slips over the platform and cover it with a tissue culture plate lid to prevent moisture drops from falling over the cover slips. Close the bath and incubate over night.

3.7. Washing

1. Float the cover slips by adding droplets of washing buffer around the edge of each cover slip on the parafilm. In a few moments the liquid is drawn in by capillary action and the cover slip begins to float.

2. Pick the cover slips, remove any excess hybridization solution, and place them in the wells containing washing buffer with the cell-side facing up.

3. Leave the dish shaking gently on a rotary shaker for 30 min at room temperature (~25°C). Change the wash medium once followed by another 30 min wash.

3.8. Mounting

1. Replace the washing buffer from the wells holding the cover slips with equilibration buffer and allow the cover slips to equilibrate for at least 2 min.

2. Place a 10 μl droplet of deoxygenated mounting solution (see Note 5) on a clean glass slide. Pick the cover slip from the

well, wipe its underside on a tissue, and place the cover slip over the droplet with cells facing down. Remove excess medium by laying several layers of tissue over the cover slip and pressing gently.

3. Apply a thin coat of nail polish around the edges of the cover slip, while ensuring that the cover slip does not slide and the nail polish does not enter into the medium underneath the cover slip. The entry of nail polish solvent will cause high background signal (see Note 6).

3.9. Imaging

1. Obtain 10–30 z-sections with 0.2 μm spacing with the 100× oil objective. The range of z-section should cover the entire thickness of the cell.

2. The spot size would be from 0.25 to 0.30 μm (see Note 7). The exposure times range from 0.2 to 2 s. Figure 3 shows a typical set of images that are obtained.

3.10. Image Processing

Simplest form of data for presentation is maximum intensity merges of z-sections that are appropriately contrasted. In addition, using our MATLAB program it is possible to identify the spots corresponding to individual mRNA molecules and count them. The program takes input of optical slices as TIFF images, removes pixel noise, and enhances spots by passing them through a Laplacian of Gaussian filter, presents the user a 3-D representation of spots so that a threshold can be entered, overlays circles corresponding to the identified spots on the TIFF image, and provides a count of all the molecules identified (see Note 8).

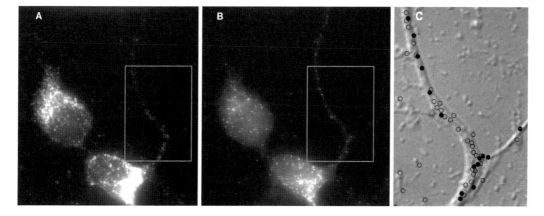

Fig. 3. An example of single molecule FISH imaging. Rat Hippocampal neurons were imaged with respect to β-actin mRNA (**a**) and *MAP2* mRNA (**b**). Probes for β-actin mRNA were labeled with Alexa 594 and the probes for *MAP2* mRNA were labeled with TMR. (**c**) Spots corresponding to both mRNA species were identified using a computer program and overlaid on a portion of a DIC image of the same neurons. The locations of β-actin mRNA molecules are depicted by *open circles* and the locations of *MAP2* mRNA molecules are depicted by *closed circles*.

4. Notes

1. We employ several strategies to maximize our signals in addition to using multiple probes to target the same mRNA. Among them are using a wide-field microscope, high numerical aperture oil objectives, cells attached to thin cover slips, long exposure times, cooled CCD camera, and the mounting medium that suppresses photobleaching. Confocal microscopes also yield detectable spots, although the signals are weaker.

2. Although we utilize 3'-amino oligonucleotides, other chemistries such as the 5'- and the internal amino labels, sulfhydryl label, and oligonucleotides coupled directly to the DNA should work as well.

3. Although this protocol is described for cover slips, cells growing over other thin media such as glass-bottom dishes or specialized chambered cover glass (Lab-Tek, Nalgene Nunc, Rochester, NY) are equally suitable. These chambers allow all steps including the cell growth to be carried out in the same vessel. A clean cover slip, fit to size, is used to cover the cells during hybridization to prevent drying of the cells. The cells in suspension can also be hybridized using the same procedure in a tube and then spread over a glass slide or chambered glass for imaging. It should be noted, however, that if cells are attached to the slide rather than to the cover slip, poorer signals result (due to increased distance between the target and objective surface).

4. The concentration of formamide used sets the stringency of the hybridization. Ten percent formamide works well for most specimens. However, higher percentage of formamide (20% or higher) can be used to increase the stringency of binding, especially for mRNAs of high GC content. It is advisable to start with low stringent conditions and then work up from there.

5. We need to intensely illuminate for 10–30 s to acquire the z-stacks. Most fluorophores photobleach during this time (with the notable exception of TMR), therefore in order to do multiplex imaging it is necessary to somehow prolong the lives of fluorophores. It has been found that oxygen is required in the pathways that lead to light mediated degradation of fluorophores. Yildiez et al. (7) described a procedure in which a cocktail of glucose deoxygenase and catalase catalytically remove molecular oxygen from the medium while consuming glucose. Use of this mixture as mounting medium enhances the half-lives of many fluorophores more than tenfold (3).

6. The purpose of nail polish is to prevent the cover slip from sliding, entry of oxygen from the edge of cover slip, and drying. However, if the acetone from the nail polish bleeds into the mounting medium, then very high background signals will be seen. In place of nail polish, silicon grease can also be used.

7. The size of spots should always be the same irrespective of the expression level. There are usually large cluster of spots seen in the nucleus, which represent the transcription sites where many nascent mRNAs accumulate. It is often not possible to distinguish single particles amongst such clusters. If there are larger spots in the cytoplasm, careful controls should be performed to eliminate possibility of artifacts.

8. Although the number of spots that the spot detection algorithm identifies is relatively insensitive to the threshold, the identification of spots should be verified at least for the first set of images in a series.

Acknowledgements

This work was supported by a grant from the National Institutes of Mental Health (MH079197). Authors acknowledge help from Salvatore A.E. Marras in probe labeling and purification.

References

1. Femino, A.M., Fay, F.S., Fogarty, K. and Singer, R.H. (1998) Visualization of single RNA transcripts in situ. *Science*, **280**, 585–590.
2. Lu, J. and Tsourkas, A. (2009) Imaging individual microRNAs in single mammalian cells in situ. *Nucleic Acids Res*, **37**, e100.
3. Raj, A., van den Bogaard, P., Rifkin, S.A., van Oudenaarden, A. and Tyagi, S. (2008) Imaging individual mRNA molecules using multiple singly labeled probes. *Nat Methods*, **5**, 877–879.
4. Raj, A., Peskin, C.S., Tranchina, D., Vargas, D.Y. and Tyagi, S. (2006) Stochastic mRNA synthesis in mammalian cells. *PLoS Biol*, **4**, e309.
5. Raj, A., Rifkin, S.A., Andersen, E. and van Oudenaarden, A. (2010) Variability in gene expression underlies incomplete penetrance. *Nature*, **463**, 913–918.
6. Banker, G. and Goslin, K. (ed.) (1998) *Culturing Nereve Cells*. MIT Press, Cambridge, MA.
7. Yildiz, A., Forkey, J.N., McKinney, S.A., Ha, T., Goldman, Y.E. and Selvin, P.R. (2003) Myosin V walks hand-over-hand: single fluorophore imaging with 1.5-nm localization. *Science*, **300**, 2061–2065.

Chapter 2

FISH and Immunofluorescence Staining in *Chlamydomonas*

James Uniacke, Daniel Colón-Ramos, and William Zerges

Abstract

Here we describe how to use fluorescence in situ hybridization and immunofluorescence staining to determine the in situ distributions of specific mRNAs and proteins in *Chlamydomonas reinhardtii*. This unicellular eukaryotic green alga is a major model organism in cell biological research. *Chlamydomonas* is well suited for these approaches because one can determine the cytological location of fluorescence signals within a characteristic cellular anatomy relative to prominent cytological markers. Moreover, FISH and IF staining offer practical alternatives to techniques involving fluorescent proteins, which are difficult to express and detect in *Chlamydomonas*. The main goal of this review is to describe these powerful tools and to facilitate their routine use in *Chlamydomonas* research.

Key words: *Chlamydomonas reinhardtii*, Cell, FISH, Algae, Plant, Chloroplast, Flagella, Pattern formation, mRNA, Localization, Fluorescence microscopy, In situ hybridization, Confocal

1. Introduction

The unicellular eukaryotic alga *Chlamydomonas reinhardtii* is a model organism for research into a variety of cell biological processes, e.g., photosynthesis, chloroplast biogenesis, and flagella-based motility (1). Researchers benefit from a powerful set of methods, and the sequenced and annotated genomes of the chloroplast and nucleus (2, 3). Yet the tools involving fluorescence microscopy have been underexploited, in large part due to strong interference from chlorophyll autofluorescence and potent silencing of transgenes encoding GFP-tagged proteins (4–8).

Fluorescence in situ hybridization (FISH) and indirect immunofluorescence (IF) staining can be used to detect the intracellular localization of endogenous mRNAs and proteins with fluorescently labeled oligonucleotide probes or antibodies, respectively. These techniques are not hampered by the difficulties encountered in the use of fluorescent proteins because they localize endogenous mRNAs and proteins, respectively. Chlorophyll autofluorescence can be eliminated when cells are chemically fixed and detergent-permeabilized. These steps required for FISH probes and antibodies to reach their intracellular targets. However, these techniques cannot reveal the dynamics of mRNA or protein localization in living cells and real-time, the major advantage of techniques involving GFP and other fluorescent protein tags.

Chlamydomonas has characteristic cellular anatomy and polarity (Fig. 1a) (9). The apical (anterior) pole is marked by the pair of flagella. The basal (posterior) cytosol is occupied mostly by a globular domain of the chloroplast from which a few finger-like lobes extend to the apical pole to cup a central nuclear-cytoplasmic region, containing the nucleus surrounded by cytoplasm, the endoplasmic reticulum and the Golgi apparatus (10–16). The chloroplast contains a prominent spherical body, the pyrenoid, which is located near the basal pole of every cell (Fig. 1a, f) (17). This highly stereotyped cellular anatomy facilitates the use of *Chlamydomonas* as a model organism for investigations of pattern formation at the cellular level (18).

This cellular anatomy also allows the researcher to identify the location(s) of a FISH or IF signal relative to prominent cytological landmarks (16, 18). The flagella, chloroplast, and pyrenoid are visible by light microscopy (Fig. 1b, c). The chloroplast also can be revealed by chlorophyll autofluorescence, when the researcher chooses not to eliminate it as described in Subheading 3.1, step 2 (19, 20). The nucleus (and chloroplast nucleoids) can be stained with DAPI (Fig. 2a, b), mitochondria with MitoTracker Green FM (Invitrogen, Carlsbad, CA, USA), and vacuoles with neutral red or MDY-64 (21).

Routine use of FISH and IF staining in *Chlamydomonas* promises to facilitate characterizations of the intracellular localization of proteins and mRNAs and bring unexpected discoveries. For example, the first FISH studies revealed surprisingly complex

Fig. 1. (continued) except for the cortical section in E (*lower row*). This cortical section was taken close to the cell perimeter and, therefore, it transects the lobe closest to the viewer. (**c**) The FISH signal of the nuclear–cytoplasmic *LhcII* mRNA, the IF signal of an r-protein of the cytoplasmic ribosome, and a DIC image of the same cell. (**d**) The *LhcII* mRNA FISH signal and the IF signal of the endoplasmic reticulum protein, protein disulfide isomerase (PDI). (**e**) The FISH signal of the chloroplast *psbA* mRNA and the IF signal of a thylakoid membrane protein, PsaA. *Arrows* indicate a gap in chloroplast lobe where it curves out of the optical section. Other cytosolic compartments extend into these gaps, as seen for PDI (and therefore endoplasmic reticulum) in *row d*. The cortical section shows that the chloroplast encloses the nuclear region as lobes rather than like the continuous rim of a cup. (**f**) The FISH of the chloroplast *rbcL* mRNA and the IF signal of Rubisco, a protein in the pyrenoid and chloroplast stroma. These results have been reported previously (16, 22, 23) (bar = 1 μm).

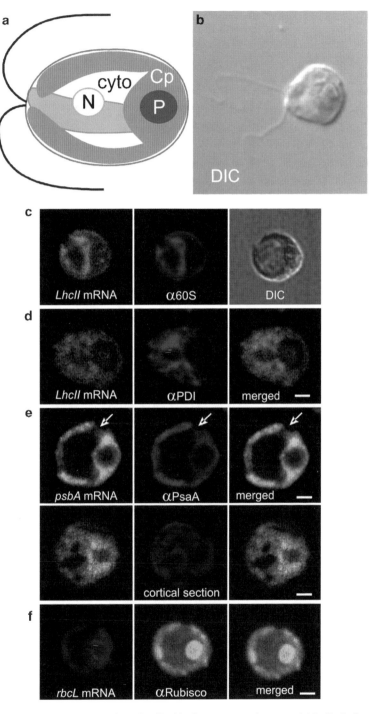

Fig. 1. *Chlamydomonas* cytology visualized by fluorescence microscopy. (**a**) An illustration of a cell oriented with its apical–basal (anterior–posterior) axis from left to right and showing the locations of the flagella, the nucleus (N), cytosol (cyto), the chloroplast (Cp), and the pyrenoid (P). (**b**) An image of a chemically fixed cell obtained by DIC microscopy. (**c–f**) Examples of FISH and IF signals (left-most and central columns, respectively) detected with a confocal laser-scanning microscope (Leica TCS SP2) in 0.2-mm longitudinal optical sections. Most optical sections were taken from the center of the cell,

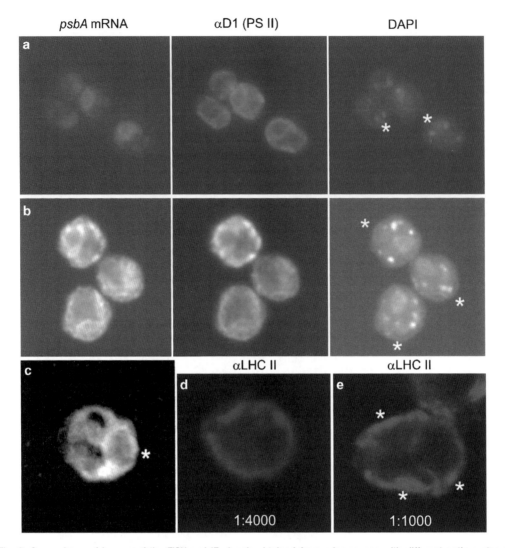

Fig. 2. Comparisons of images of the FISH and IF signals obtained from microscopes with different optic systems. (a–c) The FISH signal is from the *psbA* mRNA, the IF signal is from the photosystem II subunit D1, and DNA was stained with DAPI. (a) An axioplan fluorescence microscope (Zeiss). (b) Aristoplan microscope (Leitz) with Nomarski differential interference contrast (DIC) or epifluorescence optics. (c) A reconstructed cell image from a complete series of serial optical sections obtained with a confocal laser-scanning microscope (Leica TCS SP2). The basal (posterior) poles of most cells are marked with an *asterisks*. (d, e) The IF signal obtained from two different dilutions of antisera against the light harvesting complex II subunits (LHC). *Asterisks* in (e) indicate artifactual punctate IF signal seen when high antibody concentrations were used.

mRNA localization patterns for cellular pattern formation, protein targeting, and oxidative stress response (16, 18, 22, 23). Here we describe FISH and IF-staining protocols in detailed, yet streamlined fashion to facilitate their widespread use among *Chlamydomonas* researchers and cell biologists.

2. Materials

All chemicals and reagents should be analytical, bacteriological, or molecular biology grade.

2.1. Chlamydomonas Culture Media (24, 25)

1. Beijerinck salts (16 g NH_4Cl, 2 g $CaCl_2 \cdot 2H_2O$, and 4 g $MgSO_4 \cdot 7H_2O$ per liter, stored at 4°C).
2. Phosphate solution: 1.0 M KPO_4, pH 7 (250 mL of 1.0 M K_2HPO_4, approximately 170 mL of 1.0 M K_2HPO_4 to titrate to pH 7.0, stored at 4°C).
3. 40× Tris–Acetate Phosphate (TAP) medium stock; 96.8 g Tris-Base (Sigma), 40 mL 1 M of $KHPO_4$ to titrate to pH 7.0 with ca. 44 mL glacial acetic acid.
4. TAP medium: 25 mL of 40× TAP medium stock, 25 mL Beijerinck salts, and 1.0 mL Trace elements solution, per liter.
5. High-salt medium: 25 mL Beijerinck salts, 6.5 mL of 1 M $(K)PO_4$ pH 7.0, 1 mL of Trace elements solution.
6. Erlenmeyer flasks (50–200 mL).
7. Agar, bacteriological grade.
8. Ultra-pure water (marketed for high pressure liquid chromatography).

2.2. Trace Elements Solution

1. H_3BO_3.
2. $ZnSO_4 \cdot 7H_2O$.
3. $MnCl_2 \cdot 4H_2O$.
4. $FeSO_4 \cdot 7H_2O$.
5. $CoCl_2 \cdot 6H_2O$.
6. $CuSO_4 \cdot 5H_2O$.
7. $(NH_4)_6Mo_7O_{24} \cdot 4H_2O$.
8. Na_2EDTA.
9. 100 mL 20% KOH.

2.3. Cell Fixation and Permeabilization

1. Paraformaldehyde (toxic, purchased as a 20% (v/v) stock from Electron Microscopy Sciences, Hatfield, PA, USA; store at room temperature).
2. Methanol (Fisher), maintained at –20°C prior to use but can be stored at room temperature (toxic, inflammable).

2.4. FISH Probe Labeling

1. Synthetic oligonucleotides of 50 nt are designed and ordered, four for each target mRNA. Published protocols call for them to hybridize across exon junctions to promote specificity to the mature mRNA over unspliced precursors or genomic

DNA (16). We have used probes that hybridize within exons of a chloroplast mRNA that lack introns and these did not give a signal from chloroplast nucleoids, the structures that contain multiple copies of the chloroplast genome. The oligonucleotide sequences should be selected to have five T residues interspaced by 6–10 nt. At these positions, the oligonucleotide synthesis company is instructed to incorporate amine-modified C6-dT residues (26) (http://www.singerlab.org/protocols).

2. Amine-reactive fluorescent dyes as carboxylic acid, succinimidyl ester mixed isomers (Invitrogen). Fluorophores are selected based according to the compatibility of their excitation wavelengths with available microscope filters and lasers. We use Alexa Fluor 488 and Alexa Fluor 555 and Alexa Fluor 633 (see Notes 1 and 2). Shield from light whenever possible and store at −20°C.

3. 0.1 M Sodium bicarbonate at pH 9.0. Prepare 1 mL aliquots and store them at −20°C.

4. DMSO, store at room temperature.

5. 5.0 M NaCl, store at room temperature.

2.5. FISH Probe Purification by Denaturing Polyacrylamide Gel Electrophoresis

1. 0.1% Bromophenol blue in 50% formamide, prepared fresh.
2. 30% Acrylamide solution (19:1 acrylamide:bisacrylamide), store at 4°C.
3. 10% Ammonium persulfate, prepared fresh.
4. N,N,N,N'-Tetramethylethylenediamine (TEMED), store at room temperature.
5. 5× TBE electrophoresis buffer (445 mM Tris Base, 445 mM Boric acid, 10 mM EDTA [pH 8.0]), store at room temperature.
6. Vertical electrophoresis apparatus, with glass plates of ca. 30 cm (width) × 40 cm (height), a comb and spacers (1 mm thickness).
7. High voltage power supply.
8. Hand-held UV light source.
9. Elution Buffer (0.1% SDS, 10 mM magnesium acetate, and 0.5 M ammonium acetate), phosphate buffer with salt (PBS) (137 mM NaCl, 2.7 mM KCl, 10 mM Na_2HPO_4, and 2 mM KH_2PO_4), PBS-Mg (PBS with 5 mM $MgCl_2$), prepared fresh.

2.6. Fluorescence In Situ Hybridization

1. Fingernail polish, store at room temperature.
2. 37°C Hybridization oven, such as a slide hybridization oven for microarray hybridizations or a convection oven with precise temperature control.
3. 0.1% Poly-L-lysine (Sigma), store at room temperature.

4. 1 M HCl, store at room temperature.
5. 95% Ethanol, store at −20°C.
6. 70% Ethanol, store at room temperature.
7. Hemocytometer.
8. Coplin jars and lids.
9. ProLong Gold Anti-fade reagent (Molecular Probes).
10. 20× SSC (3.0 M NaCl, 300 mM sodium citrate, pH 7.0), store at room temperature.
11. 2.0 mg/mL Sheared salmon sperm DNA (Sigma). Store at −20°C.
12. 2.0 mg/mL *Escherichia coli* tRNA (Sigma). Store at −20°C.
13. Hybridization Buffer (4× SSC, 10 mM vanadyl ribonucleoside complex [VRC]).
14. 4.0 mg/mL bovine serum albumin (BSA), prepared fresh.
15. Prehybridization Buffer (2× SSC, 50% formamide), prepared fresh.
16. Posthybridization buffer (1× SSC, 50% formamide), prepared fresh.

2.7. Immunofluorescence Staining

1. IF staining requires the materials listed above, under Subheading 2.6, items 1–9.
2. Primary antibody. We have used rabbit polyclonal antibodies, as crude antisera and a mouse monoclonal antibody. Store in tightly sealed tubes at 4°C with 0.02% sodium azide (w/v) to prevent microbial contamination.
3. Secondary antibody is against the primary antibody and coupled to a fluorophore with an excitation wavelength that is compatible with filters or lasers on the available microscope(s). We use anti-rabbit or anti-mouse IgG–TRITC or IgG–FITC (Sigma).
4. PBS (137 mM NaCl, 2.7 mM KCl, 10 mM Na_2HPO_4, and 2 mM KH_2PO_4), store at room temperature. Blocking Solution: 0.1% BSA, and 2 mM VRC in 1× PBS, prepared fresh.
5. Primary and secondary antibody solutions (Blocking Solution with 10 mM VRC instead of 2 mM), prepared fresh.

3. Methods

The following control experiments should be considered. To determine the degree of signal specificity from a FISH probe set or antibody, cells deficient for the target mRNA or protein (e.g., due to mutation, RNAi knock-down, or a known abiotic factor) can be analyzed in parallel to the experimental cells. If the preimmune serum is available, it can be tested as a negative control. None of several different commercial secondary antibodies that we have used generated an above-background fluorescence signal in the absence of primary antibody. Nevertheless, the secondary antibody alone could be excluded as a negative control. To reveal any background autofluorescence, the FISH probe or antibodies can be excluded from a sample. It should be noted that excitation at 633 nm generates a punctuate autofluorescence within the pyrenoid. Most excitation wavelengths generate chlorophyll autofluorescence if the steps in Subheading 3.3, step 5 are not properly carried out.

Fluorescence microscopy is beyond the scope of this article (27). However, we wish to note that optical sectioning with a confocal microscope allows the researcher to normalize signal intensity across compartments with different volumes. For example, in whole cell images obtained with an epifluorescence microscopy, the greater depth of the field can make FISH and IF signals falsely appear highly concentrated in the chloroplast basal region due to its greater volume, relative to the chloroplast lobes (data not shown). Figure 2a–c show examples of images obtained by epifluorescence, Nomarski differential interference contrast (DIC), and confocal laser scanning microscopes.

3.1. Chlamydomonas Cell Culture (9, 24, 25)

The use of a cell wall (CW) mutant strain is advantageous because cells are generally larger and rounder than wild type and some pharmacological agents are ineffective in wild-type cells. We use CC-503 which carries *CW94*. However, it is advisable to also use wild type to control for potentially aberrant localization patterns in CW mutants. Cells can be cultured under photoautotrophic conditions on HSM medium in the light, mixotrophic conditions on acetate-containing TAP medium in the light. These conditions generate very different cell types (28). For analyses of localization patterns of mRNAs or proteins related to photosynthesis, we advise the use of cells cultured under photoautotrophic conditions whenever possible.

When water quality is of doubt or the cells seem to be in suboptimal condition, it may be helpful to use ultra-pure water (for high pressure liquid chromatography) for liquid media and for solutions to which cells are exposed:

1. Media are prepared with ingredients listed in Subheading 2.1 and sterilized by autoclaving.

2. Growth conditions are at 24°C with illumination by one or more 20 W fluorescent bulbs marketed for plants at a distance of 10–20 cm.

3. Strains are maintained on medium solidified with 1.5% bacteriological grade agar (wt/vol) in petri plates and transferred

to fresh plates every 3 days prior to inoculation of the liquid culture(s).

4. A liquid culture of 25–50 mL in 100 mL Erlenmeyer flask is grown over 1–2 days from an initial density of ca. 10^4 cells/mL to a final density of $2–5 \times 10^6$ cells/mL, as determined with a hemocytometer.

3.2. Trace Elements Solution (25)

1. In 550 mL H_2O, dissolve in order; 11.4 g H_3BO_3, 22 g $ZnSO_4 \cdot 7H_2O$, 5.06 g $MnCl_2 \cdot 4H_2O$, 4.99 g $FeSO_4 \cdot 7H_2O$, 1.61 g $CoCl_2 \cdot 6H_2O$, 1.57 g $CuSO_4 \cdot 5H_2O$, and 1.1 g $(NH_4)_6Mo_7O_{24} \cdot 4H_2O$.

2. Dissolve 50 g Na_2EDTA in 250 mL H_2O by heating to near 100°C.

3. Mix the two solutions and bring the resulting solution to boiling.

4. Cool to 80–90°C and adjust to pH 6.5–6.8 with ca. 100 mL 20% KOH using a pH meter calibrated at 75°C.

5. Adjust volume to 1 L with H_2O and incubate at room temperature for approximately 2 weeks in a 2 L-Erlenmeyer flask loosely plugged with cotton. A precipitate forms.

6. Filter through filter paper (Whatman).

7. Make aliquots of 50–200 mL. The working aliquot is stored at 4°C. Reserves are stored at –20°C.

3.3. Cell Fixation and Permeabilization

1. To reduce autofluorescence of the microscope slides, they are boiled in 1 M HCl for 15 min, air-dried, covered with aluminum foil, and incubated at least overnight at room temperature. Take appropriate safety precautions for manipulation of acid.

2. To enhance adherence of the cells to slides in the next step, 10 µL of 0.1% Poly-L-lysine is dispensed near one end of each slide and then smeared across its length using the edge of another slide. Treated slides are stored at room temperature in a slide rack covered with aluminum foil for 3–7 days.

3. An aliquot of ca. 500 µL containing ca. 10^6 cells is dispensed to the center of a poly-L-lysine-coated microscope slide. Cells are allowed to adhere to the slide for 45 s. One must keep track of the side of the slide with the cells prior to mounting in step 3 (Subheading 3.1) because ink labels are not resistant to the intervening steps.

4. Cells are fixed by incubating the slide for 10 min at room temperature in a Coplin jar containing 4% paraformaldehyde (v/v) freshly diluted in PBS (see Note 2).

5. Slides are incubated twice in methanol for 10 min at –20°C, again in a Coplin jar (see Note 3).

6. Slides are given two 10 min washes in PBS-Mg at room temperature.

7. Cell permeabilization involves incubating the slide in freshly prepared 2% (v/v) Triton X-100 in PBS for 10 min at room temperature (see Note 4).

8. Slides are given two additional 10 min washes in PBS-Mg at room temperature.

9. Slides are ready for FISH (Subheading 3.6) or IF staining (Subheading 3.7).

3.4. Labeling of FISH Probes

These procedures for labeling and purification of oligonucleotide FISH probes have been reported previously (26) (http://www.singerlab.org/protocols, http://probes.invitrogen.com/media/pis/mp00143.pdf) (see Note 5):

1. Labeling reactions contain 4 μL of 25 μg/μL oligonucleotide (with the five amine-modified C6-dT residues), 14 μL with 250 μg of amine-reactive fluorophore (resuspended in DMSO), 75 μL 0.1 M sodium bicarbonate buffer (pH 9.0), and 7 μL deionized water (see Note 6).

2. This labeling reaction is incubated overnight, in the dark and at room temperature.

3. The oligonucleotide is precipitated by the addition of 10 μL 5 M NaCl solution and 250 μL cold 95% ethanol, mixed, incubated at −20°C for 30 min, and finally centrifuged at maximum speed, e.g., $14,000 \times g$ for 30 min at 4°C.

4. The pellet is washed with 75% ethanol and centrifuged at maximum speed, e.g., $14,000 \times g$ for 2 min.

5. The pellet is dried by leaving the tube open for ca. 10 min.

3.5. FISH Probe Purification by Denaturing Polyacrylamide Gel Electrophoresis

The labeled oligonucleotide is purified from unlabeled oligonucleotide and free dye by denaturing polyacrylamide gel electrophoresis:

1. Mix 13 mL of a 30% acrylamide/bisacrylamide solution, 6 mL of 5× TBE buffer, 10.68 mL of water, 100 μL ammonium persulfate, and 24 μL TEMED.

2. Immediately dispense this solution between the glass plates with a 25 mL pipet.

3. Insert comb and allow the gel to polymerize for at least 30 min.

4. Prepare TBE electrophoresis buffer by diluting 200 mL of 5× TBE buffer in 800 mL water.

5. After the gel has polymerized, carefully remove the comb and use a 3-mL syringe fitted with a 22-gauge needle to rinse the wells with TBE electrophoresis buffer.

6. Attach the gel to the electrophoresis apparatus.
7. Add the TBE electrophoresis buffer to the upper and lower reservoirs.
8. The gel is exposed for 30 min to a power of 30–50 W and adjusted to maintain the glass plates at approximately 55°C. Exercise caution appropriate for high voltages.
9. Labeled oligonucleotide probes, from Subheading 3.4, step, are resuspended in 0.1% bromophenol blue in 50% formamide to the volume of a well of the gel (e.g., 50 µL).
10. Samples are incubated at 55°C for 5 min and loaded into lanes of the gel.
11. Electrophoresis is carried out at the wattage determined in Subheading 3.5, step 8 and until the bromophenol blue has migrated 60–70% of the gel. The labeled oligonucleotide should be faintly visible as a colored band during electrophoresis.
12. The power supply is turned off.
13. The gel is removed from the apparatus.
14. The glass plates are separated to expose the gel.
15. The band with the labeled oligo is visualized in a dark room with a hand-held UV source and excised from the gel with a razor blade. Labeled oligonucleotide will have migrated approximately to the middle of the gel while the free dye will have migrated with the bromophenol blue.
16. The excised gel fragment is transferred to a 1.5 mL microfuge tube, manually crushed (e.g., with a micropipette tip), and incubated overnight in elution buffer at 37°C with agitation. The probe is shielded from light whenever possible to prevent photobleaching of the fluorophore. Our probe solutions have 50–100 ng oligonucleotide/mL and a frequency of incorporation of 50–70 flour molecules per 1,000 bases, 3.1 flour molecules/probe (see Note 7).

3.6. Fluorescence In Situ Hybridization (see Notes 8 and 9)

1. The amount of FISH probe used may have to be determined empirically for each target mRNA. For each hybridization reaction, we prepare a solution containing 30–50 ng of probe and 2 µg each of sheared salmon sperm DNA and *E. coli* tRNA (from stock solutions of 2 µg/µL, stored at −20°C).
2. This solution is lyophilized at 43°C to dryness, resuspended in 10 µL of 100% ultrapure formamide, and then incubated at 85°C for 5 min.
3. A 10 µL aliquot of this hybridization solution is dispensed to the center of a cover slip situated on a piece of parafilm long enough to hold all the microscope slides. The probe mixture is then dispensed into this droplet of hybridization buffer.

4. Each slide is gently blotted with KimWipes and placed with the cell-side down onto a cover slip. Ideally, hybridizations are carried out in a slide hybridization oven with a piece of moist paper towel to maintain humidity. If a convection oven is used, slides are placed on a glass plate, which is then sealed with parafilm to prevent the hybridization solutions from drying. If the shelf of the convection oven has holes, something is placed on them to prevent air circulation in the immediate vicinity of the slides. We generally carry out hybridizations at 37°C when probing for chloroplast mRNAs and at 42°C for nuclear–cytosolic mRNAs. It may be necessary to test a range of hybridization temperatures for each target mRNA to optimize the signal to background ratio. Hybridizations are carried out overnight.

5. For each five-slide Coplin jar, 100 mL of posthybridization buffer (1× SSC, 50% formamide) is freshly prepared and incubated for approximately 30 min at 37°C.

6. Slides are washed twice for 20 min in this buffer at 37°C.

7. Slides are washed for 10 min at room temperature in 0.5× SSC.

8. Slides are washed for 10 min at room temperature in 0.25× SSC.

9. Slides are washed for 10 min at room temperature in 1× PBS, 5 mM $MgCl_2$.

10. Slides are gently blotted dry with KimWipes and 20 μL of ProLong Gold Anti-fade reagent is dispensed onto the cells.

11. A cover slip is placed on the slide and the edges are sealed with fingernail polish. The cells can viewed immediately, however, an overnight incubation at room temperature in the dark seems to improve fluorescence signal strength. Slides can be stored for at least 1 year at –20°C without noticeable loss of signal.

3.7. IF Staining

IF staining can be initiated after Subheading 3.3, step 9 or when used in combination with FISH, immediately after Subheading 3.6, step 9:

1. A 25 μL aliquot of blocking solution (PBS, 0.1% BSA, and 2 mM VRC) is dispensed onto a cover slip.

2. The slide with the fixed and permeabilized cells is dried by gentle blotting with KimWipes and then placed cell-side down onto the cover slip with the blocking solution.

3. The slide is incubated for 30 min at room temperature.

4. The cover slip is removed and cells are then incubated under a new cover slip in primary antiserum diluted with blocking buffer (containing 10 mM VRC) at 37°C for 75 min. The optimal dilution factor must be determined empirically for each primary antibody. We dilute most primary antisera at

1:1,000 although some must be diluted at 1:4,000 (see Note 10, Fig. 2d, e).

5. The cover slip is removed and slide is then washed twice in PBS for 10 min at room temperature.

6. Cells are incubated under a new cover slip in a 1:200 dilution of secondary antibody in blocking buffer (with 10 mM VRC) at room temperature for 45 min.

7. The cover slip is removed and the slide is washed in PBS for 10 min at room temperature.

8. Slides are blotted dry with KimWipes and a 25 µL drop of ProLong Gold Anti-fade is dispensed at the center of the slide and a new coverslip is placed over the drop.

4. Notes

1. Samples can be FISH-probed concurrently for multiple mRNAs using fluorescent dyes with as widely separated emission/excitation spectra as possible. The number of dyes that can be used is limited by the available laser lines. We have simultaneously probed for three mRNAs using Alexa Fluor 488, 555, and 633 dyes.

2. Fixation with glutaraldehyde results in high background autofluorescence.

3. The methanol treatments must be carried out entirely at −20°C to eliminate chlorophyll autofluorescence. These treatments can be omitted to retain chlorophyll autofluorescence in order to "stain" the chloroplast.

4. Poor permeabilization by Triton X-100 can prevent IF staining (Fig. 2b). We found that a Triton X-100 concentration of 2% (v/v) is required.

5. Other FISH probe systems exist, which we have not used. One is reported to allow quantitative expression analysis, single-copy mRNA detection, and does not require the laborious steps in Subheading 3.1, step 1 (QuantiGene ViewRNA, Affymetrix, Fremont, CA) (29). Another system has been used in *Chlamydomonas*. Oligonucleotide probes are labeled with digoxigenin or biotin, which are detected by immunofluorescence (30).

6. The labeling reaction must not contain Tris because it reacts with amine reactive dyes. Invitrogen's labeling protocol calls for 0.1 M sodium tetraborate (pH 8.5) instead of 0.1 M sodium bicarbonate (pH 9.0) (Invitrogen Manual "Amine Reactive Probes" at http://probes.invitrogen.com/media/pis/mp00143.pdf).

7. The NanoPhotometer (Implen) determines the oligonucleotide concentration and FOI of a probe using an aliquot of only 1 μL. See also Invitrogen's Manual "Amine Reactive Probes" (http://probes.invitrogen.com/media/pis/mp00143.pdf).

8. FISH signals of chloroplast mRNAs are stronger than those of even very abundant nuclear–cytosolic mRNAs.

9. A minority of experimental trials using FISH give only a weak signal that is dispersed throughout the cell. This problem is more commonly encountered when probing for nuclear–cytosolic mRNAs than for chloroplast mRNAs.

10. Excess antibody, particularly those against abundant proteins, can block central IF staining and generate artifactual punctate staining patterns (Fig. 2, compare d and e). Therefore, the primary antibody is titred by analyzing the signal strength and in situ localization patterns obtained with a range of primary antibody dilutions (e.g., from 10^{-2} to 10^{-4}). Subsequent experiments use the lowest concentration that gives an adequate quantifiable signal.

Acknowledgements

The authors thank Alisa Piekny for critical review of the manuscript and Marc Champagne and Julio Vazquez for assistance with microscopy. JU and WZ used the confocal microscopes supported by the Centre for Structural and Functional Genomics, Concordia University, the National Science and Engineering Council, and the Canadian Foundation for Innovation. WZ is funded by an operating grant (217566-08) from the National Science and Engineering Council of Canada.

References

1. Harris, E. H. (2001) Chlamydomonas as a Model Organism *Annu Rev Plant Physiol Plant Mol Biol* **52**, 363–406.
2. Merchant, S. S. et al. (2007) The Chlamydomonas genome reveals the evolution of key animal and plant functions *Science* **318**, 245–50.
3. Maul, J. E., Lilly, J. W., Cui, L., dePamphilis, C. W., Miller, W., Harris, E. H., and Stern, D. B. (2002) The *Chlamydomonas reinhardtii* plastid chromosome: islands of genes in a sea of repeats *Plant Cell* **14**, 2659–79.
4. Fuhrmann, M., Oertel, W., and Hegemann, P. (1999) A synthetic gene coding for the green fluorescent protein (GFP) is a versatile reporter in *Chlamydomonas reinhardtii Plant J* **19**, 353–61.
5. Schoppmeier, J., Mages, W., and Lechtreck, K. F. (2005) GFP as a tool for the analysis of proteins in the flagellar basal apparatus of Chlamydomonas *Cell Motil Cytoskeleton* **61**, 189–200.
6. Pittman, J. K., Edmond, C., Sunderland, P. A., and Bray, C. M. (2009) A cation-regulated and proton gradient-dependent cation transporter from *Chlamydomonas reinhardtii* has a role in calcium and sodium homeostasis *J Biol Chem* **284**, 525–33.
7. Neupert, J., Karcher, D., and Bock, R. (2009) Generation of Chlamydomonas strains that

7. efficiently express nuclear transgenes *Plant J* **57**, 1140–50.
8. Franklin, S., Ngo, B., Efuet, E., and Mayfield, S. P. (2002) Development of a GFP reporter gene for *Chlamydomonas reinhardtii* chloroplast *Plant J* **30**, 733–44.
9. Stern, D. (2009) *Introduction to Chlamydomonas and Its Laboratory Use*, Vol. II, second ed., The Chlamydomonas Sourcebook; Organellar and Metabolic Processes Elsevier Academic Press, Amsterdam.
10. Ohad, I., Siekevitz, P., and Palade, G. E. (1967) Biogenesis of chloroplast membranes. II. Plastid differentiation during greening of a dark-grown algal mutant (*Chlamydomonas reinhardtii*) *J Cell Biol* **35**, 553–84.
11. Goodenough, U. W., and Levine, R. P. (1970) Chloroplast structure and function in ac-20, a mutant strain of *Chlamydomonas reinhardtii*. 3. Chloroplast ribosomes and membrane organization *J Cell Biol* **44**, 547–62.
12. Chua, N. H., Blobel, G., Siekevitz, P., and Palade, G. E. (1976) Periodic variations in the ratio of free to thylakoid-bound chloroplast ribosomes during the cell cycle of *Chlamydomonas reinhardtii J Cell Biol* **71**, 497–514.
13. Bourque, D. P., Boynton, J. E., and Gillham, N. W. (1971) Studies on the structure and cellular location of various ribosome and ribosomal RNA species in the green alga *Chlamydomonas reinhardtii J Cell Sci* **8**, 153–83.
14. Schmidt, R. J., Richardson, C. B., Gillham, N. W., and Boynton, J. E. (1983) Sites of synthesis of chloroplast ribosomal proteins in Chlamydomonas *J Cell Biol* **96**, 1451–63.
15. Schotz, F., Bathelt, H., Arnold, C.-G., and Schimmer, O. (1972) Ergebnisse der Elektronenmikroskopie von Serienschnitten und der daraus resultierenden dreidimensionalen Rekonstruktion *Protoplasma* **75**, 229–254.
16. Uniacke, J., and Zerges, W. (2007) Photosystem II Assembly and Repair Are Differentially Localized in Chlamydomonas *Plant Cell* **19**, 3640–54.
17. Michael, R., McKay, L., and Gibbs, S. P. (1991) Composition and function of pyrenoids: cytochemical and immunocytochemical approaches *Canadian Journal of Botany* **69**, 1040–1052.
18. Colon-Ramos, D. A., Salisbury, J. L., Sanders, M. A., Shenoy, S. M., Singer, R. H., and Garcia-Blanco, M. A. (2003) Asymmetric distribution of nuclear pore complexes and the cytoplasmic localization of beta2-tubulin mRNA in *Chlamydomonas reinhardtii Dev Cell* **4**, 941–52.
19. Nishimura, Y., Misumi, O., Kato, K., Inada, N., Higashiyama, T., Momoyama, Y., and Kuroiwa, T. (2002) An mt(+) gamete-specific nuclease that targets mt(-) chloroplasts during sexual reproduction in *Chlamydomonas reinhardtii Genes Dev* **16**, 1116–28.
20. Levitan, A., Trebitsh, T., Kiss, V., Pereg, Y., Dangoor, I., and Danon, A. (2005) Dual targeting of the protein disulfide isomerase RB60 to the chloroplast and the endoplasmic reticulum *Proc Natl Acad Sci U S A* **102**, 6225–30.
21. Garcia, R. J., Kane, A. S., Petullo, D., and Reimschuessel, R. (2008) Localization of oxythtracycline in *Chlamydomonas reinhardtii Journal of Phycology* **44**, 1282–1289.
22. Uniacke, J., and Zerges, W. (2008) Stress induces the assembly of RNA granules in the chloroplast of *Chlamydomonas reinhardtii J Cell Biol* **182**, 641–6.
23. Uniacke, J., and Zerges, W. (2009) Chloroplast protein targeting involves localized translation in Chlamydomonas *Proc Natl Acad Sci U S A* **106**, 1439–44.
24. Harris, E. H. (1989) The Chlamydomonas sourcebook: a comprehensive guide to biology and laboratory use, Academic Press, San Diego.
25. Rochaix, J. D., Mayfield, S. P., Goldschmidt-Clermont, M., and Erickson, J. (1988) The molecular biology of Chlamydomonas, in *Plant Molecular Biology; a practical approach* (Shaw, C. H., Ed.) pp 253–275, Oxford University Press, USA, Washington, D.C.
26. Femino, A. M., Fogarty, K., Lifshitz, L. M., Carrington, W., and Singer, R. H. (2003) Visualization of single molecules of mRNA *in situ Methods Enzymol* **361**, 245–304.
27. French, A. P., Mills, S., Swarup, R., Bennett, M. J., and Pridmore, T. P. (2008) Colocalization of fluorescent markers in confocal microscope images of plant cells *Nat Protoc* **3**, 619–28.
28. Heifetz, P. B., Forster, B., Osmond, C. B., Giles, L. J., and Boynton, J. E. (2000) Effects of acetate on facultative autotrophy in *Chlamydomonas reinhardtii* assessed by photosynthetic measurements and stable isotope analyses *Plant Physiol* **122**, 1439–45.
29. Taylor, A. M., Berchtold, N. C., Perreau, V. M., Tu, C. H., Li Jeon, N., and Cotman, C. W. (2009) Axonal mRNA in Uninjured and Regenerating Cortical Mammalian Axons *J. Neurosci.* **29**, 4697–4707.
30. Aoyama, H., Hagiwara, Y., Misumi, O., Kuroiwa, T., and Nakamura, S. (2006) Complete elimination of maternal mitochondrial DNA during meiosis resulting in the paternal inheritance of the mitochondrial genome in Chlamydomonas species *Protoplasma* **228**, 231–42.

Chapter 3

High Resolution Fluorescent In Situ Hybridization in *Drosophila*

Eric Lécuyer

Abstract

Tissue-specific gene expression is a major determinant in the elaboration of cells with distinctive phenotypes and functions, which is crucial for the development and homeostasis of multicellular organisms. Fluorescent in situ hybridization (FISH) is a powerful method for assessing the expression and localization properties of RNA at subcellular resolution in whole mount organism and tissue specimens. This chapter describes a high-resolution FISH protocol for the detection of RNA expression and localization dynamics in embryos and tissues of the fruit fly, *Drosophila melanogaster*. The approach utilizes tyramide signal amplification (TSA) for enhanced sensitivity and resolution in the detection of coding and noncoding RNAs, for the codetection of different RNA species or of RNA and a protein marker of interest. Furthermore, the protocol outlines details for conducting FISH in microtiter plates, which greatly enhances the throughput, practicality, and economy of the procedure.

Key words: *Drosophila*, Embryos and tissues, Fluorescent in situ hybridization, FISH, Tyramide signal amplification, mRNA, Noncoding RNA, RNA–RNA, RNA–protein costaining

1. Introduction

Determining the spatio-temporal expression pattern of a gene, be it protein-coding or noncoding, as well as the subcellular compartmentalization properties of the expressed RNA product, is an important feature for understanding its biological function (1). RNA in situ hybridization is the standard method used to assess RNA distribution patters at the organismal, cellular, and subcellular levels. The technique is applicable to a wide range of organisms and tissues, and is facilitated by the fact that labeled antisense RNA probes complementary to a target RNA of interest can be easily produced based on primary sequence information alone. Once a labeled probe is generated by in vitro transcription, it is hybridized to the target RNA in situ, and is then detected immunologically with antibodies that specifically recognize the probe label. Traditionally, the most common probe detection strategy has involved immunological detection of digoxigenin (DIG)-labeled probes with antibodies coupled to the alkaline phosphatase (AP) enzyme, which is then reacted with chromogenic substrates (2). While this approach adds an enzymatic amplification step to increase detection sensitivity, the diffusion of the AP-generated dyes away from their site of synthesis diminishes the subcellular resolution of the technique. As an alternative, the use of fluorochrome-conjugated secondary antibodies has enabled the elaboration of fluorescent in situ hybridization (FISH) procedures with enhanced subcellular resolution (3), although the sensitivity of such approaches has often been limited due to the absence of an enzymatic amplification step. In contrast, the inclusion of tyramide signal amplification (TSA) in the FISH detection procedure has significantly enhanced both the sensitivity and resolution of the technique (4–6). TSA involves probe detection with peroxidase-coupled antibodies that catalyze the formation of transiently reactive fluorochrome-conjugated tyramide molecules, which in turn become covalently bound to aromatic residues of proteins in close proximity to the probe. These improvements in sensitivity and resolution, and the ability to conduct sequential tyramide-reactions following peroxide quenching steps, have proven valuable in gaining new insights into mRNA localization pathways and for increasing the multiplexing capabilities of FISH detection strategies (5, 6).

This chapter details optimized methods for conducting high-resolution FISH on *Drosophila* embryos and dissected tissues, and for performing FISH in high-throughput 96-well plate format. Instructions are provided for making RNA probes, for harvesting and fixing *Drosophila* embryos and tissues, and for probe hybridization and TSA-based detection steps. Variations of the procedure for RNA–RNA and RNA–protein costaining are also detailed.

2. Materials

2.1. RNA Probe Preparation (see Note 1)

1. 1.5-mL Microcentrifuge tubes or standard 96-well V-bottom microplates.
2. QIAquick Gel Extraction Kit (QIAGEN Inc., Mississauga, ON, Canada; Cat. No. 28706) or 96-well Whatman Unifilter microplates (Whatman Inc., Piscataway, NJ, USA; Cat. No. 7700-1303).
3. RNAse-free water.
4. RNA Polymerases (T7, T3, or SP6), 20 U/µL.

5. Transcription buffer supplied with the RNA polymerase (10× concentration).

6. Digoxigenin (DIG) or Biotin labeled nucleotide mixes. Labeled nucleotide mixtures can be prepared more affordably by individually purchasing either DIG-11-UTP (Roche Applied Science, Laval, QC, Canada; Cat. No. 11209256910) *or* Biotin-16-UTP (Roche Applied Science; Cat. No. 11388908910), which are then combined with nonlabeled nucleotides. The final mix contains 3.5 mM of DIG- or Biotin-labeled UTP, 10 mM ATP, 10 mM GTP, 10 mM CTP, and 6.5 mM UTP (see Note 2).

7. RNAse inhibitor, 40 U/μL.

8. 3 M Sodium acetate.

9. Ice-cold 100 and 70% ethanol.

10. Table top centrifuge with rotors for spinning microcentrifuge tubes or 96-well plates.

2.2. Collection and Fixation of Drosophila Embryos/Tissues

2.2.1. Embryo Collection and Fixation

1. Standard *Drosophila* collection cylinder cages (Genesee Scientific, San Diego, CA, USA; Cat. No. 59-101).

2. Apple juice agar plates: 40% apple juice, 4% sucrose, 3.5% agar, 0.3% methyl paraben, and yeast paste prepared by mixing dried baker's yeast powder with tap water.

3. Chlorine bleach solution diluted 1:1 with room temperature tap water (3% final concentration).

4. Small paint brush and a collection basket. These baskets can be made with a Nitex mesh and the top part of a 50-mL polypropylene tube that has been cut in half and for which a hole has been cut in the cap. The mesh is then screwed into place in between the cap and tube.

5. 20-mL glass scintillation vials (Thermo Fisher Scientific, Ottawa, ON, Canada; Cat. No. 03-337-15).

6. Freshly prepared 37% formaldehyde solution: In a 20-mL scintillation vial, add 3.7 g of granular paraformaldehyde (Electron Microscopy Sciences, Hatfield, PA, USA; Cat. No. 19208), with 7.3 mL of Milli-Q water and 70 μL of 2 N KOH (10 mL total volume). Dissolve the paraformaldehyde by heating the solution to 100°C on a stirring hot plate for 3–5 min, with constant mixing using a mini stir bar. Cool the solution on ice for 5 min, filter through a 0.45-μm filter and store at room temperature (see Note 3).

7. 1× Phosphate-buffered saline (PBS) at pH 7.4.

8. PBSF solution: 1× PBS and 4% formaldehyde.

9. Heptane.

10. Methanol.

2.2.2. Dissection and Fixation of Drosophila Tissues for FISH

1. 1× PBS at pH 7.4.
2. Collection baskets (as detailed in Subheading 2.2.1).
3. Two fine dissection forceps (Almedic, Montreal, QC, Canada; No. 5 forceps).
4. Deep-well depression slides (VWR International, Ville Mont-Royal, QC, Canada; Cat. No. 48333-002).
5. Standard dissection microscope.
6. 37% formaldehyde solution, freshly prepared (Subheading 2.2.1).
7. PBSF solution (see Subheading 2.2.1).
8. PBT solution: 1× PBS and 0.1% Tween-20.

2.3. Single FISH on Drosophila Embryos/Tissues

2.3.1. Postfixation, Hybridization and Posthybridization Washes

1. 5-mL Polypropylene tubes, 1.5 Microfuge tubes and 0.2 mL half-skirted 96-well PCR plates (Thermo Fisher Scientific, Ottawa, ON, Canada; Cat. No. AB-0900).
2. Microplate sealing foil (Ultident Scientific, Saint-Laurent, QC, Canada; Cat. No. 24-PCR-AS-200).
3. PBT solution (see Subheading 2.2.2).
4. 37% Formaldehyde solution, freshly prepared (Subheading 2.2.1).
5. PBTF solution: PBT and 4% formaldehyde.
6. 20 mg/mL proteinase K (Sigma–Aldrich Canada Ltd, Oakville, ON, Canada; Cat. No. P2308). Dissolve in Milli-Q water and store aliquots (50–100 µL) at –20°C.
7. PBTK solution: 1× PBS and 1 µg/mL proteinase K.
8. PBTG solution: PBT and 2 mg/mL glycine.
9. RNA hybridization buffer (RHB): 50% formamide, 5× SSC, 100 µg/mL heparin, 100 µg/mL sonicated salmon sperm DNA, and 0.1% Tween-20. Filter through a 0.2-µm filter and store at –20°C (stable for several months).
10. Heating block or water bath adjustable to 56, 80, and 100°C, or a PCR machine.
11. Nutating mixer.

2.3.2. Development of FISH Signal

1. 1× PBS at pH 7.4.
2. PBT solution (see Subheading 2.2.2).
3. PBTB solution: PBT and 1% milk powder.
4. Detection of DIG-labeled probe (see Note 4):
   (a) Biotinylated anti-DIG antibody followed by streptavidin-HRP, recommended for strongest signal:
   Biotin-conjugated mouse monoclonal anti-DIG (Jackson ImmunoResearch Laboratories Inc., West Grove, PA, USA; Cat. No. 200-062-156) *and* streptavidin-HRP

conjugate (Invitrogen Canada Inc., Burlington, ON, Canada; Cat. No. S-911).

(b) HRP-conjugated anti-DIG antibodies, suitable for strongly expressed genes or for double labeling experiments:
HRP-conjugated mouse monoclonal anti-DIG (Jackson ImmunoResearch Laboratories Inc.; Cat. No. 200-032-156) *or* HRP-conjugated sheep polyclonal anti-DIG (Roche Applied Science; Cat. No. 11633716001).

5. Tyramide signal amplification (see Note 4):
Cy3-tyramide (Perkin Elmer Life Sciences, Waltham, MA, USA; Cat. No. SAT704A001EA) *or* Alexa488-tyramide conjugate (Invitrogen Canada Inc.; Cat. No. T-20932).

6. 100× 4′,6-diamidino-2-phenylindole (DAPI) solution (0.1 mg/mL). Dissolve DAPI in methanol and store in the dark at −20°C.

2.3.3. Storage, Mounting, and Viewing of Samples

1. Mounting solution (MS): 70% glycerol, 2.5% DABCO (1,4-diazabicyclo [2.2.2.] Octane; Sigma–Aldrich; Cat. No. D-2522). In a 50-mL Falcon tube, dissolve 1.25 g of DABCO crystals in 15 mL of 1× PBS lightly shaking the tube for 1 min. Then add 35 mL of glycerol and mix on a nutator until the solution becomes homogeneous. Store at −20°C.

2. Microscope slides.

3. Coverslips (22 × 22 mm).

4. Transparent nail polish.

5. Fluorescence or confocal microscope.

2.4. Double FISH on Drosophila Embryos/Tissues

1. Reagents for postfixation of embryos, probe hybridization, probe detection, and mounting of samples as described in Subheadings 2.3.1–2.3.3.

2. Quenching solution: PBT and 1% peroxide.

3. Detection of DIG-labeled probe with HRP-conjugated antibodies (see Note 4):
HRP-conjugated mouse monoclonal anti-DIG (Jackson ImmunoResearch Laboratories Inc.; Cat. No. 200-032-156) *or* HRP-conjugated sheep polyclonal anti-DIG (Roche Applied Science; Cat. No. 11633716001).

4. Detection of Biotin-labeled probe:
Streptavidin-HRP (Invitrogen Canada Inc.; Cat. No. S-911).

5. Tyramide signal amplification (see Note 4):
Cy3-tyramide (Perkin Elmer Life Sciences, Boston, MA, USA; Cat. No. SAT704A001EA) *and* Alexa488-tyramide conjugate (Invitrogen Canada Inc.; Cat. No. T-20932).

2.5. RNA–Protein Double Labeling on Drosophila Embryos/Tissues

1. Reagents for post-fixation of embryos, probe hybridization, probe detection, and mounting of samples as described in Subheadings 2.3.1–2.3.3.
2. Primary antibody directed against the protein of interest. To prevent antibody cross-detection, make sure that the species origin of this antibody differs from that of the anti-DIG antibody used to detect the FISH probe.
3. Select a fluorochrome-conjugated *or* HRP-conjugated secondary antibody directed against the species of the primary antibody.

3. Methods

3.1. RNA Probe Preparation

1. The first step in generating probes for FISH is to clone a sequence from a gene of interest into a plasmid multi-cloning site flanked by bacteriophage promoter sequences (e.g., T3, T7, or Sp6). Once the gene segment has been sequence verified, the plasmid can then be used as a template for PCR using primers that overlap the promoter sequences, thus generating a DNA fragment with promoter elements at each extremity. The PCR reaction is run on an agarose gel and the appropriately sized fragment is extracted and purified using commercially available gel-extraction kits. Alternatively, the base plasmid can be linearized by restriction digestion at a site downstream from the promoter, then gel purified as described above. In either scenario, care should be taken to work in RNAse-free conditions (see Note 1), as these purified DNA fragments are then used as templates for in vitro transcription reactions, as described in step 2.

   The PCR-based approach is particularly useful when templates for several genes are prepared simultaneously, as most sequences can be amplified in parallel using universal primers that overlap the T7, Sp6, and/or T3 sequences. Indeed, this strategy was employed for the preparation of RNA probes for high-throughput in situ hybridization studies (6, 7) using the *Drosophila Gene Collection* (DGC) cDNA libraries (8, 9). These libraries harbor full-length cDNAs for the majority of *Drosophila* genes, which are cloned into base vectors containing bacteriophage promoter elements. Since they are organized as 96-well format bacterial glycerol stocks, bacterial cultures (100–200 µL of Luria Broth supplemented with the appropriate selection antibiotic) are seeded from these libraries and 5 µL of an overnight culture is used to seed batch PCR reactions to systematically amplify library cDNAs. The PCR products are then bulk purified by centrifugation over 96-well Whatman Unifilter plates according to the manufacturer's recommendations, concentrated by ethanol precipitation in V-bottom 96-well plates, and resuspended in 25–50 µL of RNAse-free water.

2. Transcription reactions to synthesize labeled antisense RNA probes are performed essentially as described on the specification sheets for the DIG/Biotin RNA labeling kits (Roche Applied Science). Working in RNAse-free conditions (see Note 1), mix 0.5–1 µg of DNA template with 2 µL DIG or Biotin RNA labeling mix, 2 µL 10× transcription buffer, 1 µL RNAse inhibitor (40 U/µL), 2 µL RNA polymerase (20 U/µL), and RNAse-free water to a final volume of 20 µL. Incubate at 37°C for 2–4 h.

   For PCR templates prepared in 96-well format, probes can be bulk synthesized in RNAse-free V-bottom microplates in a total reaction volume of 10 µL. In each well, 5 µL of DNA template is combined with 5 µL of prealiquoted transcription reaction mixture containing: 1 µL 10× transcription buffer, 0.5 µL DIG labeling mix, 0.5 µL RNAse inhibitor (40 U/µL), 1 µL RNA polymerase (20 U/µL), and 2 µL RNAse-free water. Plates are then covered with adhesive sealing foil and incubated for 2–4 h at 37°C.

3. Once the transcription reactions are complete, briefly spin the samples down and add RNAse-free water to bring the total sample volume up to 50 µL, then add 5 µL of 3 M sodium acetate and 125 µL of ice-cold 100% ethanol (see Note 5), and precipitate the samples overnight at −70°C. The next morning, centrifuge the samples at maximum speed in a cooled table top centrifuge, eliminate the supernatant and wash the samples once with 150 µL of 70% ethanol. Dry the RNA pellets for ~15 min by inverting the tubes on a clean tissue, then resuspend the probes in 50–100 µL of RNAse-free water. To verify the quality and to quantify the probe, run a 1–3 µL sample on a standard 1–2% agarose gel stained with ethidium bromide. Store probes at −70°C.

3.2. Collection and Fixation of Tissues for FISH

The FISH protocol described herein has been found to work efficiently in a variety of *Drosophila* tissues. Specific instructions for harvesting and fixing embryos and simple dissected tissues are given, although specialized harvesting and permeabilization steps may be required for particular types of tissues, such as ovaries (10).

3.2.1. Embryo Collection and Fixation

Embryo harvesting can be performed in small or large scale depending on the number of embryos needed and the size of the population cages used. The following reagent volumes are for collections performed in 900 cm^3 collection cylinders using 100 mm apple juice collection plates, which typically give 100–250 µL of settled embryos in a 4-h collection interval. To improve the quality of yields, keep the flies in optimal environmental growth conditions (i.e., 25°C and 70% humidity) and well-fed by adding a coin-sized portion of yeast paste onto the middle of the apple juice plate:

1. Prepare fresh 37% formaldehyde stock solution prior to embryo dechorionation.

2. Recover the apple juice plate from the cage, making sure to replace it with a fresh food plate, and collect the embryos by adding a bit of room temperature (RT) tap water to the plate and delicately brush the embryos off with a small paint brush. Transfer the embryos into a collection basket and rinse thoroughly with tap water.

3. Dechorionate the embryos by bathing the collection baskets in the chlorine bleach solution for approximately 90 s. Rinse the embryos thoroughly with room temperature tap water or with embryo rinse solution (0.7% NaCl and 0.03% Triton X-100) to remove residual bleach.

4. Disassemble the collection basket and, using tweezers, transfer the mesh (i.e., which contains the embryos) into a 20-mL glass scintillation vial containing a biphasic mixture of 2.5 mL PBSF (lower phase) and 7.5 mL heptane (upper phase). Shake the embryos off the mesh using tweezers, and then seal the vial and shake for 20 min at RT (see Note 6).

5. Using a Pasteur pipette, first eliminate most of the lower aqueous (PBS) phase, taking care not to draw up the embryos found at the interface. Then draw up the remaining PBS phase, the embryos, and a bit of the upper heptane phase into the pipette. In a drop wise manner, eliminate the remaining PBS from the Pasteur pipette and then drop the embryos into a 1.5-mL tube containing a biphasic mixture of 0.5 mL methanol (lower phase) and 0.5 mL heptane (upper phase). Devitellinize the embryos by shaking the samples by hand for ~30 s and let the embryos settle. Most of the embryos will sink to the bottom. Eliminate the upper heptane phase and wash the settled embryos three times with 1 mL of methanol. These fixed embryo samples can be stored in methanol at −20°C for several months to years.

3.2.2. Dissection and Fixation Drosophila Tissues

1. Prepare fresh 37% formaldehyde stock solution.

2. When aiming to harvest adult tissues, collect and maintain adult flies in food vials or cages supplied with fresh yeast paste to improve the overall yield and quality of the preparations. After a few days, the flies can be anesthetized with carbon

dioxide and transferred to a Petri dish containing 1× PBS in preparation for dissection. When studying larval tissues, the larvae are collected and rinsed with room temperature tap water using a collection basket.

3. Using fine forceps, dissect the tissues (e.g., imaginal disks, salivary glands) in PBS on deep-well depression slides. Transfer the tissues to a microcentrifuge tube containing ice-cold PBS until enough tissue is obtained for analysis.

4. Remove the PBS and add 1 mL of PBSF. Fix the samples by rocking on a nutator for 20 min at RT.

5. Remove the PBSF and wash the samples five times with 1 mL of PBS at RT for 2 min each.

6. Rinse the embryos once with 1 mL of a 1:1 mixture of methanol:PBS, then twice with 1 mL methanol. Samples can be stored at −20°C in methanol for up to several months.

3.3. Single FISH on Drosophila Embryos/Tissues

3.3.1. Post-fixation, Hybridization, and Posthybridization Washes

Hybridizations are typically performed in 0.2 mL half-skirted 96-well PCR plates, using ~10–15 µL of settled tissue/well. Using PCR plates, which can easily be cut into small sections for processing low numbers of samples, greatly simplifies sample manipulation/storage and is very useful for testing a variety of experimental conditions (i.e., antibody titrations, samples of different genotypes, chemical treatments, etc.). However, for the initial post-fixation steps 1–10 described below, it is preferable to treat the tissues in a batch format in a larger tube (e.g., 1.5- or 5-mL polypropylene tubes), before aliquoting the samples into the PCR plate at step 11 just prior to prehybridization. Treating the samples as a batch reduces experimental variability, since all the embryos/tissues are in the same tube, and makes the manipulations easier at the proteinase K digestion step where delicate mixing is required. Once the samples have been aliquoted into the PCR plates using wide-aperture tips (see Note 7), care should be taken to seal the plates appropriately with sealing foil and a roller for all the incubation and washing steps. Unless indicated otherwise, washing and incubation steps are performed on a rocking platform/nutator at RT, and the solution volumes used are: 100 µL/sample for 0.2-mL PCR tubes, 1 mL/sample for 1.5-mL tubes, and 3 mL/sample for 5-mL tubes:

1. Rehydrate the samples by rinsing once with methanol, once with a 1:1 mixture of methanol:PBT, and twice with PBT. When working with embryos and robust dissected tissues that require permeabilization with proteinase K, proceed to the next step. With delicate dissected tissues, for which proteinase K digestion may damage the sample, continue directly to step 9.

2. Post-fix the samples for 20 min in PBTF.

3. Wash three times with PBT for 2 min each.

4. Add a freshly prepared solution of PBTK to the samples and incubate for 10 min at RT without rocking, or adjust the incubation time according to the tissue (see Note 8). During this incubation, delicately mix the samples every 2 min by gently rotating the tube once or twice or by jetting with a pipetteman. Transfer the samples to ice and incubate for 1 h. This prolonged incubation on ice favors uniform penetration and action of the protease.

5. Remove the PBTK and stop the digestion by washing two times with PBTG for 2 min on nutator at RT.

6. Rinse two times in PBT.

7. Post-fix again with PBTF for 20 min, as in step 2 above.

8. Wash five times in PBT for 2 min each to remove all traces of fixative.

9. Rinse the samples with a 1:1 mixture of PBT:RHB. Replace the mixture with RHB. At this point, the embryos can be stored for days/weeks at –20°C. If the samples were processed as a large batch, distribute the tissue evenly into PCR plates using wide aperture tips, aiming for a final volume of ~10–15 µL settled tissue/well (see Note 7). When ready to hybridize, proceed to step 10.

10. In a separate tube, boil 100 µL/sample of RHB at 100°C for 5 min, then cool on ice for at least 5 min. This freshly boiled RHB will be used for sample prehybridization.

11. Remove the RHB from the aliquoted samples, then add 100 µL of cooled preboiled RHB and prehybridize the samples at 56°C in a heat block, water bath or PCR machine. Incubate at 56°C for a minimum of 2 h without rocking.

12. Prepare the probe solution by adding 50–100 ng of probe in 100 µL of RHB, heat at 80°C for 3 min, and cool on ice for at least 5 min. The probe solution can be kept on ice until the prehybridization is completed.

13. Remove the RHB used for prehybridization through aspiration with an eight-well manifold connected to a water or vacuum pump, and add the probe solution to the samples. Incubate at 56°C over night without rocking.

14. Preheat all wash solutions to 56°C. Remove the probe solution and rinse the embryos once with prewarmed RHB.

15. Wash with prewarmed RHB, and then with 3:1, 1:1, and 1:3 mixtures of RHB:PBT, for 15 min each at 56°C without rocking.

16. Wash four times with prewarmed PBT for 5 min each at 56°C. Cool the samples to RT.

3.3.2. Development of FISH Signal

Unless otherwise indicated, the wash volumes used below are 100 µL/sample and all incubation steps are performed at RT with rocking. At this point, the PCR plate is placed in a small box, which is then fastened vertically on the rocking platform/nutator using large elastics. Placing the plate in a box also serves to shield the samples from ambient light in order to protect photosensitive fluorochrome-coupled reagents:

1. Block samples by incubating with PBTB for 10 min (see Note 9).
2. Incubate with appropriate anti-DIG antibody (diluted 1/400 in PBTB) solution for 1.5–2 h at RT (see Note 4).

   If an HRP-conjugated antibody is used in step 2, rinse the samples once with PBTB following the antibody incubation, then perform a nuclear counter stain by incubating for 10 min with PBTB containing 1× DAPI, then proceed directly to step 6.
3. Perform six washes for 8 min each with PBTB.
4. Incubate samples with streptavidin-HRP (diluted 1/100 in PBTB) for 1–1.5 h.
5. Rinse once with PBTB, then perform a nuclear counter stain by incubating for 10 min with a PBTB containing 1× DAPI.
6. Wash six times for 8 min each with PBTB, then once with PBT, and twice with PBS for 5 min each.
7. Prepare 1/50 dilution of the appropriate tyramide conjugate with the amplification buffer supplied in the tyramide kit (see Note 4). Remove the last PBS wash from the samples, add 50 µL/sample of tyramide solution, and incubate 1.5–2 h.
8. Wash six times for 8 min each with PBS.

3.3.3. Storage, Mounting, and Viewing of Samples

1. Remove the PBS and add 125 µL of MS to the samples. Allow the samples to equilibrate for a few hours before mounting (see Note 10). The samples can be stored for months/years at 4°C in a light shielded receptacle, with minimal loss of the tyramide stain intensity. In contrast, the DAPI nuclear counter stain may fade over time, but can be recovered by adding 1–2 µL of 100× DAPI solution to the sample.
2. Delicately resuspend the sample with a wide aperture tip and transfer a ~30 µL aliquot onto a clean slide, then cover with a 22×22-mm coverslip. Seal the edges with transparent nail polish. Slides can be stored for a few weeks at 4°C in the dark. The DAPI stain tends to diffuse away after a few weeks on slides. Therefore, it is better to mount a fresh sample aliquot for reanalysis at a later date.
3. Analyze samples by fluorescence or confocal microscopy.

The images shown in Fig. 1a–d exhibit the diverse varieties of subcellular localization patterns observed by high-resolution FISH using probes to different types of coding and noncoding RNAs in *Drosophila* syncytial stage embryos.

3.4. Double FISH on Drosophila Embryos/Tissues

When conducting double FISH experiments to simultaneously visualize different RNA species, various commercially available nucleotide labels can be used (see Note 2). The following steps are recommended for the successive detection of DIG and Biotin labeled probes using TSA:

1. Generate two probes, each with a different label, as described in Subheading 3.1.
2. Collect and fix tissues as described in Subheading 3.2.
3. Perform the hybridization with both probes simultaneously; all other pre- and posthybridization washes are as described in Subheading 3.3.1.
4. Block the samples with PBTB for 10 min with constant mixing.
5. Incubate samples with the appropriate HRP-conjugated anti-DIG antibody (diluted 1/400 in PBTB) for 1.5–2 h (see Note 4).
6. Wash six times for 8 min each with PBTB, then once with PBT and twice with PBS for 5 min each.
7. Prepare a 1/50 dilution of the first tyramide conjugate using the amplification buffer supplied in the tyramide kit (see Note 4). Remove the last PBS wash, add 50 µL of tyramide solution, and incubate for 1.5–2 h. This and all of the following steps should be carried out in a light shielded receptacle.
8. Wash six times for 8 min each with PBS.
9. Inactivate the first tyramide reaction by incubating with quenching solution for 15 min (see Note 11). Wash once with PBS and twice with PBT for 5 min each.
10. Block with PBTB for 10 min, as in step 4.
11. Incubate samples for 1–1.5 h with streptavidin-HRP (diluted 1/100 in PBTB).
12. Rinse samples once with PBTB, then perform a nuclear counter stain by incubating for 10 min with a PBTB solution containing 1× DAPI.
13. Wash five to six times for 8 min each with PBTB, then once with PBT, and twice with PBS for 5 min each.
14. Prepare 1/50 dilutions of the second tyramide conjugate with the amplification buffer supplied in the tyramide kit (see Note 4). Add 50 µL tyramide solution and incubate for 1.5–2 h.

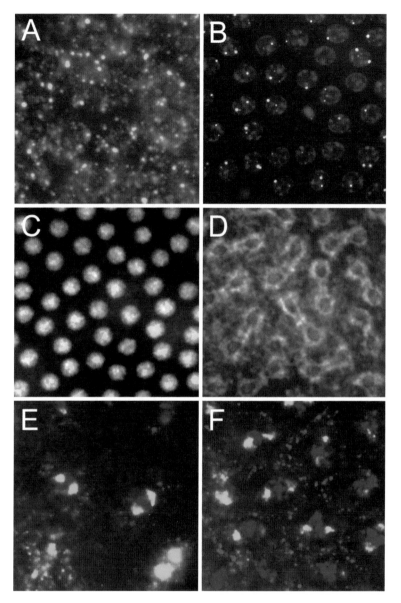

Fig. 1. Examples of subcellular localization patterns observed for different mRNAs and noncoding RNAs (**a–d**), as well as RNA–protein codetection (**e, f**). FISH was performed using DIG-labeled antisense probes, and hybridized probes were detected and visualized though consecutive incubations with a biotinylated anti-DIG antibody, streptavidin-HRP, and Cy3-tyramide, whereas nuclear DNA is revealed using DAPI. For samples (**a–d**), the Cy3-tyramide and DAPI signals are false colored in *blue* and *red*, respectively. (**a**) *Smaug* mRNA localizes within cytoplasmic foci surrounding the nuclei. (**b**) Using a probe to the *mir309-6* cluster of micro RNAs reveals staining within nuclear foci, which likely represent unprocessed primary nascent transcripts. (**c**) The *U6* noncoding small nuclear RNA, which is involved in the splicing of pre-mRNAs, exhibits a broad localization throughout the nucleoplasm. (**d**) The RNA component of the signal recognition particle, *7SL* RNA, shows an enriched localization around anaphase chromosomes and at the spindle midzone. For samples (**e, f**), in addition to the RNA and DNA signals, which are colored in *red* and *blue* respectively, immunostaining for alpha-Tubulin (*green*) was performed after the RNA detection and quenching steps, using a monoclonal rat anti-Tubulin antibody (clone Yol 1/34; Millipore, Billerica, MA, USA; Cat. No. CBL270), followed by an HRP-conjugated anti-rat secondary and Alexa488-tyramide. These images show codetection of alpha-Tubulin with (**e**) the centrosome-localized *CG1962* mRNA, or (**f**) with *Canoe* mRNA, which is targeted to forming cell junction structures.

15. Wash six times for 8 min each with PBS.
16. Mount and view samples as described in Subheading 3.3.3.

3.5. RNA–Protein Double Labeling in Drosophila Embryos/Tissues

1. Collect, process, and hybridize samples essentially as described in Subheadings 3.2 and 3.3.1; with the exception of the proteinase K step, which may have to be adapted for optimal immunostaining (see Note 7).
2. Take care to select non-crossreactive detection reagents (i.e., antibodies generated in different host species). Add the primary antibody against the protein of interest, along with the appropriate probe detection reagent; HRP- or Biotin-conjugated anti-DIG antibodies, or streptavidin-HRP, for DIG- and Biotin-labeled probes, respectively. Incubate samples with appropriate antibody solution (diluted in PBTB) for 1.5–2 h.
3. Perform six washes for 8 min each with PBTB.
4. Add secondary detection reagents (fluorochrome-conjugated secondary antibodies and streptavidin-HRP). Perform incubations, washes, DAPI staining, and TSA reaction as in Subheading 3.3.2 (see Note 12).
5. Mount samples as described in Subheading 3.3.3.

Examples of RNA–protein double labeling are shown in Fig. 1e, f, in which different mRNAs, which localize to centrosomes (*CG1962*) and forming cell junctions (*Canoe*), respectively, are codetected with alpha-Tubulin.

4. Notes

1. When working with RNA, it is imperative to limit potential contamination by RNAse enzymes by thoroughly cleaning all bench surfaces and equipment (i.e., pipettors, electrophoresis equipment, etc.), either with 70% ethanol or commercially available RNAse decontamination solutions. Care should be taken to always wear clean gloves when manipulating reagents and equipment needed for the RNA work, and certified RNAse-free tips should be used for pipetting all solutions. It may also be useful to set aside a specific set of pipettes for RNA work, or alternatively to designate a set of pipettes for procedures involving the use of RNAses, such as plasmid preparations.
2. While DIG and Biotin labels are utilized in this protocol, a variety of other labeled nucleotides (e.g., fluoresceine-, dinitrophenyl-, or Alexa-conjugated nucleotides) can be used to make labeled probes. Optimized conditions for the use and detection of these labels for FISH in *Drosophila* have been described previously (5).

3. Preparing smaller batches of fresh formaldehyde solutions as needed ensures consistent and strong fixation of samples.

4. Different combinations of antibodies can be used for the detection of DIG-labeled probes. To obtain the strongest FISH signal, the use of the biotinylated anti-DIG antibody in combination with streptavidin-HRP is recommended, as this strategy brings more HRP molecules to the vicinity of the probe compared to the HRP-coupled anti-DIG antibodies. However, the later antibodies are particularly useful for double FISH experiments, especially if Biotin is used as a secondary probe label, or for RNA–protein codetection experiments, when antibody cross-reactivity is an issue. In addition to Cy3- and Alexa 488-tyramide conjugates, a variety of other fluorochrome-conjugated tyramide reagents are commercially available from Perkin Elmer Life Sciences and Molecular Probes (Invitrogen Canada Inc.). For RNA–protein double labeling, the use of Cy3-tyramide is recommended for detecting the RNA probe, since we have found this tyramide to be more stable and resistant to the many washing steps required for the double labeling procedure. The optimal antibody and tyramide working dilutions recommended in this procedure may require empirical optimization in different laboratories due to variability in research environments, product stocks, or the types of specimens to be analyzed.

5. While lithium chloride is sometimes recommended as a salt for selective RNA precipitation, comparative analysis in our hands revealed that sodium acetate offers higher probe precipitation efficiency. In addition, we have found it unnecessary to remove the DNA template through DNAse I treatment or to perform carbonate degradation of the probes, as these treatments can sometimes compromise probe quality. Depending on the permeability of the tissue of interest, it may be useful to perform carbonate degradation of long probes (i.e., >1,000 nucleotides). Alternatively, one can generate a pool of small probes (i.e., 200–1,000 nucleotides) complementary to different sections of a target mRNA. This strategy is especially useful for detecting low abundance targets.

6. While the samples can be shaken by hand, a better alternative is to wrap individual vials with hand paper and to place these in a small box that is then fastened onto a vortex with elastics. The samples are shaken by setting the vortex with constant agitation at low speed.

7. When pipetting embryo/tissue samples, wide aperture tips should be used to avoid damaging the embryos. Wide aperture tips can be purchased from a variety of suppliers. If these are not available, simply cut off the ends of traditional tips. When aliquoting the samples into PCR plates, care should be taken to eliminate any bubbles that may have formed in the

layer of tissue at the bottom of the wells, otherwise the samples may tend to clump up during the hybridization steps and upon addition of the mountant for long-term storage.

8. Proteinase K digestion is an important parameter for optimal probe entry into the embryo or tissue of interest. Over digestion can damage the tissue, while under digestion can hinder probe accessibility to its target. While traditional protocols suggest short incubation at high proteinase K concentrations, we typically perform these digestions for a longer period of time at lower proteinase K concentrations, followed by an extended incubation on ice, which was found to improve staining uniformity. When preparing and testing new proteinase K stocks, or when working with new types of tissue, one should titrate the concentration of proteinase K in order to find the optimal working concentration. Some tissues, such as dissected larval tissues, tend to be more sensitive to proteinase K digestion; as a result, we sometimes omit the proteinase K digestion step. It may also be necessary to reduce proteinase K levels when performing RNA–protein double labeling (Subheading 3.5), since over digestion can perturb epitope recognition, as has been documented previously (11).

9. The concentration of milk or other blocking agents (e.g., bovine serum albumin, serum, commercial blocking agents) used to block nonspecific antibody binding should be empirically tested to improve staining specificity for different antibodies.

10. Samples that have been precociously mounted often show a hazy background that dissipates a few hours after the mountant has been added to the embryos.

11. While quenching, the first tyramide reaction with 1% hydrogen peroxide works efficiently when performing successive tyramide reactions, treatment with 0.01 M HCl for 10 min or heating at 70°C for 15 min have been suggested as alternative treatments for inactivating the first HRP reaction (5, 12). For some target tissues that have strong endogenous peroxidase activity, it may be useful to perform a quenching step before starting the probe detection procedure. For this, the samples should be incubated for 15 min with 1% hydrogen peroxide in PBT at RT on a nutator, then washed three times with PBT before starting the PBTB blocking step.

12. In RNA–protein costaining experiments, to visualize a protein marker of interest by immunofluorescence, one can either utilize a fluorochrome- or HRP-conjugated secondary antibody raised against the species of the primary antibody FC fragment. Secondary antibodies from Jackson ImmunoResearch Laboratories Inc., which have the multi-labeling (ML) specification, are particularly recommended for these applications.

When an HRP-conjugated secondary antibody is used, a TSA reaction is then performed, which can sometimes reveal staining features that are not obviously apparent with fluorochrome-conjugated secondary antibodies. In contrast, the enhanced signal amplification obtained with TSA may in some cases lead to undesirable or aberrant staining features. Therefore, we suggest testing both strategies for each primary antibody to identify the conditions that provide the best results.

Acknowledgements

I wish to acknowledge the support and guidance of Dr. Henry Krause and members of the Krause laboratory where many aspects of these procedures were first optimized.

References

1. Lécuyer, E., and Tomancak, P. (2008) Mapping the gene expression universe. *Curr Opin Genet Dev* **18**, 506–12.
2. Tautz, D., and Pfeifle, C. (1989) A non-radioactive *in situ* hybridization method for the localization of specific RNAs in *Drosophila* embryos reveals translational control of the segmentation gene hunchback. *Chromosoma* **98**, 81–5.
3. Hughes, S. C., and Krause, H. M. (1999) Single and double FISH protocols for *Drosophila*. *Methods Mol Biol* **122**, 93–101.
4. Wilkie, G. S., and Davis, I. (1998) Visualizing mRNA by *in situ* hybridization using high resolution and sensitive tyramide signal amplification. *Elsevier Trens J., Technical Tips Online*, T014458.
5. Kosman, D., Mizutani, C. M., Lemons, D., Cox, W. G., McGinnis, W., and Bier, E. (2004) Multiplex detection of RNA expression in *Drosophila* embryos. *Science* **305**, 846.
6. Lécuyer, E., Yoshida, H., Parthasarathy, N., Alm, C., Babak, T., Cerovina, T., Hughes, T. R., Tomancak, P., and Krause, H. M. (2007) Global analysis of mRNA localization reveals a prominent role in organizing cellular architecture and function. *Cell* **131**, 174–87.
7. Tomancak, P., Beaton, A., Weiszmann, R., Kwan, E., Shu, S., Lewis, S. E., Richards, S., Ashburner, M., Hartenstein, V., Celniker, S. E., and Rubin, G. M. (2002) Systematic determination of patterns of gene expression during *Drosophila* embryogenesis. *Genome Biol* **3**, RESEARCH0088.
8. Rubin, G. M., Hong, L., Brokstein, P., Evans-Holm, M., Frise, E., Stapleton, M., and Harvey, D. A. (2000) A *Drosophila* full-length cDNA resource. *Science* **287**, 2222–4.
9. Stapleton, M., Liao, G., Brokstein, P., Hong, L., Carninci, P., Shiraki, T., Hayashizaki, Y., Champe, M., Pacleb, J., Wan, K., Yu, C., Carlson, J., George, R., Celniker, S., and Rubin, G. M. (2002) The *Drosophila* gene collection: identification of putative full-length cDNAs for 70% of *D. melanogaster* genes. *Genome Res* **12**, 1294–300.
10. Lécuyer, E., Nećakov, A. S., Cáceres, L., and Krause, H. M. (2008) High-resolution fluorescent *in situ* hybridization of *Drosophila* embryos and tissues. *CSH Protocols*, doi:10.1101/pdb.prot5019.
11. Nagaso, H., Murata, T., Day, N., and Yokoyama, K. K. (2001) Simultaneous detection of RNA and protein by *in situ* hybridization and immunological staining. *J Histochem Cytochem* **49**, 1177–82.
12. Speel, E. J., Ramaekers, F. C., and Hopman, A. H. (1997) Sensitive multicolor fluorescence *in situ* hybridization using catalyzed reporter deposition (CARD) amplification. *J Histochem Cytochem* **45**, 1439–46.

Chapter 4

Localization and Anchorage of Maternal mRNAs to Cortical Structures of Ascidian Eggs and Embryos Using High Resolution In Situ Hybridization

Alexandre Paix, Janet Chenevert, and Christian Sardet

Abstract

In several species, axis formation and tissue differentiation are the result of developmental cascades which begin with the localization and translation of key maternal mRNAs in eggs. Localization and anchoring of mRNAs to cortical structures can be observed with high sensitivity and resolution by fluorescent in situ hybridization coupled with labeling of membranes and macromolecular complexes. Oocytes and embryos of ascidians – marine chordates closely related to vertebrates – are ideal models to understand how maternal mRNAs pattern the simple ascidian tadpole. More than three dozen cortically localized maternal mRNAs have been identified in ascidian eggs. They include germ cell markers such as *vasa* or *pem-3* and determinants of axis (*pem-1*), unequal cleavage (*pem-1*), and muscle cells (*macho-1*). High resolution localization of mRNAs, proteins, and lipids in whole eggs and embryos and their cortical fragments shows that maternal mRNA determinants *pem-1* and *macho-1* are anchored to cortical endoplasmic reticulum and segregate with it into small posterior somatic cells. In contrast, mRNAs such as *vasa* are associated with granular structures which are inherited by the same somatic cells plus adjacent germ cell precursors. In this chapter, we provide detailed protocols for simultaneous localization of mRNAs and proteins to determine their association with cellular structures in eggs and embryos. Using preparations of isolated cortical fragments with intact membranous structures allows unprecedented high resolution analysis and identification of cellular anchoring sites for key mRNAs. This information is necessary for understanding the mechanisms for localizing mRNAs and partitioning them into daughter cells after cleavage.

Key words: Ascidian eggs and embryos, mRNA localization, In situ hybridization, Fluorescent detection, Cortex isolation, RNA probe

1. Introduction

In yeast and in the eggs and embryos of *Drosophila*, *Xenopus*, several ascidian species and the cnidarian *Clytia*, subcellular localization and polarization of mRNAs are key to axis formation and cell differentiation (1–4). It is thought that anchorage to cortical structures mediates the segregation of maternal mRNAs into specific blastomeres and the spatiotemporal regulation of their translation during early development.

Ascidians (*Ciona intestinalis*, *Ciona savignyi*, *Phallusia mammillata*, *Halocynthia roretzi*, and *Botryllus schlosseri* species) provide a good model for studying mRNA localization and cell type segregation since many mRNAs (called *postplasmic/PEM* RNAs) are localized and polarized in the cortex of oocytes and embryos (5–7).

Ascidians are marine chordates (subphylum Urochordata) considered to be the closest ancestors of the vertebrates (8). Ascidian embryos develop quickly into simple tadpoles. They have been known for more than a century for their "mosaic" type of development, whereby distinct parts of the egg periphery contain determinants for different embryonic tissues (9). Small numbers of cells in the embryo undergo stereotyped morphogenetic processes – asymmetric cell division, gastrulation, and notochord formation – which pattern the ascidian tadpole (7).

After fertilization and three mitotic divisions, many maternal mRNAs are concentrated in a macroscopic cortical structure, the centrosome attracting body (CAB). These localized mRNAs and the CAB segregate into small posterior blastomeres during a further series of cleavages between the eight-cell and 64-cell stages (7, 10). This family of posteriorly localized cortical mRNAs is called *postplasmic/PEM* and comprises 39 mRNAs in the ascidian species *C. intestinalis* (11). It has been shown that two *postplasmic/PEM* mRNAs (*macho-1* and *pem-1*) are determinants for muscle formation and asymmetric division (5, 12, 13).

In situ hybridization (ISH) on fixed cells after permeabilization with detergents and solvents is commonly used to study mRNA localization and gene expression (Fig. 1). Combining ISH with labeling of proteins and cell structures has provided new insights about the localization and segregation of *postplasmic/PEM* RNAs in germ and somatic cells of ascidian embryos as well as their zygotic expression (5, 11, 14–16). Anchorage of mRNA to cellular structures can be further analyzed using ISH on isolated cortices preparations, an "open cell" preparation which precludes using detergents and solvents (Fig. 1).

Postplasmic/PEM RNAs are localized to the oocyte periphery during oogenesis, and then become polarized along the vegetal hemisphere during maturation (10, 17, 18) (Fig. 2a, b, d). Following fertilization, *postplasmic/PEM* RNAs are further concentrated in the vegetal cortex via an actomyosin driven contraction (16, 19, 20) (Fig. 2d). After meiosis completion, *postplasmic/PEM* RNAs are relocalized to the posterior pole of the zygote (Fig. 2d), and finally concentrated at the eight-cell stage in the

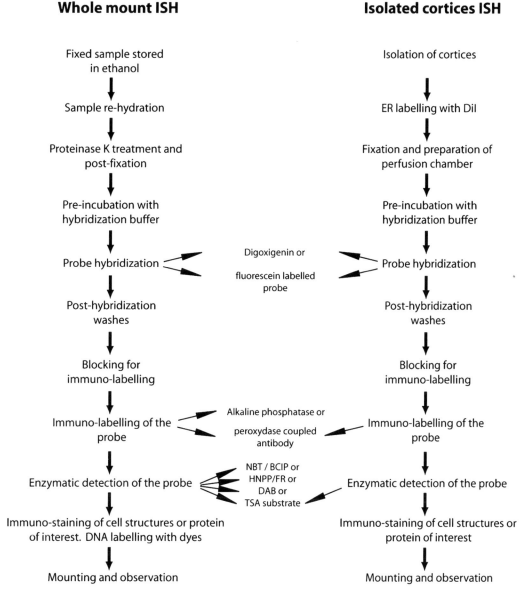

Fig. 1. Summary of ISH protocols for whole mount (*left*) and isolated cortices (*right*).

CAB (Fig. 2a–g). The CAB is responsible for a series of three unequal divisions generating two small cells (7, 21). An additional unequal division at the time of gastrulation partitions some mRNAs (such as *macho-1* and *pem-1*) into cells with somatic fate (B8.11 blastomeres) while other *postplasmic/PEM* RNAs such as the germ line marker *vasa* are also distributed into primordial

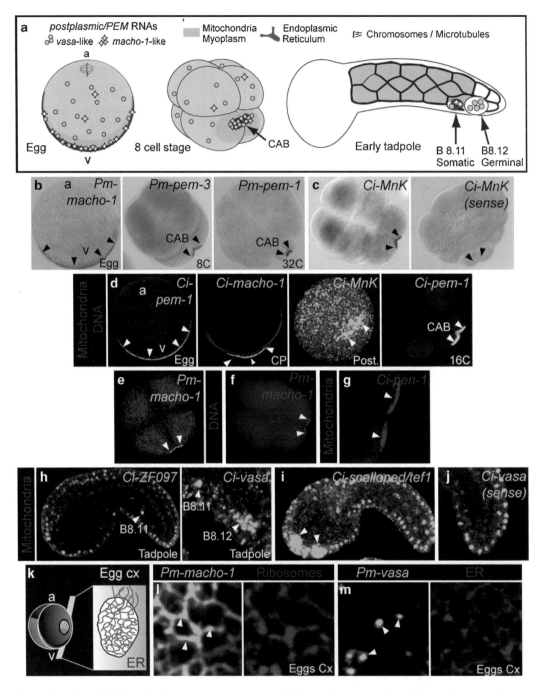

Fig. 2. Examples of mRNA localization and expression pattern. (**a**) Schematic representation of the localization, anchorage, and cell type distribution of *postplasmic/PEM* RNAs. Localized maternal mRNAs of *macho-1*-type (*yellow stars* associated with cER) and *vasa*-type (*blue dots* representing granules) are shown in an egg (*a* animal, *v* vegetal), an eight-cell embryo and an early tadpole ("tailbud"). Mitochondria-rich myoplasm is shown in *green* and ER in *red*. Derived from drawings in ref. 11. (**b**) Localization of *Pm-macho-1*, *Pm-pem-3*, and *Pm-pem-1* (*arrowheads*) assayed by ISH and detected using NBT/BCIP colorimetric substrate precipitation (8C: eight-cell stage embryo; 32C: 32-cell stage embryo). *Pm-pem-1* is detected with a probe of 290 nucleotides in length. (**c**) ISH of *Ci-MnK* mRNA using an antisense probe (*left*) and a control sense probe (*right*), detected using the NBT/BCIP substrate precipitation method. *Arrowheads* indicate position of the CABs.

germ cells (PGCs: B8.12 blastomeres) (Fig. 2a, h) (11, 14, 16). Using isolated cortical fragments and techniques for colocalizing mRNAs with proteins and organelles (Fig. 2k–m), we were able to show that *postplasmic/PEM* RNAs attach to two different subcellular structures: one population of mRNAs which includes the determinants *macho-1* and *pem-1* adhere to a subdomain of endoplasmic reticulum (the cortical endoplasmic reticulum, cER) (19, 22) whereas others such as *vasa* are localized in granular structures (putative germ granules) (11). These distinct anchorage sites likely determine the final destination of *postplasmic/PEM* RNAs in germ and/or somatic cells (Fig. 2a).

Here we have grouped all the methods useful for visualization of mRNAs in *Ciona* and *Phallusia* eggs and embryos, with emphasis on recent improvements in ISH on whole cells and isolated cortices, fluorescent labeling, and simultaneous detection of mRNAs, proteins and cell structures.

Fig. 2. (continued) No signal is observed in the CAB with the sense negative control probe (*right*). (**d**) Relocalization pattern of *postplasmic/PEM* RNAs after fertilization. *Ci-pem-1*, *Ci-macho-1* and *Ci-MnK* are detected using the TSA method (*arrowheads*, Alexa488-Tyramide, in *green*). Mitochondria were immunostained (*magenta*) and DNA labeled with Hoechst (*blue*) after ISH process. Images taken with a confocal microscope. In egg, *postplasmic/PEM* RNAs polarize along the vegetal cortex. Following fertilization, they are further concentrated in the vegetal Contraction Pole ("CP"; first phase of cytoplasmic reorganization). After meiosis completion, they are relocalized to the posterior pole of the zygote ("Post.", second phase of cytoplasmic reorganization). Finally, after the eight-cell stage they are concentrated in the CAB, which drives a series of three unequal divisions between the 8 and 64-cell stages (16C; 16-cell stage embryo). (**e**) Localization of *Pm-macho-1* (*arrowheads*) assayed by ISH and detected directly with a fluorescently labeled antisense probe (*green*). No enzymatic amplification was performed (see Note 5). Image taken with a confocal microscope. (**f**) Localization of *Pm-macho-1* (*arrowheads*) assayed by ISH and detected using HNPP/FR colorimetric substrate precipitation (*red*). DNA was labeled with Hoechst (*blue*) after ISH process. Image taken with a fluorescence microscope. (**g**) Localization of *Ci-pen-1* (*arrowheads*) assayed by ISH and detected using the TSA method (Cy3-Tyramide, in *red*). Mitochondria were immunostained (*magenta*) after ISH process. Image taken with a confocal microscope. (**h**) *Postplasmic/PEM* RNA segregation into B8.11 (somatic) or B8.11 and B8.12 (germinal) blastomeres at tadpole stages. *Ci-ZF097* and *Ci-vasa* (*arrowheads*) were detected using TSA method (Alexa488-Tyramide, in *green*). Mitochondria were immunostained (*magenta*) after the ISH process. Images taken with a confocal microscope. *Ci-ZF097* segregates only in B8.11 cells whereas *Ci-vasa* is also segregated into B8.12 cells. Small *green spots* are background nuclear staining due to TSA method (*see* Note 3). (**i**) Zygotic expression of *Ci-scalloped/tef1* in neurons in the palps of the tadpole head (*arrowheads*), detected using the TSA method (Alexa488-Tyramide, in *green*). Image taken with a confocal microscope. Small *green spots* are background nuclear staining due to TSA method (*see* Note 3). (**j**) ISH with control probe corresponding to *Ci-vasa* sense strand, detected using the TSA method (Alexa488-Tyramide, in *green*). Image taken with a confocal microscope. Nuclear background is observed with TSA reaction when confocal laser power is set very high (see Note 3). (**k**) Schematic drawing of an egg and the corresponding isolated cortex fragment (Egg cx) showing the ER (*red*), which lines the vegetal cortex. Derived from drawings in ref. 11. (**l**) *Pm-macho-1* localization in a cortex isolated from an egg, assayed by ISH using the TSA detection method (Alexa488-Tyramide, in *green*). After ISH, the cortex was immunostained with an antibody, which recognizes the ribosome component P-S6 (*magenta*). *Pm-macho-1* (*left, arrowheads*) has a reticulated localization pattern corresponding to the distribution of ER (*right*). (**m**) *Pm-vasa* localization in a cortex isolated from an egg, assayed by ISH using the TSA detection method (Alexa488-Tyramide, in *green*). Unlike *Pm-macho-1*, *Pm-vasa* (*left, arrowheads*) has a granular localization pattern, which does not correspond to the ER (*right*, labeled with DiI in *red* before fixation and ISH).

2. Materials

2.1. Ascidians

The reader should consult the "Tunicate portal" (http://www.tunicate-portal.org/) and the ANISEED website (Ascidian Network of In Situ Expression and Embryological Data) (http://crfb.univ-mrs.fr/aniseed/index.php) to find resources for ascidian species and laboratories, genomes and cDNA sequences, cell lineages, and markers. Videos of maturation, fertilization and development are available on our BioMarCell website (http://biodev.obs-vlfr.fr/recherche/biomarcell/). Basically five main species of ascidians (the solitary species *C. intestinalis*, *C. savignyi*, *P. mammillata*, and *H. roretzi*; and the colonial species *B. schlosseri*) are used for developmental, and cell and molecular biology research. In all of these species, techniques for localizing *postplasmic/PEM* RNAs, determining mRNA expression using ISH, and assessing the roles of genes by introduction of antisense Morpholino Oligonucleotides, antisense RNAs or siRNAs (5, 6, 13, 23) have been successfully applied. Since *C. intestinalis* is found on temperate coasts worldwide, it is the species of reference although it can show marked seasonal variations in terms of oocyte production and quality of development. The genome of *C. intestinalis* was sequenced in 2002 (24). Cultivation and transgenic lines are now established as well as a limited number of mutants (25, 26). There are interesting comparisons between *C. intestinalis* and other ascidian species. *H. roretzi* is a Japanese species with very large (about 280 μm) but opaque eggs and embryos ideal for micromanipulations and blastomere isolation. It also has marked reproductive seasons. Our preferred ascidian is *P. mammillata*, a European species with very transparent eggs and embryos. It can be easily kept and fed in aquariums and provides abundant oocytes and well-developing embryos all year. It is best for high resolution and live imaging and to prepare isolated cortical fragments. The procedures for dechorionation and fertilization of ascidian oocytes can be found in previous publications (27), as well as in Noriyuki Satoh's reference book "Developmental Biology of Ascidians" (28), a chapter from Billie J. Swalla in Methods in Cell Biology (29) and our recent chapter on embryological methods in ascidians in another volume of Methods in Molecular Biology (30).

2.2. Synthesis of RNA Probes

The following reagents are used: MinElute Reaction Cleanup Kit (Qiagen #28204), nuclease-free water (Promega #P119), 5× transcription optimized buffer (Promega #P1181), 20 U/μl RNA polymerase (Promega #P207 for T7 and #P208 for T3 polymerase), 10× DIG RNA labeling Mix (Roche Applied Science #11277073910), 100 mM DTT (Promega #P1171), 40 U/μl recombinant RNAsin (Promega #N251), 50 U/μl RQ1 DNAse

(Promega #M6101), 10× RQ1 DNAse Buffer (Promega #M198A), and Illustra ProbeQuant G-50 Micro Column (GE Healthcare Life Sciences #28-9034-08).

2.3. Whole Mount ISH

1. PBS (10×): 1.37 M NaCl, 26.8 mM KCl, 100 mM Na_2HPO_4, and 17.6 mM KH_2PO_4 at pH 7.5. Autoclave and store at room temperature (RT). Dilute with H_2O to obtain a 1× solution (store at 4°C for 1–3 days). In order to obtain PBS/Tween-20 solution (PBS-Tw), add Tween-20 detergent at a final concentration of 0.1%. In order to obtain PBS/formaldehyde solution, add formaldehyde (without trace methanol, see below for stock solution and Note 1) to the PBS solution to a final concentration of 4% (see Note 2 for avoiding RNAse contamination).

2. Formaldehyde stock solution (20%): 10 g of paraformaldehyde powder in 50 ml of H_2O. Add 80 µl of 10 N NaOH. Keep at 50°C overnight (ON) and mix until completely dissolved (see Note 1). Pass through a 0.4-µm filter and store in aliquots at −20°C. Thawed aliquot must be warmed at 50°C for 1 h before use. For convenience, formaldehyde solution can also be purchased commercially (32% stock solution without methanol, Electron Microscopy Sciences #15714).

3. ISH fixative: 4% formaldehyde and 0.5 M NaCl in 100 mM MOPS buffer at pH 7.6. Store at 4°C for 1 month maximum or at −20°C for longer.

4. Proteinase K: 1 mg/ml stock solution (Sigma–Aldrich #P6556). Store at −20°C in aliquots.

5. Denhardt's solution (100×): 2% w/v Ficoll 400, 2% w/v polyvinylpyrrolidone (PVP360), and 2% w/v bovine serum albumin (BSA, Cohn Fraction V). Warm at 37°C and mix until completely dissolved, and store at −20°C until use.

6. Yeast RNA stock solution (10 mg/ml): 1 g of yeast total RNA extract (Roche Applied Science #10109223001) in 50 ml of H_2O. Bring the pH to 7 with NaOH. Warm at 37°C and mix until completely dissolved, and store at −20°C until use.

7. SSC solution (20×): 3 M NaCl and 300 mM citric acid, pH 7. Autoclave and store at RT. For convenience, 20× SSC solution can also be purchased commercially.

8. Hybridization Buffer (HB): 50% formamide (see Note 1), 6× SSC (from a 20× stock solution), 5× Denhardt's (from a 100× stock solution), 1 mg/ml yeast RNA (from a 10 mg/ml stock solution) and 0.1% Tween-20. Warm at 65°C until complete mixing of the reagents, and store at −20°C until use.

9. ISH washing Solution A (Solution-A): 50% formamide, 5× SSC, and 1% SDS. Store at −20°C until use.

10. ISH washing Solution B (Solution-B): 50% formamide, 2× SSC, and 1% SDS. Store at −20°C until use.

11. ISH washing Solution C (Solution-C): 2× SSC and 0.1% Tween-20. Store at −20°C until use.

12. ISH washing Solution D (Solution-D): 0.2× SSC and 0.1% Tween-20. Store at −20°C until use.

13. Blocking reagent stock solution (10%): dissolve 5 g of blocking reagent (Roche Applied Science #11096176001) in 50 ml of 1× maleic acid buffer (100 mM maleic acid/150 mM NaCl, pH 7.5). Warm at 50°C and mix until completely dissolved. Autoclave and store at −20°C.

14. ISH blocking buffer: 150 mM NaCl, 0.5% blocking reagent (from a 10% stock solution) and 100 mM Tris–HCl at pH 8. Store in aliquots at −20°C.

15. Enzymatically coupled antibodies raised against digoxigenin: alkaline phosphatase-coupled antibody (Roche Applied Science #11093274910) or peroxidase-coupled antibody (Roche Applied Science #11207733910).

16. NBT/BCIP washing solution: 50 mM $MgCl_2$, 100 mM NaCl, 0.1% Tween-20, and 100 mM Tris–HCl at pH 9.5. Store at −20°C.

17. NBT/BCIP reaction mix: make fresh solution of NBT (nitro blue tetrazolium chloride, at 0.4 mg/ml) and BCIP (5-bromo-4-chloro-3-indolyl phosphate, at 0.19 mg/ml) in 50 mM $MgSO_4$ and 100 mM Tris–HCl at pH 9.5. NBT/BCIP reaction mix can be conveniently prepared by dissolving one ready-to-use NBT/BCIP tablet (Roche Applied Science #11697471001) in 10 ml H_2O.

18. HNPP/FR washing solution: 10 mM $MgCl_2$, 100 mM NaCl, 0.1% Tween-20, and 100 mM Tris–HCl at pH 8. Store at −20°C.

19. HNPP/FR reaction mix: mix just before use, 10 μl of HNPP solution (2-hydroxy-3-naphthoic acid-2′-phenylanilide phosphate, 10 mg/ml in DMF) with 10 μl of FR solution (Fast Red, 25 mg/ml in H_2O) in 1 ml of buffer containing 10 mM $MgCl_2$, 100 mM NaCl, and 100 mM Tris–HCl at pH 8. HNPP solution and FR powder are provided in the HNPP Fluorescent Detection Set (Roche Applied Science #11758888001).

20. TSA reaction mix: we recommend the Alexa488-TSA kit (Molecular Probes #T-20922), which gives very strong green fluorescence signal without background (for other fluorophores or haptens TSA kits, see Note 3). Dilute 1 μl of 30% H_2O_2 in 200 μl of amplification buffer (provided in the kit). Next, add 1 μl of the intermediate H_2O_2 dilution (in order to have a H_2O_2 final concentration of 0.0015%) and 1 μl of

Alexa488-tyramide (reconstituted as manufacturer recommends) to 100 μl of amplification buffer, mix and apply immediately onto the samples.

2.4. ISH on Isolated Cortices

1. EMC: 480 mM NaCl, 9.4 mM KCl, and 23.6 mM EGTA. Bring pH to 8 with NaOH, autoclave, and store at RT.

2. Buffer X: 350 mM K-aspartate, 130 mM taurine, 170 mM betaine, 50 mM glycine, 19 mM $MgCl_2$, and 10 mM of HEPES buffer. Equilibrate pH at 7 with KOH. Sterilize with 0.2-μm filter and store at −20°C in 10 ml aliquots.

3. CM-DiI [$C_{16}(3)$]: fixable DiI is purchased in small vial package (Molecular Probes #C7000). Before ER labeling, reconstitute a vial of fixable DiI in 20 μl of ethanol. This stock solution (2.5 mg/ml) can be stored at 4°C for several days. Next, 1.7 μL of the stock solution is diluted in 0.5 ml of Buffer X and emulsified using a syringe. It is immediately applied to the isolated cortical fragments to label membranes (DiI micelles fuse with ER membranes and diffuse throughout the ER tubes and sheets).

4. CIM and CIM fixative solutions: 800 mM Glucose, 100 mM KCl, 2 mM $MgCl_2$, 5 mM EGTA, and 10 mM of MOPS buffer. Bring pH to 7 with KOH. Sterilize with a 0.2-μm filter and store at −20°C in aliquots. For CIM fixative, add to CIM solution, 3.7% of formaldehyde (without methanol) and 0.1% of glutaraldehyde (see Note 1).

5. cortex Hybridization Buffer (cHB): 50% formamide (see Note 1), 6× SSC, 5× Denhardt's solution, and 1 mg/ml yeast RNA. Mix and warm at 42°C for 1 h. Store at −20°C in aliquots.

6. cortex ISH washing Solution A (cSolution-A): 50% formamide and 4× SSC. Store at −20°C in aliquots.

7. cortex ISH washing Solution B (cSolution-B): 50% formamide and 2× SSC. Store at −20°C in aliquots.

8. cortex ISH washing Solution C (cSolution-C): 50% formamide and 1× SSC. Store at −20°C in aliquots.

9. RNase A buffer: 500 mM NaCl, 5 mM EDTA, and 10 mM Tris–HCl at pH 8. Store at −20°C in aliquots.

10. RNase A: 10 mg/ml stock solution (Novagen #70856). Store at −20°C in aliquots.

2.5. Protein Immuno-Localization and DNA Labeling After ISH

1. Permeabilizing and blocking solutions: for PBS-Tw/BSA, add Tween-20 and BSA (Cohn Fraction V) at a final concentration of 0.1 and 0.5%, respectively. For PBS/BSA, add BSA at a final concentration of 1%.

2. Primary antibodies: many antibodies can be used after ISH processes in order to compare mRNA and protein localization.

In the example described in the method chapter (see Subheadings 3.5 and 3.10), we used an antibody which recognizes a mitochondrial ATP synthase subunit (personal communication of T. Nishikata; NN18 mouse monoclonal antibody, Sigma–Aldrich #N5264) and an antibody raised against a phosphorylated residue of the ribosomal protein S6 (P-S6 rabbit polyclonal antibody, Cell Signaling Technology #2211).

3. Secondary antibodies: secondary antibodies raised against rabbit and mouse immunoglobulin are purchased from Jackson ImmunoResearch laboratories. In the example described in the method chapter (see Subheadings 3.5 and 3.10), we used Cy5-conjugated goat anti-mouse (#115-175-062) and Cy5-conjugated goat anti-rabbit (#111-175-045) antibodies.

4. Hoechst 33342 stock solution: Hoechst is purchased from Sigma–Aldrich (#B2261) and reconstituted at 10 mg/ml in H_2O. Store at −20°C in aliquots. Aliquots are diluted 1:50 in H_2O before use and can be stored for several months at 4°C as a 100× solution.

5. Antifade mounting medium: Citifluor AF1 (Electron Microscopy Sciences #17970).

3. Methods

Since the pioneering mRNA localization studies using sections of ascidian eggs and embryos and isotopically labeled DNA probes (31, 32), ISH techniques have been greatly improved. Non-radioactive ISH on whole egg and embryo preparations using RNA probes labeled with small molecules and detected via enzymatic reaction has now become widespread as a means to study mRNA localization and zygotic expression (5, 33–35). This quick and reliable method has been used extensively in ascidians for both large scale ISH screens and for unraveling the gene networks underlying cell differentiation (23, 36). More recently, methods of fluorescent detection and isolated cortices have been applied to ISH to determine subcellular localization of mRNAs at the highest possible resolution (11, 19, 22).

The first step for determining mRNA localization or expression pattern is to synthesize an RNA probe complementary to the mRNA of interest using in vitro transcription. This probe will contain small molecules/haptens (usually digoxigenin) attached to the nucleotides (UTPs). The next step is to perform ISH on fixed ascidian oocytes or embryos (Fig. 1). This requires 3 days: 1 day for sample treatment and probe hybridization, 1 day for posthybridization wash and immunolabeling of the probe, and

1 day for the detection of the hybridized probe using precipitating or fluorescent reagents.

Several kinds of signal amplification can be used for RNA probe detection (Fig. 1). As for many organisms, colorimetric substrate precipitation with NBT/BCIP is the standard RNA probe detection system used with Ascidians (Fig. 2b, c). This inexpensive detection method allows for long reaction times and easy monitoring of the probe signal with visible light. Substrate precipitation detection systems have a relatively low spatial resolution, and are well-suited to detect the presence of a transcript when the purpose is to monitor promoter activity and gene expression at the tissue or whole cell level. For the higher resolution required to determine spatial localization within a cell, one should use fluorescent detection of the RNA probe and confocal microscopy (Fig. 2d, g). Tyramide System Amplification (TSA, see Note 3) allows sensitive and high resolution imaging of the mRNA localization at the subcellular level. A variety of fluorophores are available for TSA detection (see Note 3). In addition, TSA labeling of RNAs can be used in combination with techniques for the localization of proteins and the labeling of cellular structures (membranes, microfilaments, and chromosomes) using fluorescent dyes and immunostaining (Fig. 2d, g, h) (11, 14).

A still higher resolution for the localization of mRNA can be achieved using isolated cortical fragments (Figs. 1 and 2k–m). Cortices have been previously isolated from sea urchin, frog, mollusc, and mouse eggs (37). In both *Xenopus* (38) and ascidians, mRNAs remain attached to isolated cortical fragments, but it is only in ascidians that high resolution ISH has been achieved (11, 17, 19, 22). Cortices isolated on coverslips look like flat pancake-like imprints ranging from 20 to 30 μm in diameter for *Ciona* or *Phallusia* (about 1/30 of the surface of one egg). Isolated cortices of ascidian oocytes and embryos are comprised essentially of the plasma membrane layer to which a monolayer of cER, microfilaments, particles, and microtubules adhere (19, 37, 39). Many *postplasmic/PEM* RNAs remain attached onto these isolated cortical fragments and it is easy to determine what cellular components they are associated with. Unlike what happens in fixed and permeabilized cells, the continuity of the cER is retained intact in isolated cortices due to the use of a low concentration of glutaraldehyde and the elimination of solvents or detergents which are typically included in fixation and ISH protocols where permeabilization is necessary.

We describe below how to synthesize RNA probes and how to perform ISH on whole cells and isolated cortex preparations (Fig. 1) together with membrane, DNA and protein labeling. Protocols involving RNA require clean laboratory practices and solutions (see Note 2). It is also important to work quickly in order to avoid temperature variations during posthybridization washes.

3.1. Synthesis of RNA Probes

1. Digest ON 5 µg of plasmid containing the cDNA of interest (see Note 4 for probe length). The plasmid is linearized by digestion with restriction enzymes (blunt end or 5′ overhang cut), which cut in 5′ position of the cDNA for antisense probe (complementary to the mRNA of interest) and in 3′ for sense (negative control) probe. *C. intestinalis* cDNAs are obtained from the *Ciona* gene collection (in pBluescript 2 SK-vector) (40). For this vector, T3 and T7 promoters are used for synthesizing sense and antisense probes respectively. If the cDNA does not contain any PvuII restriction sites, we recommend using this enzyme in order to obtain a DNA fragment containing the cDNA of interest and both T3/T7 promoters.

2. Purify the digestion reaction using a MinElute Reaction Cleanup Kit and elute in 10 µl Nuclease-free water in order to obtain template DNA at a final concentration of 0.5 µg/µl.

3. Synthesize the probe using RNA polymerase and digoxigenin labeled UTPs. Although we usually use digoxigenin labeled RNA probe for ISH, other haptens could be used (see Note 5). Mix the following reagents:
   (a) Linearized DNA (0.5 µg/µl in Nuclease-free water): 2 µl
   (b) 5× Transcription optimized buffer: 4 µl
   (c) 20 U/µl RNA polymerase: 1 µl
   (d) 10× DIG RNA labeling Mix: 2 µl
   (e) 100 mM DTT: 2 µl
   (f) 40 U/µl Recombinant RNAsin: 0.5 µl

   Complete with nuclease-free water to reach a total volume of 20 µl.

   Incubate at 37°C for 2 h. Add 1 µl of RNA polymerase and incubate at 37°C for one more hour.

4. After the transcription reaction is completed, eliminate the template DNA by adding 2 µl of 50 U/µl RQ1 DNAse, 5 µl of 10× RQ1 DNAse Buffer, 0.5 µl of 40 U/µl Recombinant RNAsin, and 22.5 µl of Nuclease-free water. Incubate for 1 h at 37°C.

5. Purify the labeled RNA probe by applying the 50-µl reaction mix onto an Illustra ProbeQuant G-50 Micro Column following the manufacturer's procedure.

6. Estimate the probe's concentration by analyzing 1 µl on a 1% agarose gel with ethidium bromide. A standard reaction using this procedure yields 50 µl of probe with a concentration of 50 ng/µl, corresponding to a 100× probe for ISH performed on whole mount oocytes and embryos or on isolated cortices. Store the probe at −20°C until use.

3.2. Fixation of Eggs and Embryos for Whole Mount ISH

1. Fix 100–500 dechorionated eggs or embryos by depositing them in 1.3 ml of ISH fixative (in a 1.5-ml conical tube with screw cap) using a 200-μl pipette tip (or any glass capillary or plastic pipette tip with a diameter greater than 150 μm). Fixation duration can range from 2 h at RT to ON at 4°C. Use moderate shaking (20 rpm).

2. Tubes containing fixed oocytes and embryos are set vertically in a rack to allow eggs and embryos to settle to the bottom of the tube. Wash three times with PBS, one time with 25% ethanol in H_2O, one time with 50% ethanol, one time with 75% ethanol, and finally two times with pure ethanol. Store at −20°C until use.

3.3. Probe Hybridization and Immunolabeling During Whole Mount ISH

1. Rehydrate fixed ascidian eggs and embryos in a 1:1 ethanol/PBS-Tw so.ion, followed by three washes in PBS-Tw. Transfer the eggs and embryos to a 0.5-ml conical tube.

2. Digest samples for 25 min at RT with PBS-Tw solution containing 2 μg/ml proteinase K. Do not incubate longer. Wash three times in PBS-Tw.

3. Post-fix with PBS/formaldehyde solution for 1 h at RT with shaking (20 rpm). Wash three times in PBS-Tw.

4. Incubate samples for 10 min at RT in a 1:1 HB/PBS-Tw solution, followed by 10 min incubation in HB. Next, incubate for 1 h in HB at 65°C. During this last incubation, keep the HB at 65°C.

5. Replace HB with 100 μl of new HB containing 0.5 ng/μl of antisense probe and incubate for 14–18 h at 65°C. Use sense probe as a negative control (Fig. 2c, j). During ON hybridization, place Solutions-A, B, C, and D at 65°C.

6. Wash two times for 30 min at 65°C in Solution-A (the HB containing the probe can be kept at −20°C for reuse, see Note 6). Wash two times for 15 min at 65°C in Solution-B. Wash two times for 15 min at 65°C in Solution-C. Wash two times for 15 min at 65°C in Solution-D. Finally, wash three times in PBS-Tw at RT.

7. Block samples with ISH blocking buffer for 30 min at RT with shaking (20 rpm).

8. Incubate ON at 4°C with shaking (20 rpm) in ISH blocking buffer containing enzymatically coupled antibody raised against digoxigenin (or other haptens, depending on the labeled probe, see Note 5). Use alkaline phosphatase-coupled antibody at 1:2,000 dilution for detection by substrate precipitation with NBT/BCIP or HNPP/FR system. Use peroxidase-coupled antibody at 1:1,000 dilution for fluorescent covalent labeling with the TSA system (substrate precipitation

labeling can also be used with peroxidase, see Note 7). Note that abundant mRNAs can be detected without amplification (see Note 5 and Fig. 2e).

3.4. Detection of the Hybridized Probe in Whole Mount ISH

1. Wash eggs and embryos five times with PBS-Tw. After the last wash, samples can be stored several days at 4°C before detection is carried out.

2. Transfer samples to a multi-well plate (see Note 8) placed inside a humid chamber (see Note 9).

3. Detect the hybridized probe using the enzymatic reaction.

    (a) When using alkaline phosphatase and NBT/BCIP (Fig. 2b, c), wash one time in NBT/BCIP washing buffer. Next, incubate with the NBT/BCIP reaction mix at RT in a humid chamber (mix samples and NBT/BCIP frequently using a pipette). Check labeling (which should appear in the form of dark-blue precipitate) using a stereo-microscope. Depending on the abundance of the mRNA, the reaction time can range from a few minutes to several hours (see Note 10).

    (b) When using alkaline phosphatase and HNPP/FR (Fig. 2f), wash one time in HNPP/FR washing buffer. Next, incubate with HNPP/FR reaction mix at RT in a humid chamber (mix samples frequently and HNPP/FR using a pipette). Reaction time is about the same as for NBT/BCIP. The reaction forms a reddish precipitate emitting a red fluorescent signal which can be visualized using a fluorescent or confocal microscope using excitation/emission settings similar to those used for rhodamine dyes.

    (c) When using peroxidase and TSA reaction (Fig. 2d, h, i), wash one time in the TSA buffer provided by the manufacturer. Next, incubate with 50 µl of TSA reaction mix (Alexa488 coupled) at RT in a humid chamber (mix samples frequently using a pipette). Reaction time using TSA system is approximately two times shorter than with substrate precipitation labeling. In our experience, the reaction time ranges from 5 min to 2 h.

4. Stop the reaction by washing three times with PBS-Tw. For NBT/BCIP and HNPP/FR reactions, the samples must be postfixed with PBS/formaldehyde solution for 1 h and then washed again three times with PBS-Tw.

5. Mount the samples between slide and coverslip for imaging as described in Subheading 3.6.

If necessary, labeling of DNA with Hoechst dye and of proteins with antibodies can be performed after the mRNA detection step (see Subheading 3.5 below and Note 11).

3.5. Immunostaining of Proteins After TSA Reaction in Whole Mount ISH

It is possible to label proteins with antibodies after whole mount ISH. Here we describe how to immunostain mitochondria (Fig. 2d, h):

1. Primary antibody labeling of mitochondria: dilute NN18 antibody at 1:400 in PBS/BSA solution (see item 2, Subheading 2.5 and Note 11) and add to sample. Incubate ON at RT with shaking (20 rpm). Resuspend the samples from time to time.

2. Secondary antibody labeling: wash the samples five times in PBS-Tw. Dilute an appropriate secondary antibody in PBS/BSA solution (coupled with Cy5, see item 3, Subheading 2.5) and add to sample. Incubate for 4 h at RT with shaking (20 rpm). Resuspend the samples from time to time.

3. DNA labeling: wash samples in PBS-Tw. Incubate with Hoechst 33342 in PBS-Tw for 15 min at RT with shaking (20 rpm) (Fig. 2d). Wash three times in PBS-Tw.

3.6. Mounting of Whole Mount ISH Samples

1. Wash labeled oocytes and embryos three times with PBS.

2. Mount the samples on a glass microscope slide: pipette 20 μl of Citifluor AF1 on the slide. Pipette samples (approximately 20 μl) into the Citifluor drop. Gently cover the Citifluor drop containing eggs or embryos with a 22×22-mm glass coverslip (#1), which has a tad of putty on each corner. Apply pressure on the coverslip corners with forceps in order to affix the coverslip onto the slide and hold eggs and embryos in place. Seal with nail polish.

3. Observe whole mount ISH on an appropriate transmission, fluorescence, or confocal microscope.

3.7. Isolated Cortex Preparations and Endoplasmic Reticulum Labeling for High Resolution Localization of mRNAs

1. Isolated cortices of oocytes and embryos up until the 16-cell stage can be prepared, as described in our previous publications (11, 19, 22, 39), on the BioMarCell Web site (http://biodev.obs-vlfr.fr/recherche/biomarcell/ascidies/method.html) and in a separate volume of Methods in Molecular Biology (30).

    In summary, a glass coverslip (18×18 mm, #1) is successively washed in 10% Tween-20, H_2O, 100% ethanol, then dried, and two tracks of double stick tape are placed along the length of the coverslip. Next, 50–100 ascidian oocytes or synchronous embryos are deposited in a drop of artificial sea water without calcium (EMC) between the two tracks of double stick tape on the coverslip. The EMC solution is carefully replaced with the isotonic cytoplasm-mimicking Buffer X. Isolated cortices are prepared by breaking oocytes and embryos with a stream of Buffer X squirted from a Pasteur pipette. Check using a stereo microscope that all eggs or embryos are sheared. Wash the cytoplasmic debris quickly with Buffer X.

2. ER adhering to the plasma membrane of isolated cortical fragments is labeled by adding fixable CM-DiI emulsified in Buffer X to the coverslip, incubating for 1 min, then washing in Buffer X. DiI emits a red fluorescent signal which allows visualization of the cER network (use the same setting used for rhodamine dye) (Fig. 2m).

3. Fix the isolated cortices on coverslips kept in a humid chamber (see Note 9) by treatment with CIM fix solution for 30 min at RT with gentle shaking (5 rpm). Wash one time with CIM solution, followed by three washes with PBS.

4. Make a perfusion chamber by carefully laying the coverslip on the slide using the tracks of double stick tape. Be careful to avoid drying the cortices; if necessary add PBS. With forceps apply a slight pressure on the sticky tape borders of the coverslip in order to attach it firmly to the slide. Seal the two lateral sides of the coverslip to the slide with epoxy resin (Araldite, see Note 12). When the resin hardens, wash the cortices three times with PBS by passing liquid through the perfusion chamber. Store in a humid chamber. At this time, the quality of cortices preparation and cER labeling can be quickly checked using a fluorescent microscope.

3.8. RNA Probe Hybridization and Immunolabeling on Isolated Cortical Fragments

1. Perfuse the cortices in the chamber with a 1:1 dilution of cHB/PBS and incubate for 10 min at RT. Perfuse with cHB and incubate for 10 min at RT. Next, incubate for 1 h with cHB at 42°C (see Note 13 for hybridization temperatures). During this last incubation, keep the cHB at 42°C.

2. Replace the cHB with 50 µl of new cHB containing 0.5 ng/µl of antisense probe (use sense probe as a negative control) for 12–14 h at 42°C. During this ON hybridization step, place the cSolutions-A, B, and C at 42°C.

3. Wash two times for 30 min at 45°C in cSolution-A. Wash two times for 15 min at 45°C in cSolution-B.

4. Wash two times in RNAse A buffer for 15 min at 37°C. Incubate at 37°C for 30 min with RNAse A at 20 µg/ml in RNase A buffer. Repeat the incubation once and wash at 37°C for 15 min in RNAse A buffer.

5. Wash one time with cSolution-B for 30 min at 45°C, and two times for 15 min at 45°C with cSolution-C. Finally, wash one time with a 1:1 PBS/cSolution-C, and five times with PBS. Block for 30 min at RT with ISH blocking buffer.

6. Incubate ON at 4°C with ISH blocking buffer containing peroxidase-coupled antibody raised against digoxigenin at 1:1,000 dilution (or other haptens, see Note 5).

3.9. Detection of the Hybridized RNA Probe on Isolated Cortical Fragments

1. Wash five times with PBS.
2. Detect the hybridized probe using the fluorescent TSA reaction (Alexa488-coupled). Wash one time in TSA buffer provided by the manufacturer. Incubate the cortices in the perfusion chamber at RT with 50 µl of TSA reaction mix. The TSA reaction on isolated cortices usually proceeds faster than on whole cell preparation. In our experience, the reaction time ranges from 5 min to 1 h depending on the mRNA (Fig. 2l, m).
3. Stop the reaction with five washes with PBS.
4. If desired, cortical proteins can be labeled using immunostaining procedures after RNA probe detection (see Subheading 3.10 below and Note 11). Otherwise, mount the samples as described in Subheading 3.11.

3.10. Immunostaining of Proteins After Fluorescent ISH on Isolated Cortices

As for whole eggs or embryos it is possible to label proteins of isolated cortices with antibodies after ISH. Here we describe how we label ribosomal proteins to visualize ribosomes on ER (Fig. 2l):

1. For primary antibody labeling of ribosomes, dilute P-S6 antibody at 1:100 in PBS/BSA solution (see item 2, Subheading 2.5 and Note 11). Perfuse the chamber containing the cortices and incubate for 1 h at RT. Wash five times by perfusing PBS through the chamber.
2. For secondary antibody labeling, dilute appropriate secondary antibody in PBS/BSA solution (coupled with Cy5, see item 3, Subheading 2.5). Perfuse and incubate for 1 h at RT. Wash five times in PBS.

3.11. Mounting and Observing Cortex Samples

1. Replace PBS by diluted citifluor AF1 (two-third of citifluor AF1 with one-third of PBS). Seal the perfusion chamber containing labeled cortices with nail polish.
2. Observe labeled cortices using a fluorescence or confocal microscope.

3.12. Conclusions

Ascidians are ideally suited to study localization and anchoring of key developmental mRNAs situated in the cortex of oocytes and embryos because it is possible to determine their subcellular localization and cell type segregation using ISH on whole cell and isolated cortices preparations (11, 14, 19, 22). This strategy could be used on mammalian, amphibian, or invertebrate oocytes and embryos provided cortices can be isolated and characterized (37), and would be particularly interesting for *Drosophila* where localized cortical mRNA determinants were discovered (2).

4. Notes

1. Formamide and formaldehyde are toxic chemicals. Any solutions containing these chemicals must be manipulated with gloves and under a fume hood. Formamide solutions (hybridization buffer and ISH washing solutions) are stored in polypropylene tubes. During ISH processes, solutions are warmed up. Since the tube cap could leak, we recommend checking the integrity of the cap and sealing the tube with parafilm. The same precautions should be taken in preparing formaldehyde stock solution.

2. RNAse contamination is usually an important concern when working with RNAs. In vitro transcription is performed in presence of RNAse inhibitor and ISH protocols in presence of formamide. RNAse may also be destroyed by proteinase K and formaldehyde fixation. Therefore, we do not operate in specific RNAse-free environment and we do not use DEPC treated solutions. We just wear gloves and use clean materials.

3. TSA is based on the activation of labeled tyramide derivatives by peroxidase and their subsequent covalent coupling to amino groups of adjacent proteins. This allows strong and high resolution labeling without signal diffusion in contrast to substrate precipitation methods which does diffuse. Many TSA kits corresponding to various fluorophores and haptens exist. Although we usually use Alexa488-TSA kit from Molecular Probes (#T-20922), we sometimes use Cy3-TSA (#NEL704A) and Cy5-TSA (#NEL705A) kits from PerkinElmer since they are cheaper and provide good red and far red fluorescent signals. In our experience, the biotin-TSA reaction followed by fluorescent streptavidin labeling does not give much better result than direct fluorophore-TSA based reactions. Also note that quenching of endogenous peroxidases is not necessary in fixed ascidian eggs and early embryos, but nuclear background could be observed with TSA reaction when confocal laser power is set very high (use sense probe as control, Fig. 2j).

4. We successfully obtain good ISH signal with probes ranging from 290 b to more than 4 kb.

5. Usually RNA probes are labeled with digoxigenin. This hapten does not generate significant background. RNA probes could also be labeled with biotin or fluorescein (Roche Applied Science #11685597910 and #11685619910). We do not recommend biotin labeling since detection with streptavidin gives background labeling of mitochondria in whole cell preparations of ascidian. Only the most abundant

mRNAs can be detected with probes labeled by fluorescein without amplification (Fig. 2e). Amplification of signal is possible using anti-fluorescein antibodies (Roche Applied Science #11426338910 and #11426346910), which do not cause significant background.

6. After the ON hybridization, the HB solution containing the probe can be stored at 20°C and reused several times without significant decrease of the ISH signal.

7. Peroxidase and substrate precipitation labeling can also be performed using DAB reaction (3,3 -diaminobenzidine tetrahydrochloride) which produces a blackish precipitate (Sigma–Aldrich #D0426). The reaction time is approximately the same as NBT/BCIP system.

8. We recommend that for developing the enzymatic reaction and subsequent immunostaining, let the samples be put in multi-well plastic plates. Cut the plate in order to have the appropriate number of wells. Flat-bottom plates (Falcon flexible plate flat-bottom, #353912) are convenient to check precipitation reaction under a stereo microscope. Use a U-bottom plate (#353911) for the TSA reaction and subsequent immunostaining since it allows convenient changes of solutions and smaller volume incubations (about 50 μl).

9. Make humid chambers with Petri dishes. Place wet filter paper inside the dish and seal with parafilm.

10. The time required for detection of mRNAs with the substrate precipitation methods can range from a few minutes to several hours. For instance, *Pm-macho-1* mRNA localization in the CAB of 16-cell stage embryos is visible after 10 min using the NBT/BCIP system, whereas *Pm-POPK-1* mRNA requires 3 h to detect a signal. For long incubations, we recommend changing the substrate reaction mix every 3 h.

11. Although ISH processes expose samples to formamide, proteinase K, and high temperature treatments, we found that many antibodies which work with formaldehyde fixation and ethanol dehydration can be used after ISH. We therefore routinely performed immuno-localization after ISH and TSA reaction (11, 17). In the examples provided in this chapter, we label mitochondria on whole cell samples with the NN18 antibody and ribosomes on the cER of isolated cortex preparations with the P-S6 antibody.

12. Araldite can be purchased in general hardware stores. Use quick polymerization Araldite (applying the resin can take 4 min and complete polymerization requires approximately 2 h).

13. In order to preserve the integrity of the fragile perfusion chamber and DiI staining, ISH on isolated cortices is performed at

lower temperatures than for ISH on whole cells. To keep the stringency high, two changes are made for cortex ISH: RNase A treatment removes partially hybridized probe, and the composition of posthybridization solutions give a lower probe melting temperature than for whole cell ISH. We have found that hybridization at 42°C followed by posthybridization washes at 45°C and two RNAse A treatments provided a good compromise between sensitivity and specificity. In the case of abundant mRNAs, hybridization could be performed at 50°C and washes at 52°C after one RNAse A treatment.

Acknowledgments

We thank H. Yasuo, C. Hudson, T. Lepage, P. Dru, A. McDougall, E. Houliston (Developmental Biology unit, Villefranche-sur-Mer Marine Station, CNRS/UPMC UMR7009, France), and L. Yamada (Sugashima Marine Biological Laboratory, Graduate School of Science, Nagoya University, Japan) for helpful discussions and advice about ISH. This work was supported by grants to Christian Sardet from ANR (Agence Nationale de la Recherche), CNRS-ACI (Centre National de la Recherche Scientifique – Action Concertée Incitative), ARC (Association pour la Recherche sur le Cancer), and AFM (Association Française contre les Myopathies). Alexandre Paix was supported by CNRS, ARC, and ANR (Post-doctoral fellowship, A. McDougall laboratory).

References

1. Momose, T., and Houliston, E. (2007) Two oppositely localised frizzled RNAs as axis determinants in a cnidarian embryo, *PLoS Biol* **5**, e70.
2. St Johnston, D. (2005) Moving messages: the intracellular localization of mRNAs, *Nat Rev Mol Cell Biol* **6**, 363–75.
3. King, M. L., Messitt, T. J., and Mowry, K. L. (2005) Putting RNAs in the right place at the right time: RNA localization in the frog oocyte, *Biol Cell* **97**, 19–33.
4. Paquin, N., and Chartrand, P. (2008) Local regulation of mRNA translation: new insights from the bud, *Trends Cell Biol* **18**, 105–11.
5. Prodon, F., Yamada, L., Shirae-Kurabayashi, M., Nakamura, Y., and Sasakura, Y. (2007) Postplasmic/PEM RNAs: a class of localized maternal mRNAs with multiple roles in cell polarity and development in ascidian embryos, *Dev Dyn* **236**, 1698–715.
6. Brown, F. D., Tiozzo, S., Roux, M. M., Ishizuka, K., Swalla, B. J., and De Tomaso, A. W. (2009) Early lineage specification of long-lived germline precursors in the colonial ascidian Botryllus schlosseri, *Development* **136**, 3485–94.
7. Nishida, H. (2005) Specification of embryonic axis and mosaic development in ascidians, *Dev Dyn* **233**, 1177–93.
8. Delsuc, F., Brinkmann, H., Chourrout, D., and Philippe, H. (2006) Tunicates and not cephalochordates are the closest living relatives of vertebrates, *Nature* **439**, 965–8.
9. Conklin, E. G. (1905) The organization and cell lineage of the ascidian egg, *J. Acad. Sci. Philadelphia* **13**, 1–119.
10. Sardet, C., Paix, A., Prodon, F., Dru, P., and Chenevert, J. (2007) From oocyte to 16-cell stage: cytoplasmic and cortical reorganizations that pattern the ascidian embryo, *Dev Dyn* **236**, 1716–31.

11. Paix, A., Yamada, L., Dru, P., Lecordier, H., Pruliere, G., Chenevert, J., Satoh, N., and Sardet, C. (2009) Cortical anchorages and cell type segregations of maternal postplasmic/PEM RNAs in ascidians, *Dev Biol* **336**, 96–111.
12. Negishi, T., Takada, T., Kawai, N., and Nishida, H. (2007) Localized PEM mRNA and protein are involved in cleavage-plane orientation and unequal cell divisions in ascidians, *Curr Biol* **17**, 1014–25.
13. Nishida, H., and Sawada, K. (2001) macho-1 encodes a localized mRNA in ascidian eggs that specifies muscle fate during embryogenesis., *Nature* **409**, 724–29.
14. Shirae-Kurabayashi, M., Nishikata, T., Takamura, K., Tanaka, K. J., Nakamoto, C., and Nakamura, A. (2006) Dynamic redistribution of vasa homolog and exclusion of somatic cell determinants during germ cell specification in Ciona intestinalis, *Development* **133**, 2683–93.
15. Nakamura, Y., Makabe, K. W., and Nishida, H. (2003) Localization and expression pattern of type I postplasmic mRNAs in embryos of the ascidian Halocynthia roretzi, *Gene Expr Patterns* **3**, 71–5.
16. Yamada, L. (2006) Embryonic expression profiles and conserved localization mechanisms of pem/postplasmic mRNAs of two species of ascidian, Ciona intestinalis and Ciona savignyi, *Dev Biol* **296**, 524–36.
17. Prodon, F., Sardet, C., and Nishida, H. (2008) Cortical and cytoplasmic flows driven by actin microfilaments polarize the cortical ER-mRNA domain along the a-v axis in ascidian oocytes, *Dev Biol* **313**, 682–99.
18. Prodon, F., Chenevert, J., and Sardet, C. (2006) Establishment of animal-vegetal polarity during maturation in ascidian oocytes, *Dev Biol* **290**, 297–311.
19. Prodon, F., Dru, P., Roegiers, F., and Sardet, C. (2005) Polarity of the ascidian egg cortex and relocalization of cER and mRNAs in the early embryo, *J Cell Sci* **118**, 2393–404.
20. Roegiers, F., Djediat, C., Dumollard, R., Rouviere, C., and Sardet, C. (1999) Phases of cytoplasmic and cortical reorganizations of the ascidian zygote between fertilization and first division, *Development* **126**, 3101–17.
21. Patalano, S., Pruliere, G., Prodon, F., Paix, A., Dru, P., Sardet, C., and Chenevert, J. (2006) The aPKC-PAR-6-PAR-3 cell polarity complex localizes to the centrosome attracting body, a macroscopic cortical structure responsible for asymmetric divisions in the early ascidian embryo, *J Cell Sci* **119**, 1592–603.
22. Sardet, C., Nishida, H., Prodon, F., and Sawada, K. (2003) Maternal mRNAs of PEM and macho 1, the ascidian muscle determinant, associate and move with a rough endoplasmic reticulum network in the egg cortex, *Development* **130**, 5839–49.
23. Imai, K. S., Levine, M., Satoh, N., and Satou, Y. (2006) Regulatory blueprint for a chordate embryo, *Science* **312**, 1183–7.
24. Dehal, P., Satou, Y., Campbell, R. K., Chapman, J., Degnan, B., De Tomaso, A., Davidson, B., Di Gregorio, A., Gelpke, M., Goodstein, D. M., Harafuji, N., Hastings, K. E., Ho, I., Hotta, K., Huang, W., Kawashima, T., Lemaire, P., Martinez, D., Meinertzhagen, I. A., Necula, S., Nonaka, M., Putnam, N., Rash, S., Saiga, H., Satake, M., Terry, A., Yamada, L., Wang, H. G., Awazu, S., Azumi, K., Boore, J., Branno, M., Chin-Bow, S., DeSantis, R., Doyle, S., Francino, P., Keys, D. N., Haga, S., Hayashi, H., Hino, K., Imai, K. S., Inaba, K., Kano, S., Kobayashi, K., Kobayashi, M., Lee, B. I., Makabe, K. W., Manohar, C., Matassi, G., Medina, M., Mochizuki, Y., Mount, S., Morishita, T., Miura, S., Nakayama, A., Nishizaka, S., Nomoto, H., Ohta, F., Oishi, K., Rigoutsos, I., Sano, M., Sasaki, A., Sasakura, Y., Shoguchi, E., Shin-i, T., Spagnuolo, A., Stainier, D., Suzuki, M. M., Tassy, O., Takatori, N., Tokuoka, M., Yagi, K., Yoshizaki, F., Wada, S., Zhang, C., Hyatt, P. D., Larimer, F., Detter, C., Doggett, N., Glavina, T., Hawkins, T., Richardson, P., Lucas, S., Kohara, Y., Levine, M., Satoh, N., and Rokhsar, D. S. (2002) The draft genome of Ciona intestinalis: insights into chordate and vertebrate origins, *Science* **298**, 2157–67.
25. Sasakura, Y., Oogai, Y., Matsuoka, T., Satoh, N., and Awazu, S. (2007) Transposon mediated transgenesis in a marine invertebrate chordate: Ciona intestinalis, *Genome Biol* **8 Suppl 1**, S3.
26. Chiba, S., Jiang, D., Satoh, N., and Smith, W. C. (2009) Brachyury null mutant-induced defects in juvenile ascidian endodermal organs, *Development* **136**, 35–9.
27. Sardet, C., Speksnijder, J., Inoue, S., and Jaffe, L. (1989) Fertilization and ooplasmic movements in the ascidian egg, *Development* **105**, 237–49.
28. Satoh, N. (1994) Developmental biology of ascidians.
29. Swalla, B. J. (2004) Procurement and culture of ascidian embryos, *Methods Cell Biol* **74**, 115–41.
30. Sardet, C., McDougall, A., Yasuo, H., Dumollard, R., Hudson, C., Pruliere, G., Hebras, C., LeNguyen, N., Chenevert, J., and Paix, A. (in press) Embryological methods in ascidians: the Villefranche-sur-Mer protocols, *Methods Mol Biol*.

31. Jeffery, W. R. (1982) Messenger RNA in the cytoskeletal framework: analysis by in situ hybridization, *J Cell Biol* **95**, 1–7.
32. Jeffery, W. R. (1984) Spatial distribution of messenger RNA in the cytoskeletal framework of ascidian eggs, *Dev Biol* **103**, 482–92.
33. Yasuo, H., and Satoh, N. (1993) Function of vertebrate T gene, *Nature* **364**, 582–3.
34. Wada, S., Katsuyama, Y., Yasugi, S., and Saiga, H. (1995) Spatially and temporally regulated expression of the LIM class homeobox gene Hrlim suggests multiple distinct functions in development of the ascidian, Halocynthia roretzi, *Mech Dev* **51**, 115–26.
35. Yoshida, S., Marikawa, Y., and Satoh, N. (1996) Posterior end mark, a novel maternal gene encoding a localized factor in the ascidian embryo, *Development* **122**, 2005–12.
36. Yamada, L., Shoguchi, E., Wada, S., Kobayashi, K., Mochizuki, Y., Satou, Y., and Satoh, N. (2003) Morpholino-based gene knockdown screen of novel genes with developmental function in Ciona intestinalis, *Development* **130**, 6485–95.
37. Sardet, C., Prodon, F., Dumollard, R., Chang, P., and Chenevert, J. (2002) Structure and function of the egg cortex from oogenesis through fertilization, *Dev Biol* **241**, 1–23.
38. Alarcon, V. B., and Elinson, R. P. (2001) RNA anchoring in the vegetal cortex of the xenopus oocyte., *J. Cell Science* **114**, 1731–41.
39. Sardet, C., Speksnijder, J., Terasaki, M., and Chang, P. (1992) Polarity of the ascidian egg cortex before fertilization, *Development* **115**, 221–37.
40. Satou, Y., Kawashima, T., Shoguchi, E., Nakayama, A., and Satoh, N. (2005) An integrated database of the ascidian, Ciona intestinalis: towards functional genomics, *Zoolog Sci* **22**, 837–43.

Chapter 5

Visualization of mRNA Localization in *Xenopus* Oocytes

James A. Gagnon and Kimberly L. Mowry

Abstract

Visualization of in vivo mRNA localization provides a tool for understanding steps in the mechanism of transport. Here we detail a method of fluorescently labeling mRNA transcripts and microinjecting them into *Xenopus laevis* oocytes followed with imaging by confocal microscopy. This technique overcomes a significant hurdle of imaging RNA in the frog oocyte while providing a rapid method of visualizing mRNA localization in high resolution.

Key words: RNA localization, *Xenopus*, Oocyte, RNA transport, Confocal microscopy

1. Introduction

RNA localization is a conserved mechanism of establishing cell polarity in a variety of cell types and organisms. Such spatial regulation of gene expression can define specialized regions of the cell, and prominent examples include germ layer specification during vertebrate development and cytoskeletal rearrangements involved in cell motility (1–3).

Visualization of RNA distribution patterns has provided valuable insights into RNA transport steps and mechanisms. A number of techniques have been developed to visualize subcellular RNA localization, including in situ hybridization with digoxigenin- and fluorescently-labeled probes (4–6), molecular beacons (7), and fluorescent protein tethering (8, 9).

RNA localization has been studied extensively in *Xenopus laevis* oocytes, where RNAs are localized during oogenesis and underlie patterning along the animal–vegetal axis (3–6, 10–18). The *Xenopus* oocyte offers several significant advantages for studies of RNA transport. First, oocytes are easily obtained through nonlethal surgery. Each surgery can yield thousands of oocytes, making the system amenable to biochemical analyses. Second, oocytes are large in size, easily visible in detail under standard light microscopes, offering facile microinjection of RNA, proteins, DNA, and antibodies, which can be targeted into the nucleus or cytoplasm. Third, isolated oocytes are amenable to culture outside of the frog (19, 20). However, one disadvantage is increasing opacity as yolk protein accumulates during oogenesis, complicating imaging approaches. Here we describe a method of visualizing RNA localization in *Xenopus* oocytes that overcomes this issue while providing striking, high-resolution images of in vivo RNA transport.

2. Materials

2.1. In Vitro RNA Transcription

1. DEPC-treated deionized H_2O (DEPC-H_2O): Add one to two drops DEPC (Sigma–Aldrich) per 100 ml deionized H_2O. Incubate for 30 min at room temperature. Autoclave.

2. 10× Transcription Buffer (10× Tx): 60 mM $MgCl_2$, 400 mM Tris–HCl (pH 7.5), and 20 mM spermidine–HCl. Store as 1 ml aliquots at –20°C.

3. 20× cap/NTP mix: 10 mM CTP, 10 mM ATP, 9 mM UTP, 2 mM GTP, and 20 mM G(ppp)G Cap Analog (New England Biolabs #S1407L). Store as 25 µl aliquots at –20°C.

4. Fluorescent nucleotides: Chromatide Alexa Fluor 488-5-UTP or 546-14-UTP (Invitrogen #C11403 and #C11404, see Note 1).

5. G-50 solution: Hydrate 5 g Sephadex G-50 beads (Sigma–Aldrich) in 100 ml of deionized H_2O. DEPC-treat as detailed above. Before use, add the following: 0.5 ml 0.2 M EDTA, 1 ml of 1 M Tris–HCl at pH 8.0, 0.5 ml of 20% SDS (all solutions must be RNase-free). Store at 4°C.

6. G-50 column: Remove and discard the plunger from a 3-ml syringe (BD Biosciences) and place the barrel of the syringe into a 15-ml conical tube (Corning). Plug the syringe with a small amount of glass wool (a plug about half the size of a penny). Swirl the G-50 solution (see item 5, Subheading 2.1) to resuspend beads. Add 2 ml G-50 solution to the empty column. Spin for 1 min at $1,000 \times g$ in benchtop centrifuge. Add 200 µl of DEPC-H_2O to each column. Spin. Repeat wash twice more for a total of three washes. Remove the column to a fresh 15-ml conical tube prior to use.

2.2. Oocyte Microinjection

1. Needles: To make beveled glass needles with an outer diameter of ~0.05 mm, we pull 3.5-in. capillaries (Drummond Scientific item #3-000-203-G/X) using a Sutter Instrument Co. micropipette puller. Needles are beveled to an angle of 40° using a Narishige Co. EG-4 micropipette grinder.

2. Microinjection apparatus: Harvard Apparatus model #PLI-100.

3. Injection dish: We line a small plastic dish with 1/8-in. thick black foam rubber, cut to fit and secured to the dish with double-sided tape. The white oocytes stand out against the black foam background.

4. MBSH buffer: 88 mM NaCl, 1 mM KCl, 2.4 mM NaHCO$_3$, 0.82 mM MgSO$_4$, 0.33 mM Ca(NO$_3$)$_2$, 0.41 mM CaCl$_2$, and 10 mM HEPES (pH 7.6).

2.3. Oocyte Culture

1. Collagenase solution: 3 mg/ml collagenase (Sigma–Aldrich #C0130), 0.1 M KPO$_3^+$ (pH 7.4). Make fresh, do not store.

2. 24-Well plates (Sigma–Aldrich #CLS3527).

3. Antibiotic stocks: Nystatin (10,000 U/ml, Gibco, store in 1 ml aliquots at −20°C). Penicillin/streptomycin (10,000 U/ml, 10 mg/ml, Gibco, store in 100 μl aliquots at −20°C). Gentamicin (10 mg/ml, Gibco, store at 4°C).

4. Incomplete oocyte culture medium (In-OCM): 50% L-15 medium (Sigma–Aldrich), 15 mM HEPES (pH 7.6), 1 mg/ml insulin (Sigma–Aldrich). We typically make a 50-ml stock, which can be stored at 4°C for up to 2 months.

5. Complete oocyte culture medium (OCM): 980 μl In-OCM, 5 μl nystatin stock, 10 μl gentamicin stock, and 5 μl penicillin/streptomycin stock. Sterilize using a 0.22-μm syringe filter (Millipore #SLGP033RS). Make fresh, do not store.

2.4. Oocyte Fixation and Immunofluorescence

1. Glass vials for fixation (Fisher #03-339-26B).

2. 10× MEM Stock: 1 M MOPS (pH 7.4), 20 mM EGTA, and 10 mM MgSO$_4$. Store at room temperature.

3. MEMFA: 1 ml of 10× MEM, 1 ml of 37% formaldehyde, and 8 ml of deionized H$_2$O. Prepare fresh for each use (see Note 2).

4. Methanol: Anhydrous methanol (Alfa Aesar #41467), must be fresh.

5. Proteinase K Solution: 0.1 M Tris–HCl at pH 7.5, 10 mM EDTA, and 50 μg/ml Proteinase K (Sigma–Aldrich). Prepare fresh for each use.

6. PBT: PBS (pH 7.4), 0.2% BSA, and 0.1% Triton X-100 (Roche).

7. Alexa Fluor 633 – coupled secondary antibodies (Invitrogen #A21070 (goat anti-rabbit IgG) or #A21050 (goat anti-mouse IgG)) (see Note 1).

2.5. Confocal Microscopy

1. Murray's Clearing Medium: two parts benzyl benzoate (MD Biomedicals) and one part benzyl alcohol (Fisher). Store at room temperature.

2. Imaging dishes: FluoroDish (WPI Inc. #FD3510-100).

3. Methods

3.1. Preparation of Fluorescently Labeled RNA by In Vitro Transcription

1. Linearize plasmid DNA containing the relevant sequence and upstream promoter sites for transcription by T7, SP6, or T3 RNA polymerase.
2. Resuspend DNA at 1 µg/µl with DEPC-H_2O.
3. Add the following reagents, in order, to a sterile 1.5-ml tube:

2 µl	10× Tx buffer
1 µl	20× cap/NTP mix
11 µl	DEPC-H_2O
1 µl	0.2 M DTT
1 µl	RNasin (Promega)
1 µl	Linearized DNA template (see step 1, Subheading 3.1)
1 µl	1 mM Alexa Fluor 546-14-UTP
1 µl	(α-^{32}P)UTP (1 µCi/µl, Perkin Elmer)
1 µl	RNA polymerase (Promega)

4. Mix reagents gently and centrifuge briefly (10 s in a microcentrifuge).
5. Incubate for 2–4 h at 37°C, covering the tube with aluminum foil to prevent photobleaching of fluorophore.
6. Add 1 µl of 1 mg/ml RNase-free DNase (Promega) to degrade template DNA.
7. Incubate for 15 min at 37°C.
8. Add 79 µl of 20 mM EDTA (pH 8.0) to stop the reaction.
9. Remove 1 µl to a separate tube, labeled as "input" and save (see Note 3).
10. Spin reaction through a 1-ml G50 column, which will retain unincorporated nucleotides while excluding full length mRNA.
11. Concentrate the RNA by ethanol precipitation:
    (a) Add 250 µl of 100% ethanol, 10 µl of 7 M ammonium acetate, 1 µl of carrier RNA (yeast tRNA at 10 µg/µl) or 1 µl of glycogen (20 mg/ml).
    (b) Mix well and freeze until solid on dry ice or at –80°C for ~30 min.
    (c) Spin at maximum speed in microcentrifuge for 15 min. Remove the supernatant. Wash the pellet with 150 µl of cold 75% ethanol.
    (d) Spin for 3 min at maximum speed in microcentrifuge. Remove the supernatant and allow the pellet to dry for 3 min, at 37°C.

12. Resuspend RNA in 11 μl DEPC-H$_2$O. Remove 1 μl to a separate tube, labeled as "incorporated" (see Note 3).

13. Determine the percent incorporation (see Note 3) and bring the RNA to a final concentration of 50 nM in DEPC-H$_2$O (see Note 4 for troubleshooting RNA yield).

14. The RNA should be frozen in 5 μl aliquots, for single use (to avoid freeze/thaw cycles), and can be stored for several months at −80°C.

3.2. Oocyte Microinjection

1. RNA preparation: Thaw an aliquot of RNA (50 nM), and denature the RNA at 70°C for 3–5 min. Spin for 10 min at maximum speed in microcentrifuge to remove particulates, remove to a fresh tube and keep on ice.

2. Oocyte preparation: Surgically remove oocytes from albino *X. laevis* females (Nasco) and defolliculate by incubation in Collagenase Solution for 15 min, gently shaking (~200 rpm) at 18°C. Check to ensure ovary has released the defolliculated oocytes (see Note 5). Wash the oocytes three times with MBSH buffer. Manually sort stage III/IV oocytes (21), which are 250–400 μm in diameter.

3. Oocyte microinjection:
   (a) Calibrate needle with DEPC-H$_2$O to deliver 2 nl per injection.
   (b) Load RNA into needle.
   (c) Place sorted oocytes in injection dish in MBSH buffer.
   (d) Carefully inject each oocyte with 2 nl of RNA at 50 nM.
   (e) Expel RNA, rinse needle with DEPC-H$_2$O, and load the next RNA for injection (see Note 6).

3.3. Oocyte Culture

1. Place injected oocytes in a well of a sterile 24-well plate.
2. Remove buffer and replace with 400 μl OCM per well.
3. Incubate oocytes at 18°C for time points ranging between 8 and 48 h.
4. After culture, remove any dead oocytes; we routinely observe >90% survival (see Note 5).

3.4. Oocyte Fixation and Immunofluorescence

1. Place cultured oocytes in glass vials and rinse with MBSH.
2. If you are co-imaging RNA and protein distribution, skip to step 3, Subheading 3.4. If you are imaging RNA alone, fix oocytes as follows:
   (a) Remove MBSH and replace with 1 ml MEMFA (see Note 2).
   (b) Rock for 20 min at room temperature.
   (c) Wash oocytes once with 1 ml MBSH. Skip to step 4, Subheading 3.4.

3. For immunofluorescence, treat the oocytes with Proteinase K, followed by fixation and antibody incubation, as follows:
   (a) Remove MBSH and add Proteinase K solution. Incubate for 3 min at room temperature.
   (b) Remove Proteinase K solution and wash twice with 1 ml of MEMFA (see Note 2).
   (c) Rock vials for 1 h at room temperature in 1 ml of MEMFA.
   (d) Remove MEMFA and add 1 ml of PBT, rock at room temperature for 15 min. Repeat PBT wash twice, for a total of three washes.
   (e) Replace PBT with 500 µl of fresh PBT plus 2% BSA and 2% goat serum and rock for 2 h at room temperature (see Note 7).
   (f) Replace solution with 500 µl of PBT plus 2% BSA, 2% goat serum, and the appropriate dilution of primary antibody. (You may use 250 µl, if antibody is limiting.) Rock vials overnight at 4°C.
   (g) Replace the primary antibody solution with 500 µl of PBT and rock at room temperature for 1.5 h. Wash twice more with PBT, for a total of three washes.
   (h) Replace PBT with 500 µl of PBT plus 2% BSA, 2% goat serum, and the appropriate dilution of the secondary antibody. Rock vials overnight at 4°C.
   (i) Replace the solution with 500 µl PBT and rock at room temperature for 1.5 h. Wash twice more with PBT, for a total of three washes.
4. Dehydrate oocytes as follows:
   (a) Remove half of the volume of buffer (MBSH or PBT), and replace with an equal volume of anhydrous methanol.
   (b) Remove half of the buffer/methanol solution and replace with an equal volume of anhydrous methanol. Repeat.
   (c) Remove all of the solution, replace with anhydrous methanol.
   (d) Wash once with anhydrous methanol.
5. Oocytes can be stored in methanol at −20°C until ready to image.

3.5. Imaging of RNA and Protein Distribution by Confocal Microscopy

1. Place Murray's Clearing Medium into a Fluorodish.
2. Transfer oocytes from glass vials to an imaging dish, taking care to transfer as little methanol as possible (see Note 7).
3. Wait for several minutes for the oocytes to become optically clear. The oocytes should also sink to the bottom of the imaging dish. If not, gently tap the oocytes with forceps to break surface tension.

4. Image oocytes using an inverted confocal microscope. The entire field of oocytes can first be imaged using a 10× objective, which facilitates scoring localization for entire batches of oocytes. For high-resolution imaging of individual oocytes, a 20× or 63× objective should be used (see also, Note 8).

5. Examples of RNAs visualized using this protocol are shown in Fig. 1. An example of RNA and protein colocalization is shown in Fig. 2.

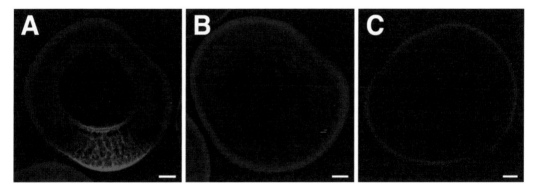

Fig. 1. Vegetal RNA localization in *Xenopus* oocytes. Stage III oocytes were injected with Alexa-546-labeled VLE RNA (a) or Alexa-546-labeled vector control RNA (b) and cultured to allow localization. Control oocytes were uninjected (c). The VLE (*V*g1 *L*ocalization *E*lement; (11)) directs RNA localization to the vegetal cortex, bottom, while the vector control RNA (of similar length, synthesized from pSP73) is uniformly distributed throughout the oocyte cytoplasm. Scale bar = 50 μm.

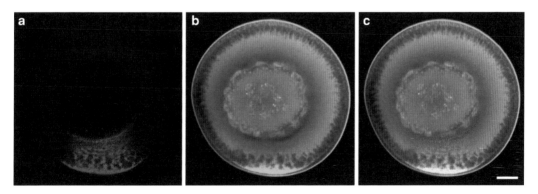

Fig. 2. RNA and protein colocalization in *Xenopus* oocytes. Stage III oocytes were injected with Alexa-546-labeled VLE RNA and cultured to allow localization of the injected RNA. Immunofluorescence was subsequently performed using antibodies directed against PTB/hnRNP I, an RNA binding protein that associates with VLE RNA and is required for VLE RNA localization (6). Vegetal localization of VLE RNA is shown in the *red* channel (a), and PTB distribution is visualized in the *green* channel (b). Colocalization of VLE RNA and PTB protein is evident in the overlay of the *red* and *green* channels, and is shown in *yellow* (c). Scale bar = 50 μm.

4. Notes

1. Autofluorescence of yolk proteins in the oocyte places limitations on the choice of fluorescent nucleotides and secondary antibodies. If possible, fluorescent nucleotides and secondary antibodies giving emission at longer wavelengths (e.g., 546, 633 nm) should be used, as autofluorescence is significant at shorter wavelengths (e.g., 488 nm).

2. The 10× MEM stock is good for several months as long as it remains colorless; store at room temperature wrapped in aluminum foil. Twenty minutes is the minimum time for fixation; however, we can also fix for several hours without negatively affecting oocyte quality.

3. To calculate RNA yield, first determine cpm in "input" (see step 9, Subheading 3.1) and "incorporated" (see step 12, Subheading 3.1) samples using a standard scintillation counter (note that the "input" represents 1% of the sample, while the "incorporated" represents 10% of the sample).

$$\% \text{ incorporation} = \frac{\text{"incorporated"}}{10 \times \text{"input"}} \times 100$$

Since 2.64 µg is maximum yield of RNA for a 20 µl transcription reaction where GTP, the limiting nucleotide, is 0.1 mM, the yield of RNA (in µg) is calculated as:

$$\text{RNA yield} = \% \text{ incorporation} \times 2.64 \text{ µg} \times 0.01.$$

Typical reaction yields are in the range of 50–100 µl of RNA at 50 nM.

4. Problems with low RNA yield could be due to one or more of the following issues:

    (a) Low template DNA concentration: Quantify template concentration to ensure addition of 1 µg of linear DNA per transcription reaction.

    (b) Nonlinear template DNA: Run a sample of template on an agarose gel to confirm complete linearization of plasmid DNA.

    (c) Impurity of template DNA: After linearization of the template DNA, treat with proteinase K, followed by phenol extraction to ensure removal of any contaminating protein. After ethanol precipitation, wash the DNA pellet with 75% ethanol to remove residual salts.

    (d) Loss of RNA during precipitation: Be careful to completely freeze precipitation reaction as described in step 11, Subheading 3.1. Do not phenol extract the RNA.

(e) RNase contamination: Verify that all enzymes are RNase-free; DEPC-treat all solutions as described in step 1, Subheading 2.1.

5. Problems with oocyte viability generally stem from extended collagenase treatment during oocyte isolation or nonsterile conditions during oocyte culture. The following precautions are necessary:

   (a) Careful collagenase treatment is essential when isolating oocytes. Ovary should be cut with scissors before placing in collagenase solution. Defolliculation is complete when the majority of ovary chunks have completely released individual oocytes. Incubation should be closely monitored especially when using a new batch of collagenase for the first time; lots can vary greatly in activity. Some lots may take up to 25 min to reduce ovary to defolliculated oocytes. However, other lots may complete this task in half the time. Buffer pH is also extremely critical; if defolliculation takes an abnormally long time the buffer pH should be checked.

   (b) OCM quality must be maintained to successfully culture oocytes. In-OCM should be a bright red/pink color. Any other color indicates that the pH has deviated from normal and the In-OCM should be discarded. Store In-OCM no longer than 2 months at 4°C. We always make complete OCM fresh immediately before use to culture oocytes. Some antibiotics lose their efficacy after repeated freeze/thaw cycles, resulting in bacterial or fungal contamination. We avoid this issue by aliquoting the antimicrobial stock solutions in small volumes, as described in step 3, Subheading 2.3.

   (c) Successfully cultured oocytes maintain their spherical shape and do not stick to the bottom of the plate well during culture. Microinjection is a stressful procedure for oocytes; approximately 10% of oocytes will not survive even under the best conditions. Greater than 30% oocyte death after culture should be a warning sign that oocytes are not healthy and the experiment may be jeopardized.

6. The steps below can help to prevent problems with needle clogging during microinjection:

   (a) Always centrifuge the RNA and carefully place on ice before loading into the needle. Centrifugation removes large particulate matter that quickly clogs the needle.

   (b) Ensure slight outward pressure in the needle. This will prevent the viscous cytoplasm from entering the needle upon piercing the oocyte.

   (c) Keep the tip of the needle immersed in liquid whenever possible. If you have to take the needle out of liquid, quickly expel a drop to cover the tip. RNA can dry rapidly and clog the needle.

7. Problems with signal detection can occur, and may arise from a number of sources:

    (a) No RNA signal/no localization: Adequate controls are critical to dissect issues with RNA signal. Negative controls include both uninjected oocytes (see Fig. 1c), and injection of a nonlocalizing control RNA (see Fig. 1b), which should be uniformly distributed in the oocyte cytoplasm. Additionally, a positive control RNA that is known to exhibit localization (see Fig. 1a) is also critical. If both the nonlocalizing RNA and the positive control show no fluorescence increase over background, there may be problems in synthesis of fluorescently labeled RNA. If, however, fluorescence of the positive control RNA is evident throughout the cytoplasm, the oocyte quality may be insufficient to support localization, and the experiment must be repeated.

    (b) No immunofluorescence (IF) signal: Antibody selection is crucial. The optimal antibody to use is one previously shown to work for IF in *Xenopus* oocytes or tissues. However, many antibodies have been shown to work for IF in other systems, but have not been tested in *Xenopus*. We have had success using such antibodies; however, when using an antibody that has not been tested for IF in any system, a series of careful controls must be used to ensure useful data. One essential negative control is the secondary only control, which is treated identically to the experiment oocytes but without incubation with primary antibody. A useful positive control when testing unknown antibodies is a proven antibody that works for IF in *Xenopus* oocytes. Finally, the blocking solution should use serum from the animal that the secondary was made; we routinely use goat serum and goat secondary antibodies.

    (c) Uncleared oocytes: Problems with oocyte clearing can result in residual opacity and inability to visualize signal deep into the oocyte. Often the nucleus cannot be visualized at all. This can be avoided by carefully dehydrating the oocytes into anhydrous methanol during fixation. Fresh anhydrous methanol and multiple washes are required for dehydration, as even trace amounts of water will prevent clearing. Also, when transferring oocytes from methanol to Murray's Clearing Medium, great care should be taken to minimize the amount of methanol transferred with the single drop containing the oocytes. This ensures that the Murray's Clear penetrates the oocytes without being diluted with methanol.

8. Additional imaging tips: We generally image with a fairly open pinhole (>1 Airy Unit), as fluorescence intensity can be quite weak. However, results may vary from batch to batch of oocytes. Additionally, some autofluorescent subcellular structures may be visualized in the 488 nm (green channel), which can be useful for orienting oocytes along the animal/vegetal axis.

Acknowledgements

We would like to thank Mowry lab members past and present, who helped to develop these methods. This work was supported by NIH grant # R01GM071049 to KLM.

References

1. Shav-Tal Y., and Singer R.H. (2005) RNA Localization. *J Cell Sci.* **118**, 4077–4081.
2. St Johnston D. (2005) Moving messages: the intracellular localization of mRNAs. *Nat Rev Mol Cell Biol.* **6**:363–75.
3. King, M.L., Messitt, T.J. and Mowry, K.L. (2005). Putting RNAs in the right place at the right time: RNA localization in the frog oocyte. *Biol. Cell.* **97**:19–33.
4. Forristall C., Pondel M., Chen L., and King M. L. (1995) Patterns of localization and cytoskeletal association of two vegetally localized RNAs, Vg1 and Xcat-2. *Development* **121**:201–208.
5. Kloc M., and Etkin L.D. (1995) Two distinct pathways for the localization of RNAs at the vegetal cortex in *Xenopus* oocytes. *Development* **121**:287–97.
6. Cote C.A., Gautreau D., Denegre J.M., Kress T.L., Terry N.A., and Mowry K.L. (1999) A Xenopus protein related to hnRNP I has a role in cytoplasmic RNA localization. *Mol. Cell* **4**:431–7.
7. Bratu D. P., Cha B. J., Mhlanga M. M., Kramer F. R., and Tyagi S. (2003) Visualizing the distribution and transport of mRNAs in living cells, *Proc Natl Acad Sci USA* **100**, 13308–13313.
8. Bertrand E. C., Pascal S., Matthias S., Shailesh M., Singer R.H., and Long R.M. (1998) Localization of ASH1 mRNA particles in living yeast. *Mol Cell* **2**:437–445.
9. Forrest K.M., and Gavis E.R. (2003) Live imaging of endogenous RNA reveals a diffusion and entrapment mechanism for nanos mRNA localization in Drosophila. *Curr Biol.* **15**:1159–68.
10. Yisraeli J.K., Sokol S., and Melton D.A. (1990) A two-step model for the localization of maternal mRNA in *Xenopus* oocytes: involvement of microtubules and microfilaments in the translocation and anchoring of Vg1 mRNA. *Development* **108**:289–98.
11. Mowry K.L., and Melton D.A. (1992) Vegetal messenger RNA localization directed by a 340-nt RNA sequence element in *Xenopus* oocytes. *Science* **21**:991–4.
12. Deshler J.O., Highett M.I., and Schnapp B.J. (1997) Localization of *Xenopus* Vg1 mRNA by Vera protein and the endoplasmic reticulum. *Science* **276**:1128–31.
13. Havin L., Git A., Elisha Z., Oberman F., Yaniv K., Schwartz S.P., Standart N., and Yisraeli J.K. (1998) RNA-binding protein conserved in both microtubule- and microfilament-based RNA localization. *Genes Dev.* **12**:1593–8.
14. Zhang J., Houston D.W., King M.L., Payne C., Wylie C., and Heasman J. (1998) The role of maternal VegT in establishing the primary germ layers in *Xenopus* embryos. *Cell* **94**:515–24.
15. Houston D.W., and King M.L. (2000) A critical role for Xdazl, a germ plasm-localized RNA, in the differentiation of primordial germ cells in *Xenopus. Development* **127**:447–56.
16. Yoon Y.J., and Mowry K.L. (2004) *Xenopus* Staufen is a component of a ribonucleoprotein complex containing Vg1 RNA and kinesin. *Development* **131**:3035–45.

17. Czaplinski K., Köcher T., Schelder M., Segref A., Wilm M., and Mattaj I.W. (2005) Identification of 40LoVe, a *Xenopus* hnRNP D family protein involved in localizing a TGF-beta-related mRNA during oogenesis. *Dev. Cell* **8**:505–15.
18. Messitt T.J., Gagnon J.A., Kreiling J.A., Pratt C.A., Yoon Y.J., and Mowry K.L. (2008) Multiple kinesin motors coordinate cytoplasmic RNA transport on a subpopulation of microtubules in *Xenopus* oocytes. *Dev. Cell* **15**:426–36.
19. Wallace R.A., Misulovin Z., and Wiley H.S. (1980) Growth of anuran oocytes in serum-supplemented medium. *Reprod Nutr Dev.* **20**:699–708.
20. Yisraeli J.K., and Melton D.A. (1988) The material mRNA Vg1 is correctly localized following injection into *Xenopus* oocytes. *Nature* **336**:592–5.
21. Dumont, J. N. (1972) Oogenesis in *Xenopus laevis* (Daudin). I. Stages of oocyte development in laboratory maintained animals. *J. Morphol.* **136**:153–179.

Chapter 6

Visualization of mRNA Expression in the Zebrafish Embryo

Yossy Machluf and Gil Levkowitz

Abstract

Examination of spatial and temporal gene expression pattern is a key step towards understanding gene function. Therefore, *in situ* hybridization of mRNA is one of the most powerful and widely used techniques in biology. Recent advances allow the reliable and simultaneous detection of mRNA transcripts, or combinations of mRNA and protein, in zebrafish embryos.

Here we describe a standard protocol for visualizing the precise expression pattern of a single transcript or multiple gene products. The procedure employs fixation and permeabilization of embryos, followed by hybridization with tagged antisense riboprobes. Excess probes are then washed and hybrids are detected by enzyme-mediated immunohistochemistry utilizing either chromogenic or fluorescent substrates.

Key words: Gene expression, mRNA detection, Whole-mount, *In situ* hybridization, Zebrafish, Embryos, Labeled riboprobes, Fluorescence, Colorimetric assay

1. Introduction

In situ hybridization (ISH) is the most prevalent method of choice for analyzing the presence and distribution of specific nucleic acid sequences in preserved cells or fixed tissue samples (1). ISH can be used to establish gene expression pattern, to define synexpression groups, and to identify tissue- and cell-specific markers important for distinct developmental stages (2). Early approaches to ISH employed radioactively labeled probes to detect transcripts on histological sections. The introduction of nonradioactive labeling systems allowed faster and easier transcript visualization in whole-mounted tissues and embryos. Subsequent improvements permitted the detection of two or three different gene products using different colors within the same sample, utilizing either chromogenic stains or fluorescent dyes (Fig. 2) (3–5). When combined with laser scanning confocal microscopy, fluorescent ISH allows detailed analysis of gene expression, including low abundant mRNAs, at a single-cell resolution (6, 7). The application of mRNA ISH to the semi-transparent zebrafish (*Danio rerio*) embryos highlights the advantages of this technique in studying gene expression and developmental processes (8–10). A comprehensive and updated collection of gene expression data in zebrafish is available online at Zebrafish Information Network (at http://zfin.org) (11).

While individual ISH protocols in zebrafish are varied, all are ultimately comprised of a few basic steps (Fig. 1). First, *in vitro* synthesis of labeled RNA probes: commonly used tags are digoxigenin (DIG, a plant steroid), fluorescein, and biotin (vitamin B_7). In parallel, zebrafish embryos are prepared by dechorionation, fixation, permeabilization, and re-fixation. Then, embryos are incubated with the labeled probe(s), followed by removal of unbound probe(s). Hybridized probes are detected by anti-tag antibodies (Ab) conjugated to an enzyme, usually alkaline phosphatase (AP) or peroxidase (POD) [also referred as horseradish peroxidase (HRP)]. Finally, the Ab-probe-transcript complex is visualized by addition of either chromogenic [such as nitroblue tetrazolium (NBT)/5-bromo-4-chloro-3-indolyl phosphate (BCIP) or 3,3′-diaminobenzidine (DAB), respectively] or fluorescent phosphatase substrates [Fast Red or enzyme-labeled

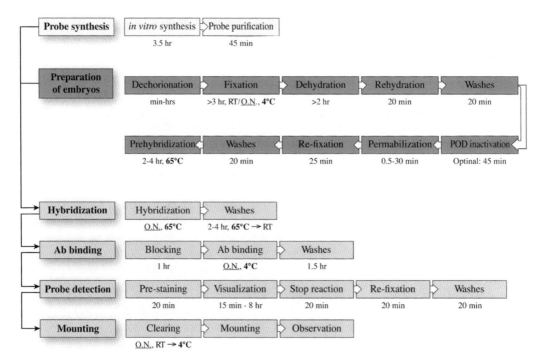

Fig. 1. Flow scheme of the *in situ* hybridization (ISH) procedure. The different steps of the protocol are shown. Information about the time and temperature required at each step is indicated. *O.N.* overnight, *RT* room temperature.

fluorescence (ELF) (12) and lately even NBT/BCIP (13, 14) for AP, tyramide signal amplification (TSA) for POD (6, 15, 16)]. These substrates are then catalyzed by the respective Ab-conjugated enzymes to produce a visible color. Depending on the enzyme-conjugated antibodies used, the multicolor reaction can be performed simultaneously or sequentially. ISH protocols can be combined with immunostaining or transgenic fluorescent lines that express fluorescent proteins such as green fluorescent protein (GFP) or red fluorescent protein (RFP), allowing labeling of desired cell types or structures.

This article provides detailed step-by-step instructions for the application of whole-mount mRNA ISH to zebrafish embryos.

2. Materials

It is recommended that all solutions be made using RNase-free reagents and diethyl pyrocarbonate (DEPC)-treated double distilled water (DDW) (see Note 1).

2.1. Preparation of Labeled Probes

This step is used to prepare large amounts of labeled single-stranded RNA probes (see Note 2).

1. 1 μg of purified linearized DNA containing the gene of interest and RNA polymerase priming site (T3, T7 or SP6) (see Note 3).
2. Sterile, RNase-free, high-purity water.
3. RNA polymerases: SP6 (Roche, cat. no. 10810274001), T7 (Roche, cat. no. 11881767001) or T3 (Roche, cat. no. 11031163001).
4. 10× polymerase transcription buffer (Roche, cat. no. 1465384).
5. 10× RNA labeling mix, containing either DIG-11-uridine-5′-triphosphate (UTP) (Roche, cat. no. 11277073910), fluorescein-12-UTP (Roche, cat. no. 11685619910) or biotin-16-UTP tag (Roche, cat. no. 11685597910).
6. RNase inhibitor (Roche, cat. no. 03335399001).
7. DNase I, RNase-free (Roche, cat. no. 10776785001).
8. 0.2 M ethylenediaminetetraacetic acid (EDTA) (J.T. Baker, cat. no. 11-8993-01) at pH 8.0.
9. Agarose (Sigma, cat. no. A9539).
10. Tris–acetate–EDTA (TAE) (50× solution, Biological industries, cat. no. 01-870-1A) agarose gel running buffer.
11. Ethidium bromide 10 mg/ml (Bio-Rad cat. no. 161-0433).

2.2. Preparation of Embryos

Zebrafish embryos undergo serial actions which prepare them for hybridization, including dechorionation, fixation, permeabilization, and prehybridization.

1. Danieau buffer: 1.74 mM NaCl, 0.21 mM KCl, 0.12 mM $MgSO_4$, 0.18 mM $Ca(NO_3)_2$, 0.15 mM HEPES. Adjust pH to 7.4 by drop-wise addition of 1 M NaOH. Filter through 0.22-μm filter. Add ~0.05% volume/volume (v/v) methylene blue before use.

2. Dulbecco's phosphate-buffered saline (PBS) modified, sterile, without calcium chloride and magnesium chloride (Invitrogen, cat. no. 14200-067). Although the use of commercial PBS is favored, it can be prepared according to a standard lab protocol. For 10× PBS, dissolve the following components in 900 ml of DDW: 2 g potassium chloride, 80 g sodium chloride, 2 g potassium phosphate (monobasic, anhydrous), 11.5 g sodium phosphate (dibasic, anhydrous). Adjust pH to 7.2 with sodium hydroxide. Complete volume to 1 l with DDW. Sterilize by autoclaving.

3. Pronase E (Sigma, cat. no. P5147) 1 mg/ml in Danieau buffer. Prepare stock solution of 10 mg/ml in DDW and store at −20°C.

4. 4% weight/volume (w/v) paraformaldehyde (PFA) (Sigma, cat. no. P6148) in 1× PBS at pH = 7.2 (see Note 4). *Caution*: noxious substance. Avoid inhalation and contact with eyes and skin.

5. 100% methanol.

6. PBST: 0.1% (v/v) Tween-20 (Sigma, cat. no. P1379) in PBS.

7. Proteinase K (Roche, cat. no. 03115828001) 10 mg/ml in PBST (1:2,000 dilution), store at 4°C.

8. 20× standard saline citrate (SSC) buffer (also known as saline sodium citrate, or sodium chloride sodium citrate): 3 M NaCl, 0.3 M trisodium citrate (Sigma, cat. no. C0909), in DEPC-treated DDW. Adjust to pH = 7.0 by drop-wise addition of 1 M HCl.

9. Prehybridization mix (preHM).
   For *colorimetric assay* use: 50% (v/v) deionized formamide (Sigma, cat. no. F9037, store at 4°C), 5× SSC at pH = 7.0, 50 μg/ml heparin (Sigma, cat. no. H3393, store at −20°C), 500 μg/ml yeast tRNA (Sigma, cat. no. R6625) (see Note 5), 0.1% (v/v) Tween-20, 0.01 M citric acid (Sigma, cat. no. 27490) (bring to pH 6.0) in sterile DDW.
   For *fluorescent (TSA) assay* use the same solution with the following modifications: 5 mg/ml yeast tRNA, citric acid is not required.
   Store at −20°C.

2.3. Hybridization

RNase-free conditions are no longer obligatory, and dilutions can be done using DDW. 0.1% (v/v) Tween-20 is required to prevent embryos sticking to each other.

1. SSCT: SSC buffer containing 0.1% (v/v) Tween-20.
2. Formamide (J.T. Baker, cat. no. 4028) dilutions in 2× SCC solution containing of 0.1% (v/v) Tween-20.
3. Tris buffered saline plus Tween (TBST): 100 mM Tris–HCl at pH 7.5, 150 mM NaCl, 0.5% (v/v) Tween-20.

2.4. Antibody Binding

1. Blocking solutions:

    For *colorimetric assay* use: PBST containing 2% (v/v) lamb serum (Invitrogen, cat. no. 16070096), 2 mg/ml BSA (Sigma, cat. no. A7906).

    For *fluorescent (TSA) assay*: 10× blocking stock solution containing 10% (w/v) blocking reagent (Roche, Cat. no. 11096176001) dissolved in maleic acid buffer [100 mM maleic acid (Sigma, cat. no. M0375), 150 mM NaCl, pH 7.5]. Autoclave and store at −20°C. For preparation of working solution, dilute 1:10 in TBST buffer.

2. Antibody solutions: dilute Ab in the appropriate blocking solutions.

3. Common antibodies: anti-DIG-AP Fab fragments (Roche, cat. no. 11093274910), anti-DIG-POD Fab fragments (Roche, cat. no. 11207733910), anti-Fluorescein-AP Fab fragments (Roche, cat. no. 11426338910), anti-Fluorescein-POD Fab fragments (Roche, cat. no. 11426346910), anti-biotin-AP Fab fragments (Roche, cat. no. 1426311910), and anti-biotin-POD Fab fragments (Roche, cat. no. 1426303910). All Abs are polyclonal from sheep.

2.5. Probe Detection

1. Pre-staining solutions:

    (a) Low-salt alkaline-Tris buffer for *colorimetric assay*: 100 mM Tris–HCl at pH = 9.5, 50 mM $MgCl_2$, 100 mM NaCl, and 0.1% (v/v) Tween-20 in sterile DDW.

    (b) High-salt alkaline-Tris buffer for *colorimetric/fluorescent (Fast Red) assay*: 100 mM Tris–HCl at pH = 8.2, 400 mM NaCl, and 0.1% (v/v) Tween-20 in sterile DDW. High salt is necessary to reduce background staining.

    (c) No pre-staining solution is used for *fluorescent (TSA) assay*.

2. Substrate solutions:

    (a) NBT/BCIP buffer for *colorimetric assay*: add 10 µl of NBT/BCIP solution (Roche, cat. no. 11681451001) to 1 ml pre-staining solution (low-salt alkaline-Tris buffer at pH 9.5). Staining buffer is yellowish. Cover with Aluminum foil to protect from light.

(b) Fast Red buffer for *colorimetric/fluorescent (Fast Red) assay*: Dissolve 1 tablet of Fast Red (Roche, cat. no. 11496549001) in 2 ml pre-staining solution (high-salt alkaline-Tris buffer at pH 8.2). It is recommended to filter the solution using Whatman No. 2 paper to omit undissolved particles. Alternatively, spin down solution and use the clear supernatant liquid.

(c) TSA buffer for *fluorescent (TSA) assay*: Commercial TSA kit (Invitrogen, cat. no.T20932 for tyramide labeled with Alexa Fluor 488) (see Note 6). Dilute tyramide stock solution 1:100 in amplification buffer containing 0.0015% (v/v) H_2O_2.

(d) DAB for *colorimetric assay* and for *immunostaining*: Dissolve 1 tablet of DAB (Sigma, cat. no. D9292) and 1 tablet of urea hydrogen peroxide (Sigma, cat. no. U1380) in 5 ml of sterile DDW, and add 0.1% (v/v) Tween-20. Substrate solution should be used within 1 h. Handling and storage with DAB requires special *caution* (see Note 7).

3. AP inactivation solution: 0.1 M glycine hydrochloride at pH 2.2, 0.1% (v/v) Tween-20.

3. Methods

Wearing gloves and lab coat is obligatory throughout the whole procedure.

3.1. Basic Protocol: One-Color ISH

All steps up to Ab binding (Subheading 3.1.4) are carried out in RNase-free conditions (see Note 1), and are performed at room temperature (RT) if not mentioned otherwise. All steps are performed in 1.7- or 2-ml microcentrifuge tubes, each containing up to 25–30 embryos, and placed horizontally on their side so that the embryos are spread as much as possible, to allow maximal surface contact with solutions. Gentle agitation (~30–50 rpm) during washes is not recommended when embryos are younger than 24 h post-fertilization (hpf), since they may disintegrate. When necessary, pre-warm solutions.

3.1.1. Preparation of Labeled Probes

1. *In vitro* synthesis reaction of labeled RNA probe consists of the following ingredients and should not exceed a total volume of 20 μl: 2 μl 10× polymerase transcription buffer, 2 μl 10× RNA labeling mix, 1 μl RNase inhibitor, 1 μg linearized plasmid, 1 μl (20 units) RNA polymerase, and sterile RNase-free water.
2. Mix gently and spin down.
3. Incubate for 2 h at either 37°C (T3 or T7 polymerases) or 40°C (SP6 polymerase).

Optional: Add additional 1 μl (20 units) of RNA polymerase and incubate further for 1 h (this step is only required when higher yield of probe is desired or low amount of template was used).

4. Add 2 μl of RNase-free DNase I and incubate at 37°C for 15 min (this step is aimed to eliminate template DNA, which may interfere with the subsequent hybridization step).

5. Stop reaction by adding 2 μl of 0.2 M EDTA at pH 8.0.

6. Purify the resulting probe using a spin-column-based RNA purification kit [e.g., RNeasy mini kit (Qiagen, cat. no. 74104)].

7. Test the quality of the probe by running 1 μl through an agarose gel (see Note 8).

8. Probe can either be kept in −80°C or diluted with an equal volume (1:1) of preHM and stored at −20°C.

3.1.2. Preparation of Embryos

Since pigmentation may interfere with staining, it is highly recommended to block pigment production during embryonic development [by application of 1-phenyl-2-thiourea (PTU) to Danieau buffer at ~18 hpf (see Note 9)]. This method is, to our experience, superior to bleaching pigmented embryos during the course of ISH (using hydrogen peroxide, following rehydration).

Preferably, embryos are dechorionated prior to fixation (see Note 10). Enzymatic dechorionation is possible only prior to fixation, as opposed to manual dechorionation.

The dehydration and rehydration processes are essential, even if embryos are used immediately, rather than stored.

Usually, all processes are performed in 1.5–2-ml microcentrifuge tubes, each containing up to 25–30 embryos.

1. *Dechorionation*: Place embryos in Petri dish filled with PBST. Removal of chorion can be done manually or enzymatically (see Note 11).

2. *Fixation*: Incubate embryos in fresh 4% (w/v) PFA for at least 3 h at RT, or overnight (O.N.) at 4°C.

    Optional: Pigmented embryos can be bleached in 3% H_2O_2 (see Note 12).

3. *Dehydration*: Wash embryos with 100% methanol for 10 min, replenish with fresh methanol and place in −20°C for at least 2 h. Embryos can be stored under these conditions for up to few months.

4. *Rehydration*: Incubate embryos for 2–5 min (at least until embryos sink to the bottom of the tube) in successive dilutions (v/v) of methanol in PBST: 75, 50, and 25%.

5. *Washing*: Rinse once with PBST, and then wash three times, 5 min each, with PBST.

Optional: When POD-conjugated Ab is used at Subheading 3.1.4, bleach embryos in 1% hydrogen peroxide for 30 min in PBST, to inactivate endogenous peroxidase activity, and wash twice, 5 min each, with PBST.

6. *Permeabilization*: Treat embryos with a fresh dilution of 10 μg/ml Proteinase K in PBST. The exact duration of this treatment depends upon enzyme activity (batch), age of embryos, and expression pattern of analyzed transcript (deep structures vs. surface tissues) (see Note 13).

7. *Re-fixation*: Stop reaction by briefly rinsing with PBST (twice), and re-fix embryos in 4% (w/v) PFA for 20 min.

8. *Washing*: Remove residual PFA by washing embryos three times, 5 min each, with PBST.

9. *Pre-hybridization*: Incubate embryos in ~300 μl preHM for 2–4 h at 65°C (place tubes on their sides, in either heated water bath or incubator).

3.1.3. Hybridization

In general, it is recommended to separately optimize hybridization conditions for each probe. High-stringency hybridization and washing conditions are aimed at preventing cross-hybridization and nonspecific background, respectively.

10. *Hybridization*: Discard preHM and add 250 μl of preHM preheated to 65°C containing 100–150 ng (see Note 14) of antisense labeled RNA probe (from Subheading 3.1.1). Hybridize O.N. at 65°C.

    Optional: Use sense labeled probe, as control.

11. *Washing*: Remove the hybridization solution (see Note 15), and wash embryos as followed:

    For *colorimetric assay*:

    66% preHM: 33% 2× SSC (v/v), for 5 min at 65°C

    33% preHM: 66% 2× SSC (v/v), for 5 min at 65°C

    2× SSC + 0.1% (v/v) Tween-20, for 5 min at 65°C

    0.2× SSC + 0.1% (v/v) Tween-20, for 20 min at 65°C

    Wash twice with 0.1× SSC + 0.1% (v/v) Tween-20, for 20 min each at 65°C

    66% 0.1× SSC: 33% PBST (v/v), for 5 min at RT

    33% 0.1× SSC: 66% PBST (v/v), for 5 min at RT

    PBST, for 5 min at RT.

    For *fluorescent (TSA) assay*:

    PreHM, for 20 min at 65°C

    Wash twice with 50% (v/v) formamide in 2× SSCT, for 20 min each at 65°C

    Wash once with 25% (v/v) formamide in 2× SSCT, for 20 min at 65°C

Wash twice with 2× SSCT, for 20 min each at 65°C

Wash three times with 0.2× SSCT, for 30 min each at 65°C

TBST, for 5 min at 65°C.

3.1.4. Antibody Binding

Pre-blocking is required to saturate nonspecific binding sites for the Ab, in order to reduce background. Avoid agitation when embryos are younger than 24 hpf.

12. *Blocking*: Incubate in 200 µl blocking solution for 1 h at RT with gentle agitation.

13. *Antibody binding*: Incubate in 200 µl Ab solution O.N. at 4°C with gentle agitation. Antibodies are directed against either DIG or Fluorescein and conjugated to either AP or POD enzyme (see Note 16).

14. *Washing*: Remove the Ab solution, rinse embryos twice, and wash 5 timed, for 15 min each.

    Colorimetric assay washes are performed with PBST, while *Fluorescent (TSA) assay* washes are carried out with TBST buffer.

3.1.5. Probe Detection

Staining reaction may utilize various substrates which are catalyzed by the enzymes conjugated to the Ab, resulting in the localized precipitation of a colored product at the site of hybridization (Fig. 2a, b).

15. *Pre-staining*: Incubate three times, 5 min each, in 200 µl of pre-staining buffer.

16. *Visualization*: Remove the pre-staining buffer and add 500 µl of freshly prepared substrate solution (see Note 17). Incubate in the dark to protect from photoreactivity. Monitor the color reaction periodically (usually every 20 min) (see Note 18).

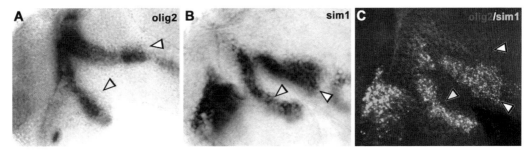

Fig. 2. Whole-mount *in situ* hybridization (ISH). Analyses of *oligodendrocyte transcription factor 2* (*olig2*) (**a**, **c**) and *single minded 1* (*sim1*) (**b**, **c**) expression in zebrafish embryos (lateral view, anterior to the left). (**a**, **b**) One-color colorimetric ISH: embryos were hybridized with specific digoxigenin-labeled probes directed against either *olig2* (**a**) or *sim1* (**b**). Expression was visualized using alkaline phosphatase (AP) and BCIP/NBT staining. (**c**) Two-color fluorescent ISH: embryo was simultaneously hybridized with differentially labeled probes. Expression of *olig2* transcripts (*red*) was visualized using a digoxigenin-labeled probe, AP and Fast Red staining. Localization of *sim1* transcripts (*green*) was detected using a fluorescein-labeling probe, POD and Alexa Fluor 488-based TSA staining. Mutually exclusive expression domains are marked by *white triangles*, whereas a co-expression domain is indicated by a *yellow triangle*. Lateral view, anterior to the left.

17. *Stop reaction*: When the desired signal vs. background intensity staining is accomplished, discard the substrate solution, rinse twice and wash three times, 5 min each.

    As in step 14, *Colorimetric assay* washes are performed with PBST, while *Fluorescent (TSA) assay* washes are carried out with TBST buffer.

18. *Re-fixation*: Re-fix embryos with 4% (w/v) PFA for 20 min.

19. *Washing*: Remove residual PFA by washing the embryos three times, 5 min each, with PBST.

3.1.6. Mounting

Embryos are gradually transferred from water to glycerol, a clearing reagent that conveys a higher degree of transparency to the biological sample and allows for long-term storage at 4°C.

20. *Clearing*: Equilibrate embryos (for at least 5 min) in successive dilutions (v/v) of glycerol in PBST: 25, 50, and 75%. Leave stained embryos in 75% glycerol in PBST O.N. at 4°C. Embryos can be stored at 4°C for several months or more (see Note 19).

21. *Mounting*: Dissect, de-yolk, and flat-mount embryos in 75% glycerol in PBST on a microscope slide.

3.2. Combined Protocol: Two-Color ISH

Simultaneous analysis of expression patterns of two (or more) genes within the same embryo is required for visualizing the relative positions of the respective genes (Fig. 2c).

To perform double ISH, the antisense probes must be differentially labeled. The most common labeling combination utilizes DIG and fluorescein tags. Biotinylated probes give high background in zebrafish and other vertebrate embryos. Labeled probes are detected by antibodies, conjugated to either AP or POD. To determine the optimal experimental detection strategy, it is advised to initially perform separated ISH with single antisense probes labeled with either tags, and to use antibodies conjugated to different enzymes which catalyze diverse substrates. These experiments also serve as references when co-localized signals are expected. Preferably, the probe that produces a stronger signal should be labeled with fluorescein and visualized first, since fluorescein-labeled probes are generally less stable and yield weaker signal compared to DIG-labeled probes. If the different antibodies are conjugated to different enzymes then staining can be performed simultaneously, whereas utilization of antibodies conjugated to the same enzyme necessitates sequential staining of probes, which includes an intermediate inactivation step (see below).

3.2.1. Multicolor Colorimetric ISH

This method allows the detection of two different transcripts in the same embryo. It utilizes two probes that are labeled differently, AP-conjugated antibodies and various AP chromogenic substrates. Importantly, while the different probes are hybridized simultaneously, they are detected in sequential rounds of Ab binding and staining reaction (Subheadings 3.1.4 and 3.1.5), separated by a crucial step: removal or inactivation of the previously applied AP, either by heating or low pH treatment.

This protocol describes detection of fluorescein-labeled probe with Fast Red, followed by visualization of DIG-labeled probe with NBT/BCIP. One can use biotin-labeled probe with 5-Bromo-6-chloro-3-indoxylphosphate(Magenta-phos)/2-(4-iodophenyl)-3-(4-nitrophenyl)-5-phenyltetrazolium chloride (INT) substrate to visualize a third transcript (colored in yellow).

3.2.1.1. Experimental Procedure (Based on the Basic Colorimetric ISH Protocol)

1. Preparation of labeled probes and embryos (Subheadings 3.1.1 and 3.1.2).
2. Hybridization (Subheadings 3.1.3) – hybridization solution contains a mix of the differentially labeled probes.
3. Detection of the fluorescein-labeled probe (Subheadings 3.1.4 and 3.1.5) – use anti-fluorescein-AP Fab fragments and Fast Red-related solutions.
4. Removal of anti-fluorescein-AP Fab fragments – incubate for 10 min in AP inactivation solution at RT (see Note 20). Wash four times, 5 min each, with PBST.
5. Detection of the DIG-labeled probe (Subheadings 3.1.4 and 3.1.5) – use anti-DIG-AP Fab fragments and NBT/BCIP-related solutions. Omit blocking (it is not required).
6. Mounting (Subheading 3.1.6).

An alternative method for multicolor colorimetric ISH (3) employs anti-tag antibodies conjugated to different enzymes, thus allowing simultaneous binding of antibodies and sequential staining reaction without the need for an inactivation step.

In most protocols, DAB is the recommended substrate for POD, producing a dark brown precipitate. It is usually used to detect the fluorescein-labeled probe, while DIG-labeled probe is visualized by AP supplemented with either NBT/BCIP or Fast Red substrates, giving dark blue/purple or bright red precipitates, respectively.

3.2.2. Multicolor Fluorescent ISH

This robust and reliable procedure describes the detection of fluorescein-labeled probe with AP-conjugated Ab, which catalyzes Fast Red to produce red signal. Following AP inactivation, DIG-labeled probe is visualized by POD-conjugated Ab and TSA which results in green fluorescence (Fig. 2c).

3.2.2.1. Experimental Procedure (Based on the Basic Fluorescent ISH Protocol)

1. Preparation of labeled probes and embryos (Subheadings 3.1.1 and 3.1.2) – inactivation of endogenous POD is an obligatory step in this protocol.
2. Hybridization (Subheadings 3.1.3) – hybridization solution contains a mix of the differentially labeled probes.

3. Detection of fluorescein-labeled probe (Subheadings 3.1.4 and 3.1.5 until step 16) – use anti-fluorescein-AP Fab fragments (diluted 1:500) and Fast Red-related solutions. Rinse twice with PBST.

4. Inactivation of first AP – incubate in AP inactivation solution for 30 min and then thoroughly wash in PBST. Continue to re-fixation and washing steps (Subheading 3.1.5, steps 18 and 19).

5. Wash three times, 5 min each, with TBST.

6. Detection of DIG-labeled probe (Subheadings 3.1.4 and 3.1.5) – incubate in blocking solution for 1 h, use anti-DIG-POD Fab fragments (diluted 1:400) and TSA substrate solution. Staining reaction lasts about 1 h.

7. Mounting (Subheading 3.1.6).

Alternatively, one can utilize POD-conjugated antibodies, and a combination of multi-fluor TSA kits for a convenient multi-target detection, which allows a flexible experimental design due to wide range of fluorescence emission (5, 13) (see Note 21).

3.3. Simultaneous Localization of Multiplex Signals: Localization of Tissue Antigen

The diverse ISH procedures can be combined with additional techniques to visualize multiple types of signals within a single zebrafish embryo in a cellular resolution.

Often, one desires to compare the localization of mRNA transcript with its own translation product or with other tissue antigen. It is recommended to first perform the ISH procedure (without mounting), in order to reduce the likelihood of transcript degradation. On the other hand, the tissue antigen to be recognized by the antibody may be affected by the harsh conditions of the ISH or susceptible to protease digestion. Therefore, each antibody should be examined independently for its capacity to identify the antigen after ISH. In some cases, modifications of the ISH protocol should be implemented, in order to allow the sequential antigen immunostaining, mainly with regard to permeabilization (e.g., using acetone instead of proteinase K treatment), fixation reagent, and hybridization conditions (lower temperature, shorter duration). Attention should be given to the choice of antibodies utilized by both methods. For example, since sheep antibodies are employed in ISH to detect the labeled probes, none of the antibodies used for detection of tissue antigen should be raised against sheep immunoglobulins (4).

3.3.1. Colorimetric Antigen Immunostaining

In our hands, best results were obtained when using the ABC method (using the appropriate VECTASTAIN Elite ABC Kit, Vector Laboratories). Primary Ab against the chosen antigen is recognized by biotinylated secondary Ab, which is then tightly bound by avidin-conjugated POD that is supplemented by DAB substrate to produce brown precipitates.

3.3.1.1. Experimental Procedure (Based on the Basic Colorimetric ISH Protocol)

1. Perform ISH as in Subheadings 3.1.1–3.1.5 until step 18. Use DIG-labeled probe, Ab conjugated to AP, and stain with NBT/BCIP.

2. Wash twice, 10 min each, with PBDTT [PBS, 1% (w/v) DMSO, 0.1% (v/v) Tween-20, 0.5% (v/v) Triton X-100].

3. For early larva (≥72 hpf) only: repeat permeabilization and fixation. Incubate larva in 10 µg/ml of proteinase K in PBDTT for the indicated time (see Note 13). Wash twice, 5 min each, with PBDTT. Re-fix larva in 4% (w/v) PFA for 1 h at RT. Wash three times, 5 min each, with PBDTT.

4. Incubate in the appropriate dilution of primary antibody in blocking solution [1% (w/v) BSA and 2% (v/v) goat serum in PBDTT], for 5 h at RT or O.N. at 4°C.

5. Rinse twice and wash four times, 30–60 min each, with PBDTT (see Note 22).

6. Incubate in blocking solution for 1 h.

7. Incubate in the appropriate secondary biotinylated Ab (Vector Elite kit) diluted 1:1,000 in blocking solution, for 5 h at RT or O.N. at 4°C.

8. Rinse twice and wash four times, 30–60 min each, with PBDTT (see Note 22).

9. Prepare AB complex: add equal volumes of reagent A (Avidin DH solution) and reagent B (biotinylated enzyme) to 1 ml of PBDTT (usually 4–15 µl of each, should be calibrated a priori). Mix and incubate for 45 min at RT.

10. Incubate embryos in 250 µl of AB complex solution for 45–60 min.

11. Wash four times, 15–30 min each, with PBDTT (see Note 22).

12. Incubate in DAB solution (see Subheading 2.5, step 2) until sufficient staining signal is obtained (see Note 7).

13. Rinse twice and wash three times, 5 min each, with PBDTT.

14. Mount embryos as described in Subheading 3.1.6, using PBDTT rather than PBST.

3.3.2. Fluorescent Antigen Immunostaining

Various procedures are available for the detection of antigens fluorescently. Visualization via standard POD/TSA-mediated fluorescent immunohistochemistry is widely spread. Nevertheless, we usually use fluorescently labeled secondary antibody. In the following protocol, we describe the detection of GFP in zebrafish reporter line.

3.3.2.1. Experimental Procedure (Based on the Basic Fluorescent ISH Protocol)

1. Perform ISH as in Subheadings 3.1.1–3.1.5 until step 18.
2. Incubate for 30 min in blocking solution [PBS, 10% (v/v) goat serum, 1% (w/v) DMSO, and 0.3% (v/v) Triton X-100].
3. Incubate O.N. at 4°C with a primary mouse anti-GFP Ab (Millipore, cat no. MAB3580) diluted 1:200 in blocking solution.
4. Rinse once and wash three times, 30–60 min each, with PBS-Triton (see Note 22).
5. Incubate with secondary goat anti-mouse Ab conjugated to Alexa 488 (green, Invitrogen, cat. no. A-11029) or Alexa 594 (red, Invitrogen, cat. no. A-11032) diluted 1:200 in blocking solution (see Note 23), for 2–3 h at RT or O.N. at 4°C.
6. Rinse once and wash three times, 30–60 min each, with PBS-Triton (see Note 22).
7. Mounting (Subheading 3.1.6) – use PBDTT rather than PBST.

4. Notes

1. Most dry reagents can be purchased as certified RNase-free. DEPC-treated water is prepared by adding 1 ml of DEPC to 1,000 ml of DDW, followed by 1 h incubation at RT. The solution is then autoclaved and allowed to cool before use. For the treatment of other instruments and tools, RNase decontamination solutions are commercially available.
2. Usually, sense probes of the corresponding antisense probes are used as control for background staining. However, it is advised to assess specificity by comparing the expression patterns revealed by different non-overlapping probes of the same gene. When designing probes, we recommend directing them against ~350 nucleotides of the 3′-UTR, as these sites are usually less conserved sequences. Such probes are less likely to cross hybridize with other gene family members.
3. Linearized DNA is generated by digesting 5–10 μg of a plasmid containing RNA polymerase promoter site [e.g., pCR2.1 (Invitrogen), pBluescript II (Stratagene) or pGEM-T-based vectors (Promega)], using the appropriate restriction enzyme. The restriction enzyme should cut at a unique site located at the 5′ end of the gene sequence when preparing an antisense probe and at the 3′ end of the gene sequence when preparing a sense control probe. Importantly, if restriction enzymes that produce 3′ overhangs (such as ApaI, BglI, KpnI, SacI, SacII, etc.) are used, the ends of the template should be blunted before transcription, by using a T4 DNA polymerase kit, in order to avoid

RNA polymerases false priming at the site (17) which may lead to a reduced efficiency of the *in vitro* transcription reaction (18). Alternatively, linearized DNA can be produced by PCR utilizing a cDNA library or plasmid DNA of gene of interest and specific primers directed to a unique fragment in the gene of interest, wherein the reverse primer would include the RNA polymerase promoter site. Care must be taken to ensure that the direction of the insert of interest relative to the RNA polymerase promoter site is correct. For antisense probe production, RNA polymerase promoter site is placed in a reverse orientation (3′→5′) relative to the template cDNA, whereas for control sense probes, RNA polymerase promoter site should be placed upstream and in the forward orientation (5′→3′). The resulting linearized DNA can be analyzed by agarose gel electrophoresis to ascertain complete digestion or existence of single PCR product, and then purified by either standard phenol–chloroform extraction followed by ethanol precipitation or by a commercially available column-based DNA purification kit [e.g., PCR purification kit (Qiagen, cat. no. 28104)]. Linearized DNA can be stored at −20°C.

4. Preparation of PFA solution is potentially hazardous and should only be done in a chemical hood. For preparation of a 1 l solution of 4% (w/v) PFA add 40 g of PFA to 850 ml DDW (if possible it is advised to use DEPC-treated DDW, but DDW also works fine). Heat the solution to 50°C while stirring (it is recommended to use preheated DEPC-treated DDW). Adjust pH to 7.2 by adding 10–12 drops of 1.0 M NaOH (~1 drop per 100 ml volume) and monitoring using pH-indicating paper. Importantly, do not overheat the solution (above 70°C) since PFA is degraded at higher temperatures, leading to acidification of the solution due to formation of formic acid. Do not use electric pH-meter as it may be damaged by the solution. After the solution becomes clear and the powder has completely dissolved, allow to cool then add 100 ml 10× PBS solution and DEPC-treated DDW to complete to 1 l volume. Place on ice. Store aliquots at −20°C. Due to critical importance of the initial fixation, as opposed to subsequent re-fixations, only use freshly thawed PFA for embryo fixation. Alternatively, commercial PFA solution (16%) is also available (Electron Microscopy Science, cat. no. 15710).

5. RNA from torula yeast type VI. Lyophilized powder is reconstituted in sterile DDW at a concentration of 50 mg/ml. To separate nucleic acids from proteins, tRNA is usually extracted by phenol/chloroform to avoid protein contamination. Alternatively, boil, spin down and use only upper liquid fraction. Aliquot are stored at −20°C. Notably in some protocols the preHM used in washes (but not hybridization) does not contain tRNA and heparin (7), but we obtained similar results using the non-modified preHM.

6. Tyramide Signal Amplification (TSA) is an enzyme-mediated detection method that utilizes the catalytic activity of POD to generate high-density labeling of a target protein or nucleic acid sequence *in situ*, in standard immunohistochemistry and ISH protocols, respectively. It confers enhanced sensitivity, greater resolution, flexibility in experimental design due to multiple choices in labeling, amplification rounds and color detection, and a short reaction time. In the described procedure, the transcript-probe complex is detected by POD-labeled antibody, which catalyzes the deposition of multiple copies of a fluorophore (Alexa Fluor 488)-labeled tyramide derivative in the vicinity of the POD–target interaction site.

7. DAB is a hazardous and toxic chemical. Avoid skin or eye contact with the tablets or solution, as well dust inhalation or ingestion. Keep away from heat. Work in chemical hood using gloves and a coat. Tablet' container should be tightly closed and kept in $-20°C$.

8. The overall integrity and quality of the labeled riboprobe may be assessed by electrophoresis under denaturing conditions. An easy alternative to standard denaturing agarose gels is to add 1 μl of labeled probe to 9 μl of formamide containing 0.7 ng/μl ethidium bromide, spin down, incubate at 65°C for 10 min, immediately chill on ice, and run through a 1% (w/v) Agarose/TAE gel alongside 25 and 100 ng samples of DIG-labeled RNA standard (Roche, cat. no. 11373099910). Ideally, a single riboprobe fragment should appear on the gel. However, often multiple RNA bands may be observed. Usually, these probes are still useable, though smear indicates RNA degradation. It is recommended to assess the usefulness of these probes by performing a preliminary single color ISH and comparing the resulting expression pattern to previous findings or literature data. Finally, determine probe concentration using a spectrophotometer (such as NanoDrop products, Thermo Scientific).

9. PTU is an inhibitor of tyrosinase, an enzyme required for melanin synthesis. Application of PTU to zebrafish larvae prevents melanin pigment formation and facilitates detection of ISH staining. Since PTU also affects early development, mainly of catecholaminergic neurons, add PTU only to 18–26 somite-stage embryos.

10. It is advised to dechorionate embryos before fixation. Yet, fixation of embryos while still in their chorion would not affect early embryos (until tailbud stage), but may induce body bending of late embryos (18 hpf and older). Therefore, analysis of transcript distribution within the tail of late embryos necessitates dechorionation prior to fixation.

11. Manual dechorionation of small quantity of embryos: chorions can be removed easily using two watchmaker forceps (no. 55 or no. 5, it is critical that tips are sharp and are aligned to each other), under a stereomicroscope. Care should be taken not to damage embryos, especially younger embryos, which are more fragile. Alternatively, large numbers of embryos can be enzymatically dechorionated. Incubate embryos in small volume Petri dish in 1 mg/ml Pronase E in Danieau buffer for a brief time (typically up to 10 min, depending on the embryonic stage). For dechorionation of less than 24 hpf embryos, use Petri dishes covered with 1% (w/v) agarose in Danieau buffer. Release embryos from chorion by gently passing them through a Pasteur pipette. When one to two embryos start to come out of the chorion, immediately pour embryos into large volume of clean Danieau buffer. Repeat rinsing for at least six times. Transfer embryos into new 2-ml microcentrifuge tube. Once dechorionated, embryos are very fragile and should be treated gently. It is recommended to transfer embryos using a fire-polished glass Pasteur pipette as they may stick to polypropylene surfaces.

12. Incubate embryos in 3% (v/v) H_2O_2 and 0.5% (w/v) KOH in PBST until pigmentation is completely disappeared (approximately 30–60 min). Fresh hydrogen peroxidase solution should be used. Stop reaction by washing for 5 min with PBS, and progressively dehydrate embryos by washing for 5 min with the following dilutions (v/v) of methanol in PBST: 25, 50, 75, and 100%.

13. Different batches of Proteinase K may vary in their activity. It is therefore recommended to determine the exact incubation time empirically for each Proteinase K Batch. Approximated incubation times with Proteinase K solution are indicated in the table below:

			Somites			hpf				
Developmental stage		≤ bud	1–8	9–14	15–26	24	36	48	72	96
Proteinase K treatment (min)	All but TSA	0.5	1[a]	3[a]	5	10	15	17	25	30
	TSA only	0.5	1[a]	3[a]	5	15	22	30	45	60

[a]Not compulsory, may affect yolk integrity

Permeabilization of embryos is a prerequisite for the penetration of RNA probes into the embryos. Under-digestion will cause insufficient probe accessibility, whereas over-digestion will alter the morphology of the embryo, and may make it prone to disintegration.

14. Embryos should be completely submerged in hybridization solution. Excessive amounts of labeled RNA probe may increase background staining.

15. A given batch of probe can be reused several times (usually ~3 times), depending on its signal intensity, transcript abundance, tissue accessibility, etc.

16. Dilution of an Ab in blocking solution depends on the specific Ab, the expression pattern of the gene of interest, experimental conditions etc. Often, dilutions of 1:1,000 (and up to 1:10,000) and 1:200 (and up to 1:1,000) are used for *colorimetric* and *fluorescent* assays, respectively.

17. In order to avoid background staining, use only freshly prepared substrate buffers, kept in the dark.

18. Staining duration of colorimetric assays can be readily determined, as signal is monitored visually. However, staining period of fluorescent assays must be determined empirically for each probe, as a non-informative ubiquitous fluorescence throughout the sample is observed prior to probe removal. Reaction time of colorimetric assays is within the range of 20 min (for highly abundant transcripts) to 8 h (for weakly expressed RNAs), where majority of gene products are usually detected after 1–1.5 h. If required, replenish with fresh substrate solution every 1.5 h. To reduce background staining, it is recommended to perform staining reaction at 4°C, which might prolong the reaction time (O.N. in case of weak probes).

19. The yolk produces a significant background autofluorescence and can also display photosensitivity leading to progressive change of its color from yellow to brown when exposed to light. Both may affect visualization of gene expression. It is therefore recommended to dissect the yolk prior to mounting. Alternatively, incubation in an acidic buffer (PBST at pH 3.0) before clearing in glycerol may prevent photosensitivity of the yolk. Yet, this treatment affects tissue morphology and can only be used for early stage embryos (up to 15-somite stage).

20. An alternative heat treatment (at 65°C for 20 min or at 80°C for 10 min) does not appear to completely inactivate the conjugated AP (3).

21. The fluorescent dyes Cy5 and Cy3 are sensitive to methanol/H_2O_2 treatment employed during POD inactivation. Therefore, in a multicolor fluorescent ISH, Cy5 and Cy3 TSA reactions should only be used in the last staining reaction (15).

22. The duration of incubations with antibody solution, blocking, and washes is variable and depends on the enzyme characteristics (mainly specificity and background signal).

23. Equally successful, one can use Cy2-, Cy3-, Cy5- (or any other fluorophore) conjugated goat anti-mouse Ab (we use antibodies from Jackson Immunoresearch Laboratories, cat. no. 115-225-003, 115-165-003, 115-175-003, respectively).

Acknowledgments

We are grateful to Nataliya Borodovsky and Amos Gutnick for stimulating discussion and comments on this manuscript. Nataliya Borodovsky also provided the ISH images. The research in the Levkowitz lab is supported by the German-Israeli Foundation (grant number 183/2007); Israel Science Foundation (grant number 928/08) and the Harriet & Marcel Dekker Foundation. G.L. is an incumbent of the Tauro Career Development Chair in Biomedical Research.

References

1. O'Keefe, H.P., Melton, D.A., Ferreiro, B. and Kintner, C. (1991) In situ hybridization. *Methods Cell. Biol.* **36**, 443–463.
2. Thisse, B., Heyer, V., Lux, A., Alunni, V., Degrave, A., Seiliez, I., Kirchner, J., Parkhill, J. P. and Thisse, C. (2004) Spatial and temporal expression of the zebrafish genome by large-scale in situ hybridization screening. *Methods Cell. Biol.* **77**, 505–519.
3. Hauptmann, G. (1999) Two-color detection of mRNA transcript localizations in fish and fly embryos using alkaline phosphatase and beta-galactosidase conjugated antibodies. *Dev. Genes Evol.* **209**, 317–321.
4. Jowett, T. (2001) Double in situ hybridization techniques in zebrafish. *Methods* **23**, 345–358.
5. Clay, H. and Ramakrishnan, L. (2005) Multiplex fluorescent in situ hybridization in zebrafish embryos using tyramide signal amplification. *Zebrafish* **2**, 105–111.
6. Welten, M. C., de Haan, S. B., van den Boogert, N., Noordermeer, J. N., Lamers, G. E., Spaink, H. P., Meijer, A. H. and Verbeek, F. J. (2006) ZebraFISH: fluorescent in situ hybridization protocol and three-dimensional imaging of gene expression patterns. *Zebrafish* **3**, 465–476.
7. Thisse, C, Thisse, B. (2008) High-resolution in situ hybridization to whole-mount zebrafish embryos. *Nat. Protoc.* **3**, 59–69.
8. Broadbent, J. and Read, E. M. (1999) Wholemount in situ hybridization of Xenopus and zebrafish embryos. *Methods Mol. Biol.* **27**, 57–67.
9. Schulte-Merker, S., Ho, R. K., Herrmann, B. G. and Nüsslein-Volhard, C. (1992) The protein product of the zebrafish homologue of the mouse T gene is expressed in nuclei of the germ ring and the notochord of the early embryo. *Development* **116**, 1021–1032.
10. Thisse, C., Thisse, B., Schilling, T. F. and Postlethwait, J. H. (1993) Structure of the zebrafish snail1 gene and its expression in wild-type, spadetail and no tail mutant embryos. *Development* **119**, 1203–15.
11. Sprague, J., Bayraktaroglu, L., Clements, D., Conlin, T., Fashena, D., Frazer, K., Haendel, M., Howe, D. G., Mani, P., Ramachandran, S., Schaper, K., Segerdell, E., Song, P., Sprunger, B., Taylor, S., Van Slyke, C. E. and Westerfield, M. (2006) The Zebrafish Information Network: the zebrafish model organism database. *Nucleic Acids Res.* **34**, D581–585.
12. Paragas, V. B., Zhang, Y. Z., Haugland, R. P. and Singer, V. L. (1997) The ELF-97 alkaline phosphatase substrate provides a bright, photostable, fluorescent signal amplification method for FISH. *J. Histochem. Cytochem.* **45**, 345–357.
13. Jékely, G. and Arendt, D. (2007) Cellular resolution expression profiling using confocal detection of NBT/BCIP precipitate by reflection microscopy. *Biotechniques* **42**, 751–755.
14. Trinh, le A., McCutchen, M. D., Bonner-Fraser, M., Fraser, S. E., Bumm, L. A. and McCauley, D. W. (2007) Fluorescent in situ hybridization employing the conventional NBT/BCIP chromogenic stain. *Biotechniques* **42**, 756–759.
15. Brend, T. and Holley, S.A. (2009) Zebrafish whole mount high-resolution double fluorescent in situ. *J. Vis. Exp.* **25**, pii: 1229. doi: 10.3791/1229.
16. Tessmar-Raible, K., Steinmetz, P. R., Snyman, H., Hassel, M. and Arendt, D. (2005) Fluorescent two-color whole mount in situ hybridization in Platynereis dumerilii

(Polychaeta, Annelida), an emerging marine molecular model for evolution and development. *Biotechniques* **39**, 460–462.

17. Hargrave, M., Bowles, J. and Koopman P. (2006) In situ hybridization of whole-mount embryos. *Methods Mol. Biol.* **326**, 103–113.

18. Coverdale, L. E., Burton, L. E. and Martin, C. C. (2008) High-throughput whole mount in situ hybridization of zebrafish embryos for analysis of tissue-specific gene expression changes after environmental perturbation. *Methods Mol. Biol.* **410**, 3–14.

Chapter 7

High-Resolution Fluorescence In Situ Hybridization to Detect mRNAs in Neuronal Compartments In Vitro and In Vivo

Sharon A. Swanger, Gary J. Bassell, and Christina Gross

Abstract

The localization of specific mRNAs into dendrites and/or axons is an important mechanism to enrich proteins at their sites of function and influence neuronal development, plasticity, and repair. The fluorescence in situ hybridization (FISH) methods described here have provided high sensitivity and resolution enabling investigation into the mechanism, regulation, and function of mRNA localization in vitro and in vivo. Two methods are described in detail. The first method employs digoxigenin- or fluorophore-conjugated oligonucleotide probes for the detection of localized mRNAs in dendrites, spines, axons, and growth cones of cultured neurons. The second method employs digoxigenin-labeled RNA probes and fluorescence tyramide amplification for the detection of less abundant mRNAs localized to dendrites in vivo. Both methods enable the visualization and quantification of mRNA granules, and changes in their localization in response to various stimuli. The high-resolution FISH technology described here has broader applications beyond the study of mRNA localization. It enables the quantitative analyses of developmental and cell type-specific patterns of gene expression, and how these are modified by physiological signals or during disease states.

Key words: Fluorescence in situ hybridization, mRNA localization, Digoxigenin-labeled probe, Fluorescence tyramide amplification, Dendritic spine, Growth cone, Calcium/calmodulin-dependent protein kinase II alpha (CaMKIIα), Activity-regulated cytoskeleton-associated protein/activity-regulated gene 3.1 (Arc/Arg3.1)

1. Introduction

Post-transcriptional gene regulation through mRNA localization, translation, and degradation plays an important role in neuronal development, structure, and function. The analysis of steady state and stimulus-induced mRNA expression within distinct neuronal compartments is critical for understanding neuronal cell biology. Early experiments used radiolabeled probes to discover the presence of abundant mRNA molecules in dendrites in vivo and in vitro, such as microtubule associated protein 2 (MAP2, (1)) and CaMKIIα mRNA (2) in brain sections, and MAP2 mRNA in dendrites of cultured neurons (3). The development of new nonisotopic in situ hybridization methods using digoxigenin-labeled probes detected by fluorophore- or alkaline phosphatase-conjugated antibodies revealed the dendritic and/or axonal localization of mRNAs, in vitro (4–6) and in vivo (7), that had previously appeared confined to the cell body (8).

Fluorescence in situ hybridization (FISH) allows for high-resolution and quantitative analysis of mRNA localization within distinct subcellular compartments, such as RNA granules within axonal growth cones and dendritic spines (5, 6). This FISH method was first used to detect poly(A) mRNA in processes of cultured neurons (8). Since then, a number of technical improvements to our methodology have further enhanced sensitivity, reduced nonspecific background, and enabled quantitative analysis of digital images (5, 6, 9–11). Here, we describe our current methodology using fluorophore-conjugated antibodies to detect digoxigenin-labeled oligonucleotide probes (Fig. 1). More recently, the use of fluorophore-labeled oligonucleotide probes has enabled multiplex detection of several mRNAs in non-neuronal cells (12). Here, we describe our method using fluorophore-labeled oligonucleotide probes to visualize mRNAs within dendrites and spines of cultured neurons (Fig. 2). Both FISH methods used with high-resolution widefield microscopy and image deconvolution permit detailed analysis of mRNA granules in specialized neuronal compartments such as dendritic spines and axonal growth cones.

Although several isotopic and nonisotopic in situ hybridization methods have provided sufficient sensitivity to detect abundant mRNAs in well-defined dendritic laminas of the brain, such as in the hippocampus and the cerebellum (13), they are not suitable to detect less abundant mRNAs in dendrites, and also lack the subcellular resolution to visualize dendritic mRNAs in brain areas that do not have a defined laminar structure (e.g., cortex or midbrain). Here, we describe our FISH method for the subcellular analysis of mRNA distribution in vivo. To optimize sensitivity, this method utilizes digoxigenin-labeled RNA probes combined with a fluorescence tyramide amplification method. A similar method has been previously used to detect immediate early gene expression, e.g., Arc/Arg3.1 mRNA, in tissue with subcellular resolution (14, 15); although, the presence of Arc/Arg3.1 mRNA within dendrites was not shown. Until now, very few studies have shown low-abundance mRNAs in dendrites in vivo (16, 17), probably due to lack of sensitivity and/or compromised tissue morphology. Our modified protocol has been used to detect the well-known dendritic mRNA CaMKIIα, but also less abundant dendritic mRNAs, such as PSD-95 and GluR1 (17). Here, we show seizure-induced dendritic localization of Arc/

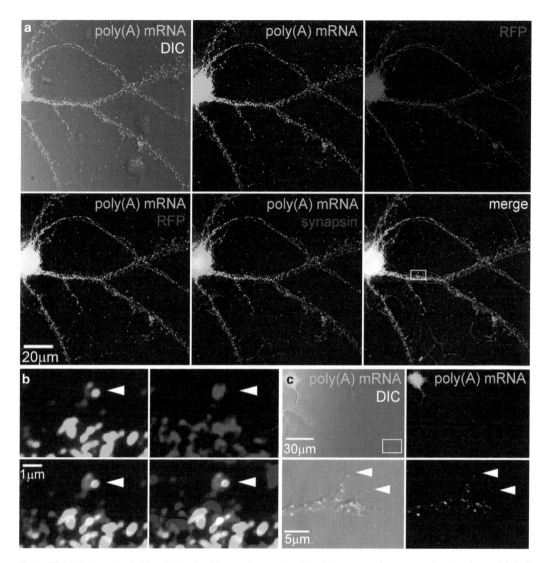

Fig. 1. FISH detection of poly(A) mRNA in dendrites and axons of cultured hippocampal neurons using digoxigenin-labeled oligonucleotide probes. (a) A specific poly(A) mRNA signal is detected in the cell body and dendrites of a 16 DIV neuron (DIC overlay, *top left*). The poly(A) mRNA signal maintains high signal-to-noise ratio in distal dendrites (>100 μm from the cell body, *top center*). 14 DIV neurons were transfected with RFP (*top right*), and can be visualized in tandem with poly(A) mRNA (*bottom left*). Concurrent synapsin immunostaining shows poly(A) mRNA granule localization at synapses (*bottom center*). (b) Poly(A) mRNA granules are detected in dendritic spines (*top left, arrowhead*). The spine is filled with RFP signal (*top right*), which overlaps with poly(A) mRNA (*bottom left*) and synapsin (*bottom right*) signals. (c) Poly(A) mRNA is detected in the cell body and neurites of a 3 DIV neuron (DIC overlay, *top left*). The poly(A) mRNA signal extends the length of the axon (*top right*) and into the growth cone palm and filopodia (*arrowheads, bottom panels*).

Arg3.1 mRNA and dendritic localization of microtubule associated protein 1b (MAP1B) mRNA in vivo (Fig. 3).

Several laboratories have recently developed improved FISH protocols that allow for analysis of mRNA localization (18). We find the FISH techniques presented here to be especially powerful to detect and analyze high- and low-abundance mRNA species abundant mRNAs in fine neuronal structures in vitro and in vivo.

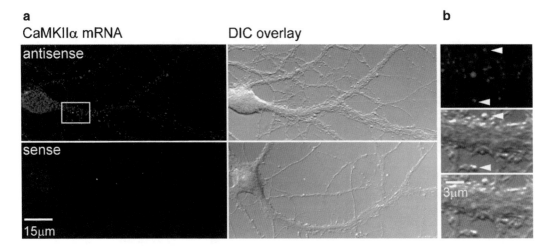

Fig. 2. FISH detection of CaMKIIα mRNA in dendrites and spines of cultured hippocampal neurons using fluorophore-conjugated oligonucleotide probes. (**a**) FISH with an antisense Cy3-labeled probe specifically detects CaMKIIα mRNA as compared to the sense CaMKIIα probe. CaMKIIα mRNA is detected in the cell body and through the dendritic arbor as seen in the DIC overlay (*top right*). (**b**) The antisense CaMKIIα probe is sufficient to detect mRNA molecules in dendritic spines (*arrowheads*). These images are magnified from the *boxed area* in (**a**).

2. Materials

If not noted otherwise, all solutions are prepared on the day of the experiment with water (H_2O) taken freshly from a Milli-Q Synthesis purification system (or similar), which produces pyrogen- and nuclease-free H_2O with a total organic carbon content of 2–5 ppb, and pyrogen levels (EU/ml) of <0.001. If these guidelines are followed, DEPC-treatment, autoclaving, or filtration is not necessary (if not otherwise noted). All solutions are prepared and stored in sterile, RNAse-free microcentrifuge and conical tubes, or autoclaved glassware. Unless otherwise stated, all chemicals are purchased from *Sigma-Aldrich, St. Louis, MO*.

2.1. Materials for Fluorescence In Situ Hybridization on Cultured Neurons

The materials needed for the hippocampal neuron dissection and culture including dissection tools, media and culture vessels, have been described previously (19). All experiments herein use hippocampal neurons harvested from E18 rat embryos.

2.1.1. Hippocampal Neuron Culture

2.1.2. Digoxigenin-Labeled Oligonucleotide Probe Preparation and Labeling

1. Oligonucleotide probes synthesized with internal T(C6)-amino 5′ modifications and purified by reverse-phase HPLC (*Biosearch Technologies, Novato, CA*).
2. Digoxigenin-3-O-methylcarbonyl-ε-aminocaproic acid-N-hydroxysuccinimide ester (Digoxigenin-NHS; *Roche Applied Science, Indianapolis, IN*).
3. Dimethyl formamide (DMF).

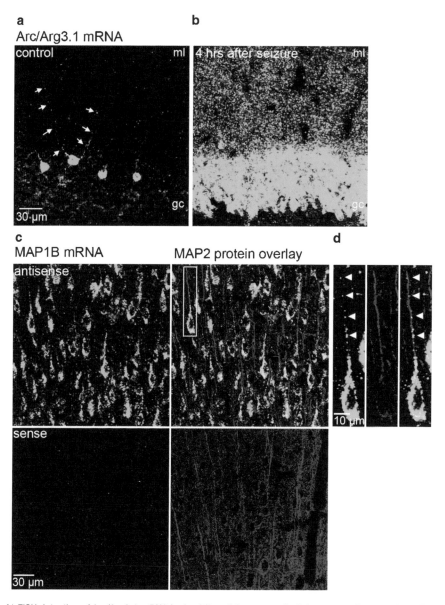

Fig. 3. (a, b) FISH detection of Arc/Arg3.1 mRNA in dendrites of the mouse dentate gyrus under control conditions (a) and 4 h following kainic acid-induced seizure (b). Note that the FISH method using RNA probes and detection with fluorescein-tyramide amplification is suitable to detect Arc/Arg3.1 mRNA specific signal in distal dendrites of single cells under control conditions ((a), indicated by *arrows*), as well as in the entire granule cell (gc) and molecular layer (ml) following induction of synaptic activity (b). (c, d) FISH detection of MAP1B mRNA in dendrites of the mouse cortex co-immunostained for the dendritic marker protein MAP2. (c) Overlay of the MAP1B FISH signal (*green*) with MAP2 protein (*red*) staining suggests that MAP1B mRNA is targeted into dendrites of the cortex (*upper panel*). A MAP1B sense probe does not detect any specific signal (*lower panels*). (d) Magnification of a single cortical neuron (*white box* in (c)) indicates the area shown in (d)) demonstrates that MAP1B mRNA positive granules can be visualized in distal regions (≥50 μm) of a dendrite.

4. 0.1 M sodium borate buffer at pH 8.8.
5. Sephadex G50 gel filtration columns (*GE Healthcare Biosciences, Pittsburgh, PA*).
6. 70% ethanol.

7. Sodium acetate solution: 3 M $NaC_2H_3O_2$ in H_2O, adjust to pH 5.2 using acetic acid.
8. DIG Nucleic Acid Detection Kit (*Roche Applied Science*).
9. Zeta-Probe blotting membrane (*BioRad, Richmond, CA*).

2.1.3. Fluorophore-Labeled Oligonucleotide Probe Preparation and Labeling

1. Oligonucleotide probes synthesized with internal T(C6)-amino 5′ modifications and purified by reverse-phase HPLC (*Biosearch Technologies*).
2. Amersham CyDye mono-reactive dye pack (Cy3 or Cy5; *GE Healthcare Biosciences*) or AlexaFluor 488 Amine Labeling Kit (*Invitrogen, Carlsbad, CA*).
3. 0.1 M sodium carbonate buffer at pH 8.8.
4. Materials listed in Subheading 2.1.2, items 5–7 are also needed.

2.1.4. Fluorescence In Situ Hybridization with Digoxigenin-Labeled Oligonucleotide Probes

1. Sterile 12-well plates, nuclease-free (*BD Biosciences, San Jose, CA*).
2. 4% paraformaldehyde in 0.1 M phosphate buffer (PB) with 5 mM $MgCl_2$ at pH 7.4 (see Note 1 for preparation).
3. 10× phosphate buffered saline (PBS: 0.01 M KH_2PO_4, 0.1 M Na_2HPO_4, 1.37 M NaCl, and 0.027 M KCl; *Roche Applied Science*).
4. 1× PBS with 5 mM $MgCl_2$.
5. Salmon sperm DNA (10 mg/ml, *Invitrogen*, store at –20°C).
6. *E. coli* tRNA solution: 10 mg/ml tRNA (*Roche Applied Science*) in H_2O. 500 µl aliquots are stored at –20°C for up to a year.
7. 20× sodium citrate buffer (SSC): 3.0 M NaCl and 0.3 M sodium citrate at pH 7.0 (*Roche Applied Science*).
8. Deionized formamide, store at 4°C.
9. Dextran sulfate solution: 50 mg/ml dextran sulfate in H_2O (see Note 2 for preparation), 1 ml aliquots can be stored at –20°C for up to a year.
10. Hybridization Buffer (HB): 200 µl dextran sulfate (50 mg/ml), 200 µl bovine serum albumin (BSA; 20 mg/ml; *Roche Applied Science*; store at –20°C), 100 µl ribonucleoside vanadyl complexes (RVC; 200 mM in H_2O; aliquoted and stored at –20°C), 100 µl 20× SSC buffer, 10 µl 10 mM PB, and 390 µl H_2O. The solution is mixed by vortexing. HB is made fresh immediately before use and kept on ice.
11. Tris-buffered saline (TBS): 50 mM Tris–HCl and 150 mM NaCl at pH 7.4.
12. TBS with 0.3% Triton (v/v).

13. TBS with 0.1% Triton (v/v).

14. Tris/Glycine buffer: 200 mM Tris-HCl at pH 7.4 and 0.75% glycine (w/v).

15. Blocking buffer: 2% BSA Fraction V (w/v; *Roche Applied Science*) and 2% FBS (v/v) in TBS with 0.1% Triton.

16. Immunofluorescence (IF) buffer: 1% BSA Fraction V (w/v) and 1% FBS (v/v) in TBS with 0.1% Triton.

17. Mouse anti-digoxigenin antibody (*Jackson Immunoresearch Laboratories, West Grove, PA*).

18. Donkey anti-mouse Cy3-conjugated antibody (optionally Cy2- or Cy5-conjugated antibodies can be used; *Jackson Immunoresearch Laboratories*).

19. 1× PBS.

20. DAPI (4′,6-diamidino-2-phenylindole).

21. Mounting media (see Note 3 for preparation).

22. Propyl gallate.

23. Superfrost glass microscope slides (*Fisher Scientific, Fair Lawn, NJ*).

2.1.5. Fluorescence In Situ Hybridization with Fluorophore-Labeled Oligonucleotide Probes

1. The materials listed in Subheading 2.1.4, items 1–10 and 19–23 are necessary to complete FISH with fluorophore-labeled oligonucleotide probes.

2. Optionally, if protein immunocytochemistry is to be conducted in addition to fluorophore-labeled oligonucleotide mRNA detection, then the materials listed in Subheading 2.1.4, items 1–23 are necessary.

2.2. Materials for Fluorescence In Situ Hybridization on Brain Tissue

2.2.1. Preparation of Tissue

1. Physiological saline: 0.9% (w/v) NaCl in H_2O, supplemented with ≥1 USP unit/ml heparin sulfate.

2. 4% paraformaldehyde: 4% paraformaldehyde, 0.2 M NaH_2PO_4, and 1 mM $MgCl_2$ at pH 7.4; filtrated through grade 1 cellulose filters (*GE Healthcare Biosciences*, see Note 4 for preparation).

3. 30% (w/v) sucrose: 30 g sucrose (nuclease-free) in 1× PBS (1:10 dilution of 10× PBS (*Roche Applied Science*).

4. Tissue-Tek OCT media (*Sakura Finetek, Torrance, CA*).

5. Superfrost Plus microscope slides (*Fisher Scientific*).

6. Cryostat/Microtome (suitable to cut 10–15 µm thick frozen sections).

2.2.2. Preparation of Riboprobes

1. Plasmid containing the cDNA of interest flanked by Sp6, T7 or T3 RNA polymerase promoters and unique restriction sites (e.g., pcDNA3 from *Invitrogen*).

2. Restriction enzymes and buffers from any manufacturer (e.g., *Fermentas, Glen Burnie, MD*), phenol/chloroform/isoamyl (25:24:1) solution, chloroform, and ethanol.

3. DIG RNA labeling Kit (Sp6/T7) (*Roche Applied Science*).

4. tRNA solution (see Subheading 2.1.4, item 6).

5. 3 M sodium acetate: 3 M $NaC_2H_3O_2$ at pH 5.2 (pH adjusted with acetic acid).

6. 0.1 M DTT dissolved in H_2O.

7. 0.4 M $NaHCO_3$ dissolved in H_2O.

8. 0.6 M Na_2CO_3 dissolved in H_2O.

9. Neutralization solution: 3 M $NaC_2H_3O_2$ at pH 6 in H_2O (pH adjusted with acetic acid).

10. Glycogen solution: 20 mg/ml glycogen in H_2O (*Roche Applied Science*).

2.2.3. Pretreatment of Brain Sections, Hybridization, and Washes

1. Glass staining dish with cover (*Fisher Scientific*).

2. SSC solutions: dilutions of 20× SSC stock solution (*Roche Applied Science*) with H_2O.

3. 0.1 M Triethanolamine-HCl: 0.1 M triethanolamine in H_2O at pH 8.0 (pH adjusted with HCl).

4. Acetic anhydride (*Fisher Scientific*).

5. Methanol/Acetone solution: 50% (v/v) Methanol and 50% (v/v) Acetone.

6. Hybridization buffer: 4× SSC, 50% (v/v) formamide, 1× Denhardts (50× Denhardt's from *Invitrogen*, contains 1% (w/v) Ficoll (type 400), 1% (w/v) polyvinylpyrrolidone, and 1% (w/v) bovine serum albumin), 10% (w/v) dextran sulfate (for preparation of dextran sulfate solution, see Subheading 2.1.4, item 9), 0.5 µg/ml herring sperm ssDNA (10 mg/ml solution from *Roche Applied Science*), and 0.25 µg/ml tRNA from *E. coli*, slowly mixed on a rotator to avoid air bubbles, stable at –80°C for 2 weeks.

7. Immunostain moisture chamber, black (*Evergreen Scientific, Los Angeles, CA*).

8. HybriSlip hybridization covers (*Invitrogen*).

9. RNAse A solution: 10 mg/ml RNAse A (Ribonuclease A from bovine pancreas) in H_2O, store in 0.5 ml aliquots at –20°C.

2.2.4. Detection

1. 3% H_2O_2 solution: 1:10 dilution of a 30% H_2O_2 solution (*Fisher Scientific*) in 1× SSC.

2. 10× TBS 100: 1.5 M NaCl and 1 M Tris–HCl at pH 7.5; autoclave, stable at room temperature for up to a month.

3. TBS 100 buffer (tris buffered saline): 1:10 dilution of 10× TBS.

4. TNB buffer: 0.5% (w/v) blocking reagent in 1× TBS 100, prepared from 10× TBS and 10% (w/v) blocking reagent (*Roche Applied Sciences*, prepared as described in the manufacturer's instructions).

5. TNT buffer: 1× TBS 100 with 0.5% (v/v) Tween 20.

6. Tyramide signal amplification (TSA) system, fluorophore-coupled (*PerkinElmer, Waltham, MA*).

7. Mounting media (see Subheading 2.1.4, items 21 and 22).

3. Methods

3.1. Methods for Fluorescence In Situ Hybridization on Cultured Neurons

The FISH methods described herein use digoxigenin-labeled oligonucleotide probes detected by fluorophore-conjugated antibodies (Fig. 1) and fluorophore-labeled oligonucleotide probes (Fig. 2) to localize mRNA molecules within subcellular compartments of cultured neurons.

3.1.1. Hippocampal Neuron Culture

Hippocampal neuron cultures are prepared as described previously (19). Briefly, hippocampal neurons are dissociated and plated at low density on 15 mm poly-l-lysine-coated glass coverslips (*No.1, Carolina Biological Supply, Burlington, NC*). Neurons are cocultured with glial feeder layers in Neurobasal media supplemented with Glutamax and B27 (*Invitrogen*) at 37°C with 5% CO_2. For mRNA detection in mature dendrites and spines, neurons are cultured for 14–21 days in vitro (DIV). To detect axonal mRNAs, neurons are cultured for 3–5 DIV.

3.1.2. Digoxigenin-Labeled Oligonucleotide Probe Preparation and Labeling

1. Oligonucleotide sequence design: Antisense oligonucleotide probes are designed to be approximately 50 nucleotides in length with 45–55% GC content, complementary to unique, non-overlapping target mRNA sequences, and with as little homology to other sequences as possible. Selected sequences should have minimal secondary structure and favorable hybridization properties assessed by software (OLIGO, *Molecular Biology Insights, Inc., Cascade, CO*). To allow precise control of hapten incorporation, five thymidine residues approximately ten bases apart are amino-modified. The amino-modification allows the direct chemical conjugation of succinimide ester compounds (e.g., digoxigenin-NHS or fluorophore-NHS) to the oligonucleotide. To increase signal detection, four distinct oligonucleotides are produced for each target mRNA. As a negative control, an oligonucleotide with a scrambled antisense sequence or the sense sequence is prepared for each target mRNA. The probes are purified by reverse-phase HPLC and resuspended in H_2O at 5–10 μg/μl.

2. 20 μg of the oligonucleotide(s) are dried in a speed vacuum, and then resuspended in 200 μl 0.1 M sodium borate buffer. If four oligonucleotides are to be used to detect a single mRNA, then 5 μg of each oligonucleotide are used. The digoxigenin-NHS is dissolved in DMF at 1.67 mg/ml, and 200 μl digoxigenin solution is added to 200 μl of the oligonucleotide suspension. The probe solution is rotated overnight at room temperature and protected from light.

3. Oligonucleotide purification: A G50 Sephadex column is used for oligonucleotide purification. The labeled oligonucleotides are concentrated using a speed vacuum and resuspended in 0.1 M sodium carbonate buffer. The oligonucleotide suspension volume must be 100 μl or less and at a concentration no greater than 1 μg/μl to be applied to the column. The column preparation, oligonucleotide application, and fraction collection are completed as per the manufacturer's instructions (*GE Healthcare*). The oligonucleotides are precipitated from each fraction by adding 100% ethanol (2.5 times the fraction volume) and 3 M sodium acetate (0.1 times the fraction volume). The samples are placed at –20°C for at least 2 h, then centrifuged in a microcentrifuge at $20,000 \times g$ for 20 min at 4°C. The DNA pellets are rinsed with 70% ethanol and centrifuged at $20,000 \times g$ for 10 min at 4°C, twice. The DNA is resuspended in H_2O at approximately 100 μg/ml, assuming full extraction of DNA from the column (i.e., if 20 μg of DNA was used for labeling, dissolve DNA in a total of 200 μl H_2O).

4. The efficiency of digoxigenin labeling is determined by an alkaline phosphatase immunoassay using the DIG Nucleic Acid Detection Kit. 1 μl of each fraction is dotted on a Zeta-Probe blotting membrane in duplicate, allowed to fully dry, and then crosslinked to the membrane using UV light. The blot is then processed for digoxigenin detection as per the manufacturer's instructions (*Roche Applied Science*). The reaction is quenched by washing with H_2O and the membrane is allowed to dry at room temperature (see Note 5).

3.1.3. Fluorophore-Labeled Oligonucleotide Probe Preparation and Labeling

1. The oligonucleotide probes are designed as described in Subheading 3.1.2, step 1.

2. Oligonucleotide preparation and labeling: 20 μg of oligonucleotide(s) are dried using a speed vacuum and resuspended in 70 μl 0.1 M sodium carbonate buffer. Dissolve 1 vial (*GE Healthcare* Cy3 or Cy5) or 1 mg (*Alexa* 488) of dye in 30 μl of DMSO. The dye and oligonucleotide solutions are combined, left at room temperature overnight with occasional vortexing, and protected from light.

3. The fluorophore-labeled oligonucleotides are purified and precipitated as described in Subheading 3.1.2, step 3.

4. The specific activity of fluorophore-labeled oligonucleotides is determined using a spectrophotometer. The resuspended oligonucleotide is diluted 1:100 in H_2O. The DNA concentration is measured by absorbance at 260 nm, and the fluorescence is determined by measuring the absorbance at the maximum absorption wavelength for the fluorophore used (e.g., 488 nm for AlexaFluor 488, 514 nm for Cy3, and 643 nm for Cy5). The oligonucleotide probe is resuspended at a final concentration of 25 ng/μl.

3.1.4. Fluorescence In Situ Hybridization with Digoxigenin-Labeled Oligonucleotide Probes

1. Hippocampal neurons cultured on glass coverslips are placed into 12-well plates containing 4% paraformaldehyde, and the neurons are fixed for 20 min at room temperature. The plate should not be moved during fixation as this can compromise the fixation process.

2. The neurons are washed three times in 1× $PBS/MgCl_2$ for 5 min. Unless otherwise noted, all washes are completed at room temperature on an orbital shaker.

3. To equilibrate the samples for hybridization, the neurons are washed in 1× SSC buffer for 10 min, and then in 1× SSC with 40% formamide for 5 min.

4. The HB is prepared while the neurons are being fixed, washed, and equilibrated. Approximately 30 μl of HB is needed for each coverslip (see Note 6).

5. Prehybridization: To minimize nonspecific oligonucleotide hybridization, the neurons are incubated with prehybridization solution for 1.5 h at 37°C. For each coverslip, 20 μg of salmon sperm DNA and 20 μg of tRNA are dried using a speed vacuum, then resuspended in 15 μl 2× SSC with 80% formamide, heated for 5 min at 95°C, and briefly cooled on ice. 15 μl of HB is added to the formamide mixture, per coverslip, and mixed well. A moisture chamber is assembled with a piece of parafilm laid flat where the coverslips will be placed for incubation. 28 μl prehybridization solution is dotted on the parafilm for each coverslip and the coverslips are placed on the solution neuron-side down using fine forceps. The chamber is covered and placed at 37°C for 1.5 h.

6. Hybridization: The probe solution for one coverslip contains 25 ng of labeled oligonucleotide probes, 20 μg of salmon sperm DNA, and 20 μg of tRNA (final probe concentration 0.5–1.0 ng/μl). The oligonucleotide and carrier molecules are dried in speed vacuum, resuspended in 15 μl of 2× SSC with 80% formamide, heated at 95°C for 5 min, and briefly cooled on ice. 15 μl of HB is added to the formamide mixture,

per coverslip, and mixed well. On a new piece of parafilm, 28 μl of probe solution is dotted for each coverslip. The coverslips are carefully lifted off of the parafilm after prehybridization using fine forceps and blotted with a laboratory tissue to remove excess prehybridization solution (see Note 7). The coverslips are immediately placed onto the hybridization solution and put at 37°C for 5 h.

7. The coverslips are carefully removed from the parafilm and the neurons are washed two times with prewarmed 1× SSC with 40% formamide for 20 min at 37°C. Then, the neurons are washed briefly three times with 1× SSC, followed by two 5 min washes in 1× SSC.

8. The neurons are equilibrated in 1× PBS/$MgCl_2$ for 5 min, and then, post-fixed in 4% paraformaldehyde for 5 min at room temperature. The neurons are then washed three times in 1× PBS/$MgCl_2$ for 5 min.

9. The neurons are equilibrated in 1× TBS for 10 min at room temperature.

10. Then, the neurons are permeabilized with 0.3% Triton-TBS for 10 min, washed with Tris-Glycine buffer for 5 min, and washed with 0.1% Triton-TBS for 5 min.

11. The neurons are incubated in Blocking buffer for 1 h at room temperature.

12. The neurons are washed in immunofluorescence (IF) buffer for 5 min, and then incubated with mouse anti-digoxigenin antibody diluted 1:1,500 in IF buffer for 1 h at room temperature (if double labeling is desired, then additional primary antibodies can be added at this step). The antibody incubations are completed in a moisture chamber. 30 μl of antibody solution is dropped onto a flat piece of parafilm and the coverslips are carefully inverted onto the antibody solution with fine forceps.

13. The neurons are washed three times for 10 min in 2 ml of IF buffer.

14. The neurons are incubated with donkey anti-mouse fluorophore-conjugated antibody diluted 1:1,000 in IF buffer for 30 min at room temperature.

15. The neurons are washed three times for 10 min in 2 ml of IF buffer.

16. If DAPI stain is desired, then the neurons are washed in 1× PBS for 5 min, treated with DAPI (1:1,000 in 1× PBS) for 5 min, and washed for 5 min with 1× PBS.

17. Mounting media aliquots are thawed 2–3 h before the coverslips will be ready for mounting. Propyl gallate (0.6 mg/ml) is added to the mounting media, and the mounting media is rotated at room temperature for at least 1 h in the dark.

The mounting media is centrifuged at $20,000 \times g$ for 5 min to pellet non-dissolved propyl gallate.

18. The coverslips are rinsed briefly with H_2O to remove salts and detergents, air-dried, and then mounted on glass slides with 15 µl of mounting media.

19. The slides are dried at room temperature overnight in the dark, and then stored at –20°C.

20. For high-resolution images, neurons are visualized on a wide-field fluorescence microscope (Nikon Eclipse TE300 inverted microscope or similar). Images are captured with a cooled CCD camera (Quantix; *Photometrics, Tuscon, AZ*, or similar), and then deconvolved using a 3D blind algorithm (AutoQuant X; *Cybermetrics, Phoenix, AZ*, or similar).

3.1.5. Fluorescence In Situ Hybridization with Fluorophore-Labeled Oligonucleotide Probes

1. In vitro FISH with fluorophore-labeled oligonucleotides is conducted exactly as described in Subheading 3.1.4, steps 1–8.

2. Optionally, if protein immunocytochemistry and fluorophore-labeled oligonucleotide FISH are to be conducted, then the protocol described in Subheading 3.1.4, steps 1–20 are completed.

3.2. Methods for Fluorescence In Situ Hybridization in Brain Tissue

The here presented FISH method for brain tissue sections using digoxigenin-labeled RNA probes (riboprobes) and fluorophore-coupled tyramide signal amplification provides preservation and accessibility of the tissue, and allows for detection of high- and low-abundance mRNAs in dendrites in vivo ((17), also see Fig. 3).

3.2.1. Preparation of Tissue Sections

1. To protect tissue morphology and mRNA structure, mice at postnatal day 21 are deeply anesthetized with an inhalative anesthetic (e.g., isoflurane) and transcardially perfused (ca. 5 ml/min) with 80 ml of prewarmed (37°C) physiological saline supplemented with 1 unit/ml of heparin sulfate to reduce blood clotting. Then, mice are perfused with 120 ml of prewarmed (37°C) 4% paraformaldehyde (see Note 8).

2. The brain is removed from the skull and stored in 5 ml of 4% paraformaldehyde at 4°C overnight.

3. After 16–18 h, the tissue is placed in 10 ml of 30% (w/v) sucrose in 1× PBS and stored at 4°C for 24 h.

4. Brains are placed in Tissue-Tek and rapidly frozen using liquid nitrogen. Frozen brains are wrapped in aluminum foil and stored at –80°C.

5. At the day of experiment, desired brain regions are cut in 10–15 µm thick sections using a cryostat/microtome and mounted on Superfrost Plus microscope slides (see Note 9). Mounted sections can be stored at –80°C for several weeks to months, but best results are usually obtained if freshly cut sections are immediately processed for FISH.

3.2.2. Preparation of Riboprobes

1. For the design of efficient riboprobes, the following guidelines should be followed: The cDNA used for in vitro transcription of the riboprobe contains 1.2–1.8 kb of the target sequence with as little homology to other sequences as possible to provide sufficient sequence specificity (see Note 10). The cDNA is subcloned into a plasmid containing promoters for two different RNA polymerases (Sp6, T7 or T3). Plasmids require two unique restriction sites, one at the 5'-, and one at the 3'-end of the sequence to allow for linearization prior to in vitro transcription. Promoters for two different polymerases, as well as single restriction sites on the 5'- and 3'-end of the cDNA allows for transcription of an antisense and a sense riboprobe from the same cDNA construct. Sense riboprobes are used as control for nonspecific signal and background labeling.

2. To prepare cDNA for in vitro transcription, 20 µg cDNA-containing plasmid is linearized with the appropriate restriction enzyme according to the manufacturer's protocol (see Note 11). 2% of the reaction volume should be checked for complete linearization by DNA gel electrophoresis. If linearization is complete, the plasmids are purified by phenol/chloroform extraction and precipitated with ethanol following standard protocols. DNA precipitates are dissolved in 15 µl of H_2O and stored at −20°C.

3. Digoxigenin-labeled transcripts are synthesized using the DIG RNA labeling Kit (Sp6/T7) from *Roche Applied Science* according to the manufacturer's protocol with 2 µl (1–2 µg) of the DNA obtained in Subheading 3.2.2, step 2. If necessary, Sp6 or T7 polymerase can be substituted with T3 polymerase (not contained in the kit, but available from *Roche Applied Science*). DNase treatment (optional in the original protocol) should be performed. The in vitro transcribed RNA probes contain ca. 4–5% UTPs labeled with digoxigenin (see Note 12). Riboprobes are precipitated by adding 80 µl H_2O, 10 µl tRNA solution, 20 µl 3 M sodium acetate (pH 5.2), and 300 µl ethanol. Samples are briefly vortexed and incubated at −20°C for at least 1 h. Riboprobes are centrifuged at $20,000 \times g$ for 20 min at 4°C, the supernatant is removed, and 1 ml 75% Ethanol is added. Samples are mixed thoroughly, and centrifuged at $20,000 \times g$ for 5 min at 4°C. After removal of the supernatant, the precipitated RNA is air-dried for 20 min at 37°C (see Note 13).

4. To provide sufficient tissue penetration, size reduction by alkaline hydrolysis is performed (20). Riboprobes are resuspended in 160 µl of 0.1 M DTT. 20 µl each of 0.4 M $NaHCO_3$ and 0.6 M Na_2CO_3 are added, the samples are mixed and incubated at 60°C. The incubation time depends on the size of the original transcript (L_0) and the desired final length (L_F, 0.1 kb is a good starting length), and can be

calculated with this formula: $t[\min] = (L_0[kb] - L_F[kb])/L_0[kb] \times L_F[kb]$ (see Note 14). Following size reduction, samples are immediately put on ice, neutralized with 7 μl ice-cold neutralization solution and precipitated by adding 2 μl glycogen solution and 500 μl ice-cold ethanol. Samples are mixed thoroughly and incubated on ice for at least 30 min, followed by centrifugation at 20,000 × g for 20 min at 4°C. Precipitates are washed with 75% ethanol as described in Subheading 3.2.2, step 3. Air-dried RNA is resuspended in 20 μl H_2O, diluted with 50 μl hybridization buffer and mixed thoroughly. Riboprobes are stored at −80°C. Riboprobes in hybridization buffer are stable at −80°C for about 2 months and do not lose activity following repeated freeze and thaw cycles (up to five times).

3.2.3. Pretreatment of Brain Sections

1. *Optional*: Slides with mounted brain sections are removed from the −80°C storage and brought to room temperature. Fresh and frozen slides are air-dried at room temperature.

2. Dried slides are placed upright into a slotted glass jar. Unless otherwise noted, incubation and washing steps are conducted at room temperature on an orbital shaker. All solutions and buffers can be poured slowly in or out of the glass jar.

3. Sections are postfixed for 5 min in ice-cold 4% paraformaldehyde.

4. Sections are washed twice for 10 min with ice-cold 2× SSC.

5. To reduce nonspecific hybridization, free amino-groups are acetylated with acetic anhydride. Sections are equilibrated for 5 min in ice-cold 0.1 M triethanolamine-HCl, pH 8.0. Immediately before acetylation, 50 ml ice-cold 0.1 M triethanolamine-HCl at pH 8.0 is placed in a sealable tube and the pre-equilibration solution is poured off the sections. 750 μl of acetic anhydride is added to the tube containing 0.1 M triethanolamine-HCl at pH 8.0, mixed thoroughly for 2–3 s, immediately poured on the slides and incubated for 10 min.

6. Sections are washed briefly three times with ice-cold H_2O and incubated for 5 min in ice-cold Methanol/Acetone solution.

7. Sections are washed twice for 10 min in ice-cold 2× SSC.

8. For prehybridization, slides are placed upright outside the jar for 2–3 min to remove excess liquid, and then placed horizontally in a moisture chamber, humidified with paper tissues soaked with 50% formamide in 2× SSC. Sections are covered with 400 μl hybridization buffer and incubated in the sealed moisture chamber for at least 2 h at 55°C. Prehybridization can be extended to up to 24 h.

3.2.4. Hybridization

1. Antisense and sense riboprobes are diluted 1:5, 1:10, and 1:20 in hybridization buffer and mixed thoroughly (see Note 15). 50 µl of hybridization buffer per brain section is needed. To reduce secondary structures of riboprobes, the hybridization solutions are incubated at 95–99°C for 5 min and immediately placed on ice for 5 min.

2. Prehybridized sections are removed from the moisture chamber and placed upright on a paper tissue to allow prehybridization solution to drain off. Excess solution can be removed carefully using a tissue without touching the sections.

3. Slides are placed back in the moisture chamber; 50 µl of hybridization solution is applied carefully directly onto the sections and covered with HybriSlip hybridization covers.

4. Sections are incubated for 14–18 h at 55°C (see Note 16). The probe is hybridized in a solution containing final concentrations of 50% formamide and 4× SSC.

3.2.5. Post-hybridization Washes

1. Following hybridization, slides are placed in 2× SSC for 10 min to remove HybriSlip hybridization covers (see Note 17).

2. After removal of the hybridization covers, sections are washed again for 10 min in 2× SSC.

3. To remove non-hybridized RNA, sections are incubated for 15 min at 37°C with 10 µg/ml RNAse A in 2× SSC (pre-warmed to 37°C).

4. Sections are washed twice for 10 min at room temperature in 2× SSC.

5. Sections are equilibrated for 5 min in 0.5× SSC at room temperature, followed by 30 min incubation in 0.5× SSC at 50°C. The 0.5× SSC solution should be prewarmed to 50°C prior usage.

6. Slides are washed twice for 10 min in 2× SSC at room temperature.

3.2.6. Detection

1. To inactivate endogenous peroxidases, sections are incubated for 15 min in 3% (v/v) H_2O_2 in 1× SSC, followed by three times 5 min washes in 1× SSC.

2. Sections are then incubated for 5 min in TBS 100 buffer.

3. Nonspecific binding sites are blocked by 30 min incubation at room temperature in TNB buffer.

4. After blocking, slides are placed upright on a paper tissue for 1–2 min to allow excess TNB buffer to drain off. Antibody solution is prepared by diluting sheep anti-digoxigenin-POD, Fab fragments (*Roche Applied Science*) in TNB buffer (starting dilution 1:1,000, needs to be optimized for each lot of antibody individually, see Note 18). Slides are placed in a sealable moisture chamber humidified with wet paper tissues,

covered with approximately 100–200 μl of antibody solution each and incubated for 2 h at room temperature (see Note 19). *Optional*: If codetection for a specific protein is desired, the first antibody specific to the respective protein can be included here.

5. Sections are washed five times 10 min in TNT buffer. [*Optional*: If simultaneous immunohistochemistry is performed, a 1 h incubation in secondary antibody (1:200 in TNB buffer, reactive to the species the first antibody was generated in and coupled to a fluorophore other than the one used in the following tyramide signal amplification) can be included, followed by five times 10 min washes in TNT buffer. Incubation in secondary antibody, as well as washes and all following steps should be performed in the dark.]

6. The tyramide signal amplification (21) is performed in a horizontal moisture chamber, protected from light. Slides are placed upright on a paper tissue for 1–2 min to allow TNT buffer to drain off. Slides are then placed in the horizontal chamber and sections are covered with 50 μl fluorophore-coupled tyramide working solution (*PerkinElmer*, tyramide reagent diluted 1:50 in amplification buffer, according to the manufacturer's instructions). Sections are incubated in TSA for exactly 8 min and slides are placed immediately in TNT buffer (see Note 20).

7. Sections are washed five times 10 min in TNT buffer.

8. Slides are briefly washed in H_2O to remove salts and detergents, air-dried, mounted with glass coverslips using mounting media as described in Subheading 3.1.4, step 17. After drying overnight at room temperature in the dark, microscope slides can be stored at −20°C.

9. For high-resolution visualization, tissue sections are imaged as z-stacks using a confocal laser scanning microscope.

4. Notes

1. To prepare 4% paraformaldehyde in 0.1 M PB with 5 mM $MgCl_2$, paraformaldehyde powder is added to 0.1 M PB, then heated and stirred. Sodium hydroxide is added to aid dissolution of paraformaldehyde. The pH is brought to 7.4, and then 1 M $MgCl_2$ is added to a final concentration of 5 mM $MgCl_2$. The solution is brought to the final volume with 0.1 M PB, and filtered.

2. Due to poor solubility of dextran sulfate in H_2O, the 50 mg/ml solution is heated to 80°C and inverted frequently (ca. 1 h).

Insoluble particles are pelleted by centrifugation at $1,000 \times g$ for 10 min at 4°C, then the solution is brought to the final volume with H_2O.

3. To prepare mounting medium, 25 g polyvinyl alcohol is added to 100 ml 1× PBS slowly while stirring and protected from light. The stirring is continued overnight. After the polyvinyl alcohol is dissolved, the pH is adjusted to 7.2 using pH indicator strips as the solution is very viscous. 50 ml glycerol is added and the solution is stirred overnight, again. Once the solution is homogenous, it is aliquoted and stored at −20°C.

4. Due to poor solubility of PFA in H_2O, 4% PFA solution should be prepared using the following protocol. To prepare 250 ml, 1.1 g NaOH and 10 g paraformaldehyde are added to 200 ml H_2O and stirred until dissolved (ca. 10 min, residual undissolved solids may still be visible). 6 g anhydrous NaH_2PO_4 is added (final concentration 0.2 M) and the solution is stirred until dissolved, followed by 250 μl of 1 M $MgCl_2$ (final concentration 1 mM). The pH is adjusted to 7.4 using 10 N NaOH, and the volume is raised to 250 ml with H_2O, followed by filtration through grade 1 cellulose filters.

5. Dots appear quickly and will saturate. So, samples are checked every 30 s after addition of detection buffer and the reaction is quenched with water prior to saturation. Make your judgment or image the blot prior to drying in order to assess which fractions contain efficiently labeled probes.

6. Approximately 10–15% extra HB is prepared for each experiment because the HB solution is viscous due to the high concentration of dextran sulfate, and an appreciable amount will be lost during pipetting. Oligonucleotide probes designed to specific mRNA sequences have approximately 50% GC content, whereas an oligonucleotide probe to detect poly(A) mRNA is composed of only thymidine bases. Given that probes rich in G-C base pairs have higher melting temperatures than those rich in A-T pairs, the hybridization buffer stringency must be reduced. In all washes and hybridization buffers, 15% formamide is used instead of 40% and the length of hybridization can be shortened to 3 h at 37°C.

7. High surface tension holds the coverslips to the parafilm. It may be helpful to put a drop of 1× SSC buffer on the coverslip to aid in the gentle removal using the fine forceps. The careful removal of coverslips from parafilm is crucial; if you drag the coverslips the neuronal morphology will be comprised.

8. To ensure good tissue preservation, efficient perfusion of the mice with physiological saline should be monitored, e.g., by

change of the color of the liver from dark red to pale. Fixation can be monitored by checking the stiffness of the mouse body (e.g., the neck or tail).

9. Thinner sections (10 μm) are beneficial for detection of dendritic mRNAs, especially those with low abundance. However, if the morphology of the tissue suffers, up to 15 μm thick sections can be used.

10. Riboprobes spanning large parts of the desired sequence are preferred, as they ensure sequence specificity and circumvent detection difficulties in vivo, where parts of the sequence might be masked with e.g., associated proteins. However, if homologies to other sequences prevent the use of larger cDNA fragments, or if probes specific to certain parts of the mRNA are required (e.g., 5′- or 3′-UTR or the open reading frame), the size of the cDNA to make the riboprobes might be reduced to 400–500 nt.

11. For linearization of plasmids, enzymes without endogenous star activity are preferred. Digestion of plasmids should then be performed with an excess of restriction enzyme for ca. 3 h. If linearization is incomplete, more enzyme (usually 0.5–1 μl) can be added and the incubation time should be increased to 4–5 h.

12. Riboprobes can be checked for digoxigenin incorporation with a dot blot as described in the manufacturer's protocol. To assure the right size of the transcript, 2% of the sample can be analyzed on an agarose formaldehyde RNA gel (22). This is especially important if riboprobes are being synthesized for the first time.

13. In our experience, additional purification of the in vitro transcribed riboprobe by gel filtration using Sephadex G50 columns to remove unincorporated nucleotides is not necessary, and should be avoided as any additional handling of the probe might compromise its quality.

14. Although size reduction of the riboprobe increases tissue penetration, for some probes it might also increase background staining. In this case, final fragment size could be up to 0.2 kb, or probes can be used without size reduction.

15. The optimal dilution of the riboprobe can vary depending on the probe sequence, and needs to be adjusted for every newly generated riboprobe.

16. Most riboprobes work at a hybridization temperature of 55°C. If the GC content of a specific probe is very high (>60%) or very low (<40%), hybridization temperature should be increased (for high GC content) or decreased (for low GC content) in 3°C increments.

17. Usually, HybriSlip hybridization covers come off very easily during the first wash in 2× SSC. If not, it is generally better to

remove hybridization covers by lifting them up using forceps as opposed to "sliding" them off, which might compromise tissue morphology.

18. To provide sufficient tissue penetration of the antibody, it is necessary that anti-digoxigenin Fab fragments are used.

19. The same moisture chamber can be used for hybridization and antibody incubation. After hybridization, however, the moisture chamber has to be cleaned carefully to remove any residual formamide. For antibody incubation and the tyramide reaction, tissues soaked in H_2O should be used to humidify the chamber.

20. Incubation time with the tyramide amplification solution is crucial, all sections should be incubated in the tyramide solution for the exact same time, to allow for comparison between different dilutions of the riboprobe, as well as assess background staining of the sense probe. To reduce background staining or increase signal intensity, incubation time can be adjusted.

Acknowledgment

The authors thank several past and current members of the Bassell lab for their efforts to refine and optimize this FISH technology. This work was supported by MH085617 and HD055835 to GJB, a postdoctoral fellowship and Conquer Fragile X research grant from the National Fragile X Foundation to C.G., and predoctoral fellowships F31NS063668, T32GM0860512 and T32NS007480, and the Epilepsy Foundation and Lennox & Lombroso Trust Fund to S.A.S.

References

1. Garner, C. C., Tucker, R. P., and Matus, A. (1988) Selective localization of mRNA for cytoskeletal protein MAP2 in dendrites, Nature 336, 674–679.
2. Burgin, K. E., Waxham, M. N., Rickling, S., Westgate, S. A., Mobley, W. C., and Kelly, P. T. (1990) In situ hybridization histochemistry of Ca2+/calmodulin-dependent protein kinase in developing rat brain, J. Neurosci. 10, 1788–1798.
3. Kleiman, R., Banker, G., and Steward, O. (1990) Differential subcellular localization of particular mRNAs in hippocampal neurons in culture, Neuron 5, 821–830.
4. Litman, P., Barg, J., Rindzoonski, L., and Ginzburg, I. (1993) Subcellular localization of tau mRNA in differentiating neuronal cell culture : Implications for neuronal polarity, Neuron 10, 627–638.
5. Bassell, G. J., Zhang, H., Byrd, A. L., Femino, A. M., Singer, R. H., Taneja, K. L., Lifshitz, L. M., Herman, I. M., and Kosik, K. S. (1998) Sorting of beta-actin mRNA and protein to neurites and growth cones in culture, J Neurosci 18, 251–265.
6. Tiruchinapalli, D. M., Oleynikov, Y., Kelic, S., Shenoy, S. M., Hartley, A., Stanton, P. K., Singer, R. H., and Bassell, G. J. (2003) Activity-dependent trafficking and dynamic localization of zipcode binding protein 1 and beta-actin mRNA in dendrites and spines of hippocampal neurons, J Neurosci 23, 3251–3261.

7. Paradies, M. A., and Steward, O. (1997) Multiple subcellular mRNA distribution patterns in neurons: a nonisotopic in situ hybridization analysis, *J Neurobiol* **33**, 473–493.
8. Bassell, G. J., Singer, R. H., and Kosik, K. S. (1994) Association of poly(A) mRNA with microtubules in cultured neurons, *Neuron* **12**, 571–582.
9. Zhang, H. L., Eom, T., Oleynikov, Y., Shenoy, S. M., Liebelt, D. A., Dictenberg, J. B., Singer, R. H., and Bassell, G. J. (2001) Neurotrophin-induced transport of a beta-actin mRNP complex increases beta-actin levels and stimulates growth cone motility, *Neuron* **31**, 261–275.
10. Antar, L. N., Afroz, R., Dictenberg, J. B., Carroll, R. C., and Bassell, G. J. (2004) Metabotropic glutamate receptor activation regulates fragile x mental retardation protein and FMR1 mRNA localization differentially in dendrites and at synapses, *J Neurosci* **24**, 2648–2655.
11. Dictenberg, J. B., Swanger, S. A., Antar, L. N., Singer, R. H., and Bassell, G. J. (2008) A direct role for FMRP in activity-dependent dendritic mRNA transport links filopodial-spine morphogenesis to fragile X syndrome, *Dev Cell* **14**, 926–939.
12. Levsky, J. M., Shenoy, S. M., Pezo, R. C., and Singer, R. H. (2002) Single-Cell Gene Expression Profiling, *Science* **297**, 836–840.
13. Bramham, C. R., and Wells, D. G. (2007) Dendritic mRNA: transport, translation and function, *Nat Rev Neurosci* **8**, 776–789.
14. Guzowski, J. F., and Worley, P. F. (2001) Cellular compartment analysis of temporal activity by fluorescence in situ hybridization (catFISH), *Curr Protoc Neurosci* **Chapter 1**, Unit 1 8.
15. Guzowski, J. F., McNaughton, B. L., Barnes, C. A., and Worley, P. F. (1999) Environment-specific expression of the immediate-early gene Arc in hippocampal neuronal ensembles, *Nat Neurosci* **2**, 1120–1124.
16. Miyashiro, K. Y., Beckel-Mitchener, A., Purk, T. P., Becker, K. G., Barret, T., Liu, L., Carbonetto, S., Weiler, I. J., Greenough, W. T., and Eberwine, J. (2003) RNA cargoes associating with FMRP reveal deficits in cellular functioning in Fmr1 null mice, *Neuron* **37**, 417–431.
17. Muddashetty, R. S., Kelic, S., Gross, C., Xu, M., and Bassell, G. J. (2007) Dysregulated metabotropic glutamate receptor-dependent translation of AMPA receptor and postsynaptic density-95 mRNAs at synapses in a mouse model of fragile X syndrome, *J Neurosci* **27**, 5338–5348.
18. Levsky, J. M., and Singer, R. H. (2003) Fluorescence in situ hybridization: past, present and future, *J Cell Sci* **116**, 2833–2838.
19. Kaech, S., and Banker, G. (2006) Culturing hippocampal neurons, *Nature protocols* **1**, 2406–2415.
20. Cox, K. H., DeLeon, D. V., Angerer, L. M., and Angerer, R. C. (1984) Detection of mrnas in sea urchin embryos by in situ hybridization using asymmetric RNA probes, *Dev Biol* **101**, 485–502.
21. Speel, E. J., Hopman, A. H., and Komminoth, P. (2006) Tyramide signal amplification for DNA and mRNA in situ hybridization, *Methods Mol Biol* **326**, 33–60.
22. Sambrook, J., and Russell, D. W. (2001) *Molecular cloning: a laboratory manual*, **3rd ed**., Cold Spring Harbor Laboratory Press, Cold Spring Harbor, NY.

Chapter 8

Localization of mRNA in Vertebrate Axonal Compartments by In Situ Hybridization

José Roberto Sotelo-Silveira, Aldo Calliari, Alejandra Kun, Victoria Elizondo, Lucía Canclini, and José Roberto Sotelo

Abstract

The conclusive demonstration of RNA in vertebrate axons by in situ hybridization (ISH) has been elusive. We review the most important reasons for difficulties, including low concentration of axonal RNAs, localization in specific cortical domains, and the need to isolate axons. We demonstrate the importance of axon micro-dissection to obtain a whole mount perspective of mRNA distribution in the axonal territory. We describe a protocol to perform fluorescent ISH in isolated axons and guidelines for the preservation of structural and molecular integrity of cortical RNA-containing domains (e.g., Periaxoplasmic Ribosomal Plaques, or PARPs) in isolated axoplasm.

Key words: ISH, Axons, Confocal ISH, PARPs, Trans-acting factors, TAF, RNA binding proteins

1. Introduction

Since the first communications describing the localization of an rRNA in an oocyte (1, 2), uses for in situ hybridization (ISH) have widened and the protocols improved to answer particular questions. Advances in detecting low copy number transcripts and spatial organization of the transcriptome helped advance the evolution of ISH techniques (3). Increasing sensitivity was one of the main goals in efforts to improve the different protocols used especially in cells where structural issues and the subcellular distribution of mRNA was not the main issue in question. In the case of neuronal cells, ISH was broadly used to detect and even semi-quantify the levels of a variety of mRNAs in the cell perikaryon. Focusing on the cell body alone would leave more than 95% of the neuronal cytoplasm, comprised in many cases mainly by dendrites and/or axons (4), uncharted. Furthermore, since mRNAs are of high abundance in the cell body, setting up protocols to detect mRNAs in these areas will lead to a default decrease in sensitivity for neuronal projections where mRNA localization is not uniform and is less concentrated. In the latter, if the user employs "cell body" ISH labeling times needed to accumulate detectable color precipitates or "cell body" parameters to detect fluorescently labeled probes, it will be difficult to obtain reliable signals. On the other hand, very long labeling times are not recommended to avoid increase background in axons.

Distribution of the protein synthesis machinery and its activity pointed out that mRNAs were being transported and localized in dendrites (and the postsynaptic densities) and axons (5, 6). mRNAs were first demonstrated to be in invertebrate unmyelinated axons or vertebrate myelinated axons by ISH (7–9). Despite a broad range of evidence showing protein synthesis activity in myelinated vertebrate axons, mRNA localization studies supporting these data only appeared in the last decade of the century (6–10). Among the main reasons for these late findings we could find the need for adaptation of ISH methods to reach sensitivity and spatial resolution to cope with the neuron and nervous tissue structural complexity. Regarding axons in the peripheral nervous system, our group demonstrated the presence of the mRNA coding for the small subunit of neurofilaments in axons of the sciatic nerve using colorimetric ISH methods (8) and the presence of beta actin mRNA with its RNA binding protein in motor axons of the lumbar ventral roots (9), but also ribosomal RNA in sciatic nerve (11). Different studies showed that mRNAs could be also detected in axons of the central nervous system like the hypothalamic–hypophyseal tract (12) where oxytocin- and vasopressin-coding mRNAs were found, or in axons of the olfactory tract where mRNAs coding for olfactory receptors were localized (13). In several studies, neuronal cell culture models were chosen to isolate axons from their complex surroundings (14, 15). Bassell and colleagues showed that axons in culture could localize beta actin mRNAs into growth cones by fast axonal transport of mRNAs embedded in ribonucleoprotein particles (RNPs) (16, 17). These studies used different approaches to gain sufficient resolution and sensitivity to answer the question of mRNA distribution in neurons. When examining an intact tissue, a balance of structural preservation coupled with the penetration of large labeled nucleic acids into the cryo-sections, combined with digoxigenin detection using alkaline phosphatase, is required (8). As the fluorescent techniques are more available and reliable, the use of laser confocal microscopy combined with either fluorescent oligos or amplification of the fluorescent signal with enzymatic methods is now preferred (9, 11).

It is possible to increase probe accessibility and resolution of axons if the user selects culture systems to perform ISH, however the main disadvantage of this approach is that the distribution of mRNAs in the animal model is not accurately assessed in this way, since axons do not fully develop in culture and the glial microenvironment is not preserved in such conditions. To solve this problem, the user can choose conventional cryo-sections or use microdissected axons from adult animals. Using methods formerly developed by Koenig and colleagues (18–23), is possible to isolate the giant axon (50–100 μm) of the Mauthner neuron, a myelinated axon from the central nervous system derived from a single cell that runs through the spinal cord of goldfish and other bony fishes. A variation of the method could be used to dissect motor neuron and sensory axons of ventral and dorsal roots of different mammals like mouse, rat, or rabbit. In both cases myelin is not present in the preparation, which allows for the fast penetration of probes and clear visualization of different probes in a stretch of up to 1,000 μm of axoplasm. In this chapter we describe methods used to detect mRNAs in micro-dissected axons from different vertebrate species and locations. mRNA may be transported as RNPs, structures also named endoaxoplasmic ribosomal plaques (EARPs) (10), via core cytoskeletal elements and then localized at cortical actin-rich Periaxoplasmic ribosomal plaque (PARP) domains (9, 18, 20, 23, 24). Therefore special attention will be given on how to preserve cortical axonal areas. Since different sources of evidences are indicating that the RNA complexity (25), either in immature (26), or mature (27) axons, is higher than previously expected, we consider ISH in single microdissected axons as a valuable tool for confirmation and/or discovery of the intra-axonal distribution and interactions of messenger RNA in neuronal projections.

2. Materials

2.1. Whole Mount Axonal Preparation in Goldfish, Rats, Mice, or Rabbits (see Note 1)

1. Cortland solution: 132 mM Na-gluconate, 5 mM KCl, 20 mM HEPES, 10 mM glucose, 3.5 mM MgSO4, and 2 mM EGTA (ethylene glycol tetraacetic acid) at pH 7.2, stored at 4°C.
2. Ammonium acetate 0.15 M at pH 4.0, Tween-20 0.01%, and NaN$_3$ 5 mM.
3. Ethyl m-aminobenzoate (MS-222).
4. Denaturing solution: 30 mM zinc acetate and 0.1 M Tricine (N-tris(hydroxymethyl) methylglycine).
5. Axon pulling solution (stock solution): (a) 0.2 M aspartate acid, and (b) 0.192 M Tris–HCl at pH 5.5.

6. Axon pulling solution (working solution): prepare a series of 30–90 mM aspartate solutions by diluting the stock aspartate solution (2.1.5) with RNAse free water as needed.

7. 1% 3-aminopropyltriethoxysilane (Polysciences, Warrington, PA) in 100% ethanol.

8. 3.75% Paraformaldehyde in PBS.

9. Tween-20.

10. #1 coverslips.

11. #5 Forceps.

12. Petri dishes (35 mm).

13. Stereoscopic microscope with fiber optics illumination source.

14. Eyebrow tool (An eyebrow attached to the tip of a Pasteur pipette).

15. YOYO-1 iodide (491/509) or POPO-1 iodide (434/456) (Invitrogen). Stock solution of YOYO-1 or POPO-1 in DMSO (1:10).

16. Epifluorescence or Confocal Microscope.

2.2. Probe Preparation

1. DIG RNA Labeling Kit (SP6/T7) (Boehringer-Mannheim, Cat. No. 1 175 025). Concentrated RNA Labeling Mix contains 10 mM each of ATP, CTP, and GTP; 6.5 mM UTP; 3.5 mM Digoxigenin-UTP; pH 7.5 (20°C).

2. Transcription Buffer (400 mM Tris–HCl (pH 8.0, 20°C), 60 mM $MgCl_2$, 100 mM Dithiothreitol (DTT), and 20 mM spermidine).

3. RNA Polymerases (SP6, T7, or T3).

4. RNasin ribonuclease inhibitor (Promega).

5. Restriction enzymes (single cut 5′ and 3′ of insert).

6. Phenol and chloroform.

7. LiCl 4 M.

8. Ethanol 70, 95, and 100%.

9. Water – RNAse/DNAse Free (with no traces of DEPC).

10. Agarose electrophoresis system.

11. Plasticware RNAse/DNAse Free.

2.3. In Situ Hybridization

1. Phosphate Buffered Saline (PBS).

2. 3.75% Paraformaldehyde in PBS.

3. 0.1 mol/L sodium diethylmalonate buffer at pH 7.0, with 0.1% Tween 20.

4. 20× SSC stock solution (0.3 M NaCl and 0.3 M sodium citrate).

5. 50% formamide.

6. 4× SSC.

7. 0.2× SSC.

8. Hybridization solution (4× SSC, 500 μg/μL salmon sperm DNA, 250 μg/μL yeast tRNA, 1× Denhardt, and 10% (w/v) dextran sulfate).

9. Bovine serum albumin.

2.4. Detection

1. Antibody anti-digoxigenin HRP conjugated antibody (Cat. No. 1-207-733 Boehringer-Mannheim).

2. 0.3% H_2O_2 in PBS.

3. TSA Plus Fluorescence Systems NEL744 (Cyanine 3) (Perkin Elmer).

4. DMSO (dimethyl sulfoxide – molecular biology or HPLC-grade).

5. Blocking reagent (Perkin Elmer Cat. No. FP1020).

6. TNT Wash Buffer (0.1 M Tris–HCl, pH 7.5, 0.15 M NaCl, and 0.05% Tween® 20).

7. TNB Blocking Buffer (0.1 M Tris–HCl, pH 7.5, and 0.15 M NaCl 0.5% Blocking Reagent).

8. Sodium peroxide.

9. 10% of dextran sulfate.

10. Fluorophore tyramide, Plus Amplification Diluent (NEN Life Science Products).

11. Prepare fluorophore tyramide Stock Solution. Each vial must be reconstituted with 0.15 mL DMSO before use. The fluorophore tyramide stock solution, when stored at 4°C, is stable for at least 3 months.

12. Prepare fluorophore tyramide working solution. Before each procedure, dilute the Fluorophore Tyramide Stock Solution 1:50 using 1× Plus Amplification Diluent to make the fluorophore tyramide working solution (28). Approximately 100–300 μL is required per slide. Discard any unused portion of working solution.

13. ProLong antifade (Invitrogen).

3. Methods

3.1. Probe Preparation

1. Linearize DNA with an appropriate restriction enzyme that does not leave 3′ end overhangs.

2. Extract with phenol–chloroform and precipitate. Resuspend in sterile water at 1 mg/mL.

3. Add the following, in the order shown:

- 1 μg of purified, linearized plasmid DNA or 100–200 ng of purified PCR fragment.
- 2 μL of 10× concentrated DIG RNA Labeling Mix.
- 2 μL of 10× concentrated Transcription Buffer (see Note 2).
- 2 μL of RNA polymerase (SP6, T7, or T3).
- Enough sterile, redistilled RNAse free water to make a total reaction volume of 20 μL.

4. Mix gently, do not vortex, spin down drops in microfuge (see Note 3).
5. Incubate at 37°C for 2–6 h.
6. Check 1 μL on a 1% agarose gel (see Note 4), and in the meantime, purify the probe using precipitation: Add 100 μL water, 10 μL 4 M LiCl, 300 μL ethanol, and 1 μL glycogen (4 mg/mL). Place at –80°C for at least 1 h. Spin and resuspend the pellet in 30 μL of water; add 15 μL of 7.8 M ammonium acetate, 1 μL glycogen and 100 μL of ethanol. Place at –80°C for at least 1 h. Spin and resuspend pellet in 30 μL of water. Store at –80°C (see Note 5).

3.2. Whole-Mount Axon Preparation

3.2.1. Mauthner Axons: Tissue Dissection

1. Anesthetize common goldfish, *Carassius auratus* (6–8 cm in length), for 30 min in ice/water bath containing MS-222 (0.01%). The isolation of brain and spinal cord tissue requires about 5 min.
2. The lower portion of the brain is exposed and severed in situ along the anterior border of the cerebellum.
3. The spinal cord and lower brainstem are rapidly dissected and placed on ice-cold Cortland medium.
4. As a next step, the brainstem is cut about midway through cerebellum; dorsal structures, including vagal lobes, facial lobe, and cerebellar crests are then excised in order to expose the rostral course of M-cell axons. The M-cell axons run superficially in the floor of the ventricle and their course can be visualized under a dissecting microscope either by reflected light or transillumination (see Fig. 1a, b).
5. Approximately 5 mm portions of brainstem and spinal cord are cut and left in ice-cold Cortland solution in a 35-mm Petri dish, waiting for denaturation step.

3.2.2. Mauthner Axon: Axon Pulling

1. CNS segments are transferred to a dish containing denaturing solution and incubated for 10 min at room temperature.
2. The spinal cord segments are transferred to a Petri dish containing axon pulling solution and a #1 coverslip previously coated with 1% 3-aminopropyltriethoxysilane.
3. After 1 min of incubation use #5 forceps to start pulling out the axoplasm while observing through a stereoscopic microscope

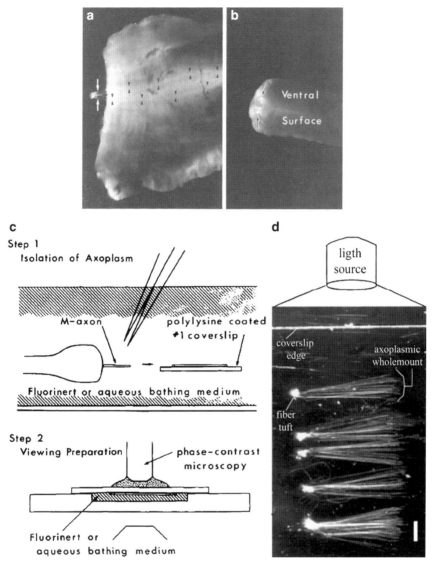

Fig. 1. Locating M-cell axons in the brainstem and spinal cord. (a) An isolated brainstem after trimming, showing two ensheathed M-cell axons, partially translated out of their in situ location. Axons lie superficially in the floor of the fourth ventricle and can be visualized in a dissecting microscope with reflected or transilluminated light. (b) M-cell axons assume a position in the ventral quadrant of the spinal cord, and can be made to protrude from the cut surface by slight compression applied to the lateral surfaces of the cord stump. Two ensheathed M-cell axons are shown protruding (*arrows*). (c) Diagrammatic description of the isolation of M-cell axons in preparation for viewing. Step 1: M-cell axons are translated out of their in situ locations with a pair of no. 5 watchmaker's tweezers in an appropriate bathing medium, and deposited fully extended on a polylysine-coated coverslip with the aid of an eyebrow hair. Step 2: after axons are attached to the coverslip, the coverslip is removed, the bottom surface is wiped, and it is inverted over the well of a chamber, whereupon the space is filled with the appropriate bathing medium. The chamber is made from two glass histological slides cemented together (courtesy of Dr. Edward Koenig, modified from Brain Research (19)). (d) A dark-field view through a dissecting microscope of five axon sprays isolated from rabbit ventral nerve root fibers that were attached to a coverslip surface. A spray is defined as multiple isolated axoplasmic whole mounts originating in a nerve fiber tuft remnant that was used to grasp the spray. *Scale bar*, 1 mm (courtesy of Dr. Edward Koenig, modified from *Journal of Neuroscience* (18)).

(see Note 6). Different concentrations of axon pulling solution should be tested for optimal recovery of cortical axoplasm (see Note 7). Once the best recovery is assessed it's possible to pull out axoplasm from the remaining segments with at least 50% chances of success with respect of cortical located structures.

4. Once the axon is out of the spinal cord, it should be moved to the coverslip surface with the aid of an eyebrow tool to attach first the end and then the middle segment of the axon (Fig. 1c).

3.2.3. Isolation of Axoplasmic Whole-Mounts from Mammalian Myelinated Spinal Root Fibers

1. Dissect lumbar spinal nerve roots from either euthanized rat, mice, or rabbit depending on the model you consider using.

2. Suspend the tissues (several nerve root/rootlet) in ice-cold Cortland solution.

3. A nerve root/rootlet, 3–5 mm, is immersed in denaturing solution for 10 min.

4. Then the tissue is placed in a 35-mm plastic culture dish containing 2 mL pulling solution and a #1 coverslip previously coated with 1% 3-aminopropyltriethoxysilane.

5. Grasp a small portion of the tip of the ventral root with #5 forceps (while holding the other end with another forceps) and pool out to obtain a spray of axons attached to a nerve fiber tuft (whole mounts). Attach isolated axoplasmic whole-mounts with the aid of eyebrow hair tools to #1 coverslips (Fig. 1d).

3.2.4. YOYO-1 and POPO-1 Staining of Axoplasmic Whole-Mounts

1. Recovery of PARPs (see Note 8) can be assessed by staining of the wholemounts with YOYO-1 iodide (491/509) (see Note 9).

2. One microliter of a 1:10 stock solution of YOYO-1 or POPO-1 in DMSO was added to 2 mL of the pulling medium for 15 min.

3. YOYO-1 or POPO-1 were washed out by brief immersion in 0.15 M ammonium acetate and 0.01% Tween 20, NaN_3 5 mM.

4. For fluorescence microscopy the coverslip with axoplasmic sprays was mounted on a flow-through chamber filled with 0.15 M ammonium acetate solution (see Note 10).

5. After verifying the presence or absence of YOYO-1 staining (epifluorescence or confocal microscope) the coverslips are returned to the Petri dishes and stored at 4°C to be used shortly thereafter for ISH.

3.3. In Situ Hybridization (see Note 11)

1. Fix isolated Zn-denatured axoplasmic whole-mounts in 3.75% paraformaldehyde buffered with PBS with 0.1% Tween 20, or if no salts are preferred, use 0.1 mol/L sodium diethylmalonate buffer at pH 7.0, with 0.1% Tween 20, for 3 h at 20°C (see Note 12).

2. Wash twice for 10 min with ammonium acetate solution at room temperature.

3. Dehydrate in 70, 95, and 100% alcohols and air dry.

4. Pre-incubate specimens in hybridization solution (i.e., 4× SSC, 500 μg/μL of salmon sperm DNA, 250 μg/μL of yeast tRNA, 1× Denhardt, and 10% (w/v) dextran sulfate) for 15 min at 42°C in a humid chamber.

5. Denature sense or antisense RNA probe and mix it with hybridization solution at a final concentration of 2 ng/μL and incubate for 3 h at 42°C in humid chamber (longer incubation time will affect PARP structures).

6. Wash coverslips once in 4× SSC for 20 min, at room temperature.

7. Wash two times with 0.2× SSC and 0.2% bovine serum albumin, at 50°C with agitation.

3.4. Detection by Tyramide Amplification

1. Quench endogenous peroxidase using 0.3% H_2O_2 in PBS for 10 min.

2. Block coverslips with 100–300 μL of TNB Buffer in a humidified chamber for 30 min at room temperature.

3. Incubate coverslips for 30 min at room temperature in a humidified chamber 1/500 of anti-DIG-HRP conjugated antibody. Using 100–300 μL of antidigoxigenin-HRP diluted in TNB Buffer.

4. Wash the slides three times for 5 min each in TNT Buffer at room temperature with agitation.

5. TSA Plus Fluorescence Systems Amplification. Pipet 100–300 μL of the Fluorophore Tyramide Working Solution onto each slide. Incubate the slides at room temperature for 3–10 min.

6. Wash the slides three times for 5 min each in TNT Buffer at room temperature with agitation in the presence of 10% of dextran sulfate to improve localization of tyramide (28).

7. Follow desired fluorescence observation by mounting the coverslip on the flow through chamber (see Subheading 3.2.2) or by mounting the coverslip upside down on a slide containing a small drop of ProLong antifade (Invitrogen). Results of ISH and detection of RNA binding proteins in axonal wholemounts can be seen in Figs. 2 and 3.

4. Notes

1. Solutions used in direct contact with axonal wholemounts should be filtered (0.45 μm) to avoid adhesion of fine particulate material.

2. To avoid precipitation caused by spermidine, reagents must be at room temperature before starting.

3. Do not use water treated with DEPC, since small traces of it will inhibit RNA polymerases, resulting in low yields of labeled transcripts.

4. Quantification of the labeled probe could be carried out by dot blots of serial dilutions (provided in the kit) or by capillary gel electrophoresis on Bioanalyzers (Agilent).

5. This extra precipitation step helps to decrease background because it improves removal of unincorporated DIG-nucleotides.

6. In the spinal cord, the axons are located in the ventral quadrants and can frequently be made to protrude by gentle compression (using number 5 forceps) of the spinal cord stump. This is accomplished by grasping the exposed cut ends of the axons with the tweezers and pulling them out in a deliberate fashion. Experience will aid in determining an optimum rate for pulling. Once the tips of the axons are visible, one can hold and later pull out the protruding end of the axon with the aid of forceps. Use extreme caution to avoid bending of the tips of forceps, since this will reduce the ability to grasp the axons.

7. It is imperative to use different concentrations of pulling solutions (30–50 mM aspartate) to establish a critical permissive concentration. This concentration is empirically defined as the one at which PARPs, PARP markers, and/or YOYO-1 staining is optimized (18), since PARPs are not always recovered at the same aspartate concentration.

8. The integrity of PARP domains and the presence of nucleic acids in the isolated axons can be verified by phase contrast and fluorescence microscopy (see Figs. 2 and 3). When preservation of PARPs is optimal, it is possible to observe phase-dense regions, which can be used as landmarks to find PARP domains after ISH development (Fig. 2). Do not stain with YOYO-1 after ISH, because hybridization buffers include blocking non-specific probe binding with an excess of nucleic acids. Instead, detect ribosomes to observe PARP distribution. It is of special value to validate a particular mRNA localization to localize an RNA-binding protein known to bind the mRNA being assessed. That can be seen in Fig. 3 in which ZBP1, a known binding partner of beta-actin mRNA (9), was localized to PARP domains in which the same message was observed by ISH (Fig. 2).

9. If one wishes to detect nucleic acids at a different wavelength, it is possible to use POPO-1 iodide (434/456), although it fades more rapidly than YOYO-1.

10. The chamber was constructed by inverting the coverslip over spacers (0.5–1 mm thick) made of Silastic elastomer (Dow

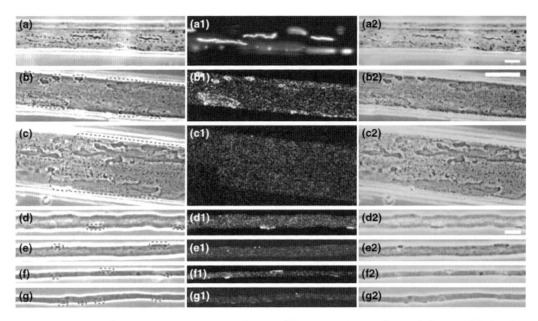

Fig. 2. Localization of β-actin mRNA by in situ hybridization (ISH) in periaxoplasmic ribosomal plaque (PARP) domains, identified by phase structural correlates (outlined by *green dash ovals*). (**a**) Structural correlates of PARP formations are revealed in a phase micrograph of an isolated goldfish Mauthner axoplasmic whole-mount. (**a1**) RNA labeled with YOYO-1, a high affinity RNA binding fluorescent dye, shows a discrete localization and distribution along the length of the whole-mount. (**a2**) A merge of RGB-rendered phase and fluorescence images in (**a**) and (**a1**) confirms that RNA fluorescence localizes in PARP formations identified by phase correlates. Representative examples of ISH labeling in axoplasmic whole-mounts isolated from (**b–b2** and **c–c2**) Mauthner, (**d–d2** and **e–e2**) rabbit, and (**f–f2** and **g–g2**) rat nerve fibers. The size and morphology of PARPs in Mauthner axons are highly variable, in contrast to the unremarkable simple morphology of PARPs in rabbit and rat whole-mounts. The b-actin mRNA antisense probe is localized in delimited domains in (**b1**) Mauthner, (**d1**) rabbit, and (**f1**) rat whole-mount ISH images. In contrast, the sense probe ISH yields diffuse background in (**c1**) Mauthner, (**e1**) rabbit, and (**g1**) rat whole-mounts. Merges of RGB-rendered phase images in (**b–g**) and corresponding (**b1–g1**) ISH images in (**b2–g2**), respectively, show that there is a close correspondence between PARP structural correlates and the localization and distribution density of antisense β-actin mRNA ISH signals. *Grayscale* images shown in (**a–g**) and (**a1–g1**) were assigned *blue-green* and *red* colors, respectively, from a lookup table to highlight correspondence (**a2, b2, d2,** and **f2**), or no correspondence (**c2, e2,** and **g2**) in merged images, using NIH ImageJ. *Calibration bars*: **a–c2**, 50 μm and **d–g2**, 10 μm, see Sotelo-Silveira et al. (9), with permission of the *Journal of Neurochemistry* (*Wiley-Blackwell*).

Corning, Midland, MI) attached to a large glass coverslip (35×50 mm) taped to a thin, U-shaped metal plate.

11. Effectiveness of the probes in detecting message can be assessed in parallel using positive control tissues (prepared by cryotomy) in which neuronal cell bodies are abundant and mRNA concentration is high (e.g., motor neurons in spinal cord).

12. Since PARP domains are exposed directly to hybridization solutions, different fixing times and hybridization length were tested to avoid PARP removal/destruction during hybridization. A combination of 3 h fixation and 3 h hybridization was determined by following YOYO-1 staining in the presence of 50% formamide, 4× SSC.

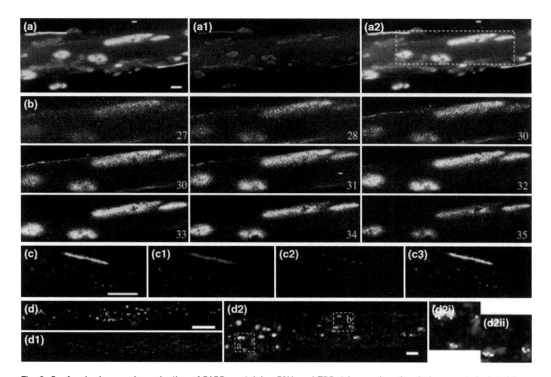

Fig. 3. Confocal microscopic evaluation of PARP-containing RNA and ZBP-1 in axoplasmic whole-mounts isolated from Mauthner and rat ventral root fibers. (**a–a2**) Low magnification (×20) merges of 39 optical slices (1.45 μm) of a Mauthner whole-mount (~55 μm in diameter), in which (**a**) RNA fluorescence labeling of PARPs with YOYO-1 in the *green* channel and (**a1**) ZBP-1 immunofluorescence in the *red* channel are shown localized in PARP formations (**a2**). A merged image of (**a**) and (**a1**) reveals co-distributions of PARP RNA fluorescence and ZBP-1 immunofluorescence. (**b**) NIH ImageJ software was used to analyze a series of nine consecutively numbered slices (b27–35) in a region outlined in (**a2**) from the outer one-third of the whole-mount in order to evaluate co-localization of RNA and ZBP-1 fluorochrome pixels. Apparent co-localization is indicated by *bright* signals, in which the intensity ratio of YOYO-1 and anti-ZBP-1 fluorochromes is ≥80%, and the overlap ratio is >0.9. Note that while inspection of (**a–a2**) indicates considerable overlap of YOYO-1 fluorescence and ZBP-1 immunofluorescence in PARPs, co-localization signals are not uniformly distributed within PARP domains. Rather, colocalization signals tend to distribute outside of the compact portions of the PARP domain. (**c–c3**) Co-distribution of (**c**) RNA (*green*), labeled by YOYO-1, (**c1**) ribosomes (*blue*), labeled with ribosome anti-P protein, and (**c2**) immunofluorescence of anti-ZBP-1 (*red*) in a surface PARP of an axoplasmic whole-mount from a rat ventral root fiber; (**c3**) a merged image of (**c–c2**). Putative "in-transit" RNA/ZBP-1-containing "particles" within the axoplasmic core of a whole-mount from a rat ventral root fiber is suggested from inspection of (**d**) RNA (*green*) (**d1**) anti-ZBP-1 immunofluorescence (*red*), and (**d2**), an enlarged merged image, which shows occasional areas containing co-localization signals (see *dash rectangles*) that are further enlarged in (**d2i** and **d2ii**). *Calibration bars*: **a–a2**, 12.2 μm and **c–d1**, 10 μm, see Sotelo-Silveira et al. (9), with permission of the *Journal of Neurochemistry* (*Wiley-Blackwell*).

Acknowledgments

The authors thank John Mercer for a critical reading of this manuscript and Edward Koenig for the use of figures presented in Fig. 1. This project was partially funded by CSIC, PEDECIBA, ANII, MEyC, PEW fellowship to JRSS, NIH grant #1R03 TW007220-01A2.

References

1. John H.A., Birnstiel M.L. Jones K.W. (1969). RNA-DNA hybrids at the cytological level. *Nature* **223**, 582–7.
2. Gall J.G., Pardue M.L. (1969). Formation and detection of RNA-DNA hybrid molecules in cytological preparations. *Proc. Natl. Acad. Sci. USA* **63**, 378–83.
3. Femino A.M., Fay F.S., Fogarty, K., Singer, R.H. (1998). Visualization of single RNA transcripts in situ. *Science* **280**, 585–90.
4. Craig A.M., Banker, G. (1994). Neuronal polarity. *Annu. Rev. Neurosci.* **17**, 267–310.
5. Sotelo-Silveira J.R., Calliari, A., Kun, A., Koenig, E., Sotelo, J.R. (2006). RNA trafficking in axons. *Traffic* **7**, 508–15.
6. Giuditta, A., Kaplan, B.B., van Minnen, J., Alvarez, J., Koenig, E. (2002). Axonal and presynaptic protein synthesis: new insights into the biology of the neuron. *Trends Neurosci.* **25**, 400–4.
7. Giuditta, A., Menechini, E., Perrone, Capano. C., Langella, M., Castigli, E., Kaplan BB. (1991). Active polysomes in the axoplasms of the squid giant axon. *J. Neurosci. Res.* **28**, 18–28.
8. Sotelo-Silveira J.R., Calliari, A., Kun, A., Benech, J.C., Sanguinetti, C, Chalar, C., Sotelo J.R. (2000). Neurofilament mRNAs are present and translated in the normal and severed sciatic nerve. *J. Neurosci. Res.* **62**, 65–74.
9. Sotelo-Siveira, J.R., Crispino, M., Puppo, A., Sotelo, J.R., Koenig, E. (2008). Myelinated axons contain beta-actin mRNA and ZBP-1 in periaxoplasmic ribosomal plaques and depend on cyclic AMP and F-actin integrity for in vitro translation. *J. Neurochem.* **104**, 545–57.
10. Koenig E. (2009). Organized ribosome-containing structural domains in axons. *Results Prob. Cell Differ.* **48**, 173–91.
11. Kun, A., Otero, L., Sotelo-Silveira, J.R., Sotelo, J.R. (2007). Ribosomal distributions in axons of mammalian myelinated fibers. *J. Neurosci. Res.* **85**, 2087–98.
12. Mohr, E., Richter, D. (1993). Complexity of mRNAs in axons of rat hypothalamic magnocellular neurons. *Ann NY Acad Sci* **689**, 564–6.
13. Vassar, R., Chao, S.K., Sitcheran, R., Nunes, J.M., Vosshall, L.B., Axel, R. (1994). Topographic organization of sensory projections to the olfactory bulb. *Cell* **79**, 981–91.
14. Eng, H., Lund, K., Campenot, R.B. (1999). Synthesis of beta-tubulin, actin, and other proteins in axons of sympathetic neurons in compartmented cultures. *J. Neurosci.* **19**, 1–9.
15. Zheng, J.Q., Kelly, T.K., Chang, B., Ryazantsev, S., Rajasekaran, A.K., Martin, K.C., Twiss, J.L. (2001). A functional role for intra-axonal protein synthesis during axonal regeneration from adult sensory neurons. *J. Neurosci.* **21**, 9291–303.
16. Kiebler, M.A., Bassell, G.J. (2006). Neuronal RNA granules: movers and makers. *Neuron* **51**, 685–90.
17. Bassell, G.J., Zhang, H., Byrd, A.L., Femino, A.M., Singer, R.H., Taneja, K.L., Lifshitz, L.M., Herman, I.M., Kosik, K.S. (1998). Sorting of beta-actin mRNA and protein to neurites and growth cones in culture. i. **18**, 251–65.
18. Koenig, E., Martin, R., Titmus, M., Sotelo-Silveira, J.R. (2000). Cryptic peripheral ribosomal domains distributed intermittently along mammalian myelinated axons. *J. Neurosci.* **20**, 8390–400.
19. Koenig E. (1986). Isolation of native Mauthner cell axoplasm and an analysis of organelle movement in non-aqueous and aqueous media. *Brain Res.* **398**, 288–97.
20. Koenig, E, Martin R. (1996). Cortical plaque-like structures identify ribosome-containing domains in the Mauthner cell axon. *J. Neurosci.* **16**, 1400–11.
21. Koenig, E., Repasky E. (1985). A rational analysis of alpha-spectrin in the isolated Mauthner neuron and isolated axons of the goldfish and rabbit. *J. Neurosci.* **5**, 705–14.
22. Koenig E. (1979). Ribosomal RNA in Mauthner axon: implications for a protein synthesizing machinery in the myelinated axon. *Brain Res.* **174**, 95–107.
23. Sotelo-Silveira, J.R., Calliari, A., Cárdenas, M., Koenig, E., Sotelo, J.R. (2004). Myosin Va and kinesin II motor proteins are concentrated in ribosomal domains (periaxoplasmic ribosomal plaques) of myelinated axons. *J. Neurobiol.* **60**, 187–96.
24. Muslimov, I.A., Titmus, M., Koenig, E., Tiedge, H. (2002). Transport of neuronal BC1 RNA in Mauthner axons. *J Neurosci.* **22**, 4293–301.
25. Yoon, B.C., Zivraj, K.H., Holt, C.E. (2009). Local translation and mRNA trafficking in axon pathfinding. *Results Prob. Cell Differ.* **48**, 269–88.
26. Willis, D., Li, K.W., Zheng, J.Q., Chang, J.H., Smit, A., Kelly, T., Merianda, T.T., Sylvester,

J., van Minnen, J., Twiss, J.L. (2005). Differential transport and local translation of cytoskeletal, injury-response, and neurodegeneration protein mRNAs in axons. *J. Neurosci.* **25**, 778–91.

27. Hanz, S., Perlson, E., Willis, D., Zheng, J.Q., Massarwa, R., Huerta, J.J., Koltzenburg, M., Kohler, M., van-Minnen, J., Twiss, J.L., Fainzilber, M. (2003). Axoplasmic importins enable retrograde injury signaling in lesioned nerve. *Neuron* **40**, 1095–104.

28. van Gijlswijk, R.P., Zijlmans, H.J., Wieganet, J., Bobrow, M.N., Erickson, T.J., Adler, K.E., Tanke, H.J., Raap, A.K. (1997). Fluorochrome-labeled tyramide: use in immunocytochemistry and fluorescence in situ hybridization. *J. Histochem. Cytochem.* **45**, 375–82.

Part II

Visualizing mRNAs In Vivo Using Molecular Probes or Reconstituted Fluorescent Proteins

# Chapter 9

Tiny Molecular Beacons for *in vivo* mRNA Detection

Diana P. Bratu, Irina E. Catrina, and Salvatore A.E. Marras

Abstract

The molecular beacon technology is an established approach for visualizing native mRNAs in living cells. These probes need to efficiently hybridize to accessible RNA regions in order to spatially and temporally resolve the dynamic steps of the RNA life cycle. A refined method using two computer algorithms, *mfold* and RNAstructure, is described for choosing shorter, more abundant target regions for molecular beacon binding. The probes are redesigned as small hairpins and are synthesized from 2′-*O*-methyl RNA/LNA chimeric nucleic acids. These tiny molecular beacons are stable in the cellular environment and have a high affinity for binding to target RNAs. The user-friendly synthesis protocol and ability to couple to a variety of fluorophores make tiny molecular beacons the optimal technology to detect less abundant, highly structured RNAs, as well as small RNAs, such as microRNAs. As an example, tiny chimeric molecular beacons were designed to target regions of *oskar* mRNA, microinjected into living *Drosophila melanogaster* oocytes and imaged via spinning disc confocal microscopy.

Key words: Molecular beacons, 2′-*O*-methyl RNA, LNA, Fluorescence live cell imaging, mRNA localization, mRNA transport

1. Introduction

Significant advances have been made over the last decade for visualizing and tracking of individual mRNAs within distinct mRNP complexes in subcellular space in real time (1, 2). Such approaches have provided a live glimpse of specific biological processes that have remained opaque thus far (3). Among them is the molecular beacon technology, which involves probes that fluoresce only upon hybridizing to specific complementary mRNA sequences. Introduced as an innovative and general approach, molecular beacons have been used in a variety of cell types, detecting mRNA at various levels of expression (4). Coupled with fast 3D imaging over time, molecular beacon technology has enabled highly time-resolved studies of RNA–protein interactions *in vivo* providing details of the dynamically orchestrated relationship between an mRNA and various proteins involved in its transport.

Molecular beacons are internally quenched hairpin-shaped oligonucleotide probes that fluoresce upon hybridization with their target sequence (Fig. 1a) (5, 6). Target-bound probes fluoresce as much as 100 times more intensely than background levels of unbound probes, enabling highly sensitive detection (Fig. 3b). Molecular beacons are designed using the *mfold* RNA-folding software (7, 8). The most stable predicted structure **must** reflect a hairpin, ensuring that the fluorophore and quencher are within minimum distance from each other (9). Due to their stem, the recognition of targets by molecular beacons is so specific that if the target differs even by a single nucleotide, the probe does not bind to it. Molecular beacons can be synthesized from modified nucleic acids (i.e., 2′-O-methyl-ribonucleic acids, LNA – locked nucleic acids) (Fig. 1b) to ensure greater stability of the probe and the probe–target hybrid, and be labeled with a wide variety of fluorophores and respective quenchers (Tables 1 and 2), enabling detection of multiple targets simultaneously (6).

For live cell imaging, molecular beacons have been synthesized from 2′-O-methyl ribonucleotides, thus assuring resistance to cellular nucleases, and hybrid stability with RNA, thus evading degradation by Ribonuclease H (9).

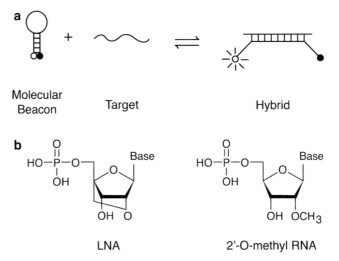

Fig. 1. (**a**) Scheme of molecular beacon operation. In the absence of a complementary target, these molecules are nonfluorescent, because the stem hybrid keeps the fluorophore in close proximity to the quencher. In the presence of target, the probe sequence in the loop hybridizes to the target, forming a rigid double helix inducing a conformational reorganization that separates the quencher from the fluorophore leading to an increase in fluorescence. (**b**) Molecular structures of LNA vs. 2′-O-methyl RNA. Locked Nucleic Acid (LNA) is a nucleic-acid modification where the 2′ oxygen and the 4′ carbon atoms in the furanose ring are bridged via a methylene moiety.

Table 1
Fluorophore labels for molecular beacon probes

Fluorophore	Alternative fluorophore	Excitation (nm)	Emission (nm)
TMR	Alexa 546[a], Cy3[b]	555	575
Texas Red	Alexa 594[a]	585	605
Cy5[b]	Alexa 647[a]	650	670

[a]Alexa fluorophores are available from Invitrogen
[b]Cyanine dyes are available from Amersham Biosciences

Table 2
Quencher labels for molecular beacon probes

Quencher	Absorption maximum (nm)
Deep Dark Quencher I[a]	430
Dabcyl	475
Eclipse[b]	530
Iowa Black FQ[c]	532
Black Hole Quencher 1[d]	534
Black Hole Quencher 2[d]	580

[a]Deep Dark Quenchers are available from Eurogentec
[b]Eclipse quenchers are available from Epoch Biosciences
[c]Iowa quenchers are available from Integrated DNA Technologies
[d]Black Hole Quenchers are available from Biosearch Technologies

Incorporation of LNA bases into oligonucleotides leads to exceptionally high-affinity binding to complementary sequences. Their negatively charged backbone confers good solubility, making these RNA derivatives easily synthesized using standard DNA/RNA synthesis methods (10).

Here, we show that introducing LNAs into 2′-O-methyl RNA oligonucleotides enables the design of smaller hairpins that match the stabilities of longer 2′-O-methyl RNA variants when bound to target. We describe how to design and synthesize these tiny LNA/2′-O-methyl RNA chimeras as well as introduce a refined approach for target region selection. We demonstrate that these tiny molecular beacons are stable, bright, highly specific, and effective in binding multiple mRNA target sites. The implementation of these changes in molecular beacon design will make this technology more practical for targeting highly structured mRNAs, thus enabling the simultaneous visualization of several mRNAs. Moreover, their application can extend to the detection of small RNA targets, such as microRNAs or piRNAs in living cells.

2. Materials

2.1. Target mRNA Selection and Design of Tiny Molecular Beacons

1. *mfold*: access to server http://mfold.rna.albany.edu/.
2. RNA folding program: http://mfold.rna.albany.edu/?q=mfold/RNA-Folding-Form.
3. RNAstructure program: Free download from the Mathews Lab at the University of Rochester Medical Center: http://rna.urmc.rochester.edu/rnastructure.html.

2.2. Tiny Molecular Beacon Synthesis

1. 394 DNA/RNA Synthesizer (Applied Biosystems, Foster City, CA).
2. Locked Nucleic Acid (LNA) Phosphoramidites (Exiqon, Woburn, MA).
3. 2′-O-methyl RNA Phosphoramidites (Glen Research, Sterling, VA).
4. Dabcyl-linked controlled-pore glass synthesis column (Biosearch Technologies., Novato, CA).
5. 5′-thiol modifier C6 (Glen Research) (see Note 1).
6. High-pressure liquid chromatograph equipped with C18 reverse-phase column and dual wavelength detector: System Gold (Beckman Coulter, Brea, CA).
7. HPLC Buffer A: 0.1 M Triethylammonium acetate at pH 6.5, filtered and degassed.
8. HPLC Buffer B: 0.1 M Triethylammonium acetate in 75% (v/v) Acetonitrile at pH 6.5, filtered and degassed.
9. Ethanol.
10. 3 M Sodium acetate at pH 5.2.
11. 0.15 M Dithiothreitol (DTT).
12. 0.15 M Silver nitrate.
13. Tetramethylrhodamine-5-iodoacetamide (Invitrogen, Carlsbad, CA) (see Note 1).
14. Texas Red C5 bromoacetamide (Invitrogen).
15. *N,N*-Dimethylformamide (DMF).
16. 0.2 M Sodium bicarbonate at pH 9.0.
17. TE buffer: 1 mM EDTA and 10 mM Tris–HCl at pH 8.0.
18. Nuclease-free water (Ambion Inc., Austin, TX).

2.3. In Vitro Characterization of Tiny Molecular Beacons

1. Hybridization buffer: 10 mM Tris–HCl at pH 8.0, 50 mM KCl, and 1 mM $MgCl_2$.
2. Molecular beacon. Dissolve molecular beacons for stock solutions in TE buffer and store at −20°C. Dilute working

solutions in nuclease-free water, keep protected from light, and store at −20°C up to 1 month.

3. Oligonucleotide target complementary to the probe sequence of the molecular beacon. Dissolve in TE and store at −20°C.

4. Spectrofluorometer: QuantaMaster (Photon Technology International, Birmingham, NJ).

5. Thermal cycler with a capacity to monitor fluorescence in real time: iCycler iQ5. (Bio-Rad, Hercules, CA).

2.4. In Vivo mRNA Imaging with Tiny Molecular Beacons

1. Injection buffer: 50 mM Tris–HCl at pH 7.5, 100 mM NaCl, and 1.5 mM $MgCl_2$.

2. Molecular beacon. Dilute stock solution to 100 and 200 ng/μL in injection buffer. Keep protected from light, and store at −20°C up to 1 month.

3. Halocarbon Oil 700 (Sigma-Aldrich, St. Louis, MO).

4. Dupont tweezers No. 5 (World Precision Instruments, Sarasota, FL).

5. Microscope cover slips (22×40 mm) (VWR, West Chester, PA).

6. Microinjection apparatus (Eppendorf Inc., Hauppange, NY).

7. Fluorescence microscope (inverted) (Leica Microsystems Inc., Bannockburn, IL).

8. Image acquisition/analysis/processing software: Volocity (Perkin Elmer, Waltham, MA) and ImageJ (freeware from NIH): http://rsbweb.nih.gov/ij/.

3. Methods

3.1. Selection of mRNA Target Regions and Design of Tiny Molecular Beacons

Theoretically, any sequence within a target RNA can be chosen as a site for molecular beacon binding. The endless possibilities give one the confidence that such regions are easily identified. However, the extent of target accessibility is primarily a consequence of complex secondary and tertiary intramolecular structures, which are not easy to predict and can mask many of these regions.

3.1.1. Selection of RNA Target Regions

1. Fold mRNA sequence using RNA folding program *mfold* (7, 8).

2. Using default settings at 37°C, 1 M NaCl, and no divalent ions, obtain an immediate output for RNA sequences <800 nt in length. A "batch job" is otherwise submitted, which gets completed in 15 min on average, depending on the server's availability. One can also receive an email notification that the job is complete and a link to the result posted on the web server.

3. Select to view the first MFE structure (e.g., jpg file), that represents the most thermodynamically stable secondary structure predicted. Use default parameter settings for suboptimal structures, which are sufficient for this analysis.

4. Analyze the entire ensemble of structures using two parameters: ss-count and P-num. ss-count defines the probability of a nucleotide to be single-stranded. P-num denotes the total number of different base pairs that can be formed by a particular base within the full set of structure results. These values are assigned to each nucleotide in the mRNA sequence and can be represented via a color-coded map. To view the folded RNA structure representing ss-count values, choose the "ss-count" annotation, and click on the image to redraw the structure. Each nucleotide will become colored with warm or cold colors depending on the respective ss-count values. Warm colors represent high values, indicating single-stranded regions. Change the annotation to "P-num", and the colors will now reflect the values that indicate the stability of structures formed (warm colors represent well determined base pairings). Evaluate the ss-count and P-num color-annotated secondary structures (see Note 2).

5. To winnow down the number of candidate sites, employ a second algorithm (11). Input the RNA sequence (in CAPS) in the RNAstructure program as "New Sequence" and select to "Fold as RNA." The file will be saved as a "Sequence File" (.seq), and a "CT File" (.ct) will automatically be generated. Use the default Suboptimal Structure Parameters. Start fold. The calculation will take a few minutes, depending on the length of the RNA (approx. 30 min for 3 kb using 8GB RAM at 1066 MHz) and if the suboptimal structure parameters are changed.

6. Select RNA OligoWalk module. This scans the folded RNA sequence for regions to which various-length oligonucleotides are capable of binding. The "Input File" is the ".ct" file saved above. This file contains the sequence with base pair information. Use the default mode, "Break Local Structure", for calculating the free energy of intramolecular structure of the oligonucleotide and of the local RNA structure predicted. In this mode, the target structure breaks wherever the oligonucleotide binds ($\Delta G_{break\ target}$), oligonucleotides lose pairs in self-structures ($\Delta G_{oligo\text{-}oligo}$ and $\Delta G_{oligo\text{-}self}$) and gain pairs in oligonucleotide-target binding (ΔG_{duplex}). This algorithm calculates the equilibrium affinity of a set-length complementary oligonucleotide to the RNA target and predicts its overall free energy of binding while taking into account stability of the newly formed duplex, local secondary structure within the target RNA, as well as intermolecular and intramolecular secondary structures formed by the oligonucleotide (12).

7. Run OligoWalk calculation for an oligonucleotide concentration of 100 ng/μL, ranging in length from 10 to 15 nucleotides, including suboptimal target structures. The calculation will take just a few minutes, depending on the length of the oligonucleotide and target RNA.

8. The ΔG bar graph generated will facilitate the "walk" of the oligonucleotide on the RNA sequence, one nucleotide at a time, thus obtaining information about each oligonucleotide/RNA hybrid formed.

9. Choose a target region that generates the most negative (−) kcal/mol values for ΔG_{duplex} and $\Delta G_{overall}$, with a difference between them of ≤ −10 kcal/mol (see Note 3).

10. Correlate these potential target regions with the ss-count and P-num annotation structures and make an educated "guess" for an RNA target region.

3.1.2. Design of Tiny Molecular Beacons

Molecular beacons synthesized from 2′-O-methyl-RNA are normally designed to have 4–5 nt stems (with a low G/C rich content) and 18–25 nt in the loop/probe region of the hairpin which ensures >10°C difference between melting temperature of the probe–target hybrid and the target detection temperature (25 or 37°C) (http://www.molecular-beacons.org).

With the incorporation of LNA bases in the loop/probe region of the hairpin, the stability of the probe/target hybrid increases (see Note 4), thus enabling the design of probes with shorter hairpin loops and stems. Tiny molecular beacons are chimeras with 2′-O-methyl RNA/LNA loops and 2′-O-methyl RNA stems. They have 3–4 nt length stems (with GC variations only) with 9–12 nt in the hairpin loop (see Note 5–7).

3.2. Synthesis of Tiny Molecular Beacons

Molecular beacon probes possessing a backbone chemistry containing a combination of LNA and 2′-O-methyl RNA nucleotides can be synthesized using standard automatic DNA chemistry using quencher labeled solid supports, a 5′ fluorophore or 5′ terminal modifier phosphoramidites, LNA, and 2′-O-methyl RNA phosphoramidites (13). The 5′ terminal modifier phosphoramidite is utilized if a fluorophore reporter is required for which no phosphoramidite is available (see Note 1). For the synthesis of the molecular beacon probes described in this chapter, we utilized a controlled-pore glass column to introduce dabcyl at the 3′ end of the oligonucleotide during the automated synthesis. Compared to standard automated DNA synthesis protocols, and as recommended by the phosphoramidite manufacturers, the coupling step and oxidation step in each synthesis cycle was doubled. At the 5′ end of the oligonucleotide, a trityl-protected sulfhydryl modifier was introduced for a subsequent manual coupling of Texas Red C5 bromoacetamide or Tetramethylrhodamine-5-iodoacetamide. Standard post-synthesis protocols were followed after the oligonucleotide synthesis was completed. Before the conjugation of the fluorophore, the oligonucleotide was first purified by HPLC to remove non-full-length oligonucleotides that do not contain a 5′ trityl-protected sulfhydryl group. The protective trityl moiety was then removed from the 5′-sulfhydryl group and a fluorophore was introduced in its place using an iodoacetamide or bromoacetamide derivative. This conjugation was followed by a second HPLC purification to remove loose fluorophores and unconjugated oligonucleotides.

1. After the automated synthesis and post-synthesis steps, dissolve the oligonucleotides in 500 μL of HPLC Buffer A.

2. Purify the oligonucleotides on a C-18 reverse phase HPLC column, utilizing a linear elution gradient of 20–70% HPLC Buffer B in HPLC Buffer A and run for 25 min at a flow rate of 1 mL/min. Monitor the absorption of the elution stream at 260 nm (absorption of nucleotides) and 491 nm (absorption of dabcyl). Fig. 2a shows a typical HPLC chromatogram of the purification of an oligonucleotide labeled with a trityl-protected sulfhydryl group, a mixture of LNA and 2′-O-methyl RNA nucleotides, and a 3′ dabcyl group. Collect peak B that absorbs in both wavelengths. Due to the increased hydrophobicity of this trityl containing oligonucleotide, its HPLC retention time is longer than non-full-length oligonucleotides.

3. Precipitate the collected material with ethanol and sodium acetate for at least 2 h at −20°C and spin in a centrifuge for 10 min at 8,000 × g. Discard the supernatant, dry the pellet, and dissolve the oligonucleotide in 250 μL of HPLC Buffer A.

4. In order to remove the trityl moiety, add 10 μL of 0.15 M silver nitrate and incubate for 30 min. Add 15 μL of 0.15 M DTT to this mixture and incubate for 5 min. Spin for 2 min at 8,000 × g and transfer the supernatant to a new tube. Dip the point of a pipette tip in the bottle containing the fluorophore and dissolve a small amount of tetramethylrhodamine-5-iodoacetamide or Texas Red C5 bromoacetamide in 10 μL of DMF (see Note 8). Add the fluorophore solution to 250 μL of 0.2 M sodium bicarbonate at pH 9.0. Incubate the mixture for 120 min. Each of the above mentioned solutions should be prepared just before use.

5. Precipitate the fluorophore–oligonucleotide mixture with ethanol and sodium acetate for at least 2 h at −20°C and spin in a centrifuge for 10 min at 8,000 × g. Discard the supernatant, dry the pellet, and dissolve the oligonucleotide in 500 μL of HPLC Buffer A.

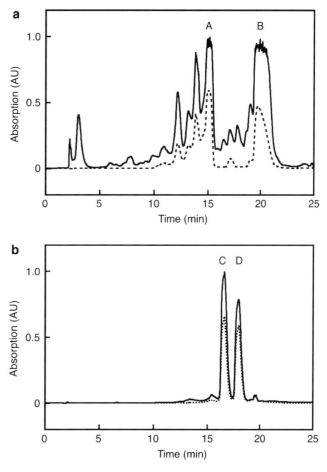

Fig. 2. HPLC purification chromatograms of oligonucleotides coupled to dabcyl before (a) and after conjugation (b) to Texas Red bromoacetamide. The *solid line* in both plots represents absorption at 260 nm and the *dotted line* in plot represents absorption at 491 nm (plot a) and 584 nm (plot b). The oligonucleotides in peak A do not contain trityl moieties, whereas the oligonucleotides in peak B are protected by trityl moieties. Peak B should be collected. Oligonucleotides in peak C and D are labeled with Texas Red and should be collected.

6. Purify the fluorophore labeled oligonucleotides on a C-18 reverse phase HPLC column, utilizing a linear elution gradient of 20–70% HPLC Buffer B in HPLC Buffer A and run for 25 min at a flow rate of 1 mL/min. Monitor the absorption of the elution stream at 260 nm (absorption of nucleotides) and 555 nm (absorption of Tetramethylrhodamine) or 584 nm (absorption of Texas Red). Fig. 2b shows a typical HPLC chromatogram of the purification of an oligonucleotide labeled with a Texas Red bromoacetamide group, a mixture of LNA and 2'-O-methyl RNA nucleotides, and a 3' dabcyl group. Collect peaks C and D that absorbs in both wavelengths.

The Texas Red bromoacetamide derivative is a mixture of two isomeric sulfonamides, which results in a slightly different HPLC retention time for each isomer, and hence two peaks are observed. The tetramethylrhodamine-5-iodoacetamide derivative is a single isomer product and only one HPLC peak will be observed.

7. Precipitate the collected material with ethanol and sodium acetate for at least 2 h at −20°C and spin in a centrifuge for 10 min at 8,000 × g. Discard the supernatant, dry the pellet, and dissolve the molecular beacon in 50 µL TE buffer. Determine the absorbance at 260 nm and estimate the yield (1 OD_{260} ~ 33 µg/mL).

3.3. In Vitro Characterization of Tiny Molecular Beacons

After the synthesis and purification, the molecular beacon is characterized in two short *in vitro* experiments to determine if the purity of the molecular beacon preparation is sufficient to avoid background fluorescence generation during live cell imaging and for the ability of the molecular beacon to form stable hybrids with target nucleic acids.

The thermodynamic characteristics of the molecular beacons are obtained by measuring the denaturing profile in a real-time PCR instrument. Ideally, the melting temperature of the molecular beacon probe–target nucleic acid hybrids should be at least 7–10°C higher than the detection temperature, which is usually room temperature or 37°C. This is also true for the stem hybrid melting temperature of the molecular beacon. Molecular beacons that increase their fluorescence intensity at least 20 times upon hybridizing to their target nucleic acid are considered good. When designed correctly and after a successful HPLC purification, most molecular beacon preparations increase their fluorescence intensity ranging from 30 to 100 times. Fig. 3 represents an example of a denaturation profile and signal-to-background ratio determination. The molecular beacon contains a combination of LNA and 2′-O-methyl RNA nucleotides, and was labeled with 5′ Texas Red and 3′ dabcyl. Fig. 3a shows the denaturation profile of this molecular beacon in presence of a complementary nucleic acid target containing a backbone chemistry of deoxyribonucleotides (DNA, dashed line) and 2′-O-methyl ribonucleotides (2′-O-methyl RNA, solid line), or no nucleic acid target (dotted line). This example shows that although a DNA target could be used (probe – target melting temperature is about 35°C), the molecular beacon forms a much more stable hybrid with the 2′-O-methyl RNA target (probe – target melting temperature is about 70°C), which is closely related to RNA in structural properties. The melting temperature of the stem hybrid is 80°C. Fig. 3b shows the result of the signal-to-background ratio experiment of the molecular beacon determined with the DNA target (open circles) and 2′-O-methyl RNA target (closed circles). The signal-to-background ratio

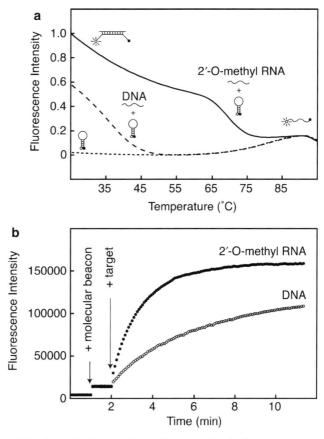

Fig. 3. (a) Plot shows the denaturation profile of a molecular beacon in presence of a complementary nucleic acid target containing a backbone chemistry of deoxyribonucleotides (DNA, *dashed line*) and 2'-O-methyl ribonucleotides (2'-O-methyl RNA, *solid line*), or no nucleic acid target (*dotted line*). (b) Plot shows the spontaneous fluorogenic response of molecular beacons to the addition of target (DNA, *open circles*; 2'-O-methyl RNA, *closed circles*). The first segment of the data is due to the fluorescence of the buffer, the second segment is due to the fluorescence of the buffer and closed molecular beacons, and the third segment shows the increase in fluorescence that occurs upon addition of the target oligonucleotides.

determined with the DNA target is about 20 and with the 2'-O-methyl RNA target is about 30 (see Note 9). Both results are sufficient, however, the plots also show that the kinetics of hybridization between the molecular beacon and its DNA target are slower than the kinetics of hybridization between the molecular beacon and its 2'-O-methyl RNA target. The results in this figure shows that, if possible, it will benefit to utilize a 2'-O-methyl RNA target in order to obtain a better characterization profile of the molecular beacon used in live cell mRNA imaging.

1. Signal-to-background ratios of molecular beacons are determined in a spectrofluorometer using the optimal excitation and emission wavelength for the reporter fluorophore.

2. Determine the fluorescence (F_{buffer}) of 150 µL of hybridization buffer using (as in above mentioned example) 584 nm as the excitation wavelength and 603 nm as the emission wavelength.

3. Add 10 µL of 1 µM molecular beacon to this solution and record the new level of fluorescence (F_{closed}).

4. Add a twofold molar excess of the oligonucleotide target and monitor the rise in fluorescence until it reaches a stable level (F_{open})

5. Calculate signal-to-background ratio as $(F_{open} - F_{buffer})/(F_{closed} - F_{buffer})$.

6. Thermal denaturation profiles of molecular beacons are determined in a thermal cycler with a capacity to monitor fluorescence in real time using the wavelength of excitation and emission specific for the fluorophore.

7. Prepare two tubes containing 25 µL of 200 nM molecular beacon dissolved in hybridization buffer and add the oligonucleotide target to one of the tubes at a final concentration of 400 nM.

8. Determine the fluorescence of each solution as a function of temperature. Decrease the temperature of these tubes from 95 to 25°C in 1°C steps, with each hold lasting 30 s, while monitoring the fluorescence during each hold.

3.4. In Vivo Imaging of oskar mRNA

We have previously used 2′-O-methyl RNA molecular beacons to directly visualize the endogenous expression of the maternal gene *oskar* in the *Drosophila melanogaster* oocytes (4). Following our report, others have confirmed the target specificity of molecular beacons within various cellular contexts (14–16).

Here we show *in vivo* accessibility to previously selected *oskar* mRNA target regions of tiny LNA/2′-O-methyl RNA chimera molecular beacons designed and synthesized as described in Subheadings 3.1 and 3.2 (Fig. 4, see Note 10).

In *Drosophila melanogaster* egg chambers, mRNAs are transcribed in the nurse cell nuclei throughout oogenesis and are transported into the oocyte via connecting ring canals. *Oskar* mRNA localizes during mid-oogenesis at the posterior pole of

Fig. 4. Tiny molecular beacon detects *oskar* mRNA *in vivo*. (**a**) Co-visualization of a tiny molecular beacon and an established 2′-O-methyl RNA probe. A solution containing 200 ng/µL of each molecular beacon (osk2209-2′-OMe-Cy5/BHQ2 (*red*) and osk1227-chimera-TxRed/dabcyl (*green*)) was injected in egg chambers of various stages. The top two rows show a stage 8 egg chamber. Images represent Z-projections of 7 × 1 µm sections at 10 and 35 min following microinjection. The bottom row represents a single section of a stage 10 egg chamber at 60 min after microinjection. (**b**) Simultaneous, *in vivo* targeting of two regions within *oskar* mRNA with two tiny molecular beacons labeled with the same fluorophore (TMR). A cocktail of osk1227-chimera-TMR/dabcyl and osk2213-chimera-TMR/dabcyl, containing 100 ng/µL each, was injected into stage 8 egg chambers (see Note 14). A DIC image of the oocyte 30 min after microinjections, followed by Z-projections of 13 × 0.5 µm sections at 0, 10, and 30 min. All images were acquired with a spinning disc microscope set-up, using 40× oil objective of 1.25 NA. The scale bars represent 20 µm.

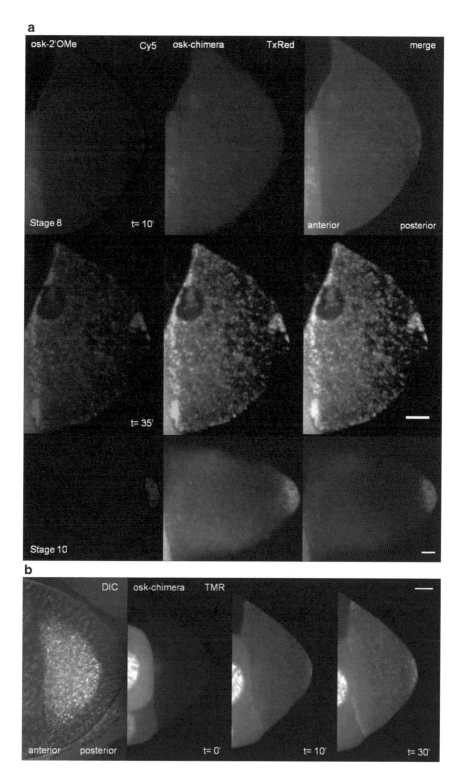

the oocyte, while briefly anchoring at the anterior cortex in the earlier stages.

1. Feed newly hatched wild type female flies with fresh yeast paste for 2–4 days.

2. Dissect ovaries in Halocarbon Oil 700 directly on glass coverslip and tease apart each ovariole, thus separating individual egg chambers.

3. Mount the glass coverslip onto the spinning disc microscope stage (see Note 11) and microinject the molecular beacon solution (2′-O-methyl RNA and/or chimera) into a nurse cell most proximal to the oocyte. Select egg chambers that are of stage 8 or older.

4. Begin recording immediately, selecting a Z-stack (6–14 slices of 0.5–1 μm each), while acquiring data every 30 or 60 s for 15 min, up to 1 h (see Note 12).

5. Acquire images using Volocity and export the file of interest as OpenLab liff file format (see Note 13).

6. Open liff files using ImageJ. Adjust images for brightness/contrast. The merged images are created using "Merge Channels".

4. Notes

1. 5′ sulfhydryl (or thiol) modifiers are used for the conjugation of iodoacetamide, bromoacetamide, or maleimide derivatives of fluorophores. 5′ amino modifiers are used for the conjugation of succinimidyl ester derivatives of fluorophores.

2. Unlike predictions of local hairpins, long-range interactions or multibranch junctions remain poorly determined. Such predicted structures could provide insights into regions of potential structural plasticity within an RNA molecule and thus reveal potential accessible sites for antisense probes. Therefore, when considering probes for targeting an RNA sequence, it is very helpful to pay close attention to the information offered by a secondary structure fold.

3. It is possible for a longer region of the target to be chosen as favorable, not just the region comprising the length of the oligonucleotide. $\Delta G_{oligo\text{-}oligo}$ and $\Delta G_{oligo\text{-}self}$ are disregarded, as the molecular beacon structure is analyzed separately using *mfold* (see Subheading 3.1.2).

4. The effect of LNA substitutions is approximately additive when LNA nucleotides are spaced by at least one 2′-O-methyl nucleotide (17). On average, an internal LNA substitution for a 2′-O-methyl RNA makes duplex stability more favorable

by −1.4 kcal/mol at 37°C (17). This corresponds to a tenfold increase in binding constant.

$$\Delta G°_{37}(\text{chimera/RNA}) = \Delta G°_{37}(2'\text{-}O\text{-MeRNA/RNA}) - 1.10 n_{iAL/UL} - 1.6 n_{iGL/CL},$$

where $\Delta G°_{37}$ (2'-O-MeRNA/RNA) is the free energy change at 37°C for duplex formation in the absence of any LNA nucleotides, $n_{iAL/UL}$ and $n_{iGL/CL}$ are the numbers of internal LNAs in AU and GC pairs, respectively.

5. Avoid stretches of 3 or more Gs or Cs.
6. Keep the loop GC content between 30 and 60%.
7. Do not include all LNAs in loop sequence.
8. Some fluorophores are directly soluble in aqueous solutions and do not need to be dissolved in DMF. Refer to the manufacturer application note for the particular fluorophore derivative.
9. If the manual conjugation of a fluorophore to an oligonucleotide resulted in a low yield, check the pH of the buffers used in the coupling reactions and use fresh dyes. The reactive dyes should be stored at −20°C in the presence of a desiccant.
10. Tiny molecular beacons:
    - The # indicates the range of nucleotides on mRNA sequence, followed by the fluorophore–quencher pair labels, and size and stability of the hairpin structure.
    - The sequence includes the stem forming nucleotides (underlined font) and the probe loop complementary to target region (CAPS font).
    - All nucleotides are 2'-O-methyl RNA with LNA substitutions in **bold** font.

      *osk*2209-2233 Cy5-BHQ2
        (stem = 5, loop = 25, ΔG = −6.7 kcal/mol)

      5'-**gc**u**gc** AAA A**G**C GGA AAA GUU UGA AGA GAA **G**ca**gc**-3'

      *osk*2213-2224 Texas Red-dabcyl
        (stem = 4, loop = 11, ΔG = −4 kcal/mol)

      5'-**c**gg**c** AAG UUU GAA GA **G**cc**g**-3'

      *osk*1227-1238 Texas Red-dabcyl
        (stem = 4, loop = 11, ΔG = −3.8 kcal/mol)

      5'-**g**cc**G** AAU CGU UGU AG **c**gg**c**-3'.
11. Confocal Microscope Setup: Leica DMI4000B inverted microscope mounted on a TMC isolation platform, Yokogawa CSU10 spinning disc head, Hamamatsu C9100-13 ImagEM EM-CCD camera, diode lasers - 491, 561, and 638 nm,

Eppendorf Patchman-Femtojet microinjector, Volocity acquisition software and ImageJ processing software.

12. Though these experiments are done at 25°C, the accessibility to the target region predicted in *mfold* at 37°C is not affected. The tiny molecular beacons are stable on target for a range of temperatures (Fig. 3b).

13. Processing can be performed with Volocity, but it is not as flexible as ImageJ for reconstructing Z-stacks.

14. Mixture of tiny molecular beacons targeting multiple RNA regions results in an increased signal per target molecule, thus benefiting the visualization of low abundance RNAs.

Acknowledgments

This work was supported by grants from the National Institute of Mental Health (MH079197) and National Institute of General Medical Sciences (SC2GM084859).

References

1. Forrest, K.M. and Gavis, E.R. (2003) Live imaging of endogenous RNA reveals a diffusion and entrapment mechanism for nanos mRNA localization in Drosophila. *Curr. Biol.* **13**, 1159–1168.
2. Fusco, D., Accornero, N., Lavoie, B., Shenoy, S.M., Blanchard, J.M., Singer, R.H. and Bertrand, E. (2003) Single mRNA molecules demonstrate probabilistic movement in living mammalian cells. *Curr. Biol.* **13**, 161–167.
3. St Johnston, D. (2005) Moving messages: the intracellular localization of mRNAs. *Nat. Rev. Mol. Cell Biol.* **6**, 363–375.
4. Bratu, D.P., Cha, B.J., Mhlanga, M.M., Kramer, F.R. and Tyagi, S. (2003) Visualizing the distribution and transport of mRNAs in living cells. *Proc. Natl. Acad. Sci. USA* **100**, 13308–13313.
5. Tyagi, S. and Kramer, F.R. (1996) Molecular beacons: probes that fluoresce upon hybridization. *Nat. Biotechnol.* **14**, 303–308.
6. Tyagi, S., Bratu, D.P. and Kramer, F.R. (1998) Multicolor molecular beacons for allele discrimination. *Nat. Biotechnol.* **16**, 49–53.
7. Zuker, M. (2003) Mfold web server for nucleic acid folding and hybridization prediction. *Nucleic Acids Res.* **31**, 3406–3415.
8. Mathews, D.H, Sabina, J., Zuker, M and Turner, D.H. (1999) Expanded Sequence Dependence of Thermodynamic Parameters Improves Prediction of RNA Secondary Structure *J. Mol. Biol.* **288**, 911–940.
9. Bratu, D.P. (2006) Molecular beacons: Fluorescent probes for detection of endogenous mRNAs in living cells. *Methods Mol. Biol.* **319**, 1–14
10. Braasch, D.A. and Corey, D.R. (2002) Cellular delivery of locked nucleic acids (LNAs). *Curr. Protoc. Nucleic Acid Chem.* Chapter 4:Unit 4.13.
11. Mathews, D.H., Burkard, M.E., Freier, S.M., Wyatt, J.R., and Turner, D.H. (1999) Predicting oligonucleotide affinity to nucleic acid targets. *RNA* **5**, 1458–1469.
12. Xia, T., SantaLucia, J., jr., Burkard, M.E. et al. (1998) Thermodynamic parameters to predict stability of RNA/DNA hybrid duplexes. *Biochemistry* **34**, 11211–11216.
13. Mullah, B. and Livak, K. (1999) Efficient automated synthesis of molecular beacons. *Nucleos. Nucleot.* **18**, 1311–1312
14. Tyagi, S. and Alsmadi, O. (2004) Imaging native beta-actin mRNA in motile fibroblasts. *Biophys. J.* **87**, 4153–4162.

15. Vargas, D.Y., Raj, A., Marras, S.A., Kramer, F.R. and Tyagi, S. (2005) Mechanism of mRNA transport in the nucleus. *Proc. Natl. Acad. Sci. USA* **102**, 17008–17013.

16. Kloc, M., Wilk, K., Vargas, D., Shirato, Y., Bilinski, S. and Etkin, L.D. (2005) Potential structural role of non-coding and coding RNAs in the organization of the cytoskeleton at the vegetal cortex of *Xenopus* oocytes. *Development* **15**, 3445–3457.

17. Kierzek, E., Ciesielska, A., Pasternak, K., Mathews, D.H., Turner, D.H. and Kierzek, R. (2005) The influence of locked nucleic acid residues on the thermodynamic properties of 2′-O-methyl RNA/RNA heteroduplexes. *Nuc. Acids Research* **33**, 16, 5082–5093.

Chapter 10

Delivery of Molecular Beacons for Live-Cell Imaging and Analysis of RNA

Antony K. Chen, Won Jong Rhee, Gang Bao, and Andrew Tsourkas

Abstract

Over the past decade, a variety of oligonucleotide-based probes have been developed that allow for direct visualization of RNA molecules in living cells. Of these, molecular beacons have garnered a particularly high degree of interest due to their simple yet exquisite unimolecular stem-loop design that allows for the efficient conversion of target recognition into a specific fluorescent signal. As a result of their favorable fluorescent enhancement and their high specificity, molecular beacons have been used for a wide range of applications, including the monitoring of RNA expression and localization in living cells, cancer cell detection, and the study of viral infections. In this chapter we describe a general methodology that can be followed for the imaging and analysis of RNA in living cells using molecular beacons. Several commonly employed methods for delivering molecular beacons into the cytosol are discussed including toxin-based cell membrane permeabilization, microinjection, and microporation. Strategies for acquiring ratiometric measurements are also described.

Key words: Molecular beacons, Live-cell imaging, Streptolysin O, Microinjection, Microporation, Ratiometric imaging, Gene expression, RNA

1. Introduction

The ability to acquire a complete spatial and temporal profile of RNA synthesis, processing, and transport could offer unprecedented insight into cell function and behavior in conditions of health and disease and in response to external stimuli. Accordingly, much effort has been devoted to developing methods to monitor gene expression in living cells. Currently, the majority of live cells imaging approaches utilize molecular beacons (MBs), which are antisense oligonucleotide probes labeled with a "reporter" fluorophore at one end and a quencher at the other end (1) (Fig. 1a). In the absence of complementary nucleic acid targets,

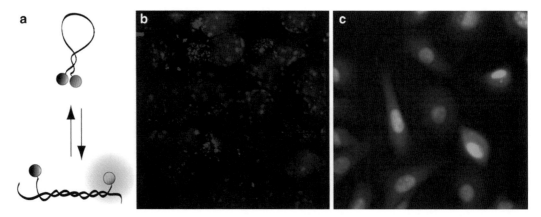

Fig. 1. A schematic of a molecular beacon and fluorescent microscopy images of living cells following the delivery of molecular beacons. (**a**) Schematic of a molecular beacon in the absence and presence of a complementary nucleic acid target. In the absence of target, molecular beacon fluorescence is quenched; however, upon hybridization, fluorescence is restored. (**b**) Fluorescence image of antisense Oct4 molecular beacons that have been delivered into undifferentiated P19 embryonal carcinoma cells using SLO. Molecular beacon reporter fluorescence is shown in *red*. The nucleus has been counterstained with DAPI, shown in *blue*. (**c**) A representative fluorescence image of molecular beacons in living MEF/3T3 cells, 4 h following microporation.

the MBs assume a stem-loop configuration. In this manner, the fluorophore and quencher are brought into close proximity and fluorescence is quenched. When MBs hybridize to complementary targets, the fluorophore is separated from the quencher and fluorescence is restored. The unique ability of MBs to convert target recognition into a measurable fluorescent signal has rendered MBs the probe of choice to study the level, distribution, transport, and processing of specific RNA molecules in living cells (2–15).

While the use of MBs for live-cell RNA analysis has become increasingly widespread, a critical requirement for detecting RNA molecules in living cells is the noninvasive and efficient delivery of MBs into the cytosol. When unassisted, oligonucleotide-based probes are not capable of efficiently passing through the cell plasma membrane. Moreover, even if MBs could enter the cells successfully, sufficient amount of probes must remain functional and be available for hybridization to the RNA of interest. Commonly utilized transfection methods for oligonucleotide delivery such as lipid and dendrimer-based delivery methods have often failed in delivering MBs into cells, as MB-transfection agent complexes do not always efficiently dissociate once internalized, leading to bright fluorescent punctate aggregates that interfere with fluorescent measurements. Further, fluid-phase endo/micropino-cytosis of the MB-transfection agent complexes can often lead to entrapment and degradation of the probes within the endosome/lysosome compartments, thus increasing background signal. The use of fusogenic peptides (16) or proton-sponged polymers such as polyethyleneimine (PEI) (17) may help the release of the probes from the endosomal compartment into the cytoplasm; however, it has been estimated that only 0.01–10% of the released oligonucleotide probes may remain functional (18). Consequently, methods whereby MBs can be directly introduced into the cytoplasmic compartments are generally preferred. We have routinely administered MBs into live cells by toxin-based membrane permeabilization using streptolysin O, microinjection, and microporation. These delivery methods are described in detail below. We also describe a general methodology that can be followed for the imaging and analysis of MB fluorescence including ratiometric imaging approaches.

2. Materials

2.1. Cell Culture (Three Examples Are Provided)

2.1.1. Example 1

1. Normal human dermal fibroblasts (HDF, Cambrex).
2. Fibroblast growth medium (Clonetics) supplemented with 2% fetal bovine serum (FBS), 1 µg/mL of human recombinant Fibroblast growth factor (hFGF), 5 µg/mL of insulin, 50 µg/mL of gentamicin, 50 µg/L of amphotericin-B.
3. Solution of 0.25% trypsin and 1 mM ethylenediamine tetraacetic acid (EDTA).

2.1.2. Example 2

1. Mouse EC cell (CRL-1825; American Type Culture Collection).
2. α-Minimum Essential Medium supplemented with 7.5% calf bovine serum and 2.5% fetal bovine serum.
3. Solution of 0.25% trypsin and 1 mM ethylenediamine tetraacetic acid (EDTA).

2.1.3. Example 3

1. MCF-7 cell (American Type Culture Collection).
2. Minimum Essential Medium with 2 mM l-glutamine and Earle's balanced salt solution (BSS) adjusted to contain 1.5 g/L sodium bicarbonate, 0.1 mM non-essential amino acids, 1 mM sodium pyruvate, and 10% fetal bovine serum.
3. Solution of 0.25% trypsin and 1 mM ethylenediamine tetraacetic acid (EDTA).

2.2. Molecular Beacon Design (Three Examples Are Provided)

2.2.1. Example 1

1. Antisense K-ras 2'-O-methyl RNA MBs, e.g., possessing a Cy5 fluorophore at the 5' end (see Note 1), and a Blackhole quencher-3, BHQ-3, at the 3' end with the sequence: 5'-Cy5-CCT ACG CCA CCA GCT CCG TAG G-BHQ3-3' (see Note 2).
2. K-ras RNA target, 5'-TTG GAG CTG GTG GCG TAG G CA-3'.
3. Nonsense 2'-O-methyl RNA MBs, 5'-Cy5-CAC GTC GAC AAG CGC ACC GAT ACG TG-BHQ-3-3' (see Note 3).

162   Chen et al.

   4. Nonsense RNA target, 5′-ATC GGT GCG CTT GTC G-3′.
   5. All MBs and RNA targets should be suspended in RNAse/DNAse free dH$_2$O to yield a stock concentration of 100 μM. The stocks should be aliquoted and stored at −20°C or below. When in use, the probe can be further diluted in buffers appropriate for cellular studies.

2.2.2. Example 2

   1. Antisense Oct-4 2′-*O*-methyl RNA MBs, e.g., possessing a Cy5 fluorophore at the 5′-end (see Note 1), and a Blackhole quencher-3, BHQ-3, at the 3′ end with the sequence: 5′-Cy5-CGC AGT CCA GGT TCT CTT GTC TCT GCG-BHQ3-3′ (see Note 2).
   2. Oct-4 RNA target, 5′-AGA GAC AAG AGA ACC TGG A-3′.
   3. Nonsense 2′-*O*-methyl RNA MBs, 5′-Cy5-CGA CGC GAC AAG CGC ACC GAT ACG TCG-BHQ-3-3′ (see Note 3).
   4. Nonsense RNA target, 5′-CGT ATC GGT GCG CTT GTC G-3′.
   5. All MBs and RNA targets should be suspended in RNAse/DNAse free dH$_2$O to yield a stock concentration of 100 μM. The stocks should be aliquoted and stored at −20°C or below. When in use, the probe can be further diluted in buffers appropriate for cellular studies.

2.2.3. Example 3

   1. Antisense c-myc 2′-*O*-methyl RNA MBs, e.g., possessing a Cy5 fluorophore at the 5′-end (see Note 1), and a Blackhole quencher-3, BHQ-3, at the 3′ end with the sequence: 5′-Cy5-GTC ACG TGA AGC TAA CGT TGA GGG TGA C-BHQ3-3′ (see Note 2).
   2. c-myc RNA target, 5′-GTC ACC CTC AAC GTT AGC TTC ACT TT-3′.
   3. Nonsense 2′-*O*-methyl RNA MBs, 5′-Cy5-GTC ACC TCA GCG TAA GTG ATG TCG TGA C-BHQ-3-3′ (see Note 3).
   4. Nonsense RNA target, 5′-GTC ACG ACA TCA CTT ACG CTG AGT TT-3′.
   5. All MBs and RNA targets should be suspended in RNAse/DNAse free dH$_2$O to yield a stock concentration of 100 μM. The stocks should be aliquoted and stored at −20°C or below. When in use, the probe can be further diluted in buffers appropriate for cellular studies.

2.3. Toxin-Based Membrane Permeabilization

   1. Streptolysin O (SLO).
   2. Tris(2-carboxyethyl)phosphine (TCEP) (Thermo Scientific).
   3. Phosphate Buffered Saline (PBS).
   4. Fibroblast growth medium (serum free; Clonetics).
   5. Clear 24-well plate.

2.4. Microinjection

1. A Femtojet and Injectman NI 2 microinjection system (Eppendorf).
2. Microinjection capillary (Femtotip I) (Eppendorf).
3. Microloader (Eppendorf).
4. Hexamethyldisilazane.
5. Microinjection buffer: 48 mM K_2HPO_4, 4.5 mM KH_2PO_4, and 14 mM NaH_2PO_4 at pH 7.2.
6. 15-mm culture dish (non-sterile).
7. Glass bottom dish (HBSt-5030, Willco Wells).

2.5. Microporation

1. OneDrop Microporator (MP-100, BTX Harvard Apparatus).
2. Dulbecco's Modified Eagle's Medium without phenol red and without antibiotics, supplemented with 10% FBS.
3. Solution of 0.1% trypsin without phenol red and 1 mM ethylenediamine tetraacetic acid (EDTA).
4. Dulbecco's Phosphate Buffered Saline (DPBS), Mg^{2+}, Ca^{2+} free.
5. Resuspension buffer R (BTX Harvard Apparatus).
6. Electroporation buffer (BTX Harvard Apparatus).
7. Electroporation Gold Tips (10 μL size) (BTX Harvard Apparatus).
8. Electroporation tube (BTX Harvard Apparatus).
9. 8-Well Lab-Tek Chambered Coverglass (155409, Nalgen Nunc).

2.6. Microscope and Imaging Software

1. Olympus IX81 Motorized inverted fluorescence microscope equipped with a back-illuminated EMCCD camera (Andor) and an X-cite 120 excitation source (EXFO).
2. A LUC PLAN FLN 40× objective, N.A. 0.9.
3. Cy5 (HQ620/60, HQ700/75, Q660lp) (Chroma Technology).
4. Alexa750 (HQ710/75, HQ810/90, Q750LP) (Chroma Technology).
5. IPLab image acquisition software.
6. ImageJ software (available from the NIH website http://rsbweb.nih.gov/ij/).

2.7. Ratiometric Imaging

1. Amino dextran (MW=10kDa, Invitrogen).
2. NeutrAvidin (Thermo Scientific).
3. Alexa750-NHS ester (Invitrogen).
4. Methyl PEOn-NHS ester (10 kDa MW, Laysan).
5. 50 mM Sodium Borate Buffer (pH 8).
6. NAP-5 gel chromatography columns (Amersham Biosciences).

7. Microcon YM-50 centrifugal devices (50,000 MW cutoff; Millipore).
8. Cary100 spectrophotometer (Varian).

3. Methods

We have previously demonstrated that streptolysin O, microinjection, and microporation can be used to efficiently deliver MBs into the cytosol of living cells. However, each method has unique advantages and disadvantages, thus the optimal method will vary depending on the application. For example, streptolysin O does not require any specialized equipment and allows MBs to be delivered into many cells simultaneously; however, a short incubation period is necessary, ~1 h, before images can be acquired. Therefore, the kinetics of MB hybridization cannot be monitored. Of course, hybridization may require more than 1 h to reach complete equilibrium. Microinjection benefits from instantaneous delivery and imaging, but is severely limited in its ability to deliver probes into a large number of cells. Further, expensive equipment is required. Microporation also requires specialized equipment (although at approximately half the cost of a microinjection system), but in this case MBs can be delivered into many cells simultaneously. Moreover, fluorescent images can be acquired within just a couple of minutes following microporation. Each of these methods is described briefly below.

3.1. Streptolysin O

Streptolysin O (SLO), is a bacterial exotoxin that reversibly forms pores on the cell surface. SLO has already been successfully used to deliver MBs into a wide variety of cell types including human dermal fibroblasts, stem cells, and cancer cells (11, 19, 20). SLO is activated using the reducing agent, TCEP. The activated SLO and MBs are then added to the cells for 10 min. Afterwards, the permeabilized cells are resealed by diluting the SLO mixture with fresh medium. Following a ~1 h incubation period, the cells are washed and imaged by fluorescence microscopy.

3.2. Microinjection

Microinjection is perhaps the most direct method for delivering MBs into the cytoplasm of living cells. Microinjection can be applied to essentially any cell type, although large cells that exhibit a more rounded morphology are generally easier to inject. Microinjection is carried out using a FemtoJet and Injectman NI 2 (Eppendorf) microinjection system fitted with Femtotips I (Eppendorf). Prior to use, Femtotips are silanized with Hexamethyldisilazane (Fluka). Further, the MB sample is centrifuged to remove any debris that might clog the needle tips. Fluorescent images are acquired immediately following microinjection.

3.3. Microporation

Microporation is a microliter-volume electroporation process that exhibits a reduction in the many harmful events often associated with electroporation, including heat generation, metal ion dissolution, pH variation, and oxide formation. We have found that microporation allows for nearly 100% MB and protein delivery with >85% viability (4, 5). A successful microporation experiment requires resuspending the cells at an appropriate concentration (~100,000 cells per 10 μL), adding the MB sample, and microporating the cells at the optimal voltage, pulse length, pulse frequency. Optimal microporation parameters for a vast number of cell lines can be found at http://www.microporator.com.

3.4. Image Analysis

Once MBs have been delivered into cells, a relative measure of gene expression can be acquired by using an image analysis software package (ImageJ, IPLabs, MetaMorph, etc.) to draw a region of interest around individual cells and measuring the total integrated fluorescence. The contribution from the background fluorescence can be removed by drawing a region of interest (ROI) outside of the cell boundary, acquiring a measure of the total fluorescence intensity and subtracting it from the measurement of total cellular fluorescence.

3.5. Ratiometric Imaging

If highly sensitive measurements are required, then ratiometric imaging can be performed. This procedure requires the administration of a "reference" dye concomitantly with the MB. The reference dye can either be attached to an oligonucleotide or to dextran. Fluorescent images of both the reference dye and the MB reporter dye are then acquired and analyzed. An alternative strategy is to attach the MBs to fluorescently labeled NeutrAvidin. This approach has the added benefit of preventing nuclear localization, which can lead to a significant improvement in sensitivity (4).

3.6. MB Delivery

3.6.1. Streptolysin O

3.6.1.1. Day 1

1. Seed cells into wells of a 24-well plate so that they will be ~70% confluent on the day of the experiment.

3.6.1.2. Day 2

2. Activate SLO by adding 5 mM TCEP to 2 U/mL of SLO, in PBS.
3. Incubate for 30 min at 37°C.
4. Dilute SLO with serum-free medium to a concentration of 0.2 U/mL (0.5 U SLO per 10^6 cells) (see Note 4).
5. Add MB stock solution to give a final concentration of ~0.2–1 μM.
6. Wash cells once with PBS.
7. Incubate cells with SLO-MB mixture for 10 min at 37°C.
8. Wash cells three times with normal growth medium.

9. Add normal growth medium (i.e., with serum) and incubate for 0.5–1 h.
10. Acquire fluorescent images of the cells using the Cy5 filter set. A representative fluorescence image of MBs, following their delivery into living cells through the use of SLO, is shown in Fig. 1b.

3.6.2. Microinjection

3.6.2.1. Day 1

1. Seed cells onto glass bottom dishes so that they will be 10–30% confluent on the day of the experiment.
2. Siliconize the microinjection capillary
   (a) Connect the microinjection capillary to the capillary holder mounted on the microinjection device.
   (b) Place a 20-mm culture dish at the center of the microscope stage and pipette Hexamethyldisilazane (HDMS) into the dish until the bottom of the dish is covered.
   (c) Lower the level of the microinjection capillary so that the tip of the microinjection capillary is immersed in the HDMS.
   (d) Set the compensation pressure (Pc) to 0 to allow HDMS to enter into the tip via capillary force.
   (e) After immersing the microinjection capillary in HDMS for least 10 min, immerse the tip in microinjection buffer that has been pipetted into a clean 20-mm culture dish.
   (f) At 0 compensation pressure, microinjection buffer will enter into the microinjection capillary via capillary force.
   (g) Wash the residual HDMS from the microinjection capillary by repeatedly pressing the "Clean Button" on the controller, while the capillary is still immersed in the buffer.
   (h) Remove the microinjection capillary from the buffer and allow it to air dry prior to usage.

3.6.2.2. Day 2

3. Dilute the MB stock solution to a concentration of 1 µM in Microinjection (MJ) Buffer.
4. Centrifuge the MB sample for 15–20 min at $16,000 \times g$ to pellet debris that might clog the microinjection capillary.
5. Load the microinjection capillary with 3–10 µL of the diluted MB solution using a microloader. These volumes are adequate to microinject several hundred cells (see Note 5).
6. Attach a filled microinjection capillary to the microscope capillary holder.
7. Position the microinjection capillary so that the tip is at the center of the field of view on the microscope (see Note 6).
8. Retract the needle by pressing the "Home" button on the microinjection system.
9. Place the cells on the microscope stage and find the focal plane with a 20× objective.

10. Press the "Home" button again the tip will return to the original position at the center of the field of view on the microscope. Slowly lower the microinjection capillary so that the tip is immersed in the media (if not already immersed) but not yet in contact with any cells (see Note 7).
11. Set the compensation pressure to be at least 15 Psi (see Note 8).
12. Set the injection parameters to Pi = 150 Psi and Ti (injection time) = 1 s (see Note 9).
13. Press "Quick Clean" several times to remove the residual air in the tip of the capillary. This can be confirmed by observing MBs being ejected from the needle tip under fluorescent light.
14. Move the tip over the top of a cell using the joystick.
15. Switch to a 40× objective for microinjection.
16. Set the Z limit (injection level).
    (a) Focus on a cell.
    (b) Lower the tip gently to touch the plasma membrane until a gentle wave passes through the cell from the site of injection.
    (c) Press the "Limit" key on the InjectMan NI 2 control board to set this height as the Z limit.
17. Raise the needle approximately 20–30 μm above the cell so that it can be moved freely around the dish without touching the cells, but with the tip still visible. This height corresponds to the search level.
18. Focus on a cell to be injected.
19. Press the injection button on the joystick to inject the cell. The needle will inject in axial direction into the cell as far as the Z limit. The needle will return to the search level after injection.
20. Move the microinjection capillary out of the field of view using the joystick.
21. Acquire fluorescent images of the cells using the Cy5 filter set.

3.6.3. Microporation

3.6.3.1. Day 1

1. Seed cells in T-25 flasks in normal growth medium with serum but without phenol red or antibiotics so that they will be 70–90% confluent on the day of the experiment.

3.6.3.2. Day 2

2. Aspirate the media from the cells and rinse the flasks using ~5 mL of DPBS (Mg^{+2}, Ca^{+2} free) (see Note 10).
3. Aspirate the DPBS and add 1 mL of phenol red-free trypsin/EDTA (see Note 11); incubate the cells for 1 min at room temperature.
4. Aspirate the trypsin and tap the flask to detach the cells from the surface.

5. Neutralize the trace amounts of trypsin by adding 1 mL of media without antibiotics (see Note 12) and without phenol red.

6. Transfer the cells to a 1.5-mL microcentrifuge tube and pellet them by spinning the tube at $1,000 \times g$ for 5 min at room temperature.

7. Aspirate the media and resuspend the cell pellet with 1 mL DPBS (Mg^{+2}, Ca^{+2} free).

8. Count the cells in DPBS (see Note 13).

9. Pellet the required number of cells necessary for experimentation (assume 110,000 cells per microporation) by centrifugation at $1,000 \times g$ for 5 min at room temperature.

10. Aspirate the DPBS and resuspend the cell pellets in resuspension buffer R (BTX Harvard Apparatus) at a concentration close to 11,000 cells per µL (see Note 14).

11. Add 1 µL of sample containing MBs to every 10 µL of cells such that the final concentration of MBs is 5 µM.

12. Gently and slowly pipette up 10 µL of the cells with MBs using the microporator pipette (see Note 15).

13. Microporate 10 µL of the cell suspension containing roughly 100,000 cells and the probes at 1,500 V with a 10 ms pulse width and three pulses total (see Note 16).

14. Following microporation wash the cells once with 1 mL of growth medium (without Phenol Red and antibiotics) supplemented with 10% serum and resuspend in another 1 mL of growth medium (without Phenol red or antibiotics) supplemented with 10% serum.

15. Seed the cells into an 8-well Lab-Tek Chambered Coverglass or glass bottom dish.

16. Acquire fluorescent images of the cells using the Cy5 filter set. A representative fluorescence image of MBs that have been microporated into living cells is shown in Fig. 1c.

3.7. Single-Cell Image Analysis

1. Open the IPLab image files in ImageJ (see Note 17).

2. Draw a region of Interest (ROI) around the cell of interest using the "Freehand Selections" tool.

3. Measure the total fluorescent intensity and area within the ROI by selecting Analyze>Measure (see Note 18). In ImageJ, the total fluorescent intensity will be reported as Integrated Density.

4. Measure the background fluorescence by drawing a ROI in an area just outside the cell of interest (see Note 19).

5. Measure the total fluorescent intensity and area within the "background" ROI.

6. Confirm that the background ROI is the same size as the cell ROI. If not, adjust the background fluorescent intensity accordingly by scaling it linearly with area.

7. Subtract the background measurement of fluorescent intensity from the cellular measurement.

3.8. Ratiometric Imaging

The above methodologies can readily be modified for ratiometric imaging by delivering a "reference" dye (i.e., fluorescently labeled dextran or NeutraAvidin) concomitantly with the MB.

3.8.1. Approach #1: MB-Dextran Mixtures

1. Prepare Alexa750-dextran (see Note 20).
   (a) Dissolve 10 mg of aminodextran (10 kDa) in 1 mL of 50 mM Sodium Borate Buffer (pH 8), giving a concentration of 10 mg/mL.
   (b) React the 10 mg/mL aminodextran solution with 1 μL of 2.5 mM Alexa750 NHS ester (i.e., a dye to dextran molar ratio of 2.5–1) for at least 4 h at room temperature.
   (c) Purify the fluorescently labeled dextrans on NAP-5 gel chromatography columns in Microinjection buffer. A maximum of 0.5 mL of sample can be loaded on each column; therefore at least two columns will be needed. The dextran will elute with the void volume.
   (d) Determine the concentration of Alexa750 fluorophore spectrophotometrically.

2. Mix Alexa750-dextran with MBs at a 2:1 molar ratio (Alexa750:MB).

3.8.2. Approach #2: MB-NeutrAvidin Conjugates

1. Prepare Alexa750-NeutrAvidin
   (a) Dissolve 10 mg NeutrAvidin (60 kDa) in 50 mM Sodium Borate buffer, pH 8, giving a concentration of 10 mg/mL.
   (b) React with the 10 mg/mL NeutrAvidin solution with 1 μL of 416.7 μM Alexa750-NHS ester (i.e., a dye to NeutrAvidin molar ratio of 2.5:1) for at least 4 h at room temperature.
   (c) Purify the fluorescent conjugates on NAP-5 gel chromatography columns in Microinjection buffer. A maximum of 0.5 mL of sample can be loaded on each column; therefore at least two columns will be needed. The dextran will elute with the void volume.
   (d) Determine the number of fluorophores per NeutrAvidin spectrophotometrically.
   (e) Synthesize pegylated NeutrAvidin conjugates by further reacting fluorescently labeled NeutrAvidins with Methyl PEOn-NHS ester at a 100:1 PEG to NeutrAvidin molar ratio.
   (f) Purify the Pegylated NeutrAvidin by repeated filtration and dilution on Microcon YM-50 centrifugal devices.

2. Conjugate biotinylated MBs to Alexa750-NeutrAvidin (see Note 21)
   (a) Mix biotinylated MBs with Alexa750-NeutrAvidin at a molar ratio of 2:1 (MBs:NeutrAvidin) for at least 4 h at room temperature.
   (b) Purified the PEG-NeutrAvidin by repeated filtration and dilution on Microcon YM-50 centrifugal devices.
   (c) Determine the relative concentration of MB and NeutrAvidin spectrophotometrically.

   The following steps can be followed for both MB-dextran mixtures and MB-NeutrAvidin conjugates.

3. Deliver MB samples into cells using any of the approaches described in Subheading 3.1 (see Notes 22 and 23).
4. Acquire fluorescent images of the cell using both the Cy5 and Alexa750 filter sets. Representative fluorescence images of MB-NeutrAvidin conjugates that have been microporated into living cells are shown in (Fig. 2).
5. Open a single pair of images in ImageJ, corresponding to the Cy5 and Alexa750 fluorescence of the same cells.
6. Stack the pair of Cy5 and Alexa750 images using the function Image>Stacks>Convert Images to Stack.
7. On the top image, draw a region of Interest (ROI) around the cell of interest using the "Freehand Selections" tool.
8. Measure the total fluorescent intensity and area within the ROI by selecting Analyze>Measure (see Note 18). In ImageJ, the total fluorescent intensity will be reported as Integrated Density. The fluorescence intensity of the top image will be listed first.

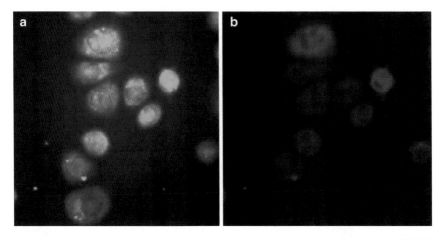

Fig. 2. Representative fluorescence images of molecular beacon-NeutrAvidin conjugates in living MEF/3T3 cells, 4 h following microporation. (**a**) Fluorescence image of the MB reporter (i.e., Cy5) and (**b**) fluorescently labeled NeutrAvidin. NeutrAvidin was labeled with the dye Alexa750.

9. On the top image, measure the background fluorescence by drawing a ROI in an area just outside the cell of interest (see Note 19).

10. Compute the total fluorescent intensity and area within the "background" ROI. Again, the fluorescence intensity of the top image will be listed first.

11. Confirm that the background ROI is the same size as the cell ROI. If not, adjust the background fluorescent intensity accordingly by scaling it linearly with area.

12. Subtract the respective background measurement of fluorescent intensity from the cellular measurement.

13. Calculate the fluorescent ratio within each cell by dividing the background subtracted fluorescent intensity of each cell from the MB image, F_{MB}, by the corresponding reference image, F_{REF}.

4. Notes

1. The use of a near- to far-infrared reporter dye (e.g., Cy5) is preferred since autofluorescence is significantly reduced at these red-shifted wavelengths.

2. When designing an antisense MB sequence, it is important to avoid RNA sequences with secondary structure. Software packages such as Beacon Designer (http://www.premierbiosoft.com) and mfold (http://www.bioinfo.rpi.edu/applications/mfold/old/dna/) can be used to assist with selection of a target sequence, but it should be noted that these programs could be inaccurate due to limitations of the biophysical models applied. Further, current computational models do not take protein–RNA interaction into account, which could also interfere with MB hybridization. Therefore, it is generally necessary to select multiple MB designs for each RNA target.

3. MBs with nonsense sequences, i.e., sequences that are not complementary to any known endogenous RNAs, are used for negative control studies.

4. The optimal amount of SLO may vary depending on the lot. Therefore, multiple concentrations of SLO may need to be tested to assess activity. This optimization procedure can be carried out using fluorescently labeled linear oligonucleotide probes to preserve MB samples.

5. Be sure to slowly pipette the MB solution into the microinjection capillary to avoid creating air bubbles.

6. Positioning of the microinjection capillary is most easily accomplished using low magnification objectives (i.e., 10× or 20×).

The magnification can then be increased step-wise to ensure that the microinjection capillary remains in site.

7. Care should be taken to avoid the contact between the tip and the bottom of the dish as this may break the tip.
8. A compensation pressure is needed to overcome capillary force. If the compensation pressure is not set high enough, the liquid loaded in the needle may never reach the tip and won't be injected. Further, without a compensation pressure the media from the culture dish can flow into the needle, diluting the probe concentration inside of the needle.
9. Suitable injection parameters vary with different cell types. The injection amount should also be adjusted as needed.
10. Mg^{+2} and Ca^{+2} should be avoided because they can reduce transfection efficiency.
11. Phenol Red should be completely removed prior to and soon after microporation as this pH indicator has a toxic effect within the cell. Phenol red can be added to the medium 24 h after microporation.
12. Antibiotics should be completely removed prior to and soon after microporation as it has a toxic effect within the cell. Antibiotics can be added to the medium 24 h after microporation.
13. A T-25 flask typically contains 200,000–400,000 cells.
14. The number of cells used per microporation experiment is cell line-specific and may affect the transfection efficiency and viability. A detailed list of cell line and the number of cells to use per microporation can be found on the product website (http://www.microporator.com).
15. Care should be taken to pipette the cells to avoid the formation of air bubbles.
16. Microporation parameters are cell line-specific; transfection efficiency and viability need to be optimized for each cell type. Detailed optimization strategies and previously optimized parameters for a wide range of cell lines can be found on the product website (http://www.microporator.com).
17. Opening IPLab files in ImageJ requires the "IPLab Reader" plugin.
18. If values for Integrated Density are not provided select Analyze>Set measurements and select the box for Integrated Density.
19. Be sure the area selected is free of cells and/or fluorescent debris. Also be sure that the background ROI that is selected does not contain any of the scattered light from the cells, i.e., the diffuse haze around the cells. Any unwanted fluorescence would lead to inaccurate quantification of MB fluorescence.

20. The reference dye was intentionally chosen to be optically distinct from the MB reporter dye to eliminate the need for spectral unmixing.

21. A biotin should be incorporated into the stem of the MB, near the quencher (e.g., 5′-Cy5-CCT ACG CCA CCA GCT CCG/iBiodT/AG G-BHQ3-3′.

22. This methodology assumes that the efficiency of delivering both the Alexa750-dextran and the MB are equivalent.

23. Based on our experience, when MB-NeutrAvidin conjugates are used, microporation and microinjection are the preferred methods of delivery.

Acknowledgments

This work was supported by the National Institutes of Health through awards CA125088 (AT), CA116102 (AT), HHSN268201000043C (GB) and CA119338 (GB); the National Science Foundation grant BES-0616031 (AT); and the American Cancer Society grant RSG-07-005-01 (AT).

References

1. Tyagi, S., and Kramer, F. R. (1996) Molecular beacons: probes that fluoresce upon hybridization. *Nat Biotechnol* **14**, 303–8.

2. Bratu, D. P., Cha, B. J., Mhlanga, M. M., Kramer, F. R., and Tyagi, S. (2003) Visualizing the distribution and transport of mRNAs in living cells. *Proc Natl Acad Sci U S A* **100**, 13308–13.

3. Chen, A. K., Behlke, M. A., and Tsourkas, A. (2007) Avoiding false-positive signals with nuclease-vulnerable molecular beacons in single living cells. *Nucleic Acids Res* **35**, e105.

4. Chen, A. K., Behlke, M. A., and Tsourkas, A. (2008) Efficient cytosolic delivery of molecular beacon conjugates and flow cytometric analysis of target RNA. *Nucleic Acids Res* **36**, e69.

5. Chen, A. K., Behlke, M. A., and Tsourkas, A. (2009) Sub-cellular trafficking and functionality of 2′-O-methyl and 2′-O-methyl-phosphorothioate molecular beacons. *Nucleic Acids Res.*

6. Dirks, R. W., Molenaar, C., and Tanke, H. J. (2001) Methods for visualizing RNA processing and transport pathways in living cells. *Histochem Cell Biol* **115**, 3–11.

7. Drake, T. J., Medley, C. D., Sen, A., Rogers, R. J., and Tan, W. (2005) Stochasticity of manganese superoxide dismutase mRNA expression in breast carcinoma cells by molecular beacon imaging. *Chembiochem* **6**, 2041–7.

8. Medley, C. D., Drake, T. J., Tomasini, J. M., Rogers, R. J., and Tan, W. (2005) Simultaneous monitoring of the expression of multiple genes inside of single breast carcinoma cells. *Anal Chem* **77**, 4713–8.

9. Rhee, W. J., Santangelo, P. J., Jo, H., and Bao, G. (2008) Target accessibility and signal specificity in live-cell detection of BMP-4 mRNA using molecular beacons. *Nucleic Acids Res* **36**, e30.

10. Santangelo, P., Nitin, N., LaConte, L., Woolums, A., and Bao, G. (2006) Live-cell characterization and analysis of a clinical isolate of bovine respiratory syncytial virus, using molecular beacons. *J Virol* **80**, 682–8.

11. Santangelo, P. J., Nix, B., Tsourkas, A., and Bao, G. (2004) Dual FRET molecular beacons for mRNA detection in living cells. *Nucleic Acids Res* **32**, e57.

12. Tyagi, S., and Alsmadi, O. (2004) Imaging native beta-actin mRNA in motile fibroblasts. *Biophys J* **87**, 4153–62.

13. Vargas, D. Y., Raj, A., Marras, S. A., Kramer, F. R., and Tyagi, S. (2005) Mechanism of mRNA transport in the nucleus. *Proc Natl Acad Sci U S A* **102**, 17008–13.

14. Wang, W., Cui, Z. Q., Han, H., Zhang, Z. P., Wei, H. P., Zhou, Y. F., Chen, Z., and Zhang, X. E. (2008) Imaging and characterizing influenza A virus mRNA transport in living cells. *Nucleic Acids Res* **36**, 4913–28.
15. Wu, Y., Yang, C. J., Moroz, L. L., and Tan, W. (2008) Nucleic acid beacons for long-term real-time intracellular monitoring. *Anal Chem* **80**, 3025–8.
16. Oliveira, S., van Rooy, I., Kranenburg, O., Storm, G., and Schiffelers, R. M. (2007) Fusogenic peptides enhance endosomal escape improving siRNA-induced silencing of oncogenes. *Int J Pharm* **331**, 211–4.
17. Thomas, M., Lu, J. J., Chen, J., and Klibanov, A. M. (2007) Non-viral siRNA delivery to the lung. *Adv Drug Deliv Rev* **59**, 124–33.
18. Dokka, S., and Rojanasakul, Y. (2000) Novel non-endocytic delivery of antisense oligonucleotides. *Adv Drug Deliv Rev* **44**, 35–49.
19. Nitin, N., Santangelo, P. J., Kim, G., Nie, S., and Bao, G. (2004) Peptide-linked molecular beacons for efficient delivery and rapid mRNA detection in living cells. *Nucleic Acids Res* **32**, e58.
20. Rhee, W. J., and Bao, G. (2009) Simultaneous detection of mRNA and protein stem cell markers in live cells. *BMC Biotechnol* **9**, 30.

Chapter 11

Genetically-Encoded Fluorescent Probes for Imaging Endogenous mRNA in Living Cells

Takeaki Ozawa and Yoshio Umezawa

Abstract

Localization of mRNAs plays pivotal roles in different cell types, including neurons and the cells in the developing stages. To visualize the dynamic movements of mRNAs in living cells, many methods have been emerged in the past decade. However, it has not been realized to visualize endogenous mRNAs with genetically encoded fluorescent probes. We recently developed fluorescent protein-based RNA probes for characterizing the localization and dynamics of mRNAs in single living cells. The probes consist of two RNA-binding domains of human PUMILIO1, each connected with split fragments of a fluorescent protein capable of reconstitution upon binding to a target mRNA. The probes are modified to specifically recognize a 16-base sequence of an mRNA of interest and to target into organelles by means of a short signal peptide. We have shown that ND6 mRNA is concentrated particularly on mitochondrial DNA (mtDNA) and movement of the mRNA is restricted in mitochondria. The probes provide a general means to study spatial and temporal mRNA localization and dynamics in intracellular compartments in living cells.

Key words: Green fluorescent protein, PUMILIO1, mRNA, Mitochondria

1. Introduction

There is much evidence that RNAs play a wide range of functions in living cells, including the transcription of genetic information, the process of translation into proteins, regulation and silencing of gene expression, and catalysts of chemical reactions. These functions are executed by specific sequences of ribonucleic acids and, in part, by spatial and temporal regulation of the expression level of RNAs in living cells (1). Efforts to obtain complete spatial-temporal profiles of the synthesis, processing, and transport of a specific RNA are therefore of importance for molecular biologists to understand complicate cellular functions in response to external stimuli. Of the RNAs, localization of messenger RNAs (mRNAs) has a key role in processes such as morphogenesis, cell migration, and memory formation. To understand these processes, imaging of intracellular mRNA has long been required in the field of molecular biology. Historically, most efforts have focused on developing nucleic acid probes, named molecular beacons, that become fluorescent upon binding to an mRNA tagged with motifs that bind to fluorescent proteins, such as MS2-GFP (2–4). The latter approach has a strong advantage because it does not necessitate microinjection of probes into each cell before imaging, and it allows for the construction of stably expressing transgenic cell lines and living organisms. However, there is a limitation in the method; the GFP is always fluorescent, whether or not it is bound to the target mRNA. This indicates that visualization of the mRNA is possible only when the unbound GFP is excluded from the area of mRNA-locating compartments or the GFP is expressed in exceedingly low amounts. In addition, a specific motif of RNA sequence, which binds to the MS2 protein, has to be inserted into the target mRNA. The artificial mRNA is expressed using techniques of cDNA transfection or homologous recombination. In both cases, the mRNA sequence is different from a native mRNA that works in living cells.

In order to overcome these limitations, we have recently developed fluorescent probes to which recognition sequences can be tailor-made based on a 16-base sequence of a target mitochondrial mRNA (mtRNA) (5). We enabled targeting of these probes into mitochondria by adding a peptide sequence corresponding to a mitochondrial targeting signal. These probes are based on the reconstitution of split fragments of a fluorescent protein (6, 7). The RNA probes consist of split fragments of enhanced green fluorescent protein (EGFP) (8) or yellow fluorescent protein (Venus) (9), each of which is connected with a sequence-specific RNA-binding domain of human PUMILIO1 (Pumilio homology domain; PUM-HD) (Fig. 1). PUM-HD is composed of eight sequence repeats and recognizes a consensus sequence, UGUANAUA (10). Each repeat acts as a molecule that specifically recognizes a single RNA base. It is of importance that specificity of PUM-HD for an RNA sequence can be altered by changing amino acid residues within the repeated sequence (11). Using the probes, we designed the PUM-HD to match the sequence of NADH dehydrogenase 6 (ND6) and visualized the mRNA in a single living cell. Upon interacting with RNA, the PUM-HDs bring the split fragments of EGFP, to which they are fused, close enough for the fragments to associate and reconstitute native EGFP structure and recover fluorescence. Monitoring the fluorescence signals allows spatial and temporal analysis of mRNA localization in single living cells. The usefulness of the

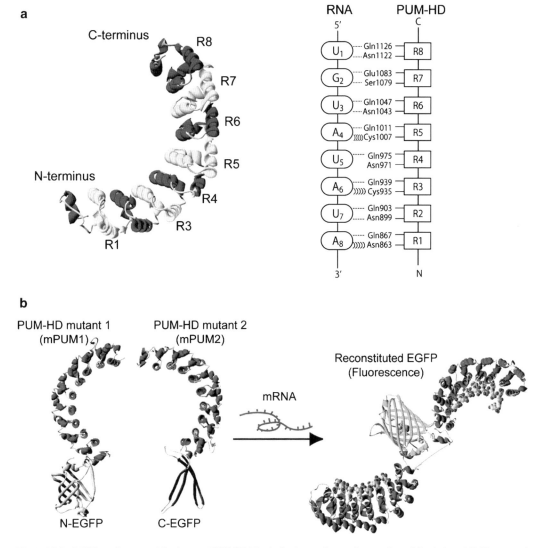

Fig. 1. (**a**) *Left*: Ribbon diagram of the human PUM-HD. The helical repeats are shown alternately *dark* and *light gray*, and are labeled as repeat 1(R1) to repeat 8(R8). Each repeat recognizes a specific base of RNA. The N and C termini are indicated. *Right*: Schematic representation of PUM-HD-RNA interaction. PUM-HD repeats are indicated by *squares* and RNA bases by *ovals* (*dashed lines*, hydrogen gonds; *parentheses*, van der Waals interactions). (**b**) Basic principle of labeling a target mRNA with a fluorescent protein. Two RNA-binding domains of PUM-HD are engineered to recognize specific sequences on a target mRNA (mPUM1-RNA and mPUM2-RNA). In the presence of the target mRNA, mPUM1 and mPUM2 bind to their target sequences bringing together the N- and C-terminal fragments of EGFP, resulting in functional reconstitution of the fluorescent protein.

RNA probes has also been shown by another group for visualization of a virus RNA in living plant cells (12). These results reported up to date demonstrate that the RNA probes have a significant potential to visualize specific natural mRNAs in living mammalian and plant cells.

2. Materials

2.1. Plasmids and Cells

1. Plasmids: GN-mPUM1, VN-mPUM1, mPUM2-GC, mPUM2-VC, and MTS-DsRed-Ex (see Note 1) (Fig. 2). Cos7 and HeLa cells (see Note 2).

2.2. Cell Culture and Transfection

1. Dulbecco's Modified Eagle's Medium (DMEM) (Gibco/BRL, Bethesda, MI) supplemented with 10% fetal bovine serum (FBS, Gibco/BRL).
2. Trypsin solution (0.25%) (Gibco/BRL).
3. OPTI-MEM medium (Gibco/BRL).
4. Lipofectamine 2000 (Invitrogen).
5. Phosphate-buffered saline, PBS. Autoclave the PBS solution before storage at room temperature.

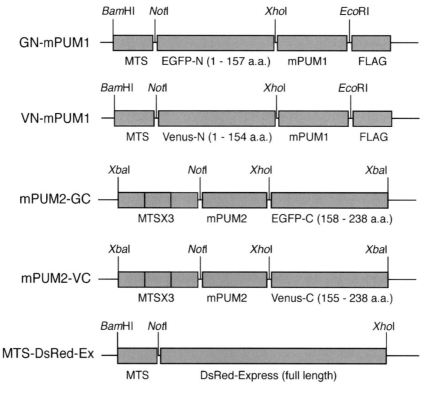

Fig. 2. Schematic structures of constructs. EGFP-N and EGFP-C indicate the cDNA sequences encoding the 1–157 and 158–238 amino acids of EGFP, and Venus-N and Venus-C are the cDNA sequences encoding the 1–154 and 155–238 amino acids of Venus, respectively. mPUM1 is a cDNA sequence of PUM-HD mutant (S863N, C935S, Q939E, Q975S, C1007N, N1043C and Y1044N). mPUM2 is a cDNA sequence of PUM-HD mutant (N1043S, Q1047E and Y1044N). FLAG, FLAG epitope; MTS, matrix-targeting signal derived from subunit VIII of cytochrome C oxidase. The cDNA is inserted into an expression vector of either pcDNA3.1(+) or pcDNA4/V5-His.

2.3. Electrophoretic Mobility Shift Assay

1. cDNAs of mutant PUM-HD (mPUM1, mPUM2).
2. Bacterial expression vector, pCold I (TAKARA Bio, Shiga, Japan).
3. BL21 *E. coli* competent cells (TAKARA Bio).
4. Isopropyl β-D-1-thiogalactopyranoside, IPTG.
5. TALON His-tag purification resin (CLONTECH, Mountain View, CA).
6. Elution buffer: 300 mM NaCl, 150 mM imidazole, and 50 mM NaH_2PO_4 at pH 7.0).
7. Ultrafiltration membrane, Centriprep YM-30 (Millipore, Billerica, MA).
8. [γ-^{32}P]ATP (GE healthcare).
9. T4 polynucleotide kinase (New England Biolabs).
10. Sephadex G-25 Quick spin column (Roche).
11. Binding buffer: 50 mM KCl, 1 mM EDTA, 0.01% (v/v) Tween-20, 0.1 mg/mL BSA, 1 mM dithiothreitol (DTT), and 10 mM Hepes at pH 7.4.
12. A TBE buffer (89 mM boric acid, 2.5 mM EDTA, and 89 mM Tris/HCl at pH 8.3).
13. A γ-ray imaging system, BAS1500 (Fujifilm, Tokyo, Japan).

2.4. Immunoprecipitation and Immunoblot Analysis

1. Lysis buffer A: 100 mM NaCl, 1 mM EDTA, 10 mM NaF, 2 mM sodium orthovanadate, 1 mM phenylmethylsulfonyl fluoride (PMSF), 10 µg/mL pepstatin, 10 µg/mL leupeptin, 10 µg/mL aprotinin, 0.1% Triton X-100, and 50 mM Tris–HCl at pH 7.4.
2. Mouse monoclonal anti-Flag antibody (Sigma, St. Louis, MO) and mouse monoclonal anti-GFP-antibody (Roche).
3. Protein Sepharose 4FF beads (GE healthcare).
4. Lysis buffer B: 10% SDS, 250 mM Tris-HCl at pH 6.8.
5. A 2× loading buffer (10% 2-mercaptoethanol, 4% SDS, 250 mM Tris–HCl at pH 6.8).
6. Primers: ND6F (5′-ATGATGTATGCTTTGTTTCT-3′), ND6R (5′-CCTATTCCCCCGAGCAATCT-3′), ND1F (5′-ATACCCATGGCCAACCTCCT-3′), and ND1R (5′-TTAGGTTTGAGGGGGAATGC-3′) (see Note 3).
7. Image analyzer (LAS-1000plus, Fujifilm).

2.5. Imaging Strategy with Microscope System

1. MitoTracker Red CMXRos (Molecular Probes).
2. DAPI (Molecular Probes).
3. Hanks' Balanced Salt Solutions (HBSS).

4. Inverted fluorescence microscope, IX71 (Olympus), equipped with 100× 1.40 NA oil objective, a 100-Watt mercury arc lamp for illumination and 50 mm-Watt Xenon lamp for bleaching with a double lamp-house system (see Note 4).

5. EM-CCD camera (iXon, ANDOR Technology) (see Note 5).

6. Meta Morph software (Universal Imaging Corporation) (see Note 6).

3. Methods

The special character of PUM-HD is its binding to stretches of a target RNA in a sequence-specific manner rather than recognizing secondary structures. The sequence of PUM-HD can be engineered to alter its target-sequence specificity and to bind to two adjacent eight-nucleotide stretches of a target sequence. To characterize the binding affinity of engineered PUM-HD to a target sequence, an electrophoretic mobility shift assay is used (Fig. 3). In addition, it is important to examine whether the direct binding of engineered PUM-HD to a target mRNA occurs in target cells (Fig. 4). For this purpose, it is suitable to use

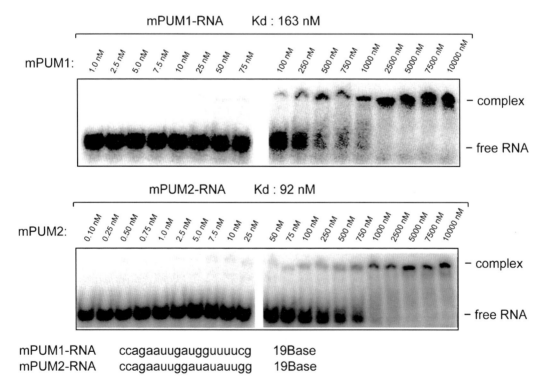

Fig. 3. Example of electrophoretic mobility shift assay. The *upper bands* are the protein–mRNA complexes and the *lower bands* are the unbound mRNA. The sequences of mRNA are indicated as mPUM1-RNA and mPUM2-RNA. When the concentration of each mPUM1 and mPUM2 increased, the proteins made a complex with its cognate mRNA.

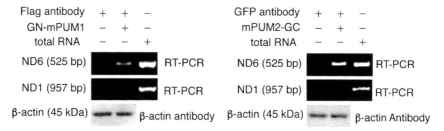

Fig. 4. Example of immunoprecipitation and RT-PCR. In order to examine a selective binding of the probes to a target mRNA, HeLa cells expressing Flag-tagged GN-mPUM1 or mPUM2-GC were lysed and each probe was immunoprecipitated with anti-Flag or anti-GFP antibody. The immunoprecipitated products were extracted and analyzed by RT-PCR. The sizes of PCR products are shown in *parentheses*.

immunoprecipitation and successive reverse transcription PCR (RT-PCR) methods for obtaining reliable results. After such biochemical analysis, imaging analysis can be performed under the fluorescence microscope.

3.1. Design of Mutant PUM-HD Proteins

There is growing evidence of the specificity and selectivity of engineered PUM-HD to RNA sequences (11, 13), which lead the following general rules for the design of PUM-HD mutants (see Note 7). Structures of the PUM-HD bound to an RNA sequence, $5'\text{-}U_1G_2U_3A_4U_5A_6U_7A_8\text{-}3'$, have revealed by X-ray crystallography (10). PUM-HD comprises eight tandem repeats (Fig. 1a). The RNA runs antiparallel to the protein such that nucleotides from U_1 to A_8 are recognized individually by the repeats from R8 to R1, respectively. Each repeat recognizes a single RNA base through three conserved side chains. Of the side chains, two chains form hydrogen bonds or van der Waals interactions with the Watson–Crick edge of an RNA base, whereas a third side chain stacks with the same based and/or the preceding base. When a mutant PUM-HD is constructed, it is only necessary to change the two residues in a repeat that interact with the Watson–Crick edge of the base; there is no need to change the third side chain used for stacking interactions.

The two amino acid residues that recognize an RNA base are named B1 and B5 from the N-terminal end. When two amino acids in a repeat of PUM-HD are mutated to match an RNA base (X), the following combination (X: B1, B5) is recommended: (U: Asn, Gln), (G: Ser, Glu), and (A: Cys or Ser, Gln). No information of the two amino acid residues is available to match the RNA base of C.

Contribution of the 8 repeat to the affinity with RNAs is not equivalent. The most important sequence is the consensus RNA recognition sequences of PUM-HD beginning with $5'\text{-}U_1G_2U_3\text{-}3'$, which recognize R8–R6. The R7 that recognizes G is sensitive to mutations. A mutation of U_3 to G also reduces the affinity of the mutant 30-fold even if the mutated PUM-HD is generated. It is therefore recommended that the consensus sequence should not be changed if at all possible. The R4 is composed of purine (A or G) whose recognition is not strict. Following the sequence, 5'-$U_5A_6U_7A_8$-3', is somewhat variable. In general, affinity of a mutant from A_6 to U_6 or A_6 to G_6 is comparable or more stable than the wild type PUM-HD. A mutation from A_8 to G_8 also keeps high affinity. There is no information about the affinity of mutation from U_5 or U_7 to another RNA base. In any cases, affinity of a mutant PUM-HD to an RNA sequence has to be examined using the electrophoretic mobility shift assay.

3.2. Electrophoretic Mobility Shift Assay

1. Prepare a cDNA, which encodes mutant PUM-HDs (mPUM1 and mPUM2) to match a target mRNA sequences (see Note 7).

2. Insert the cDNA into an *E. coli* expression vector, pCold I (see Note 8). The plasmid is dissolved in pure water (final concentration, 1 mg/mL).

3. Transform 50 µL of BL21 competent cells directly with 1–5 µg of expression vector. The cells are plated on LB agar plates and grown up to form single colonies.

4. Pick up one colony, and culture the cells in 50 mL LB medium for 8 h at 37°C.

5. Add 0.5 mM IPTG for protein expression, shaking for 24 h at 15°C (see Note 9).

6. Collect the cells by centrifugation at 500×g, and disrupt the cell walls by sonication.

7. Centrifuge at 15,000×g for 10 min and collect the supernatant.

8. Apply the supernatant onto the TALON His-tag purification resins.

9. Wash the column according to the manufacture's protocol (CLONTECH).

10. Elute the His-tagged mPUM1 with an elution buffer.

11. Concentrate the purified mPUM1 with Centriprep YM-30 up to 20 mg/mL (see Note 10).

12. Prepare different concentrations of mPUM1 diluted in a binding buffer.

13. Synthesize a [γ-^{32}P]ATP-labeled RNA oligonucleotide with T4 polynucleotide kinase according to the manufacture's protocol (New England Biolabs).

14. Remove the excess amount of [γ-^{32}P]ATP with Sephadex G-25 Quick spin column.

15. Adjust the concentration of [γ-^{32}P]ATP-labeled RNA oligonucleotides to 500 pM.

16. Incubate different concentrations of mPUM1 ranging from 1.0 nM to 10 µM with 15 pM [γ-^{32}P]ATP-labeled RNA oligonucleotides in binding buffer for 1–2 h at room temperature.

17. Load the sample onto 6% polyacrylamide gels.
18. Electrophorese samples using 0.5× TBE at a constant 100 V for 20 min at 4°C.
19. Dry the gel, expose to phosphor screen, and scan the γ-ray with BAS1500 (see Note 11).

3.3. Immunoprecipitation and RT-PCR

1. Seed HeLa cells in 6-cm culture dishes or six-well plates with the DMEM including 10% FBS and grow the cells up to 80–90% confluence.
2. Transfect the plasmids for expression of GN-mPUM1, VN-mPUM1, mPUM2-GC, and mPUM2-VC, into each dish or well using Lipofectamine 2000.
3. Incubate the HeLa cells at 37°C for 48 h in an atmosphere of 5% CO_2.
4. Lyse the cells with Lysis buffer A.
5. Centrifuge $15,000 \times g$ and keep the supernatant.
6. Add anti-Flag antibody (1/1,000 (v/v)) or anti-GFP antibody (1/2,000 (v/v)) into the solution.
7. Incubate the solution between 1 h to overnight at 4°C on a rotator.
8. Mix ProteinSepharose 4FF beads to absorb the immunoprecipitates.
9. Incubate the solution for 1 h at 4°C.
10. Wash the beads four times with Lysis buffer B.
11. Add 25–50 μL of 2× loading buffer.
12. Boil at 95–100°C for 5 min to denature the protein and mRNA in order to separate them from the beads.
13. Centrifuge at $15,000 \times g$ for 3 min and keep the supernatant.
14. Convert the extracted RNA into cDNA using a cDNA synthesis kit (Invitrogen) according to the manufacture's protocol.
15. Run PCR using the cDNA as a template. Set a pair of primers, ND6F and ND6R for mitochondrial ND6 mRNA, and ND1F and ND1R as its control experiment. In the control experiments, run PCR using cDNAs prepared from total RNA in the HeLa cells as a template (see Note 12).
16. Electrophorese the sample and measure the amount of amplified DNA at the size of 525 and 957 bp.

3.4. Cell Preparation for Fluorescence Imaging

1. Seed HeLa cells (1×10^6 cells) onto a 10-cm dish 24 h before transfection, and incubate the cells at 37°C in a CO_2 incubator (see Note 13).
2. Mix 3 plasmids; 1 μL GN-mPUM1, 1 μL mPUM2-GC, and 0.1 μL MTS-DsRed-Ex (see Note 14).

3. Transfect 2.1 µL of the mixed plasmids dissolved in 200 µL OPTI-MEM into the HeLa cells using Lipofectamine 2000.
4. Incubate the cells for 12 h at 37°C in a CO_2 incubator.
5. Harvest the cells by trypsinization, using 0.25% trypsin for 3 min at 37°C.
6. Wash the cells 3× with PBS buffer.
7. Resuspend the cells in 0.5 mL DMEM.
8. Seed the cells on a glass-bottom dish (see Note 15).
9. Incubate the cells for 12 h at 37°C in a CO_2 incubator (see Note 16).
10. Add 14 µM DAPI and MitoTracker in the medium.
11. Incubate the cells for 5 min at 37°C in a CO_2 incubator.
12. Aspirate the medium, wash the cells with PBS buffer (× 3 times), and replace it with HBSS including 5% FBS.

3.5. Microscopic Analysis

1. Set the glass-bottom dish including the HeLa cells on an inverted fluorescence microscope (IX71, Olympus), using a mercury arc lamp and a 100× 1.40 numerical aperture (NA) oil objective.
2. Adjust excitation and emission filters, and a dichroic mirror.
3. Set cooling temperature of EM-CCD camera at –50°C, and its maximal gain at 255 (see Note 17).
4. Acquire images using Meta Morph software.
5. For bleaching experiments, select a bleaching region focused with a 400-µm pinhole.
6. Irradiate in a short time interval, and acquire digital images with the EM-CCD camera (see Note 18).

4. Notes

1. The plasmids described herein can be obtained from our laboratory (e-mail address: ozawa@chem.s.u-tokyo.ac.jp).
2. This protocol can be adapted for many other cell types.
3. This primer set is used to examine the existence of mRNAs of NADH dehydrogenase subunit 6 (ND6) and its subunit 1 (ND1). When another mRNA is visualized with the PUM-HD probes, a corresponding pair of primers, which can be used to amplify a specific region of the reverse-transcribed cDNA, can be selected.
4. We have used a fluorescence microscope from Olympus because of easy handling. Numerous competitive instruments are available such as Zeiss, Leica, Nicon, etc. In the bleaching

experiments, a system of confocal laser-scanning fluorescence microscope is also useful instead of a double lamp-house system.

5. Sensitivity of EM-CCD camera is quite important to obtain fluorescence images. The observed fluorescence intensity of mRNA probes is very weak because of low copy number of mRNAs in single cells.

6. Analysis of imaging data is not always required. Analysis software is required when higher background noise is included in the obtained imaging data.

7. RNA specificity of PUMILIO-like proteins from *C. elegans* and *S. cerevisiae* been investigated (14–16). These RNA binding proteins are also available for the design of RNA imaging probes.

8. pCold I is a cold shock expression vector. The vector is designed to attach the cDNA sequence of His-tag at the 5′ end of a target sequence. The cDNA of a mutant PUM1 and PUM2 (mPUM1 and mPUM2) must be inserted in frame into the expression vector.

9. Overexpression of mPUM1 and mPUM2 results in formation of an inclusion body. In this case, reducing the shaking time may reduce inclusion body formation and improve successive protein purification.

10. Accurate quantification of the purified mPUM1 and mPUM2 is quite important. Large errors of the protein concentration result in wrong evaluation of dissociation constants between mutant PUM-HDs and target mRNA.

11. Quantification of data can also be done by scanning densitometry. Please ensure that the signal has not been saturated. Final calculation of a dissociation constant is performed with a software such as KaleidaGraph Ver. 4.0.

12. In addition to a pair of primers to amplify a target sequence, two pairs of primers for negative and positive controls are required.

13. Transfection efficiency is sensitive to culture confluence, so it is important to maintain a standard seeding protocol from experiment to experiment. Do not add antibacterial agents to the media during transfection.

14. For the purpose of observing mRNA localization, a pair of split EGFP fragments is useful for acquiring fluorescence images. In contrast, VN-mPUM1 and mPUM2-VC, which encode split fragments of the Venus protein, is used for bleaching experiments. To perform the bleaching experiments, the VN-mPUM1 and mPUM2-VC pair can be used instead of the GN-mPUM1 and mPUM2-GC pair in Subheading 3.3.

15. In order to visualize mitochondrial morphology clearly, the cell should be seeded in confluence of 30–50%.

16. The morphology of mitochondria depends on the cell type and incubation conditions. In the case of HeLa cells, mitochondria have a tubular form. However, incubation time longer than 24 h after transfection may result in fragmentation of mitochondria and result in a granular structure. Therefore, take care not to exceed the incubation time longer than 24 h. If the mitochondrial tubular structure cannot be observed

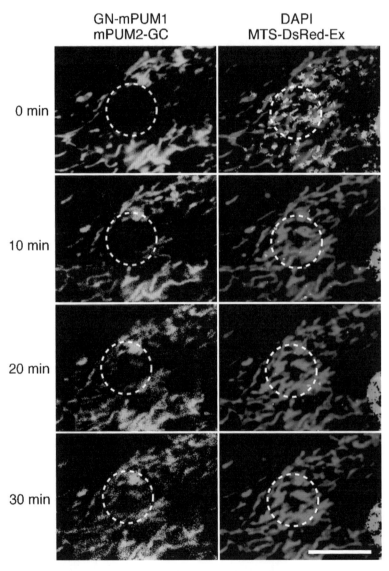

Fig. 5. Time-lapse fluorescence images of HeLa cells expressing GN-mPUM1 and mPUM2-GC upon stimulation of H_2O_2. The HeLa cells were stimulated with 1 mM H_2O_2 immediately after photobleaching, and time-lapse fluorescence images of reconstituted EGFP were acquired. Localization of mitochondria was probed with DsRed-Express (MTS-DsRed-Ex), and mtDNA was stained with DAPI. *Circles* in the images mark photobleached region. *Scale bar*, 5 μm.

under the fluorescence microscope, it is preferable to decrease in the amount of cDNAs for transfection and to repeat the cell preparation again.

17. In the case of Ixon EM-CCD camera, its cooling temperature can be reached less than −80°C if a water-cooling system is used. It is evident that lower temperature reduces the background noise, and then, we recommend to set the temperature as lower as possible.

18. When an external stimulation is added to the cells, it is essential to decide in advance an optimal condition for bleaching experiments. Bleaching time strongly depends on the power of the light source. It is then required to find the shortest time at which the fluorescence intensity is almost diminished (Fig. 5).

Acknowledgments

This work was supported by grants from Core Research for Evolutional Science and Technology (CREST) of Japan Science and Technology (JST) and the Ministry of Education, Science, and Culture, Japan.

References

1. St Johnston, D. (2005) Moving messages: the intracellular localization of mRNAs. *Nat Rev Mol Cell Biol* **6**, 363–75.
2. Tyagi, S. (2009) Imaging intracellular RNA distribution and dynamics in living cells. *Nat Methods* **6**, 331–8.
3. Bao, G., Rhee, W. J., and Tsourkas, A. (2009) Fluorescent probes for live-cell RNA detection. *Annu Rev Biomed Eng* **11**, 25–47.
4. Tyagi, S. (2007) Splitting or stacking fluorescent proteins to visualize mRNA in living cells. *Nat Methods* **4**, 391–2.
5. Ozawa, T., Natori, Y., Sato, M., and Umezawa, Y. (2007) Imaging dynamics of endogenous mitochondrial RNA in single living cells. *Nature Methods* **4**, 413–19.
6. Ozawa, T., Sako, Y., Sato, M., Kitamura, T., and Umezawa, Y. (2003) A genetic approach to identifying mitochondrial proteins. *Nature Biotechnology* **21**, 287–93.
7. Ozawa, T., Takeuchi, M., Kaihara, A., Sato, M., and Umezawa, Y. (2001) Protein splicing-based reconstitution of split green fluorescent protein for monitoring protein-protein interactions in bacteria: improved sensitivity and reduced screening time. *Anal Chem* **73**, 5866–74.
8. Zhang, G., Gurtu, V., and Kain, S. R. (1996) An enhanced green fluorescent protein allows sensitive detection of gene transfer in mammalian cells. *Biochem Biophys Res Commun* **227**, 707–11.
9. Nagai, T., Ibata, K., Park, E. S., Kubota, M., Mikoshiba, K., and Miyawaki, A. (2002) A variant of yellow fluorescent protein with fast and efficient maturation for cell-biological applications. *Nat Biotechnol* **20**, 87–90.
10. Wang, X., McLachlan, J., Zamore, P. D., and Hall, T. M. (2002) Modular recognition of RNA by a human pumilio-homology domain. *Cell* **110**, 501–12.
11. Cheong, C. G., and Hall, T. M. (2006) Engineering RNA sequence specificity of Pumilio repeats. *Proc Natl Acad Sci USA* **103**, 13635–9.
12. Tilsner, J., Linnik, O., Christensen, N. M., Bell, K., Roberts, I. M., Lacomme, C., and Oparka, K. J. (2009) Live-cell imaging of viral RNA genomes using a Pumilio-based reporter. *Plant Journal* **57**, 758–70.
13. Lu, G., Dolgner, S. J., and Hall, T. M. (2009) Understanding and engineering RNA sequence specificity of PUF proteins. *Curr Opin Struct Biol* **19**, 110–5.

14. Miller, M. T., Higgin, J. J., and Hall, T. M. (2008) Basis of altered RNA-binding specificity by PUF proteins revealed by crystal structures of yeast Puf4p. *Nat Struct Mol Biol* **15**, 397–402.
15. Wang, Y., Opperman, L., Wickens, M., and Hall, T. M. (2009) Structural basis for specific recognition of multiple mRNA targets by a PUF regulatory protein. *Proc Natl Acad Sci USA* **106**, 20186–91.
16. Zhu, D., Stumpf, C. R., Krahn, J. M., Wickens, M., and Hall, T. M. (2009) A 5′ cytosine binding pocket in Puf3p specifies regulation of mitochondrial mRNAs. *Proc Natl Acad Sci U S A* **106**, 20192–7.

Chapter 12

Visualization of Induced RNA in Single Bacterial Cells

Azra Borogovac and Natalia E. Broude

Abstract

Visualization of RNA in live cells is a challenging task due to the transient character of most RNA molecules and the lack of adequate methods to label RNA noninvasively. Here, we describe a system for regulated RNA synthesis and visualization of RNA in live *Escherichia coli* cells based on protein complementation. This method allows for labeling RNA with a relatively small protein complex that becomes fluorescent only when bound to an RNA. This method greatly reduces the high fluorescence background characteristic of methods employing intact fluorescent proteins. A short reporter RNA was shown to localize at the cell periphery in nonrandom patterns.

Key words: RNA localization, Protein complementation, Fluorescent proteins, Eukaryotic initiation factor 4A, RNA aptamer, Bacterial cells

1. Introduction

The internal organization of bacteria is a topic of high interest since the discovery of bacterial analogs of all three major eukaryotic cytoskeletal proteins: actin, tubulin, and intermediate microfilaments (reviewed in refs. 1, 2). The localization and dynamics of many cytoskeletal proteins have been thoroughly studied and linked to their function during the cell cycle and changing environment (reviewed in refs. 3–6). Chromosomal DNA and bacterial plasmids have also been shown to localize specifically and move in an orderly cycle-dependent manner (6–9). In contrast, studies on RNA dynamics in live bacterial cells have only recently begun, leaving many questions open for exploration (10–14).

We recently reported a system for RNA visualization based upon protein complementation that involves binding of a split and inactive protein complex to a short added sequence on the target RNA, called an aptamer (14, 15). Aptamers are DNA or

RNA oligonucleotides that usually have strong secondary structure and bind to proteins or small molecules with high affinity. In this system, the RNA aptamer is recognized by an RNA-binding protein, the eukaryotic initiation factor 4A (eIF4A) (16). The inactive protein complex binds to the RNA aptamer, which results in its re-association and restoration of activity.

Specifically, the protein complex consists of two different fusion proteins containing fragments of EGFP and the eIF4A protein. In the first fusion, the N-terminal fragment of split EGFP is linked to the N-terminal domain of eIF4A via a polypeptide linker. Similarly, in the second fusion, the C-terminal fragment of split EGFP is fused to the C-terminal domain of eIF4A. The target RNA is tagged with the eIF4A-binding aptamer at the 3′ end. Expression of the different labeling components in *Escherichia coli* cells generates a fluorescent signal only when the aptamer-tagged target RNA binds the two fragments of eIF4A and triggers re-association of the two split EGFP fragments (14, 15) (Fig. 1). This method offers two advantages for live cell RNA imaging: (1) low background fluorescence and (2) a relatively small labeling complex. In this chapter, we describe an extension

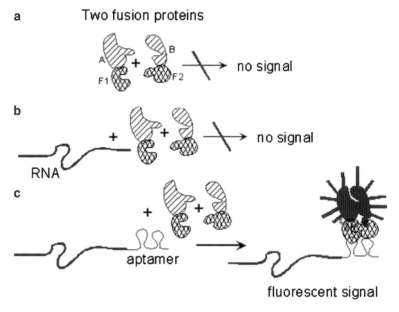

Fig. 1. Principle of the RNA imaging method based on fluorescent protein complementation. Two fusion proteins, A-F1 and B-F2, are co-expressed in a bacterial cell together with RNA tagged with an aptamer. F1 and F2 are the fragments of an RNA-binding protein, eIF4A, while A and B are the fragments of EGFP. The RNA target is tagged at the 3′ end with the eIF4A-binding aptamer. Aptamer binds two fragments of eIF4A, which trigger re-assembly of EGFP. This results in fluorescence development. (**a**) Two fusion proteins are expressed in the cell in the absence of RNA; there is no fluorescent signal. (**b**) RNA target is expressed in the cell, but it does not have the aptamer sequence tag; there is no signal. (**c**) Fluorescent signal develops in the presence of RNA tagged with the aptamer.

of our RNA-labeling system, whereby RNA synthesis and synthesis of the split fusion proteins are separately controlled. Using a TetR-regulated system, we visualized and quantified the transcription kinetics of a short untranslated reporter RNA at the single cell level (17). We note, however, that our most recent results on labeling various RNAs have revealed some caveats in the use of a tetracycline-regulated transcription system for kinetic studies (see Note 1).

2. Materials

2.1. Plasmids Construction

1. dNTP solution (10 mM) containing all four dNTPs.
2. Oligonucleotide primers.
3. Restriction enzymes, thermostable DNA polymerase, and DNA ligase.
4. DNA purification kit.
5. 1% Agarose and gel electrophoresis equipment.
6. 1× Tris–Acetate–EDTA (TAE) buffer: 40 mM Tris–acetate and 1 mM EDTA, pH 8.3.
7. 10 mg/ml ethidium bromide.

2.2. Bacterial Growth and Induction

1. Competent *E. coli* strains Bl21(DE3), *XL-10* (Stratagene), and *DH5αPRO* (Clonetech).
2. 1 M isopropyl-β-d-thio-galactopyranoside (IPTG).
3. Ampicillin (Amp), 100 mg/ml; chloramphenicol (Chp), 10 mg/ml in ethanol; and anhydrotetracycline (ATc), 250 μg/ml in ethanol.
4. Autoclaved Luria–Bertani (LB) media: 10 g tryptone/l, 5 g yeast extract/l, 5 g NaCl/l, and 1 mM NaOH.
5. Two plasmids, one encoding two protein fusions and another with regulated RNA synthesis (Fig. 2).
6. Incubators, at 37, 30, and 20°C.

2.3. Flow Cytometry Analysis

1. Cells expressing fusion proteins and RNA tagged with the eIF4A-specific aptamer before and after IPTG induction, and before and after ATc induction.
2. Cells expressing protein fusions before and after IPTG induction.
3. Cells expressing fusion proteins and an untagged RNA template before and after IPTG induction, and before and after ATc induction.

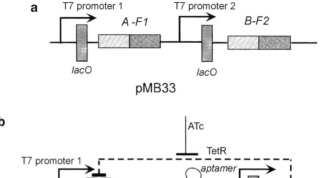
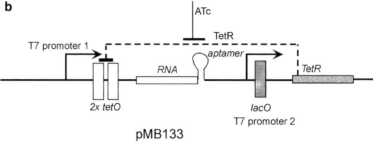

Fig. 2. Two plasmids for inducible RNA synthesis and RNA visualization in bacterial cells. (**a**) Plasmid pMB33 expressing two protein fusions from the two T7/*lacO* promoters. (**b**) Plasmid pMB133 for TetR-regulated RNA expression.

4. Phosphate-buffered saline (PBS): 150 mM NaCl, 3 mM KCl, 8 mM Na_2HPO_4, and 2 mM KH_2PO_4, pH 7.4.

5. Incubators, at 37 and 25°C.

6. Flow cytometer with 488-nm excitation filter for collection of green fluorescence.

2.4. Microscopy and Image Analysis

1. Microscope cover slips (22 × 40 × 0.15 mm) from Fisher, Pittsburgh, PA, and Lab-Tek two-well glass chamber slides from Nalge Nunc, Naperville, IL.

2. Multi-test microscope slides, 15 wells (e.g., from MP Biomedicals, LLC).

3. Piranha solution (30% H_2O_2 and H_2SO_4, 1:3) to remove organic residues from the glass slides.

4. (3-Aminopropyl)-triethoxysilane (Sigma).

5. PBS.

6. Cells expressing protein and RNA component of the complementation complex.

7. Cells expressing only protein fusions (Control 1).

8. Filter-sterilized M9 medium: 420 mM Na_2HPO_4, 220 mM KH_2PO_4, 85 mM NaCl, 50 mM NH_4Cl, 1 mM $MgSO_4 \cdot 7H_2O$, 0.1 mM $CaCl_2 \cdot 2H_2O$, and 0.003 mM Thiamine–HCl.

9. IPTG and ATc.

10. Incubators, at 37 and 25°C.

11. Inverted fluorescence microscope allowing excitation at 490–500 nm and emission detection at 525–535 nm, equipped with neutral density (ND) filters.

3. Methods

The methods described below outline: (1) tagging of an RNA of interest with an aptamer under the control of TetR protein; (2) separate induction of protein expression and of the RNA target; and the analysis of bacterial cells by (3) flow cytometry and (4) fluorescence microscopy.

3.1. Expression Plasmids

To visualize RNA in bacterial cells using fluorescent protein complementation, two compatible plasmids should be used (Fig. 2). The first plasmid (pMB33) expresses two fusion proteins from the two T7/*lacO* promoters in vector pACYCDuet1 (Novagen). The second plasmid (pMB133), based on the pETDuet1 backbone, encodes the aptamer-tagged RNA. The RNA has two copies of the 58-nt-long aptamer cloned between *Hind*III and *Not*I restriction sites. Transcription of this RNA is repressed by the TetR protein synthesized from the T7/*lacO* promoter in the same plasmid (Fig. 2). If both plasmids are expressed in *E. coli* cells, the addition of IPTG induces synthesis of the two fusion proteins from pMB33 together with the synthesis of TetR from plasmid pMB133, which represses RNA synthesis. This allows for the buildup of fusion proteins in the cell so that they can detect the aptamer-tagged RNA that is induced later by the addition of ATc, which relieves TetR repression. Both plasmids pMB33 and pMB133 are available upon request from our laboratory. Any RNA of interest can be cloned into the first multiple cloning site (MCS) in plasmid pMB133 upstream of the tandem repeat of the aptamer sequence.

3.1.1. Insertion of RNA of Interest into Plasmid pMB133

To tag an RNA of interest with the aptamer that will be recognized and labeled with the split fusion proteins, clone the RNA into the MCS1 upstream of the aptamer in pMB133. For control experiments, insert the RNA of interest and remove the aptamer. This allows for measuring background fluorescence of the RNA lacking the aptamer (see Notes 2 and 3).

3.2. Separate Induction of Protein Synthesis and of RNA Tagged with the Aptamer

Once the two vectors expressing protein fusions and the aptamer tagged RNA are available, they need to be transformed into an appropriate bacterial strain expressing T7 RNA polymerase. Induction with IPTG initiates the synthesis of the two fusion proteins and of the TetR repressor, which represses the synthesis of RNA. Once full repression is achieved, synthesis of RNA is started by the addition of ATc, which relieves TetR repression.

1. Co-transform bacterial cells, e.g., BL21 (DE3), with the plasmid pMB33 and the plasmid expressing the aptamer-tagged RNA (modified pMB133).

2. Plate-transformed cultures on LB plates supplemented with 100 μg/ml of ampicillin and 34 μg/ml of chloramphenicol for selection of both plasmids.

3. Pick a single colony from this transformation and start a day culture in 3 ml of LB medium supplemented with 100 μg/ml of ampicillin and 34 μg/ml of chloramphenicol by incubating the cultures at 37°C for 4 h with vigorous shaking. Concurrently, grow the control cells with the plasmid containing the RNA without aptamer, using the same protocol.

4. After 4 h, make multiple inoculations (1, 5, 20, and 50 μl) in 4 ml of fresh LB media supplemented with antibiotics and 1 mM IPTG to ensure that a desired optical density (OD_{600}) would be reached overnight and that the densities would be similar between the test and control strains.

5. After overnight growth, split the cell cultures with OD_{600} between 0.2 and 0.4 in half into separate tubes so that RNA synthesis can be induced in one half with the addition of ATc to 250 ng/ml, while the other half remains without ATc.

6. Analyze cell fluorescence by flow cytometry and microscopy, as described below, at different time points after ATc induction.

3.3. Flow Cytometry Analysis of RNA Labeling

This procedure allows for the assessment of the formation of the fluorescent RNP complex in vivo. The efficiency of RNP complex formation is deduced from a comparison between the fluorescence levels of ATc-induced cells and uninduced cells or cells lacking the RNA component (see Note 4). A successful experiment should result in ATc-dependent appearance of fluorescent cells, which may constitute 50–80% of all cells. The fluorescence of such cells increases with time and at the maximum (50–70 min after induction) reaches up to tenfold above the background level (see Note 5). Background fluorescence is determined by the signal that comes from cells expressing only the protein fragments, i.e., from the cells not induced with ATc. Another useful control includes cells expressing the fusion proteins and the RNA lacking the aptamer sequence (see Note 6). If the fluorescence of such cells is several folds lower than the fluorescence of cells expressing the tagged RNA, then the experiment can be considered successful, and cell fluorescence is the result of protein complementation dependent on the presence of the target RNA.

1. Prepare *E. coli* cells as described in Subheading 3.2.

2. Pellet 0.5 ml of cells by centrifugation at 8,000 rpm (3600 g) for 2 min at room temperature at different time points after

addition of ATc. Wash the cells twice with 1× PBS. Re-suspend the cells in 0.5 ml of PBS.

3. Analyze cell fluorescence using a flow cytometer. Perform excitation with a 488-nm argon laser and use a 515–545-nm emission filter (FL1). Measure fluorescence of 100,000 cells in each sample at a low flow rate. Select a small forward scatter and side scatter gate to decrease fluorescence variation due to different cell sizes.

4. Compare the fluorescence of induced cells with that of uninduced cultures. Also, measure the fluorescence of cells transformed only with the plasmid expressing the fusion proteins in the absence of an interacting RNA. As another control, measure the fluorescence of cells expressing the protein fusions and an untagged RNA (see Note 6).

3.4. Time-Lapse Analysis of Cell Fluorescence by Microscopy

One of the major advantages of protein complementation with regulated RNA synthesis is the possibility to follow RNA synthesis and distribution in single cells in real time (see Note 1 and Fig. 3). Microscopic analysis of the cells can be performed in agarose pads or on the glass cover slides. Agarose pads provide a better environment for cell growth; however, a slightly better resolution can be achieved on the glass slides. To see RNA distribution within bacterial cell, we analyzed cells immobilized on the glass surface using a custom-modified inverted microscope (Zeiss 200M) with a 100×/1.40 oil immersion objective. Excitation was performed using a solid-state 488-nm laser beam (Coherent, Sapphire), which was attenuated using a quarter wave-plate/polarizing cube combination, and beamed into the rear port of the microscope to form a nearly flat illumination field. At the imaging arm, scattered laser light and Raman emission were effectively blocked using appropriate filters (Semrock). Imaging was performed using a back-illuminated electron-multiplying CCD (iXon 897, Andor) cooled to −100°C, and hooked to the laser shutter (17).

3.4.1. Time-Lapse Image Analysis of the Cells Expressing RNA and Immobilized on Glass Surface

1. Wash cover slips in piranha solution and subsequently immerse them in (3-aminopropyl)-triethoxysilane to promote adhesion of *E. coli* cells to the surface.

2. Apply 10–20 μl of suspension of bacterial cells in M9 medium after full induction with IPTG but before induction with ATc and incubate for 5 min. Remove unbound cells by rinsing the binding surface several times with M9 media.

3. Induce RNA synthesis by adding 1–2 μl of M9 medium with 250 ng/ml of ATc.

4. Start time-lapse imaging using 0.1–0.5-s exposure times (depending on fluorescence) and 2–5-min intervals between exposures.

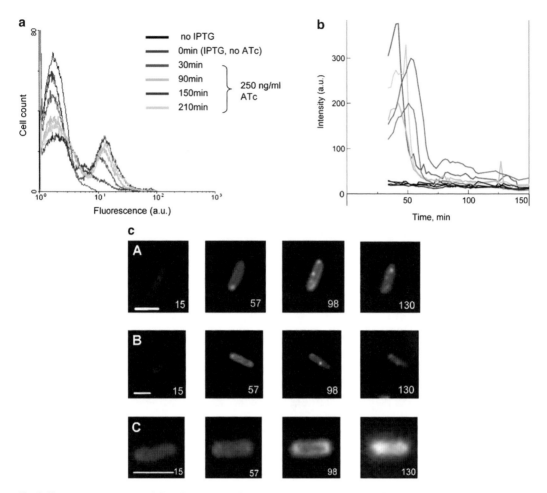

Fig. 3. Fluorescence analyses of *E. coli* cells expressing noncoding reporter RNA. (**a**) Flow cytometry of *E. coli* cells after induction of RNA synthesis with ATc. (**b**) Kinetics of total fluorescence changes in single *E. coli* cells expressing RNA. Each *curve* corresponds to fluorescence in a single cell. Delay (15–20 min) in detecting cell fluorescence is not related to a delay in RNA synthesis or a limitation of our labeling system, but is rather a technical limitation due to the handling of the cells and the microscope. (**c**) Examples of time-lapse images of the cells expressing RNA, the numbers indicate time in minutes after ATc induction.

5. Analyze total cell fluorescence and fluorescence distribution within the cell using public accessible (ImageJ) or custom software.

3.4.2. Time-Lapse Analysis of the Cells Immobilized on Agarose Pads

1. Prepare a 0.8% agarose solution in 1× PBS. Allow the solution to equalize to 50°C and add ATc to 250 ng/ml. Keep the solution warm at 50°C prior to use.

2. Remove 20 μl of agarose solution and distribute small droplets into 15 wells of a multi-test slide. Do not make the agarose pads too big as they will prop open the cover slip.

3. Immediately dispense an aliquot of cell culture with induced protein synthesis over the wells such that it is enough to cover

them. Allow the cells to settle on the pads for 4 min at room temperature.

4. Carefully aspirate excess liquid covering the pads using a vacuum aspirator. Take care not to disrupt or remove the pads from the slide.

5. Cover the slide with a long cover slip, e.g., 24×50 mm.

6. Image the cells in time-lapse format using an inverted fluorescence microscope with a 100× oil immersion objective and epifluorescence system that allows for excitation at 490–500 nm and emission detection at 525–535 nm (exposure time 100–400 ms and 5–10-min intervals). Take differential interference contrast images also to record changes in cell morphology and cell division.

7. Analyze total cell fluorescence and fluorescence distribution within the cell using public accessible (ImageJ) or custom software. Also, compare these images with those obtained with the cells expressing fusion proteins in the absence of RNA.

4. Notes

1. We measured the transcription kinetics of a short untranslated RNA in *E. coli* cells and observed pulsing (17), in accordance with earlier results from the Cluzel group (12, 13). This type of kinetics has been shown to be the result of bacterial efflux pumps, which expel the inducer, ATc, out of the cells (12, 13). Our recent results involving the labeling of different RNAs have revealed changes in cell fluorescence that were dependent on the conditions of cell culture, but not necessarily on the nature of the RNA (unpublished data). At the moment, the reasons for this variability are not completely clear; however, the most likely explanation is that the underlying mechanism entails changes in the state of the efflux pumps. Based on these results, we would not recommend using the ATc-regulated system for studying RNA kinetic studies.

2. Plasmid pMB133 contains two copies of the aptamer, which means that the RNA can bind a maximum of two copies of the fusion proteins. Earlier data on the visualization of *LacZ* mRNA and 5S ribosomal RNA that were tagged with just one copy of the aptamer showed that one copy is sufficient to visualize the signal (14).

3. The target RNA is modified at the 3′ end by inserting the eIF4A aptamer recognition sequence, a 58-nt fragment. Recent data from the Vogel laboratory showed that tagging noncoding RNAs with aptamers can affect RNA expression (18).

Therefore, Northern blots or RT-PCR should be used to test the level of target RNA expression.

4. For each particular case, especially if different cells (e.g., mammalian cells) are used, background fluorescence caused by spurious self-assembly of the fusion proteins should be tested. If there is spontaneous re-assembly resulting in significant fluorescence, several steps can be taken to reduce background: (1) change the orientation of fusion proteins; (2) decrease the concentration of fusion proteins by using a plasmid having a lower copy number or a weaker promoter; or (3) reduce the concentration of the inducer. The temperature of cell culturing also affects the background by changing the concentration of properly folded proteins: at higher temperatures (30°C), the background is lower than that at lower (20°C) temperatures (19).

5. The fluorescence of cell cultures induced with ATc should be compared to that of cells without ATc induction. Note that the fluorescence of these control cells slightly increases over time even in the absence of ATc, due to the leaky T7 promoter that also results in RNA synthesis. Additionally, ATc is autofluorescent and this fluorescence should be not confused with that of RNA.

6. The signal-to-background ratio depends upon the nature of the RNA. Structured RNAs (e.g., tRNA) consisting of complex hairpins and bulges may produce higher backgrounds and lower signal-to-background ratios than less structured mRNAs. Thus, we found that aptamer-tagged *LacZ* RNA produced a better signal-to-background ratio (e.g., four- to fivefold) compared to aptamer-tagged 5S rRNA, where the signal-to-background ratio was about twofold (14, 15). We believe that this is the case because of the nature of the eIF4A protein, which interacts with RNA via secondary structures.

References

1. Shih, Y.L., and Rothfield, L. (2006) The bacterial cytoskeleton. *Microbiol Mol Biol Rev.* **70**, 729–754.
2. Møller-Jensen, J., and Löwe, J. (2005) Increasing complexity of the bacterial cytoskeleton. *Curr. Opin. Cell. Biol.* **17**, 75–81.
3. Collier, J., and Shapiro, L. (2007) Spatial complexity and control of a bacterial cell cycle. *Curr. Opin. Biotechnol.* **18**, 333–340.
4. Jensen, R.B., Wang, S.C., and Shapiro, L. (2002) Dynamic localization of proteins and DNA during a bacterial cell cycle. *Nat. Rev. Mol. Cell Biol.* **3**, 167–176.
5. Thanbichler, M., Viollier, P.H., and Shapiro, L. (2005) The structure and function of the bacterial chromosome. *Curr. Opin. Genet. Dev.* **15**, 153–162.
6. Thanbichler, M., and Shapiro, L. (2006) Chromosome organization and segregation in bacteria. *J. Struct. Biol.* **156**, 292–303.
7. Gordon, G.S., Sitnikov, D., Webb, C.D., Teleman, A., Straight, A., Losick, R., et al. (1997) Chromosome and low copy plasmid segregation in *E. coli*: visual evidence for distinct mechanisms. *Cell* **90**, 1113–1121.

8. Pogliano, J., Ho, T.Q., Zhong, Z., and Helinski, D.R. (2001) Multicopy plasmids are clustered and localized in Escherichia coli. *Proc Natl Acad Sci U S A.* **98**, 4486–4491.

9. Ho, T.Q., Zhong, Z., Aung, S., and Pogliano, J. (2002) Compatible bacterial plasmids are targeted to independent cellular locations in Escherichia coli. *EMBO J.* **21**, 1864–1872.

10. Golding, I., and Cox, E.C. (2004) RNA dynamics in live Escherichia coli cells. *Proc Natl Acad Sci. USA* **101**, 11310–11315.

11. Golding I, Paulsson, J., Zawilski, S.M., and Cox, E.C. (2005) Real-time kinetics of gene activity in individual bacteria. *Cell* **123**, 1025–1036.

12. Le, T.T., Harlepp, S., Guet, C.C., Dittmar, K., Emonet, T., Pan, T., and Cluzel, P. (2005) Real-time RNA profiling within a single bacterium *Proc Natl Acad Sci U S A.* **102**, 9160–9164.

13. Guet, C.C., Bruneaux, L., Min, T.L., Siegal-Gaskins, D., Figueroa, I., Emonet, T., and Cluzel, P. (2008) Minimally invasive determination of mRNA concentration in single living bacteria. *Nucleic Acids Res.* **36**:e73.

14. Valencia-Burton, M., McCullough, R.M., Cantor, C.R., and Broude, N.E. (2007) RNA visualization in live bacterial cells using fluorescent protein complementation. *Nat Methods* **4**, 421–427.

15. Valencia-Burton, M., and Broude, N.E. (2007) Visualization of RNA using fluorescence complementation triggered by aptamer-protein interactions (RFAP) in live bacterial cells. *Curr Protoc Cell Biol.* Chapter 17, Unit 17.11.

16. Oguro, A., Ohtsu, T., Svitkin, Y.V., Sonenberg, N,, and Nakamura, Y. (2003) RNA aptamers to initiation factor 4A helicase hinder cap-dependent translation by blocking ATP hydrolysis. *RNA* **9**, 394–407.

17. Valencia-Burton, M., Shah. A., Sutin, J., Borogovac, A., McCullough, R.M., Cantor, C.R., et al. (2009) Spatiotemporal patterns and transcription kinetics of induced RNA in single bacterial cells *Proc Natl. Acad Sci USA*, *in press*, 2009.

18. Said N, Rieder R, Hurwitz R, Deckert J, Urlaub H, Vogel J. (2009) In vivo expression and purification of aptamer-tagged small RNA regulators. *Nucleic Acids Res.* **37**:e133.

19. Yiu HW, Demidov VV, Toran, P, Cantor CR, Broude NE. (2011) RNA detection in live bacterial cells using fluorescent protein complementation triggered by Interaction of two RNA aptamers with two RNA-binding peptides, submitted.

Part III

Visualizing mRNAs In Vivo Using Aptamers and Intact Fluorescent Proteins

Chapter 13

Visualizing mRNAs in Fixed and Living Yeast Cells

Franck Gallardo and Pascal Chartrand

Abstract

Localization of messenger RNA (mRNA) is a process used by eukaryotes to control the spatio-temporal expression of proteins involved in cellular motility, asymmetric cell division, or polarized cell growth. A better understanding of this process relies on methods to detect specifically the position of an mRNA in fixed or living cells. This chapter presents methods to visualize mRNA in both fixed and living yeast *Saccharomyces cerevisiae*. In fixed cells, position of mRNAs can be assessed by using Fluorescent In Situ Hybridization (FISH) that consists of the hybridization of fluorescent probes that target a specific transcript in situ. In living cells, dynamics of mRNAs can be monitored using a bipartite system composed of MS2 stem-loops inserted in the mRNA of interest. These stem-loops are recognized specifically by the MS2 RNA-binding protein, fused to a fluorescent protein. In vivo association between the reporter (fluorescent MS2 protein) and the MS2-tagged mRNA reconstitutes active fluorescent ribonucleoparticles that can be followed by live cell imaging. Detailed protocols for the realization of these methods are provided and several technical considerations are discussed. Together, these methods provide very robust tools to determine the intracellular position and dynamics of your mRNA of interest in yeast.

Key words: Yeast, Fluorescent in situ hybridization, MS2-GFP, Fluorescence microscopy, mRNA trafficking, Live cell imaging

1. Introduction

1.1. Fluorescent In Situ Hybridization to Visualize mRNAs in Yeast

Fluorescent in situ hybridization (FISH) is a method which consists of the hybridization of a labeled nucleic acid probe (either DNA, RNA or modified nucleic acids) on a RNA target in formaldehyde-fixed cells. Following imaging under a microscope, the position of the target RNA can be determined within a cell. The advent of fluorescent dyes, of epifluorescence and confocal microscopes, and the appearance of CCD cameras lead to an expansion of RNA cytological studies based on FISH. For instance, direct or indirect detection of a poly(dT) oligonucleotide using fluorescent dyes would reveal the whole population of polyA+ transcripts in yeast cells. Combined with yeast genetic tools, this led to the identification of mutants affecting mRNA nuclear export (1). Sequencing of the complete genome of *Saccharomyces cerevisiae* in 1996 opened the possibility of using nucleic acid sequences to determine the spatial distribution of specific transcripts within yeast. Synthetic oligonucleotide probes can now be designed, synthesized, and labeled to detect any given RNA within a cell. While little evidence suggested that the intracellular distribution of a specific mRNA would be different from the bulk population of yeast polyA+ transcripts, the discovery that some mRNAs were specifically sorted to the bud tip of yeast cells changed that perception (2). Since then, FISH has been used to study new mechanisms regulating transcription, splicing, nuclear export, and cytoplasmic localization of specific transcripts in yeast (3–5). Association of subpopulations of mRNAs to organelles like mitochondria or ER has also been studied using this technique (6, 7).

The FISH method presented here is based on the use of modified oligonucleotide probes for in situ hybridization. While these probes can be expensive, they can be designed to target specific regions of a given RNA and they can be directly labeled with fluorophore dyes, which offer an increased signal to noise ratio. For low abundance RNA, multiple probes targeting to different regions of an RNA can be designed to increase the signal. There are other types of probes for FISH and other types of probe labeling that have been developed, which are not discussed in this article. Refer to ref. 8 for details about these alternatives.

1.2. Visualization of mRNAs in Living Yeast Using Fluorescent Proteins

The development of FISH opened the possibility of visualizing specific mRNAs in yeast. However, this technique is limited by the fixation step that freezes the cell in a certain state. Efforts to develop live cell imaging of mRNAs lead to the creation of green fluorescent protein (GFP) reporter fused to specific RNA-binding proteins. While several systems have been developed over the years (9–12), the first and most widely used is based on the MS2 phage protein: the MS2-GFP system. The coat protein from the MS2 bacteriophage (MS2-CP) binds with high specificity to an RNA stem-loop structure of 19 nucleotides in the MS2 phage genome (13). Over the years, the MS2 coat protein has been engineered so that it can be fused to any protein and tethered to any RNA containing the MS2 stem-loop motif. MS2 coat protein variants have been generated, which do not multimerize and show high affinity to the RNA stem-loop, and still bind RNA as a dimer. One such variant, the V29I-dlFG mutant, shows very high affinity to the RNA stem-loop (K_d of 40 nM) (14). Moreover, a cytosine to uracil mutation at position-5 of the 19 nucleotide MS2 RNA stem-loop further increases its affinity for the MS2 coat protein (15). Combining the V29I-d1FG MS2 coat protein mutant with the cytosine-variant RNA stem-loop provides very strong binding properties for the MS2 coat protein. These MS2 coat protein and

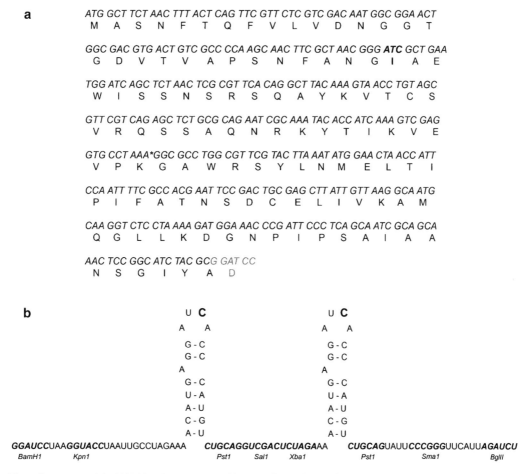

Fig. 1. Sequences of the MS2 bipartite system used in live cell imaging. (**a**) Sequence of the MS2 coat protein V29IdIFG. Position of the mutation V29I is indicated in bold. Position of the deletion of the FG domain is shown by an *asterisk*. A linker (*gray*) containing a *Bam*H1 restriction site is used to clone a fluorescent reporter in the C-terminal part of the MS2 protein. (**b**) Sequence of two MS2 stem-loops. Positions of restriction sites in the DNA counterpart for cloning are indicated. The U → C variant is shown in the loop (C in bold).

RNA stem-loop mutants are used in the MS2-GFP system in yeast and sequences of both MS2 RNA stem loops and coat protein are provided (see Fig. 1).

The MS2-GFP system involves the expression of two constructs: a fusion of the MS2 coat protein to GFP or any fluorescent protein (MS2-CP-GFP), and a multimer of the MS2 RNA stem-loop sequence in the mRNA to be visualized. When the MS2-CP-GFP protein binds to the RNA stem-loops, it acts as a fluorescent beacon that allows the detection of this mRNA within a cell by live cell microscopy. The initial studies have shown that MS2-CP-GFP protein is highly specific for mRNAs containing MS2 stem-loops in yeast (9, 10). The main advantage of using this system over the expression of a GFP-tagged endogenous RNA-binding protein is that MS2-CP-GFP is specific to the RNA containing MS2 stem-loops, while the endogenous RNA-binding protein may bind several mRNAs and reflect the behavior of all of them. Altogether, this system offers the benefits of a specific detection of MS2 stem loop-tagged mRNAs by standard epifluorescence or confocal microscopy, and the study of RNA dynamics in living yeast cells.

2. Materials

2.1. Yeast Cultures

1. YEP (yeast rich media): dilute 10 g of Yeast extract and 20 g of Bacto peptone in 900 mL of H_2O. Autoclave and add 100 mL of 20% sterile glucose solution. YC (yeast selection media) for proper selection. Add 6.7 g of Yeast Nitrogen base without amino acids, and the amino acid and bases required for selection of auxotrophic strains as described (16). Complete with 900 mL of H_2O. Autoclave and add 100 mL of 20% sterile glucose solution.
2. Carbon source: 20% D-glucose or 30% galactose, dilute in water and filter sterilize.
3. Culture tubes.
4. Erlenmeyers.

2.2. Fluorescent In Situ Hybridization

All solutions and equipment should be RNAse-free. For solutions, treat with diethylpyrocarbonate (DEPC), which can be added at 0.1 mL per 100 mL of solution, incubated overnight and autoclaved. For equipment, decontaminate by treating with 3% H_2O_2 for 30 min, then rinse ten times with DEPC-treated water. Always wear gloves to avoid RNAse contaminations.

2.2.1. Buffers

1. 1× Buffer B: 1.2 M Sorbitol and 0.1 M potassium phosphate at pH 7.5. Store at 4°C.
2. 10× PBS: 10 mM KH_2PO_4, 1.4 M NaCl, 40 mM KCl, 100 mM Na_2HPO_4 at pH 7.5. Treat with DEPC, autoclave, and store at room temperature.
3. 20× SSC: dissolve 175.3 g of NaCl and 88.2 g of sodium citrate in 800 mL of water. Adjust pH to 7.0 with HCl. Adjust volume to 1 L with water. Autoclave and store at room temperature.
4. Sodium carbonate buffer: dissolve 106 mg of Na_2CO_3 and 84 mg of $NaHCO_3$ in 10 mL of DEPC water. Adjust pH to 8.8 with HCl. Keep at −20°C.
5. Mounting medium: Dissolve 100 mg of *p*-phenylenediamine (Sigma, St-Louis, MO) in 10 mL of 10× PBS and adjust to

pH 8.0 with 0.5 M of sodium bicarbonate pH 9.0 freshly prepared. Add 90 mL of glycerol. Wrap in aluminum foil and store at −20°C. Discard when the color is too dark.

2.2.2. Enzymes and Chemicals

1. Dried Zymolyase 100T: 100 mg of Zymolyase 100T (Seikagaku, Japan) should be resuspended in 1× Buffer B, aliquoted at 60 µg per tube and lyophilized. These aliquots can be stored at −20°C in a dessicator (see Note 1).

2. 32% Paraformaldehyde (Electron Microscope Sciences, Fort Washington, PA) (see Note 2).

3. Poly-L-lysine (Sigma, St-Louis, MO). Dilute to 0.01% with DEPC treated water.

4. Cy3 monoreactive dye pack (GE Healthcare, Piscataway, NJ).

5. Formamide. Store at 4°C.

6. Pepstatin. Prepare an 850 µg/mL stock with ethanol. Store at −20°C.

7. Leupeptin. Prepare an 800 µg/mL stock with DEPC-treated water. Store at −20°C.

8. Aprotinin. Prepare a 1 mg/mL stock with DEPC-treated water. Store at −20°C.

9. Phenylmethylsulfonyl fluoride (PMSF). Prepare a 17.5 mg/mL stock with ethanol. Store at −20°C.

10. 40 U/µL of RNAse inhibitor (Fermentas, Glen Burnie, MD).

11. 200 mM Vanadium Ribosyl Complex (VRC) (New England Biolabs, Ipswich, MA).

12. 20 mg/mL RNAse-free BSA (Roche Applied Biosciences, Indianapolis, IN).

13. β-Mercaptoethanol.

14. Triton X-100.

15. 5 mg/mL solution of a 1:1 mixture of sonicated salmon sperm DNA and *E. coli* tRNA.

16. 70% ethanol prepared with DEPC-treated water. Keep at −20°C.

2.2.3. Additional Supplies

1. Coplin Jars for 22 × 22 × 0.15 mm coverslips (Fisher Scientific, Pittsburgh, PA).

2. Glass coverslips (22 × 22 × 0.15 mm) and microscope slides (72 × 25 × 1 mm), both from Fisher Scientific.

3. Sephadex G25 or G50 column, RNAse-free (Roche Applied Bioscience, Indianapolis, IN).

4. Glass plates (16 × 20 cm).

5. Parafilm M (Fisher Scientific, Pittsburgh, PA).

6. Tweezers.

7. Six-well plastic plates (Fisher Scientific, Pittsburgh, PA).

2.3. Live Cell Imaging

1. Sealing wax: 1:1:1 mixture of vaseline, lanolin and paraffin, all from Sigma (St-Louis, MO).

3. Methods

3.1. Fluorescent In Situ Hybridization

3.1.1. Probe Preparation

3.1.1.1. Probe Design

Oligonucleotide DNA probes are synthesized with amino-allyl modified thymine residues, which allows the covalent coupling of fluorescent molecules. These probes are usually 50 nucleotides long, with an amino-allyl modified thymine residue incorporated every ten nucleotide (resulting in five amino-allyl thymine residues per probe), if possible. For efficient hybridization, the total G+C content of the probe should be near 50%. If a highly expressed mRNA has to be visualized by FISH, a 5′ or 3′ end-labeled oligonucleotide can be used instead.

3.1.1.2. Probe Labeling

Probe labeling occurs when the desired activated fluorophore reacts with the terminal amino moiety of the amino-allyl modified thymine residues. We routinely use the FluoroLink Cy3 monoreactive dye, used to label antibodies, for labeling probes. However, any activated fluorophore can be used. As a result, the fluorophore is covalently linked to the probe. This step avoids the use of a fluorescent secondary antibody to detect the probe and thus, greatly increases the signal to noise ratio. Moreover, several mRNAs can be detected using different probes labeled with different fluorophores in a same experiment. We have successfully used the duo Cy3/Cy5 to detect multiple mRNAs in the same cell in situ.

1. Dry 10 µg of probe in a speed vacuum. Do not heat.

2. Resuspend probe in 35 µL of sodium carbonate buffer at pH 8.8, freshly made.

3. Dissolve the content of one vial of monoreactive fluorophore dye in 30 µL of DEPC water. See user information for more details. One vial can be used to label two probes.

4. Add 15 µL of dye to the probe and incubate at room temperature for 24–36 h in the dark, with occasional vigorous vortexing.

3.1.1.3. Probe Purification

1. Prepare a Sephadex G-25 or G-50 column. Open the upper cap and then break the bottom seal. Let the column rest for 5 min and then discard the flowthrough.

2. Spin the column for 2 min at 2,500×g in a 15-mL Falcon tube covered with aluminum foil.

3. Discard the flowthrough and use a new 15-mL Falcon tube.

4. Add the 50-μL probe reaction to the center of the column (see Note 3).

5. Spin column for 4 min at 2,500×g in a 15-mL Falcon tube covered with aluminum foil. Recover the purified probe. The probe solution should elute from the free (column bound) dye. Discard column and determine incorporation efficiency.

3.1.1.4. Calculation of Fluorophore Incorporation

This section uses the Cy3 dye as an example. The same calculation applies to any other fluorophore, using its appropriate Molecular Extinction Coefficient.

1. Dilute the labeled probe and take OD at 260 nm (DNA) and 552 nm (for Cy3).

2. Calculate the Molecular Extinction Coefficient (MEC). MEC for Cy3 fluorophore is 150,000 M^{-1} cm^{-1}. The MEC of your oligonucleotide probe is the sum of each A, G, C, and T of your probe. The MEC for adenosine is 15,400 M^{-1} cm^{-1}, for cytidine 7,400 M^{-1} cm^{-1}, for guanine 11,500 M^{-1} cm^{-1}, and thymidine 8,700 M^{-1} cm^{-1}. One can consider an average of 10,000 M^{-1} cm^{-1} per base and do the calculation with this value.

3. Calculate incorporation efficiency by using these equations:

$$[\text{Cy3dye}] = \frac{A_{552}}{150,000},$$

$$[\text{Oligo}] = \frac{(A_{260} - 0.08(A_{552}))}{\text{Oligo MEC}},$$

$$\text{Incorporation(dye / oligo)} = \frac{[\text{Cy3dye}]}{[\text{Oligo}]}.$$

While 100% incorporation can rarely be achieved, we routinely obtain 70–90% labeling efficiency.

3.1.2. Preparation of Poly-L-lysine Coated Coverslips

Yeast cells do not stick very well to glass coverslips. To maintain yeasts spheroplasts on the surface of coverslips during hybridization and washing steps, the coverslips must be pre-coated with poly-L-lysine.

1. Perform chemical stripping by boiling coverslips in 250 mL of 0.1 N HCl for 30 min in a glass beaker. Cover the beaker with aluminum foil while boiling.

2. Let the beaker cool down to room temperature and wash the coverslips ten times with distilled water.

3. Autoclave coverslips in 100 mL of distilled water. These coverslips can be stored at 4°C for several months.

4. Put one coverslip in each well of a six-well tissue culture plate. Let dry and then drop 200 µL of 0.01% poly-L-lysine DEPC on each coverslip. The poly-L-lysine solution should cover at least 75% of the coverslip surface.

5. Incubate for 2 min at room temperature, aspirate the excess of solution, and let dry at room temperature (2–3 h) (see Note 4).

6. After drying, wash each well with DEPC-treated water three times for 10 min at room temperature. While drying, coverslips may have stuck to the bottom of the well, so during washes with water, slow shaking of the six-well plate should help free the coverslips.

7. Rest each coverslip on the wall of the wells, with the coated surface on top. Aspirate excess liquid and let dry (from 2 to 3 h to overnight). If a coverslip falls into the well, it will stick to the plastic surface.

8. After drying, put the coverslips back into the wells, the coated surfaces facing up. These plates can be kept for several months at room temperature.

3.1.3. Fixation and Spheroplasting of Yeast Cells

Fixation is performed with the cross-linking agent, paraformaldehyde. Since paraformaldehyde is a very toxic reagent, it should be manipulated under the hood. Keep the cells on ice to decrease protease and RNAse activities.

1. Yeasts are grown in 50 mL of the appropriate media until they reach early to mid-log phase (OD_{600} between 0.2 and 0.4) (see Note 5).

2. Fix cells for 45 min at room temperature by adding 6.25 mL of 32% paraformaldehyde freshly prepared directly into the cell culture. Transfer into a 50-mL Falcon tube and invert gently every 10 min (see Note 6).

3. Pellet the cells by centrifuging for 4 min at $2,500 \times g$ at 4°C.

4. Wash cell pellet three times with ice-cold buffer B. Between each wash, centrifuge for 4 min at $2,500 \times g$ at 4°C. *Do not* resuspend the pellet when washing.

5. Gently resuspend the cells (do not vortex) in 1 mL of buffer B containing 20 mM vanadyl ribonucleoside complex (VRC), 28 mM β-mercaptoethanol, 0.06 mg/mL phenylmethylsulfonyl fluoride, 5 µg/mL pepstatin, 5 µg/mL leupeptin, 5 µg/mL aprotinin, and 120 U/mL of RNAse inhibitor.

6. Transfer the cells into a tube containing dried Zymolyase 100T or oxalyticase (see Note 1).

7. Incubate the cells for 16–20 min at 30°C (see Note 7).
8. Pellet the spheroplasts by centrifuging for 4 min at 2,500 × g at 4°C.
9. Wash spheroplast pellet with 1 mL of ice-cold buffer B. *Do not* resuspend cells.
10. Add 750 µL of ice-cold buffer B to the pellet and resuspend gently.
11. Drop 100 µL of spheroplast solution on each poly-L-lysine coated coverslip in a six-well plate and let them adhere to coverslips for 30 min at 4°C.
12. Wash coverslips with 3 mL of ice-cold buffer B and aspirate excess. *Do not* drop the liquid directly on the spheroplasts, as this may disperse them.
13. Dehydrate the spheroplasts by adding 5 mL of 70% ethanol prepared with DEPC-treated water. *Do not* drop the ethanol directly on the spheroplasts. Incubate at least 20 min at −20°C before performing the in situ hybridization. The spheroplasts can be kept at −20°C for a maximum of 2 months.

3.1.4. In Situ Hybridization

All the following steps must be performed in dimmed light to avoid bleaching of the labeled probes. Using several probes against the same transcript may be preferable if the expression level of the mRNA is low. We have successfully used this FISH method to detect a broad variety of RNAs expressed at endogenous levels in yeast (see Fig. 2).

3.1.4.1. Probe Preparation

For each coverslip used in the hybridization, prepare one tube of probe. We strongly suggest the use of two coverslips per experiment in order to have a duplicate if one is broken during manipulations.

1. Dilute and pool the probe(s) to a final concentration of 1 ng/µL in DEPC water.
2. Mix 10 µL of probe solution with 4 µL of a 5 mg/mL solution of a 1:1 mixture of sonicated salmon sperm DNA and *E. coli* tRNA. Lyophilize in a speed vacuum (cover with aluminum foil when drying).
3. Resuspend the probes in 12 µL of 80% formamide and 10 mM sodium phosphate at pH 7.
4. Keep covered at room temperature.

3.1.4.2. Rehydratation and Hybridization

1. Cover a 16 × 20 cm glass plate with a layer of parafilm by placing its paper side up and rubbing firmly to make it adhere to the glass. This glass plate will be used at a later step.

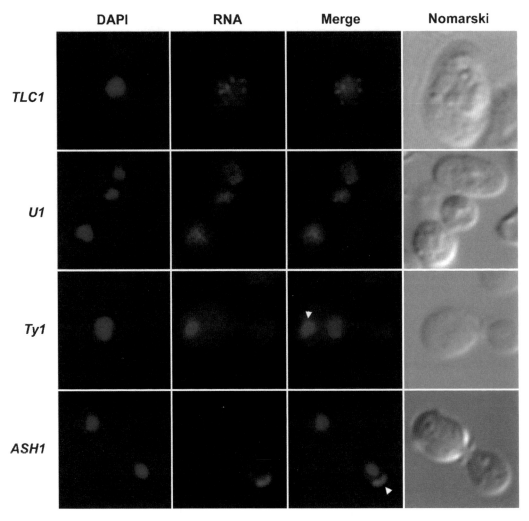

Fig. 2. Localization of various endogenous RNAs using FISH assay in the yeast *S. cerevisiae*. Examples of the localization of different RNAs are shown in the *red channel*. DAPI: nuclear staining. The FISH technique allows the detection of differentially expressed RNAs, from very low expression (*TLC1* RNA), to mid-level expression (*U1* snRNA), and high expression (*Ty1* RNA). The *arrowhead* indicates the presence of a *Ty1* RNA cluster called the T-body. The *ASH1* mRNA is an example of an asymmetric distribution of an mRNA to the bud tip (*arrowhead*). All probes used for these detections were labeled with Cy3 fluorophores.

2. Put the coverslips containing spheroplasted cells in a Coplin Jar and wash twice with 8 mL of 2× SSC for 5 min at room temperature (see Note 8).

3. Incubate in 2× SSC 40% formamide for 5 min at room temperature. During this incubation, heat your probe solution for 3 min at 95°C (see Note 9).

4. Dilute the probe solution with 12 μL of 4× SSC, 20 mM VRC, 4 μg/mL RNAse free BSA, and 50 U of RNAse inhibitor.

5. For the hybridization, use the glass plate wrapped with the parafilm sheet. The probe solution (24 μL) is dropped on

the parafilm. Lay the coverslip on the drop; the surface of the coverslip containing the spheroplasts should face the drop (see Note 10).

6. Cover the coverslips with another parafilm sheet and seal both parafilm sheets around the coverslips with a non-sharp object. *Do not* move the coverslips after they have been laid down. This creates a humidified hybridization chamber that is required to ensure good hybridization. Wrap in aluminum foil and incubate overnight in the dark at 37°C (see Note 11).

3.1.4.3. Washing

1. After hybridization, remove coverslips from the parafilm sheet and put them back in a Coplin Jar. At this step, the Coplin jar should always be wrapped in aluminum foil.
2. Wash twice with 8 mL of preheated 2× SSC 40% formamide at 37°C for 15 min.
3. Wash with 8 mL of 2× SSC 0.1% Triton X-100 for 15 min at room temperature.
4. Wash twice with 8 mL of 1× SSC at room temperature for 15 min.

3.1.4.4. Nuclear Staining and Mounting on Slides

1. Add 8 mL of 1× PBS containing 1 ng/mL of 4,6 diamidino-2-phenylindole (DAPI) (Invitrogen, Carlsbad, CA) and allow to stand for 2 min at room temperature.
2. Drop 10 µL of mounting medium on a glass slide. Lay down a coverslip on the drop, with the surface containing the spheroplasts facing the drop.
3. Remove excess liquid with Kimwipes.
4. Seal the coverslip by applying nail polish on the sides. Allow to dry at room temperature and store at −20°C (see Note 12).

3.2. In Vivo Visualization of mRNAs Using Fluorescent Proteins

3.2.1. Generating an mRNA with MS2 Stem-Loops

The first step is to insert the MS2 stem-loops in your mRNA of interest. These stem-loops can be inserted in the 5′ or 3′UTR of the mRNA, as long as they do not interfere with the expression, stability, localization, or function of the transcript (see Fig. 3a). There are different MS2 stem-loop repeats currently available, with either 2, 6, 12, or 24 repeats (see Note 13). Insertion of the MS2 stem-loops in the 5′UTR gives the advantage of being able to visualize the transcription site of the RNA (see Fig. 3b). However, it can interfere with translation. Thus, 3′UTR insertion should be favored, even if this implies the detection of only the mature form of your mRNA.

For the expression of mRNA with MS2 stem-loops, three approaches can be used. The first approach is to use a fusion reporter mRNA containing an heterologous open reading frame with MS2 stem-loops in its 3′UTR. For instance, a vector expressing a *lacZ*-6xMS2 reporter (YEP195-lacZ-6xMS2) can be used to study the role of specific RNA zipcodes or localization elements

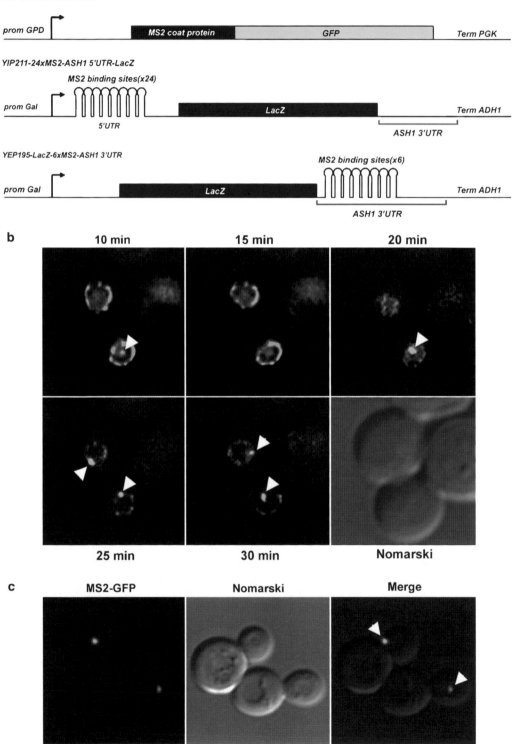

involved in mRNA trafficking. Besides the *lacZ* ORF and six MS2 stem-loops, it contains an *ADH2* 3′UTR with multiple cloning site for insertion of heterologous DNA. An example with the *ASH1* bud-localization zipcode cloned in the 3′UTR of the *lacZ-6xMS2* reporter is shown in Fig. 3c.

The second approach is to directly clone the MS2 stem-loops in the 3′UTR of the mRNA of interest, expressed from a plasmid. In this case, one can use the MS2 stem-loops from the pSL-MS2-6 plasmid (9), which contains several restriction sites for subcloning in the 3′UTR. The MS2 stem-loops can also be inserted in the 5′UTR of the gene of interest. Finally, a third approach is to directly integrate the MS2 stem-loops in the chromosome locus expressing your mRNA of interest (17). A PCR cassette containing the MS2 stem-loops and a selectable marker is integrated by homologous recombination in the 3′UTR of the gene, a few nucleotides downstream of the stop codon. The presence of *loxP* sites flanking both sides of the marker allows the Cre-mediated excision of the marker, so only the MS2 stem-loops remain in the transcript. This approach allows the visualization of a specific mRNA expressed at endogenous levels, from its own promoter.

3.2.2. Choosing the Right MS2-CP-GFP Expression Vector

For the expression of the MS2-CP-GFP protein in yeast, the original expression vector (pG14-MS2-CP-GFP) developed by the Singer lab was based on the pG14 vector (2 µm, *LEU2*) containing the MS2-CP-GFP fusion under the control of the *GPD1* promoter (9). An SV40 nuclear localization signal and an HA tag was inserted upstream of the MS2-CP-GFP open reading frame. This plasmid allows constitutive expression of a nuclear-localized MS2-CP-GFP protein. Only the MS2-fluorescent protein tethered to an mRNA containing MS2 stem-loops can be exported out of the nucleus and accumulate in the cytoplasm.

Problems with this vector are the high level of MS2-CP-GFP expression and the impossibility of detecting nuclear RNAs. New versions of this expression plasmid are now available, with lower expression levels (a CEN, *LEU2* plasmid YCP111-MS2-CP-GFP), without nuclear localization signal (YCP111-MS2-CP-GFP-ΔNLS) for detection of transcripts in the nucleus (see Fig. 3a), or with an inducible promoter (*MET25* promoter) (17). Another version, with the RedStar fluorescent protein has been recently published (18).

Fig. 3. Detection of a specific mRNA in living yeast from transcription to terminal localization using the MS2-GFP system. (**a**) Illustration of the constructions used in (**b**) and (**c**). (**b**) The MS2-CP-GFP protein allows real-time visualization of a transcription site from the chromosome-integrated galactose-inducible *24xMS2-LacZ-ASH1 3′UTR* gene. Sequential acquisition of images at 5 min interval was performed after 15 min of galactose induction. An integrated Nup49-GFP construction was used to label the nuclear membrane. *Arrowheads* indicate transcription site, which forms a larger and brighter GFP foci compared to the perinuclear Nup49-GFP signal. (**c**) Localization of the *LacZ-6xMS2-ASH1 3′UTR* mRNA tagged with MS2-CP-GFP to the bud tip. *Arrowheads* indicate RNA/GFP particle in the bud.

3.2.3. Preparing Slides for Live Cell Imaging

When cells are mounted directly on the slide, very few of them will immediately stick on the surface and imaging may be difficult due to cell movements. To get rid of this problem, cells can be placed on a small layer of agar containing the appropriate growth medium. Using this method, cells can grow for hours under the microscope, if phototoxicity is avoided (see Note 14).

1. Prepare a solution of your minimal media with a 4% carbon source as a final concentration and 1% agar.
2. Keep sterile in an eppendorf tube. Tubes can be stored for months at 4°C.
3. Heat the medium at 95°C until it melts.
4. Drop 200 µL of melted agar solution in the center of a glass slide and quickly cover with another slide creating a sandwich, with agar between the two.
5. When the agar has solidified, open the sandwich by turning the upper slide. This will release the slide and the thin layer of agar should stay on the bottom slide.

3.2.4. Live Cell Imaging

1. Both plasmids are transformed in your strain of interest and selected in the appropriate medium.
2. Grow different clones overnight in 5 mL of minimum selective media (see Note 5).
3. Dilute overnight culture in 5 mL of appropriate media and let grow until they reach OD 0.2–0.3 at A_{600}. Induce mRNA expression if necessary.
4. Drop 20 µL of cell suspension on a freshly prepared agar coated slide and cover with a coverslip.
5. Seal with melted Wax (1:1:1 mixture of paraffin, vaseline, and lanolin). Do not use nail polish (see Note 15).
6. Place slide in temperature-controlled chamber at 30°C (see Note 16).
7. Perform imaging using an epifluorescence or confocal microscope (see Note 17).

4. Notes

1. While Zymolyase 100T works fine with well expressed mRNAs, spheroplasting that yields a more sensitive FISH signal is obtained using recombinant oxalyticase purified in the lab (see supplementary data from ref. 19 for details on expression and purification of recombinant oxalyticase).

2. The quality of the formaldehyde is crucial for the preservation and detection of small details by FISH. We always use EM-grade ultrapure, single-usage sealed ampules of formaldehyde from Electron Microscope Sciences.

3. Missing the center will allow the probe preparation to sweep between the resin and the tube, resulting in free dye contamination at the end of the purification and a misleading measurement of labeling efficiency. A second purification step using a new Sephadex column can be used if free dye is still present with the probe.

4. Putting coverslips under a hood will shorten drying time.

5. To minimize the natural autofluorescence of cells, avoid growing the yeasts over mid-log phase ($OD_{600\,nm}$ 0.2–0.6). Moreover, for laboratory strains that have the *ade2* marker, adding adenine at 20 µg/mL directly to your growth media helps decrease autofluorescence due to the accumulation of an oxidized metabolite of phosphoribosyl-aminoimidazole, an adenine precursor that gives the pink color to *ade2* yeast colonies. Yeast cells grown in YPD medium, which is poor in adenine, show more autofluorescence. This is especially important when imaging at 488 nm since this pigment is fluorescent in the green emission channel and may interfere with the GFP signal.

6. Colocalization between a specific mRNA and proteins can be performed using FISH and GFP-tagged proteins expressed in yeast. However, formaldehyde fixation strongly reduces GFP fluorescence due to a decreased pH of the medium after addition of formaldehyde. For such colocalization experiments, growing cells at room temperature (allows for better folding of GFP) induces a stronger fluorescence in the GFP channel and will counteract the effect of formaldehyde. Moreover, fixing the cells in DEPC water instead of yeast media clearly helps to conserve GFP fluorescence. In this case, yeast cells can be centrifuged, resuspended in DEPC water, and fixed with formaldehyde.

7. Incubation time with lyticase may depend on the yeast strain, as some strains have thicker cell wall. Assays should be performed to optimize incubation time.

8. To remember which side of the coverslips the spheroplasted yeasts are, mark one side of the Coplin jar with a black pen and always put the coverslip with their side containing the spheroplasts facing the marked side of the jar.

9. The quality of the formamide is important in order to avoid nonspecific signal. Store the formamide stock at 4°C and do not keep for more than a year. Formamide concentration determines the stringency of the hybridization. Formamide concentration can range from 10% (for poly(dT) probe

detecting polyA+ mRNA) to 50% (for probes with high G + C content).

10. Avoid air bubbles between the probe solution and the coverslip; it will affect hybridization and induce a low FISH signal.

11. An incubation of 2–3 h can be performed if the target RNA is highly expressed and the hybridization signal is very strong. For most cases, we strongly suggest overnight incubation.

12. While the slides can be stored at –20 °C for months without fading, capture images as soon as possible, as fluorescence intensity decreases with time even in the presence of antifading agent.

13. MS2 stem-loop repeats can be deleted during amplification in bacteria. We use the HB101 bacterial strain (a rec-deficient strain) grown at 30°C to efficiently amplify the plasmids without deletion or recombination of the MS2 stem-loops.

14. Try to minimize phototoxicity when imaging. Phototoxicity is generated by illumination under the microscope, which induce reactive oxygen species and cellular damages. Cells respond to such damage, which can affect the behavior of your RNA. This may not be very important for short acquisition time (1 or 2 min), but it can be a problem if you want to image over long time period. To evaluate phototoxicity, let the cell grow after imaging. If cells have taken too much damage, they will stop dividing. Adding an antioxidant to the media can delay this effect.

15. Do not seal the slides with nail polish for live cell imaging. The solvent in the nail polish will affect GFP fluorescence and can damage cells.

16. Using a chamber with controlled temperature allows efficient growth of yeasts and is preferred when doing time-lapse imaging over long period. We can easily image several cell cycles using such chamber. Also, ATP-dependent dynamics of RNA is affected by the temperature.

17. Before starting GFP-labeled mRNAs in live cell, we strongly suggest to perform a FISH on your mRNA construction to show that the foci you observe with the MS2-CP-GFP system colocalize with the MS2-tagged mRNA in situ.

Acknowledgments

The authors thank Dr. Emmanuelle Querido for critical reading of the manuscript and Zhifa Shen for sharing results. This work is supported by a grant from the Natural Sciences and Engineering

Research Council of Canada. FG is supported by a fellowship from the Terry Fox Foundation from the National Cancer Institute of Canada. PC is a Senior Scholar from the Fond de Recherche en Santé du Québec (FRSQ).

References

1. Amberg, D. C., Goldstein, A.L., Cole, C.N. (1992) Isolation and characterization of RAT1: an essential gene of Saccharomyces cerevisiae required for the efficient nucleocytoplasmic trafficking of mRNA. *Genes Dev.* **6**, 1173–89.

2. Long, R. M., Singer, R. H., Meng, X., Gonzalez, I., Nasmyth, K., and Jansen, R.-P. (1997) Mating Type Switching in Yeast Controlled by Asymmetric Localization of ASH1 mRNA. *Science* **277**, 383–87.

3. Zenklusen, D., Larson, D. R., and Singer, R. H. (2008) Single-RNA counting reveals alternative modes of gene expression in yeast. *Nat Struct Mol Biol* **15**, 1263–71.

4. Long, R. M., Elliott, D.J., Stutz, F., Rosbash, M., Singer, R.H. (1995) Spatial consequences of defective processing of specific yeast mRNAs revealed by fluorescent in situ hybridization. *RNA* **1**, 1071–78.

5. Saavedra, C., Tung, K. S., Amberg, D. C., Hopper, A. K., and Cole, C. N. (1996) Regulation of mRNA export in response to stress in Saccharomyces cerevisiae. *Genes & Development* **10**, 1608–20.

6. Garcia, M., Darzacq, X., Delaveau, T., Jourdren, L., Singer, R. H., and Jacq, C. (2007) Mitochondria-associated Yeast mRNAs and the Biogenesis of Molecular Complexes. *Mol. Biol. Cell* **18**, 362–68.

7. Aronov, S., Gelin-Licht, R., Zipor, G., Haim, L., Safran, E., and Gerst, J. E. (2007) mRNAs Encoding Polarity and Exocytosis Factors Are Cotransported with the Cortical Endoplasmic Reticulum to the Incipient Bud in Saccharomyces cerevisiae. *Mol. Cell. Biol.* **27**, 3441–55.

8. Chartrand, P., Singer, R.H., and Long, R.M. (2000) Sensitive and high-resolution detection of RNA in situ. *Methods in Enzymol* **318**, 493–506.

9. Bertrand, E., Chartrand, P., Schaefer, M., Shenoy, S.M., Singer, R.H., and Long, R.M (1998) Localization of ASH1 mRNA particles in living yeast. *Mol. Cell* **2**, 437–45.

10. Beach, D. L., Salmon, E. D., and Bloom, K. (1999) Localization and anchoring of mRNA in budding yeast. *Current Biology* **9**, 569–78.

11. Brodsky, A. S., and Silver, P. A. (2000) Pre-mRNA processing factors are required for nuclear export. *RNA* **6**, 1737-49.

12. Daigle, N., and Ellenberg, J. (2007) λ_{N}-GFP: an RNA reporter system for live-cell imaging. *Nat Meth* **4**, 633–36.

13. Bernardi, A., Spahr, P.F. (1972) Nucleotide sequence at the binding site for coat protein on RNA of bacteriophage R17. *Proc Natl Acad Sci USA.* **69**, 3033–37.

14. Lim, F., and Peabody, D. S. (1994) Mutations that increase the affinity of a translational repressor for RNA. *Nucl. Acids Res.* **22**, 3748–52.

15. Talbot, S. J., Goodman, S., Bates, S. R. E., Fishwick, C. W. G., and Stockley, P. G. (1990) Use of synthetic oligoribonucleotides to probe RNA-protein interactions in the MS2 translational operator complex. *Nucl. Acids Res.* **18**, 3521–28.

16. Rose, M. D., Winston, F., and Hieter, P. (1990) Methods in yeast genetics. A laboratory course manual, Cold Spring Harbor Laboratory Press, Cold Spring Harbor, NY.

17. Haim, L., Zipor, G., Aronov, S., and Gerst, J. E. (2007) A genomic integration method to visualize localization of endogenous mRNAs in living yeast. *Nat Meth* **4**, 409–12.

18. Schmid, M., Jaedicke, A., Du, T.-G., and Jansen, R.-P. (2006) Coordination of Endoplasmic Reticulum and mRNA Localization to the Yeast Bud. *Current Biology* **16**, 1538–43.

19. Gallardo, F., Olivier, C., Dandjinou, A.T., Wellinger, R.J., Chartrand, P. (2008) TLC1 RNA nucleo-cytoplasmic trafficking links telomerase biogenesis to its recruitment to telomeres. *The Embo Journal* **27**, 748–57.

Chapter 14

In Vivo Visualization of RNA Using the U1A-Based Tagged RNA System

Sunglan Chung and Peter A. Takizawa

Abstract

mRNA transport is a widely used method to achieve the asymmetric distribution of proteins within a cell or organism. In order to understand how RNA is transported, it is essential to utilize a system that can readily detect RNA movement in live cells. The tagged RNA system has recently emerged as a feasible non-invasive solution for such purpose. In this chapter, we describe in detail the U1A-based tagged RNA system. This system coexpresses U1Ap-GFP with the RNA of interest tagged with U1A aptamers, and has been proven to effectively track RNA in vivo. In addition, we provide further applications of the system for ribonucleoprotein complex purification by TAP-tagging the U1Ap-GFP construct.

Key words: U1A, GFP, Tagged RNA, RNA localization, ASH1, RNP (ribonucleoprotein) complex

1. Introduction

Cell polarity is involved in a wide variety of biological processes such as differentiation, cell motility, cell fate determination, and synaptic plasticity. To generate the asymmetric distribution of proteins, several cells and organisms transport mRNA to the destined region and spatially restrict protein expression (1–3). Localizing mRNA is particularly advantageous in that it is possible to achieve the local concentration of protein with a relatively low number of transcripts. Moreover, as protein expression is limited to a restricted region, unwanted expression throughout other regions in the cell is thus inhibited.

Localized mRNA contain *cis*-acting sequences, namely localization elements or zipcodes, that are recognized by RNA-binding proteins, and are further assembled into an RNP (ribonucleoprotein) complex that actively travels to its destination by the motor proteins (4–6). In budding yeast, *ASH1* mRNA is localized to the bud tip so that Ash1 represses the expression of HO endonuclease in daughter cells. This ensures that mating-type switching does not occur in these cells and establishes different mating types in mother and daughter cells. Four localization elements are present on *ASH1* with three located in the coding region and the fourth, U3 (E3), extending into the 3'UTR (7, 8). Extensive studies on this system suggest that these zipcodes are recognized by the RNA-binding protein She2, and subsequently the myosin motor Myo4 binds in conjunction with the adapter protein She3, enabling transport of the RNP complex to the bud tip (9–12). However, this is among the only systems where all *cis*- and *trans*-acting factors required for transport have been fully established.

In order to identify proteins and *cis*-acting elements involved in mRNA transport and understand how cell polarity is achieved, it is essential to utilize a system where mRNA can be readily visualized. Although in situ hybridization still remains a powerful method for such purpose, this method only provides fixed images and does not give access to the real time movement of mRNA that generates such distribution. Thus, several methods have been developed to track RNA in live cells. We have successfully used the U1A-based tagged RNA system to detect live images of RNA in budding yeast (12). This system takes advantage of the RNA–protein interaction of U1Ap, a component of the spliceosomal U1 small nuclear ribonucleoprotein (U1 snRNP), and the 3'UTR of its own pre-mRNA (13). The U1Ap-binding site in the 3'UTR folds into a secondary structure with two internal loops that each contains a seven nucleotide recognition sequence that can recruit U1Ap with high affinity. Thus, up to two copies of U1Ap can bind to each aptamer. To visualize RNA in live cells, the RNA-binding domain of human U1Ap (1-102, ref. 14) is fused to GFP and coexpressed with the RNA of interest tagged with U1A aptamers (Fig. 1a). This embeds several molecules of U1Ap-GFP to RNA and enables the live imaging of RNA. Human U1Ap does not bind to its yeast RNA counterparts, namely U1 RNA and U1A pre-mRNA, thus, when expressed in yeast, U1Ap-GFP specifically binds to the RNA of interest (15, 16). When *ASH1* was tagged and visualized using this system in yeast, GFP-bound *ASH1* RNA was detected at the bud tip (Fig. 1b).

The bacteriophage MS2 coat protein and MS2 RNA have also been frequently used as an alternative tagged RNA system (17). However, MS2p is an obligate dimer that tends to oligomerize, even in its mutant form that prevents aggregation (18). In contrast, U1Ap (1-102) exists as a monomer and binds to its RNA target in a 1:1 ratio (19, 20). This makes it possible to control the number of U1Ap-GFP molecules bound to RNA and accurately quantify the fluorescence signal of the RNA by measuring the relative fluorescence intensity of GFP in a region of interest. In addition, as U1Ap does not multimerize, the unwanted gathering of RNP complexes via the tethering protein would not occur.

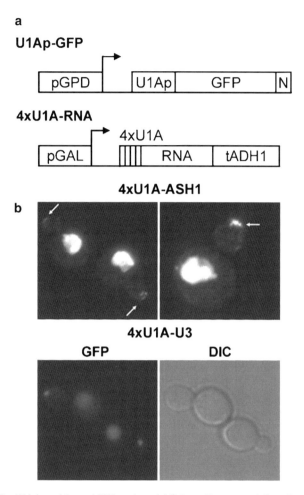

Fig. 1. The U1A-based tagged RNA system. (a) Schematic representation of constructs used to visualize RNA in vivo. The RNA-binding protein is expressed from the constitutive glycerol phosphate dehydrogenase (*GPD*) promoter, and contains U1Ap, GFP, and a nuclear localization sequence (N). The RNA of interest is tagged with U1A aptamers and coexpressed with U1Ap-GFP. Various RNA sequences can also be inserted into an RNA construct that is expressed from the inducible galactose (*GAL*) promoter, and consists of 4 repeats of U1A aptamer and an *ADH1* terminator as shown. (b) *Top*, Yeast cells coexpressing U1Ap-GFP and *ASH1* tagged with 4xU1A in the 5′UTR. GFP bound *ASH1* RNA is localized to the bud tip in large budded cells as indicated by *arrows*. The bright nuclear fluorescence is due to U1Ap-GFP not bound to RNA. *Bottom*, yeast cells coexpressing U1Ap-GFP and U3 expressed from the RNA construct shown in (a).

RNA localization is accomplished by the coordinated activity of several *trans*-acting factors. Proteins involved in the recognition and movement of RNA, and those necessary to generate an active transport complex must be assembled on RNA so that it can effectively move along cytoskeletal tracks. In addition, regulatory proteins that inhibit RNA expression during travel, and proteins that anchor the RNA at its designated site would also be required to ensure that RNA is correctly localized. In order to identify *trans*-acting factors involved in RNA localization, we have developed a method to purify an RNP transport complex by modifying the tagged RNA system. In this method, a TAP tag is added to the U1Ap-GFP construct (U1A-GFP-TAP) so that when coexpressed with 4xU1A-RNA, it is possible to pull down the RNP complex of interest using standard TAP purification (21). This method could be further used in a wide range of systems to identify proteins involved in various steps of mRNA metabolism. Here, we provide a detailed protocol on the tagged RNA system for visualization of RNA in yeast and describe the modified tagged RNA system for RNP purification.

1.1. The RNA Construct: U1A$_{tag}$-RNA

The RNA of interest is commonly tagged with four repeats of the U1Ap-binding sequence from U1A premRNA 3′UTR (13, 22; see Note 1). When using full length RNA, the location and number of the aptamers should be taken into careful consideration. With rare transcripts, it may be necessary to increase the number of aptamers to obtain detectable signals with the endogenous promoter (see Note 2). However, the insertion of excessive aptamers could result in altered mRNA stability. Furthermore, when inserted in the 5′UTR or open reading frame, translational inhibition could also occur. Thus, it is useful to screen for the best candidate among a variety of constructs. In our hands, a faint GFP signal was detected even with a single aptamer tagged to the RNA of interest when expressed on a high-copy plasmid in yeast. However, increasing the number up to four aptamers substantially improved the image quality without interfering with localization.

As a more robust solution to examine the transport of various RNA fragments, it is also possible to insert the RNA of interest into a p*GAL*-4xU1A-t*ADH1* construct (Fig. 1a). This is expressed from a high copy plasmid (2 µ) and contains the inducible galactose promoter (p*GAL*), 4 repeats of U1A aptamer, and the terminator sequence from *ADH1*. The inducible promoter allows the transient expression of RNA, providing images of RNA in motion without interfering with normal cell function. When we expressed the 77 nucleotide U3 zipcode from *ASH1* using this system, the bud-directed movement of bright GFP particles were readily detected in live yeast cells (Fig. 1b). The 5′UTR of rare transcripts can also be replaced with the *GAL* promoter and tagged with U1A aptamers in the 3′UTR to facilitate detection (see Note 3).

1.2. The RNA-Binding Construct: U1Ap-GFP

To generate a GFP fusion protein that binds to the RNA of interest, the RNA-binding domain of U1A (1-102) is fused upstream of GFP (S65T, V163A, S175G; ref. 23), and an SV40 nuclear localization signal (NLS) is added downstream of GFP to confine unbound U1Ap-GFP to the nucleus (Fig. 1a). The fusion protein is under the control of the constitutive glycerol phosphate dehydrogenase promoter (p*GPD*) and expressed from a low-copy plasmid (CEN/ARS) or integrated in the yeast chromosome (see Note 4).

1.3. Ribonucleoprotein Complex Purification

Various protein tags can be added downstream of GFP in the RNA-binding construct to pull down the ribonucleoprotein (RNP) complex of interest (see Note 5). The RNA should be expressed from a *GAL* promoter on a high copy plasmid so that an ample amount of RNA is available for purification. However, as RNA goes through multiple steps during its lifetime, the RNA to be purified should be selected with prudence. Using the minimal length RNA required for localization (or other purpose to be examined) would most likely give best results. In addition, it is essential to compare the protein profile of the purified complex with adequate controls to accurately identify proteins specifically associated to the RNA of interest. Cells expressing U1Ap-GFP-TAP only, or those coexpressing U1Ap-GFP-TAP and a control RNA tagged with U1A would serve as good sources for such purpose. The final protein profile can then be identified by mass spectrometry analysis.

We have added a TAP tag in the RNA-binding construct (p*GPD*-U1Ap-GFP-TAP-NLS) and expressed this with the 77 nucleotide U3 RNA tagged with 4xU1A (p*GAL*-4xU1A-U3-t*ADH1*) to purify the minimal RNP complex that enables bud transport. To identify proteins specifically bound to U3 RNA, protein profiles from U1Ap-GFP-TAP expressed alone, or with a non-localizing *ADH2* RNA sequence originating from the same position as U3 in *ASH1* were used as negative controls. Using this method, we were able to purify the U3 RNP complex containing Myo4, She3, and She2 with a minimal amount of nonspecific proteins (Fig. 2a).

2. Materials

2.1. Subcloning, Yeast Transformation, and Yeast Cell Culture

1. Any available plasmid to subclone and store the U1A repeats. This plasmid is not necessary when making U1A repeats directly with PCR products (see Note 6).

2. Plasmids pRS423 and pRS314 (or pRS304 for integration into yeast chromosome). This can vary depending on plasmid availability.

3. Oligonucleotide primers, U1A-1 and U1A-2 (see Note 7).

4. Yeast strain for transformation. We routinely use wild-type or strains derived from W303 (*MATa ura3-1 trp1-1 leu2-3,112 his3-11 ade2-1 can1-100 GAL*).

5. 20% glucose. Sterilized by autoclaving.

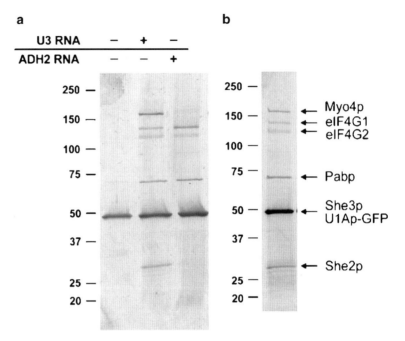

Fig. 2. Purification of U3 RNP complex using the modified tagged RNA system. (a) Extracts prepared from cells expressing U1Ap-GFP-TAP only (lane 1), U1Ap-GFP-TAP and 4xU1A-U3 (lane 2), or U1Ap-GFP-TAP and 4xU1A-*ADH2* (77 NT) (lane 3) were used for TAP purification. The purified complexes were separated on a 4–15% gradient SDS–polyacrylamide gel and silver stained. (b) Protein bands from the purified U3 RNP complex from (a) were excised from gels and identified by mass spectrometry analysis (LC–MS/MS). The modified tagged RNA system purifies a U3 RNP complex that contains the core proteins known to localize *ASH1* mRNA: Myo4, She3 and She2.

6. 20% D-(+)-raffinose. Filter-sterilized.
7. 20% D-(+)-galactose. Filter-sterilized.
8. Selective medium (500 mL): 2% glucose (should be replaced with 2% raffinose for *GAL* induction), 3.35 g yeast nitrogen base without amino acids (Difco), 0.36 g amino acid base (see Note 8), 12.5 mg adenine sulfate, 10 mg uracil, 10 mg L-histidine, 15 mg L-leucine, and 10 mg L-tryptophan. Omit ingredients as required. For plates, add 2% bacto agar.
9. YPD media (500 mL): 2% glucose (should be replaced with 2% raffinose for *GAL* induction), 10 g bacto peptone, 5 g bacto yeast extract, and 12.5 mg adenine sulfate.
10. TE: 10 mM Tris–HCl, pH 7.5, 1 mM EDTA.
11. LiAc/TE: 100 mM LiAc in 1× TE.
12. 10 mg/mL sonicated salmon sperm DNA (Stratagene), denature at 95°C for 5 min prior to use.
13. PEG/LiAc/TE: 40% polyethylene glycol (MW 3,350) and 100 mM LiAc in 1× TE.
14. Dimethyl sulfoxide (DMSO).

2.2. TAP Purification of the RNP Complex	1. Wash buffer: 25 mM Hepes–KOH at pH 7.5, 0.15 M KCl, and 2 mM $MgCl_2$.
2. Extract buffer: 25 mM Hepes–KOH at pH 7.5, 0.15 M KCl, 2 mM $MgCl_2$, 0.1% NP-40, 1 mM DTT, 5 mM ATP, 0.2 mg/mL heparin, 20 mM vanadyl ribonucleoside complexes (Sigma), 0.4 mM AEBSF (4-(2-aminoethyl) benzenesulfonyl fluoride hydrochloride), 2 μg/mL of aprotinin, leupeptin, and pepstatin.
3. HCB: 25 mM Hepes–KOH at pH 7.5, 0.15 M KCl, 2 mM $MgCl_2$, 0.1% NP-40, and 1 mM DTT.
4. TCB: 25 mM Hepes–KOH at pH 7.5, 0.15 M KCl, 0.5 mM EDTA, 0.1% NP-40, and 1 mM DTT.
5. CBB: 25 mM Hepes–KOH at pH 7.5, 0.15 M KCl, 2 mM $MgCl_2$, 1 mM imidazole, 2 mM $CaCl_2$, 0.1% NP-40, and 1 mM DTT.
6. Elution buffer: 50 mM Tris–HCl at pH 8.0, 0.15 M NaCl, 12.5 mM EDTA, and 0.1% SDS. Boil samples at 95°C for 5 min or 65°C for 10 min.
7. Calmodulin elution buffer: 25 mM Hepes, pH 7.5, 0.15 M KCl, 2 mM $MgCl_2$, 1 mM imidazole, 10 mM EGTA, 0.1% NP-40, and 1 mM DTT.
8. Porcelain mortar and pestle (CoorsTek).
9. IgG Sepharose 6 Fast Flow (GE Healthcare), Calmodulin Affinity Resin (Stratagene), AcTEV™ Protease (Invitrogen) or equivalents.

3. Methods

3.1. Preparing the Constructs	Here we describe an experimental scheme to build an RNA construct with 4 repeats of U1A expressed on pRS423 (high-copy 2 μ plasmid, selectable marker *HIS3*), and an RNA-binding construct on pRS314 (low copy CEN/ARS plasmid, selectable marker *TRP1*). In this example, the aptamers are inserted into the *Eco*RI site within the multiple cloning site of pRS423 (see Note 9). It is also possible to vary the number of U1A aptamers by ligating DNA fragments with the desired amount of repeats.
3.1.1. Making the U1A Repeats	1. Synthesize oligonucleotides with a single U1A sequence (see Note 1) or acquire a DNA template that contains U1A. Either can be used as templates for PCR with U1A-1 and U1A-2 primers so that U1A sequence is tagged with 5′-GGG-*Asc*I-*Eco*RI- and -GCGC-*Mfe*I-*Asc*I-CCC-3′ (see Notes 7 and 10).
2. Clean up the PCR, digest with *Asc*I, and ligate the purified 1xU1A into *Asc*I site of an available plasmid. This plasmid

will be used to store the single U1A sequence, thus it is not necessary to meet further requirements other than containing an available *Asc*I site.

3. Transform competent cells and isolate plasmids from colonies that contain 1xU1A.

4. Digest the 1xU1A plasmid with *Mfe*I, treat with alkaline phosphatase (CIP) to prevent self-ligation, and purify the linearized plasmid.

5. Double digest the purified 1xU1A PCR (from step 1) with *Eco*RI and *Mfe*I and purify the digested fragment.

6. Ligate 5'-*Eco*RI-1xU1A-*Mfe*I-3' (from step 5) to the *Mfe*I-cut 1xU1A plasmid (from step 4).

7. Transform competent cells and perform colony PCR with U1A-1 and U1A-2 to select colonies that contain 5'-*Asc*I-*Eco*RI-U1A-GCGC-*Mfe*I/*Eco*RI-U1A-GCGC-*Mfe*I-*Asc*I-3'. Isolate plasmids from colonies that contain 2xU1A.

8. Digest the 2xU1A plasmid with *Mfe*I, treat with alkaline phosphatase (CIP), and purify the linearized plasmid.

9. Double digest the 2xU1A plasmid with *Eco*RI and *Mfe*I, and purify the digested 2xU1A fragment (see Note 11).

10. Ligate 5'-*Eco*RI-2xU1A-*Mfe*I-3' (from step 9) to the *Mfe*I-cut 2xU1A plasmid (from step 8).

11. Transform competent cells and perform colony PCR with U1A-1 and U1A-2 to select colonies that contain 5'-*Asc*I-*Eco*RI-4xU1A-*Mfe*I-*Asc*I-3'. Isolate plasmids from colonies that contain 4xU1A.

12. Double digest the 4xU1A plasmid with *Eco*RI and *Mfe*I, and insert the purified 5'-*Eco*RI-4xU1A-*Mfe*I-3' into the *Eco*RI site of pRS423 (see Note 6).

3.1.2. Gathering the Fragments

1. *The RNA construct*: Each fragment is sequentially inserted into pRS423 using DNA fragments with appropriate restriction enzyme sites linked to both ends via PCR. The restriction map of the final RNA construct that we use is *Xho*I-p*GAL*-*Eco*RI-4xU1A-*Mfe*I/*Eco*RI-*Xma*I-*Bam*HI-*Sac*I-t*ADH1*-*Sac*I. The RNA sequence of interest can then be inserted using the *Xma*I or *Bam*HI site (see Note 12).

2. *The RNA-binding construct*: Each fragment is sequentially inserted into pRS314. The restriction map of the final RNA-binding construct we use is *Kpn*I-p*GPD*-*Xho*I-U1Ap-*Xho*I/*Sal*I-GFP-*Eco*RI-*Pst*I-*Xma*I-*Bam*HI-NLS-*Xba*I-*Sac*I. This arrangement provides several restriction sites downstream GFP, making it possible to insert protein tags for RNP purification. We inserted a TAP tag into the *Bam*HI site of this construct.

3.2. Yeast Transformation

1. Inoculate a single yeast colony in 3 mL of YPD or selective media and grow overnight at 30°C with shaking at 250 rpm.
2. Inoculate the overnight culture into 10 mL of fresh YPD and grow ~4–5 h at 30°C, 250 rpm, until the OD of the culture is ~0.5–0.7 at A_{600}.
3. Collect the cells by centrifugation at $3,000 \times g$ for 5 min at room temperature.
4. Discard the supernatant and resuspend the pellet with 1 mL LiAc/TE. Transfer the resuspended cells into a 1.5-mL tube and collect the cells by briefly spinning at $10,000 \times g$.
5. Resuspend the cells with 0.1 mL LiAc/TE.
6. Add 100 ng plasmid DNA, 50 µg denatured salmon sperm DNA, and 0.6 mL PEG/LiAc/TE to the resuspended cells and mix well by vortexing (see Note 13).
7. Incubate at 30°C for 30 min.
8. Add 70 µL DMSO and mix by inverting.
9. Heat shock at 42°C for 15–20 min.
10. Collect the cells by briefly spinning at $10,000 \times g$ and resuspend the cell pellet with 100 µL sterile water.
11. Plate ~10–50 µL cells on proper solid selective media and incubate plates in 30°C for 2–3 days (see Note 14).
12. Streak the transformed yeast colonies on fresh selective media plates and incubate in 30°C for 1–2 days.

3.3. Expressing the Constructs for Visualization

1. Inoculate cells containing the RNA construct and RNA-binding construct in 3 mL of selective media containing 2% raffinose and 200 mg/L adenine sulfate and grow overnight at 30°C with shaking at 250 rpm (see Notes 15 and 16).
2. Inoculate 10 mL of rich media containing 2% raffinose and 200 mg/L adenine sulfate with the overnight culture so that OD is ~0.5 at A_{600}.
3. Incubate for ~2 h at 30°C with shaking.
4. Add galactose to 1% and further incubate for 1–1.5 h at 30°C with shaking.
5. Collect ~5–10 µL of the cell suspension and examine by fluorescence microscopy.

3.4. RNP Complex Purification

3.4.1. Growing and Harvesting the Cells

1. Inoculate cells containing 4xU1A-RNA and U1Ap-GFP-TAP in 5 mL selective media containing 2% raffinose and grow overnight at 30°C with shaking at 250 rpm.
2. Inoculate 100 mL selective media containing 2% raffinose with the overnight culture and grow another day at 30°C with shaking at 250 rpm.

3. Inoculate prewarmed 1 L rich media containing 2% raffinose with the overnight culture so that OD is ~0.2–0.3 at A_{600} (see Notes 17 and 18).

4. Incubate at 30°C with shaking until OD is ~1.0 at A_{600}.

5. Add galactose to 1% and further incubate for 1.5 h at 30°C with shaking.

6. Collect ~5–10 μL of the cell suspension and briefly examine fluorescence to confirm RNA localization prior to purification.

7. Remove the culture and chill the cells in an ice bath.

8. Centrifuge the chilled cells at $4,000 \times g$ for 10 min at 4°C.

9. Remove the media and resuspend the cell pellet with 10 mL ice-cold wash buffer (see Note 19).

10. Collect the washed cells at $4,000 \times g$ for 10 min at 4°C.

11. Remove the buffer and resuspend the cells in half to equal pellet volume of extract buffer (see Note 20).

12. Slowly pipette the cell suspension from the tip of a pipette into a liquid nitrogen tub (see Note 21).

13. The cells can now be stored in −80°C or used immediately for purification.

3.4.2. Homogenizing the Cells

1. Use liquid nitrogen to precool a clean mortar, pestle, and spatula. This equipment should be kept cool until the lysis has been completed (see Note 22).

2. Pour liquid nitrogen into the mortar and add the frozen cell suspension.

3. Grind the cells into a fine powder. Liquid nitrogen should be occasionally added during lysis to keep the cell lysate cold and prevent the lysate from sticking to the mortar (see Note 23).

4. Collect the lysate with a cold spatula into a 15- or 50-mL conical tube.

3.4.3. TAP Purification

1. Add 1 mL of extract buffer per liter culture and thaw the cell extract.

2. Centrifuge the extract at $4,000 \times g$ for 10 min at 4°C to pellet down unbroken cells and collect the supernatant.

3. Transfer the initial lysate into 1.5- or 2-mL tubes and centrifuge at $16,100 \times g$ for 20 min at 4°C using a refrigerated tabletop centrifuge (see Note 24).

4. Collect the clear cell extract and add 100 μL of IgG sepharose equilibrated with HCB (see Note 25).

5. Incubate mixture for 2 h at 4°C on a rotator.

6. Pellet beads for 1 min at $1,000 \times g$ and remove the supernatant.

7. Wash the beads 3× with 4 mL of HCB and subsequently 1× with 4 mL of TCB.
8. Resuspend the beads with 1 mL TCB and transfer the mixture into a 1.5-mL tube.
9. Pellet beads and resuspend with 0.5 mL TCB containing 50 units of TEV protease.
10. Incubate mixture for 2 h at 16°C (or 4 h at 4°C) on a rotator or roller.
11. After collecting the eluate, rinse the beads 3× with 100 μL of TCB and combine this with the eluate.
12. Adjust the TEV eluate so that the final composition is almost identical to CBB, and add CBB to a total of 1 mL.
13. Add 50 μL of calmodulin affinity resins equilibrated with CBB and incubate for 1.5 h at 4°C on a rotator.
14. Pellet beads and wash 4× with 3 mL CBB.
15. Either elute with fresh elution buffer or calmodulin elution buffer (see Note 26).

4. Notes

1. U1A aptamer sequence (from U1A premRNA): 5′-ACAGCAUUGU-ACCCAGAGUC-UGUCCCAGA-CAUUGCACCU-GGCGCUGU-3′.
2. The use of endogenous promoters does not always lead to native level of expression as it is not guaranteed a single plasmid is present per cell.
3. The overexpression of some transcripts can possibly affect RNA transport or stability. Thus, during pilot experiments, it is advised to perform conventional in situ hybridization to confirm results obtained by the tagged RNA system.
4. In our hands, cells with excessive GFP expression occasionally showed abnormal cell shape and decreased viability. In addition, cells with less bright nuclear GFP fluorescence in the absence of RNA expression also gave better results in RNP purification experiments.
5. In order to purify the RNP complex of interest and obtain a relatively clear protein profile, a minimum of two-step purification would be necessary. The addition of a TAP tag or multiple protein tags adjacent to GFP would fulfill such requirement. However, we have experienced altered efficiencies in purification with the use of particular protein tags. This should be optimized according to the RNP of interest.

6. An alternative method to make the repeats is to directly use the PCR products for digestion and ligation. In this case, the 5′-*AscI*-*Eco*RI-1xU1A-*MfeI*-*AscI*-3′ PCR product is separately digested with *Eco*RI or *Mfe*I, and after eliminating the end fragments, *Eco*RI- and *Mfe*I-digested 1xU1A are ligated. The ligation reaction can now be used as a template for PCR with U1A-1 and U1A-2 to generate 2xU1A. This process is repeated with the 2xU1A fragments to make 4xU1A.

7. U1A-1 (5′ end primer): 5′-GGG-GGCGCGCC-GAATTC-ACAGCATTG-3′, U1A-2 (3′ end primer): 5′-GGG-GGCGCGCC-CAATTG-GCGC-ACAGC-3′. The *Asc*I site in the primers can be modified so that U1A is subcloned into an available plasmid.

8. Amino acid base: 5.0 g L-arginine, 2.5 g L-aspartic acid, 5.0 g L-glutamic acid, 2.5 g L-isoleucine, 5.0 g L-lysine, 2.5 g L-methionine, 2.5 g L-phenylalanine, 2.5 g L-serine, 5.0 g L-threonine, 2.5 g L-tyrosine, and 2.5 g L-valine.

9. It is useful to use the multiple cloning site of a vector to insert the U1A repeats. This facilitates the subsequent addition of p*GAL* and t*ADH1* and provides ample restriction sites for subcloning the RNA sequence of interest.

10. All PCRs should be performed with a high-fidelity polymerase.

11. If the plasmid is significantly larger than the 2xU1A insert, it is more practical to perform PCR using the 2xU1A plasmid with U1A-1 and U1A-2 and double digest the PCR product rather than directly use the plasmid for digestion.

12. It is also possible to subclone the RNA sequence ligated to the *ADH1* terminator. This can assist subcloning when the RNA sequence is very short. To obtain the RNA-t*ADH1* fragment, primers should be synthesized so that the 3′ end of the RNA sequence and the 5′ end of t*ADH1* contain identical (or compatible) restriction sites (A), and the remaining ends have restriction sites opted for subcloning. After digesting the PCR fragments of the RNA and t*ADH1* with enzyme A, these two fragments are purified, ligated, and the ligation reaction is used for PCR templates with the remaining outside primers.

13. When using an integrating plasmid, at least 1 μg of linearized plasmid should be used for transformation.

14. To simultaneously transform yeast cells with the RNA construct and the RNA-binding construct, both constructs are mixed together for transformation and the transformed cells are plated onto the appropriate double selective medium. In most cases, the total transformation reaction (100 μL) is

spread onto plates to obtain a reasonable number of colonies from dual transformations. In addition, it may require a slightly longer incubation in 30°C (~3 days) in order to detect full-size yeast colonies.

15. If cell growth is poor in selective media with raffinose, as an alternative, cells can be grown overnight in selective media with glucose and washed twice in rich media with raffinose prior to subsequent inoculation. However, as p*GAL* is strongly repressed by growth in glucose, it is important to completely eliminate the glucose from the overnight culture to ensure RNA expression. When using endogenous promoters, all cultures are grown in glucose media.

16. For experiments involving visualization of GFP fluorescence in *ade2* yeast, background fluorescence is often reduced by growing cells in high concentrations of adenine sulfate.

17. During pilot experiments, purifications using 500 mL cultures gave sufficient amount of U3 bound proteins for silver staining in our hands. After the initial 1 L culture, the culture volume should be scaled up or down depending on the type of analysis to be done after purification.

18. The time required for cell growth from OD 0.2 to 1.0 at A_{600} in raffinose media may vary from 5 up to 10 h. The OD at A_{600} of the initial inoculation should be increased in the case of prolonged incubations.

19. The wash buffer and extract buffer should be chilled until the solution is at 4°C prior to use. Once the cell pellet has been resuspended in wash buffer, the cells can be transferred to a smaller (15 or 50 mL) centrifuge tube.

20. The composition of the extract buffer should be adjusted according to the characteristics of the RNP complex of interest. Adequate RNase and protease inhibitors should always be included to prevent degradation.

21. Small balls will form by slowly adding the cell suspension into liquid nitrogen. It is easier to lyse cells in this form than using clumps of frozen cells.

22. A steel blender or different type of homogenizer can also be used for lysis. However, it is essential that the homogenizer is RNase-free and kept at a frozen temperature during lysis. In our hands, the mortar and pestle gave best yields for purification.

23. When lysing larger amount of cells, it is best to grind the cells in multiple proportions rather than all in once to achieve an appreciable efficiency.

24. In some cases, it may be necessary to ultracentrifuge the initial supernatant at $100,000 \times g$ for 1 h and remove the lipid layer and cell debris to improve purification efficiency.

25. We have obtained U3 RNP profiles with a minimum amount of nonspecific proteins by batch purification. Results may differ when performing column purification.

26. If the protein components of the RNP complex are to be analyzed, elution should be performed with 1× SDS–PAGE sample buffer or a mild elution buffer as that used here to maximize the yield of the purified complex. Depending on how the RNP complex is to be further analyzed, the RNP complex can also be eluted by incubating the washed beads in CEB. However, this usually results in a significantly lower yield.

References

1. Du, T. G., Schmid, M., and Jansen, R. P. (2007) Why cells move messages: the biological functions of mRNA localization. *Semin. Cell Dev. Biol.* **18**, 171–177.
2. Kloc, M., Zearfoss, N. R., and Etkin, L. D. (2002) Mechanisms of subcellular mRNA localization. *Cell* **108**, 533–544.
3. St Johnston, D. (2005) Moving messages: the intracellular localization of mRNAs. *Nat. Rev. Mol. Cell Biol.* **6**, 363–375.
4. Jambhekar, A., and Derisi, J. L. (2007) Cis-acting determinants of asymmetric, cytoplasmic RNA transport. *RNA* **13**, 625–42.
5. Martin, K.C., and Ephrussi, A. (2009) mRNA localization: gene expression in the spatial dimension. *Cell* **136**, 719–730.
6. Bullock, S.L. 2007. Translocation of mRNAs by molecular motors: think complex? *Semin. Cell Dev. Biol.* **18**, 194–201.
7. Gonsalvez, G. B., Urbinati, C. R., and Long, R. M. (2005) RNA localization in yeast: moving towards a mechanism. *Biol. Cell* **97**, 75–86.
8. Chartrand, P., Meng, X. H., Singer, R. H., and Long, R. M. (1999) Structural elements required for the localization of *ASH1* mRNA and of a green fluorescent protein reporter particle in vivo. *Curr. Biol.* **9**, 333–336.
9. Jansen, R. P., Dowzer, C., Michaelis, C., Galova, M., and Nasmyth, K. (1996) Mother cell-specific HO expression in budding yeast depends on the unconventional myosin myo4p and other cytoplasmic proteins. *Cell* **84**, 687–697.
10. Böhl, F., Kruse, C., Frank, A., Ferring, D., and Jansen, R. P. (2000) She2, a novel RNA-binding protein tethers *ASH1* mRNA to the Myo4-myosin motor via She3. *EMBO J.* **19**, 5514–5524.
11. Long, R.M., Gu, W., Lorimer, E., Singer, R. H., and Chartrand, P. (2000) She2p is a novel RNA-binding protein that recruits the Myo4p-She3p complex to *ASH1* mRNA. *EMBO J.* **19**, 6592–6601.
12. Takizawa, P. A., and Vale, R. D. (2000) The myosin motor, Myo4p, binds Ash1 mRNA via the adapter protein, She3p. *Proc. Natl. Acad. Sci. USA* **97**, 5273–5278.
13. Boelens, W. C., Jansen, E. J., van Venrooij, W. J., Stripecke, R., Mattaj, I. W., and Gunderson, S. I. (1993) The human U1 snRNP-specific U1A protein inhibits polyadenylation of its own pre-mRNA. *Cell* **72**, 881–892.
14. Scherly, D., Boelens, W., van Venrooij, W. J., Dathan, N. A., Hamm, J., and Mattaj, I. W. (1989) Identification of the RNA binding segment of human U1A protein and definition of its binding site on U1 snRNA. *EMBO J.* **8**, 4163–4170.
15. Kretzner, L., Krol, A., and Rosbash, M. (1990) *Saccharomyces cerevisiae* U1 small nuclear RNA secondary structure contains both universal and yeast specific domains. *Proc. Natl. Acad. Sci. USA* **87**, 851–855.
16. Liao, X. C., Tang, J., and Rosbash, M. (1993) An enhancer screen identifies a gene that encodes the yeast U1 snRNP A protein: implications for snRNP protein function in pre-mRNA splicing. *Genes Dev.* **7**, 419–428.
17. Bertrand, E., Chartrand, P., Schaefer, M., Shenoy, S. M., Singer, R. H., and Long, R. M. (1998) Localization of *ASH1* mRNA particles in living yeast. *Mol. Cell* **2**, 437–445.
18. Keryer-Bibens, C., Barreau, C., and Osborne, H. B. (2008) Tethering of proteins to RNAs by bacteriophage proteins. *Biol. Cell* **100**, 125–138.
19. Klein Gunnewiek, J. M. T., Hussein, R. I., van Aarssen, Y., Palacios, D., de Jong, R., van Venrooij, W. J., and Gunderson, S. I. (2000) Fourteen residues of the U1 snRNP-specific U1A protein are required for homodimerization,

cooperative RNA binding, and inhibition of polyadenylation. *Mol. Cell Biol.* **20**, 2209–2217.

20. Coller, J., and Wickens, M. (2007) Tethered function assays: an adaptable approach to study RNA regulatory proteins. *Methods Enzymol.* **429**, 299–321.

21. Puig, O., Caspary, F., Rigaut, G., Rutz, B., Bouveret, E., Bragado-Nilsson, E., Wilm, M., and Séraphin, B. (2001) The tandem affinity purification (TAP) method: a general procedure of protein complex purification. *Methods* **24**, 218–229.

22. Nagai, K. (1996) RNA-protein complexes. *Curr. Opin. Struct. Biol.* **6**, 53–61.

23. Straight, A. F., Sedat, J. W., and Murray, A. W. (1998) Time-lapse microscopy reveals unique roles for kinesins during anaphase in budding yeast. *J. Cell Biol.* **143**, 687–694.

# Chapter 15

Visualizing Endogenous mRNAs in Living Yeast Using m-TAG, a PCR-Based RNA Aptamer Integration Method, and Fluorescence Microscopy

Liora Haim-Vilmovsky and Jeffrey E. Gerst

Abstract

Localized mRNA translation is involved in cell-fate determination, polarization, and morphogenesis in eukaryotes. While various tools are available to examine mRNA localization, no easy and quick method has allowed for the visualization of endogenously expressed mRNAs in vivo. We describe a simple method (m-TAG) for PCR-based chromosomal gene tagging that uses homologous recombination to insert binding sites for the RNA-binding MS2 coat protein (MS2-CP) between the coding region and 3′-untranslated region of any yeast gene. Upon co-expression of MS2-CP fused with GFP, specific endogenously expressed mRNAs can be visualized in vivo for the first time. This method allows for the easy examination of mRNA localization using fluorescence microscopy and leaves the yeast cells amenable for further genetic analysis.

Key words: mRNA, MS2, Green fluorescent protein, Gene tagging, Cre recombinase, Yeast, *Saccharomyces cerevisiae*

1. Introduction

RNA localization and local translation is one mechanism by which proteins can be sorted asymmetrically to generate cellular polarity, morphogenesis, and cell-fate determination (1–3). This mechanism is used by a variety of organisms such as yeast, insects (*Drosophila*), amphibians (*Xenopus*), and mammals (rodents) (1, 3). However, the localization of most mRNAs in cells is still undetermined because of the use of complicated and time-consuming procedures and methodologies. Here, we describe a simple gene-tagging procedure, called m-TAG, which allows the sustained visualization of endogenous mRNAs in living yeast (4, 5).

In the m-TAG procedure (illustrated in Fig. 1), a template cassette containing a selection marker flanked on both sides by loxP sites, as well as binding sites (e.g., MS2 loops; MS2L) for the RNA-binding MS2 coat protein (MS2-CP) is amplified by PCR. The primers used for PCR contain sequences homologous to the template cassette as well as either the end of the open reading frame (ORF) of interest (e.g., the forward primer) or the beginning of its corresponding 3′-untranslated region (3′UTR) (e.g., the reverse primer). The PCR product is then transformed into yeast and the cassette is inserted into the genome by homologous recombination. This PCR-based strategy has been used before to create large-scale yeast libraries, including deletion (6), GFP- and epitope-tagged protein libraries (7, 8). However, unlike these other tagging strategies, the selection marker (used for genomic integration) is removed by Cre recombinase expression, and therefore the 3′UTR remains associated with the ORF. As the 3′UTR is often necessary for mRNA localization and stability, m-TAG allows for the proper intracellular targeting of mRNA. Finally, MS2-CP fused with GFP(×3) is expressed in cells bearing the MS2L-tagged mRNA and allows for the formation of fluorescent RNA granules and visualization of the endogenous mRNA in living yeast.

2. Materials

2.1. DNA Amplification

1. Design primers specific to the gene of interest to amplify the pLOXHIS5MS2L plasmid (4, 5).

    Primer GENE-Tag F: contains a sequence of 40 nucleotides corresponding to the 3′-end of the ORF (including the stop codon) followed by a 21-nucleotide sequence corresponding to the template plasmid pLOXHIS5MS2L, 5′-AACGCTGCAGGTCGACAACCC-3′.

    Primer GENE-Tag R: contains the reverse complement of the sequence corresponding to the first 40 nucleotides of the 3′UTR (immediately after the stop codon) followed by a 20-nucleotide sequence corresponding to the template plasmid pLOXHIS5MS2L, 5′-GCATAGGCCACTAGTGGATC-3′.

    Use polyacrylamide gel electrophoresis (PAGE)-purified primers.

2. pLOXHIS5MS2L plasmid (available from authors upon request).

2.2. DNA Precipitation

1. Ethanol 70%: Mix 0.7× volumes of ethanol with 0.3× volumes of double-distilled water (DDW). Store at 4°C.

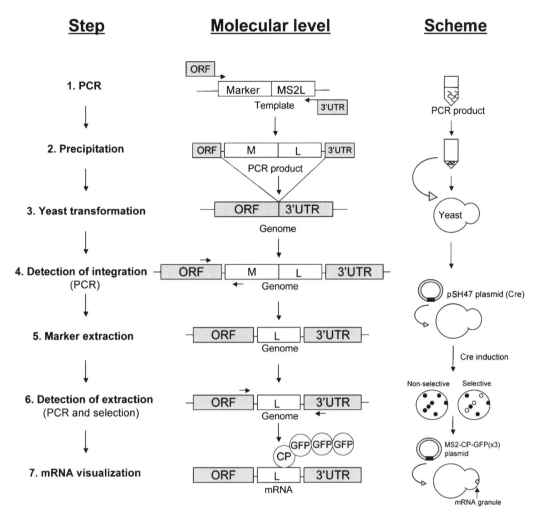

Fig. 1. An illustration of m-TAG procedure. Each step (numbered corresponding to the steps in the protocol) is illustrated at two levels: the molecular level and a scheme according to the procedure. (*Step 1*) *PCR*: Forward and reverse primers having homology to the coding region (open reading frame; ORF) and 3′UTR of the gene of interest, are used to amplify the template cassette by PCR. The template cassette contains 12 MS2 loop sequences (MS2L; L) and a selectable marker (Marker; M) flanked by loxP sites (not shown). The PCR product is placed in a tube. (*Step 2*) *Precipitation*: The PCR product containing the template cassette flanked by regions homologous to the gene of interest is concentrated into a small volume of DDW. (*Step 3*) *Yeast transformation*: The PCR product is transformed into yeast. Homologous recombination results in integration of the cassette into the gene between its ORF and 3′UTR. (*Step 4*) *Detection of integration*: Correct MS2 loop-integrated yeast are detected by PCR using the GENE-Det F and HIS3-Det R primers (shown as *small arrows*). (*Step 5*) *Marker extraction*: MS2 loop-integrated yeast are transformed with the pSH47 plasmid (designated by an annulus), which expresses Cre recombinase. Yeast cells are grown on galactose-containing medium to induce Cre recombinase expression, which results in excision of the selection marker. (*Step 6*) *Detection of marker extraction*: Recombinant yeast lacking the marker are identified by replica plating from a non-selective plate (viable yeast colonies are indicated by *black dots*) onto selective media (yeast colonies that did not grow on selective media are indicated as *open circles*). Recombinant yeast cells unable to grow on the selective plate are then verified for loss of the selection marker by PCR, using the GENE-Det F and GENE-Det R primers (shown as *small arrows*). (*Step 7*) *mRNA visualization*: Recombinant yeast cells are transformed with the pMS2-CP-GFP(×3) plasmid (designated by an annulus). MS2-CP-GFP(×3) protein binds to the MS2 loops in the mRNA. The transformed yeast cells are examined by fluorescence microscopy to visualize mRNA granules.

2.3. Yeast Growth and Transformation

1. Yeast strains: *Saccharomyces cerevisiae* lab strain, which is mutated in the *HIS3* and *URA3* genes and wild-type for the *MET* genes.

2. YPD medium: 1% (wt/vol) Bacto yeast Extract, 2% (wt/vol) Bacto Peptone, and 2% (wt/vol) d-Glucose. Autoclave and cool before use. For plates add 2% (wt/vol) bacteriological agar.

3. SC medium: The synthetic medium is prepared essentially according to the protocol of Rose et al. (9). Briefly, for 1 l of medium, add 7 g of Synthetic Dry Mix (mix composed of: 294 g ammonium sulfate; 30 g dibasic potassium phosphate; 0.3 g each of arginine, cysteine, and proline; 0.45 g each of isoleucine, lysine, and tyrosine; 0.75 g each of glutamic acid, phenylalanine, and serine; 1.0 g each of aspartate, threonine, and valine; and 100 g yeast nitrogen base lacking ammonium sulfate and amino acids) with 850 ml DDW, followed by 350 µl of 10 N NaOH, and in case of liquid medium, add either 20 g glucose or 35 g galactose (as required), followed by stirring and autoclaving in a large 2–3 l Ehrlenmeyer flask. After cooling to room temperature, 10 ml of a sterile-filtered 100× amino acid stock solution (see below) is added and mixed thoroughly. In the case of plates, 20 g bacteriological agar, instead of the sugar, should be added to the mixture prior to autoclaving. After cooling the autoclaved mixture to 55°C in a water bath, 100 ml of either prewarmed sterile 20% glucose or 35% galactose solution is then added, as required, along with 10 ml of a sterile-filtered 100× amino acid stock solution. Mix thoroughly and wait 30 min at 55°C for the bubbles to disappear, and pour into plates while warm. The 100× amino acid stock is composed of up to six amino acids/bases (e.g., adenine, histidine, leucine, methionine, tryptophan, and uracil) for the preparation of specific selective media. For example, synthetic complete medium (SC) contains all six, while synthetic medium lacking uracil (SC-U) would contain all but uracil, etc. The 100× amino acid stock solution is prepared by first adding 0.4 g of each amino acid/base required to 150 ml DDW, followed by the addition of 3 ml concentrated HCl while stirring; bringing the volume up to 200 ml; and then performing sterile filtration. Liquid media and plates can be stored at room temperature for up to 2 months. The 100× amino acid stock is stored up to 1 year at 4°C; while the Synthetic Dry Mix can be stored up to 1 year at room temperature.

4. Tris–EDTA (TE): 10 mM Tris–HCl and 1 mM EDTA in DDW at pH 7.5.

5. LiOAc 1 M: Add 51 g LiOAc to 400 ml DDW; titrate to pH 7.5 with 2 M acetic acid; and fill to 500 ml.

6. LiOAc 0.1 M: Mix 0.1× volumes of 1 M LiOAc with 0.9× volumes of TE.
7. PEG 50%: 50% PEG 3350 (wt/vol) in TE.
8. PEG 45%: Mix 0.9× volumes of 50% PEG 3350 with 0.1× volumes of 1 M LiOAc.
9. ssDNA 5.0 mg/ml: Sheared, organic extracted (i.e., phenol:chloroform (1:1), chloroform), and denatured ssDNA prepared according to standard procedures.

2.4. Detection of Proper Integration into Genome

1. Zymolase solution: Zymolase 10 mg/ml, in DDW. Store in −20°C for up to 2 months.
2. Zymolase buffer: 50 mM NaCl, 10 mM Tris–HCl, 10 mM $MgCl_2$, and 1 mM dithiothreitol (pH 7.9). Store in −20°C for up to 2 months.
3. Design a GENE-Tag F primer specific to the gene of interest. This forward primer should include 18 nucleotides corresponding to ~200 bp upstream to the termination codon of the ORF.
4. Synthesize the HIS5-Det R primer: 5′-GACTGTCAAGG AGGGTATTCTG-3′.

2.5. Marker Extraction via Cre-Mediated Recombination

1. Plasmid pSH47 (EUROSCARF, accession no. P30119).
2. SC medium containing galactose: replace 2% (wt/vol) glucose with 3.5% (wt/vol) galactose.

2.6. Detection of Proper Extraction by PCR

1. Design a GENE-Tag R primer specific to the gene of interest. This reverse primer should include 18 nucleotides corresponding to ~200 bp downstream from the beginning of the 3′UTR.

2.7. mRNA Visualization In Vivo

1. Plasmid pMS2-CP-GFP(×3) (available from authors upon request).

2.8. mRNA Visualization in Fixed Cells

1. Paraformaldehyde solution – Dissolve paraformaldehyde 4% (wt/vol) and sucrose 4% (wt/vol) in DDW. Filter sterilize and store at 4°C for up to 1 month.
2. 1 M KH_2PO_4 in DDW (filter sterilize).
3. 1 M K_2HPO_4 in DDW (filter sterilize).
4. 1 M Potassium phosphate buffer at pH 7.5 – Mix 0.834× volumes of 1 M K_2HPO_4 with 0.166× volumes of 1 M KH_2PO_4. Prepare under sterile conditions.
5. Sorbitol 2 M–36.4% sorbitol (wt/vol) in DDW. Filter sterilize.
6. KPO_4/Sorbitol solution – Mix 0.6× volumes of 2 M Sorbitol with 0.1× volumes of 1 M Potassium phosphate buffer and 0.3× volumes of DDW. Prepare under sterile conditions.

3. Methods

3.1. DNA Amplification

1. Prepare the PCR reaction in 200 μl reaction volume separated into 50 μl aliquots in four separate PCR tubes. PCR is preformed according to an optimized protocol with a high-fidelity polymerase suited for long templates under standard conditions recommended by the manufacturer (i.e., we use Bio-X-ACT Long DNA Polymerase). Final concentrations: PCR buffer 1×, 5 mM $MgCl_2$, 200 μM dNTPs, 200 μM each primer, DNA 50 ng, and 0.08 U/ml of *taq* polymerase. Prepare a control reaction that lacks the DNA template (see Note 1).

2. Amplify the template plasmid using the following cycling conditions:

   Start – 95°C for 3 min

   Melting temp – 95°C for 1 min

   Annealing temp – 75°C for 1 min

   Elongation temp – 68°C for 2.5 min

   Cycles: 40

3. Load 5 μl of the PCR reactions onto a 1% (wt/vol) agarose gel in TAE buffer; electrophorese at 120 V for ~20 min.

3.2. DNA Precipitation

1. Combine the PCR products for each separate gene into one tube.

2. Add 0.75 ml of 100% ethanol to the PCR product. Incubate for 15 min at –80°C or 1 h at –20°C.

3. Pellet at 20,000 × g for 10 min at 4°C and discard the supernatant.

4. Wash with 70% ethanol and spin at 20,000 × g for 1 min at 4°C.

5. Discard the supernatant and let the pellet dry completely at room temperature.

6. Resuspend the pellet in 8 μl of DDW. DNA can be stored at –20°C for up to 6 months or more.

3.3. Yeast Transformation

1. Inoculate an overnight yeast culture in YPD (see Note 2).

2. Dilute cells to OD = 0.2 at A_{600} and grow cells to OD = 0.6–0.8 at A_{600}. Typically, one uses 10 OD units at A_{600} per transformation (see Note 3).

3. Spin down cells at 1,000 × g for 3 min. Decant the supernatant and resuspend cells in 10 ml TE.

4. Harvest cells by centrifugation at 1,000 × g for 3 min. Discard the supernatant and resuspend in 10 ml of 0.1 M LiOAc.

5. Incubate the cells with continuous gentle shaking for 10 min at 26–30°C.

6. Spin down cells at 1,000×g for 3 min. Decant the supernatant and resuspend the cells in a minimal volume of 0.1 M LiOAc.

7. Add 10 µl of 5.0 mg/ml sterile, sonicated, and denatured ssDNA to microtubes. Also add ssDNA in control tubes, but without PCR product.

8. Add 8 µl of PCR product DNA to each tube.

9. Add 50 µl (10 OD_{600} units) of cells to the tube and mix.

10. Incubate for 10 min at room temperature.

11. Add 0.5 ml of 45% PEG 3350 and mix by vortexing.

12. Incubate at 26–30°C for 45 min.

13. Heat shock the cells for 5 min at 42°C.

14. Spin down cells at 1,000×g for 3 min. Decant the supernatant and resuspend the cells with 100 µl of sterile TE. Repeat once.

15. Plate cells onto the selective synthetic media lacking histidine.

16. Incubate plates at 26–30°C for 3 days. After growth, yeast can be stored at 4°C for about 2 weeks.

3.4. Detection of Proper Integration by PCR

1. Patch out ten individual colonies (and a wild-type strain as a control) on the plate and incubate at 26–30°C for 1 day.

2. Mix 16 µl of DDW, 2 µl of Zymolase buffer and 2 µl of Zymolase solution in a tube, for each colony patched.

3. Swirl yeast from a patch into the solution using a sterile toothpick or a sterile tip.

4. Incubate for 1 h at 37°C. This generates yeast spheroplasts.

5. Prepare the PCR reaction in 20 µl reaction volume according to optimized protocol with a standard *taq* polymerase (i.e., we used Taq Mix Purple). Final concentrations: buffer 1×, 1.5 mM $MgCl_2$, 200 µM dNTPs, 0.25 µM GENE-Det F primer, 0.25 µM HIS5-Det R primer, and 0.1 U/µl of *taq* polymerase. Add 2 µl of the spheroplasted cells. Prepare a control reaction that contains wild-type DNA.

6. Amplify the template plasmid using the following cycling conditions:

   Start – 95°C for 8 min

   Melting temp – 95°C for 45 s

   Annealing temp – 58°C for 1 min

   Elongation temp – 72°C for 1 min

   Cycles: 30

7. Load 10 µl of the PCR reactions onto a 1% (wt/vol) agarose gel in TAE buffer; electrophorese at 120 V for ~20 min. The wild-type strain should not produce a PCR product, while correct integration at

the genomic locus will generate a PCR product of 400–500 bp length, depending on the distance that the GENE-Det F primer recognizes in the ORF (see Notes 4 and 5).

3.5. Marker Extraction via Cre-Mediated Recombination

1. Transform at least two integrated individual colonies with the pSH47 plasmid (see Note 6). Plate the cells onto double selective media lacking histidine and uracil (see Note 7).
2. Patch out individual yeast colonies and incubate at 26–30°C for 1 day.
3. Inoculate yeast cells in 5 ml selective medium containing galactose as a carbon source, but lacking uracil and grow at 26–30°C to an OD of 0.5–1.5 at A_{600} (overnight) (see Note 8).
4. Dilute 5 μl of cells in 500 μl TE.
5. Plate 5 μl of diluted cells onto a YPD plate. Incubate at 26–30°C for 2 days.
6. Patch out individual yeast colonies onto a fresh YPD plate and incubate at 26–30°C for 1 day.
7. Replica the YPD plate onto a plate containing selective medium lacking histidine and a fresh YPD plate. Incubate at 26–30°C for 1 day. Integrated yeast colonies that have lost the selection marker will not grow on plates lacking histidine.

3.6. Detection of Proper Extraction by PCR

1. For each colony patched, mix 16 μl of DDW, 2 μl of Zymolase buffer, and 2 μl of Zymolase solution in a tube.
2. Swirl yeast from a patch into the solution. Use a wild-type strain as a control.
3. Incubate at 37°C for 1 h.
4. Prepared the PCR reaction in 20 μl reaction volume according to optimized protocol with a standard *taq* polymerase (i.e., Taq Mix Purple). Final concentrations: buffer 1×, 1.5 mM $MgCl_2$, 200 μM dNTPs, 0.25 μM GENE-Det F primer, 0.25 μM GENE-Det R primer, and 0.1 U/μl of *taq* polymerase. Add 2 μl of the spheroplasted cells. Prepare a control reaction that contains wild-type DNA.
5. Amplify the template plasmid using the following cycling conditions:

   Start – 95°C for 8 min

   Melting temp – 95°C for 45 s

   Annealing temp – 58°C for 1 min

   Elongation temp – 72°C for 1 min

   Cycles: 30
6. Load 10 μl of the PCR reactions onto a 1% (wt/vol) agarose gel in TAE buffer; electrophorese at 120 V for ~20 min.

The recombinant colonies should yield a PCR product larger by ~800 bp than that obtained from the wild-type cells (see Note 9).

3.7. mRNA Visualization In Vivo

1. Transform an integrated strain with the pMS2-CP-GFP(×3) plasmid. Plate the cells onto selective media lacking histidine.
2. Grow cells in 5 ml media lacking histidine at 26–30°C to an OD of 0.5–0.8 at A_{600}.
3. Spin down cells at $1,000 \times g$ for 3 min. Decant the supernatant and wash the cells with 5 ml of medium lacking methionine. Repeat once (see Note 10).
4. Incubate cells at 26–30°C for 1 h.
5. Place 3 µl of cells on a slide and cover with a coverslip.
6. Observe the cells using fluorescence microscopy (see Note 11). m-TAG integrated strains typically yield small (~50–300 nm) fluorescent granules that vary in both intensity and number. For live cell imaging, RNA granules can be easily visualized by confocal microscopy (preferentially with a 100× objective) or for time-lapse using a DeltaVision system.

3.8. mRNA Visualization in Fixed Cells

Yeast cells can also be fixed and then imaged, according to the following protocol.

Continue from step 4, Subheading 3.7.

1. Spin down cells at $1,000 \times g$ for 3 min. Decant the supernatant and resuspend the cells in 100 µl of paraformaldehyde solution.
2. Incubate tubes at 26°C for 15 min.
3. Spin down cells at $1,000 \times g$ for 3 min. Decant supernatant and wash cells with 0.5 ml KPO_4/sorbitol solution.
4. Spin down cells at $1,000 \times g$ for 3 min. Decant supernatant and resuspend cells in 100 µl KPO_4/sorbitol solution. Fixed yeast can be stored in the dark at 4°C for up to 1 month. Continue to step 5, Subheading 3.7.

4. Notes

1. High Mg^{2+} concentration (5 mM) and annealing temperature of 75°C are critical for success of the PCR reaction.
2. The yeast strain used for integration should be mutated in the *HIS3* and *URA3* genes, to allow the selection for transformed cells containing plasmids and PCR product having the *SpHis5/HIS3* and *URA3* markers. In addition, the yeast strain should be wild-type for the *MET* gene, since it will be

grown in medium lacking methionine to induce the MS2-CP-GFP (×3) expression.

3. Make sure that the yeast cells are in log phase before transformation. Use yeast cells grown to no more than OD of 1 at A_{600} and use at least 10 OD_{600} units yeast per transformation. To increase the transformation yield use well sheared, organic-extracted, and denatured ssDNA.

4. In the case of no correct integrated colonies (i.e., integration occurred elsewhere in the genome), use a longer homologous sequence (corresponding to the ORF or 3′UTR) of ~50 bp in the GENE-Tag primers. This incorrect integration may occur because of either repetitive or AT-rich sequences contained in the region homologous to the gene of interest (e.g., the 40 nucleotides corresponding to the gene) in the GENE-Tag primers.

5. The PCR can alternatively be performed with the GENE-Det-R and MS2L-Det-F (5′-GCTGGTCGCTATACTGCTG-3′) primers. This PCR reaction will reflect the number of MS2 loops left after integration and can detect instances of recombination within the repetitive loop sequence. In case of no recombination events in the MS2 loops, the expected size of the product is ~800 bp, depending upon where the GENE-Det-R primer recognizes to the 3′UTR.

6. It is important to proceed with at least two successfully integrated colonies, since internal recombination within the MS2 loop repeats may occur during the process.

7. This step can be performed at the beginning of the procedure, before transformation of the PCR product.

8. Yeast cells may grow slower on galactose-containing medium, therefore, yeast cells should be grown for 48 h.

9. In the case of a PCR product that differs by less than ~800 bp length, relative to a PCR product obtained from wild-type cells, use a different integrated colony. This may occur because of recombination within MS2 loop sequence (a repetitive sequence).

10. This step is optional, since the *MET25* promoter in pMS2-CP-GFP(×3) is leaky. Thus, MS2-CP-GFP(×3) expression without induction may be sufficient for RNA granule visualization.

11. Low MS2-CP-GFP(×3) expression can result in a weak GFP signal. Therefore, yeast cells can be grown in medium lacking methionine for 1–2 h. If the GFP signal is high, but no granules are seen, it may result from low levels of mRNA transcription. In order to induce mRNA expression, use different growth conditions, according to the gene of interest (i.e., growth on different nutrient sources, starvation conditions).

Acknowledgements

This work was supported by grants to J.E.G. from the Minerva Foundation, Germany, and the Y. Leon Benoziyo Center for Molecular Medicine, Center for Scientific Excellence, and Kahn Center for Systems Biology, Weizmann Institute of Science. J.E.G. holds the Besen-Brender Chair in Microbiology and Parasitology.

References

1. Bashirullah, A., Cooperstock, R. L., and Lipshitz, H. D. (1998) RNA localization in development *Annu. Rev. Biochem.* **67**, 335–394.
2. Gonsalvez, G. B., Urbinati, C. R., and Long, R. M. (2005) RNA localization in yeast: moving towards a mechanism *Biol. Cell* **97**, 75–86.
3. Kloc, M., Zearfoss, N. R., and Etkin, L. D. (2002) Mechanisms of subcellular mRNA localization *Cell* **108**, 533–544.
4. Haim, L., Zipor, G., Aronov, S., and Gerst, J. E. (2007) A genomic integration method to visualize localization of endogenous mRNAs in living yeast *Nat. Methods* **4**, 409–412.
5. Haim-Vilmovsky, L., and Gerst, J. E. (2009) m-TAG: a PCR-based genomic integration method to visualize the localization of specific endogenous mRNAs in vivo in yeast *Nat. Protoc.* **4**, 1274–1284.
6. Giaever, G., Chu, A. M., Ni, L., Connelly, C., Riles, L., et al. (2002) Functional profiling of the Saccharomyces cerevisiae genome *Nature* **418**, 387–391.
7. Ghaemmaghami, S., Huh, W. K., Bower, K., Howson, R. W., Belle, A., et al. (2003) Global analysis of protein expression in yeast *Nature* **425**, 737–741.
8. Huh, W. K., Falvo, J. V., Gerke, L. C., Carroll, A. S., Howson, R. W., et al. (2003) Global analysis of protein localization in budding yeast *Nature* **425**, 686–691.
9. Rose, M. D., Winston, F., Hieter, P. (1990) *Methods in Yeast Genetics, A Laboratory Course Manual*. Cold Spring Harbor Laboratory Press, Cold Spring Harbor, New York.

Chapter 16

Imaging mRNAs in Living Mammalian Cells

Sharon Yunger and Yaron Shav-Tal

Abstract

The gene expression pathway begins in the nucleus as a gene receives a cue to transcribe, and typically ends in the cytoplasm with the production of the required protein. The nuclear processes of mRNA transcription and nucleo-cytoplasmic transport are of high importance as they encompass the major control points of gene expression. While it has been possible to study the mRNA life cycle using biochemical and molecular biology approaches, the advent of methods for nucleic acid tagging in vivo, have opened up many possibilities for examining these processes in vivo. In this chapter we describe the methodology required for setting up a live-cell system for monitoring real-time mRNA dynamics in mammalian cells.

Key words: Transcription, Live-cell imaging, RNA dynamics, Nucleus

1. Introduction

Imaging-based kinetic approaches allow for probing of the nuclear compartment and underlying gene expression taking place within the natural surroundings of the living cell (1–4). Major topics of study focus on the in vivo dynamics of mRNA transcription and mRNA nucleo–cytoplasmic transport. Using a cell system in which major elements of the gene expression pathway such as DNA, mRNA, and protein were fluorescently tagged and visualized in real-time, it was possible to detect a specific gene locus, its transcribed mRNA, and the translated protein product in single living cells (5). A commonly used method for following mRNA transcripts in living cells is based on the integration of 24 MS2 sequence repeats into the gene construct, which are then transcribed as part of the 3′UTR of the gene of interest (GOI) (6). The MS2 sequence repeats originate from the MS2 bacteriophage system and therefore do not interfere or interact with any mammalian components. When these sequences are transcribed as part of mRNA transcripts, they form 24 stem-loop structures in the mature mRNA molecules. These secondary structures are specifically bound by dimers of a fluorescently tagged MS2 coat protein (MS2-CP-XFP) protein. The binding of the MS2-CP protein to the MS2 stem-loop in the mRNA is extremely efficient and specific, thereby fulfilling the task of tagging mRNA in vivo.

In this fashion it was possible to track and analyze the nucleoplasmic movements of individual mRNA–protein complexes (mRNPs) tagged with a YFP-MS2-CP protein, hence showing that mRNPs move through the nucleoplasm by a random, diffusion-based process (7). By utilizing this cellular system it has been possible to dissect the different steps undertaken by RNA polymerase II (Pol II) while transcribing a specific gene in a living cell (8). Interpretation of time-resolved fluorescence photobleaching and photoactivation data, and comparison of the experimental data to mathematical mechanistic models formalized by a series of differential equations, has led to interesting biological insights. Measuring the recruitment and residency times of RNA Pol II on an integrated transcription unit yielded the first in vivo measurements of Pol II recruitment and abortive initiation, two processes that were previously only indirectly observed. Furthermore, not only could the enzymatic kinetics of elongation be quantified but also the measurement of the kinetics of mRNA product formation in vivo, using the MS2-CP-GFP technology (7). Quantification of these reactions provided sufficient resolution for quantifying two important parameters of the transcription process: polymerase elongation speed and pausing time. This analysis demonstrated that RNA Pol II can proceed at speeds of up to 4 kb/min, much faster than detected by biochemical assays (~1 kb/min) (8, 9). Recent assays have corroborated this finding, showing that transcription moves through the genome at rates of 3–4 kb/min (10, 11). Additionally, kinetic modeling of the live-cell data indicated that RNA Pol II can pause for significant times during the elongation process. Expanding the use of such approaches will enable the formulation of precise hypotheses on how genes are regulated and will allow researchers to quantitatively test these models. Indeed, the MS2-mRNA labeling system has been useful for following mRNA transcription also in prokaryotes (12, 13) and *Dictyostelium* (14). These and other studies have demonstrated that mRNA synthesis occurs in a burst-like probabilistic manner (15, 16).

In this chapter we explain the basic requirements and the experimental steps required for setting up an experimental mammalian cell system for following mRNA in vivo. While mRNA dynamics in mammalian cells have been studied by a number of methods (17, 18), the experimental approach to be discussed herein is based on the tagging of mRNA transcripts using the MS2-CP-GFP system. This chapter will describe: (a) The construction of a gene construct containing the MS2 sequence repeats; (b) The generation of stable mammalian cell lines that contain the GOI; (c) Screening for positive clones by RNA FISH against the MS2 sequence repeats; and (d) Setting up live-cell imaging experiments to follow mRNA dynamics in living cells.

2. Materials

2.1. Cloning of 24 MS2 Sequence Repeats into a Vector with GOI

1. pSL24MS2 vector: Plasmid containing the 24 MS2 sequence repeats. Can be obtained from http://www.addgene.org/pgvec1?f=c&identifier=27120&cmd=findpl&attag=c.
2. Restriction enzymes.
3. Ligase – T4 DNA Ligase (5 u/μl, New England Biolabs, Beverly, MA).
4. Competent *Escherichia coli* bacteria for transformation (see Note 1).

2.2. Electroporation for Stable Integration of GOI-MS2 or Transient Expression of MS2-CP-GFP

1. Dulbecco's Modified Eagle's Medium (DMEM).
2. Fetal bovine serum (FBS).
3. Trypsin for detaching cells from tissue culture plate (Gibco/BRL).
4. PBS solution for washing cells.
5. MS2-CP-GFP plasmid, (1–4 μg). Can be obtained from http://www.addgene.org/pgvec1?f=c&identifier=27121&cmd=findpl&attag=c
6. Carrier DNA for electroporation: 40 μg sheared salmon sperm DNA 20 mg/ml (Cat. No. 0644, Amresco, OH).
7. Electroporator e.g., Bio-Rad Gene Pulser Xcell (Bio-Rad, Hercules, CA).
8. Gene Pulser Cuvette 0.4 cm (Cat. No. 165-2088, Bio-Rad).
9. Adherent cell line of choice. Human U2OS and HeLa cell lines are recommended (can be purchased from the ATCC).
10. Antibiotics for stable selection depending on the selection marker in the gene construct.
11. Cloning cylinders (Cat. No. 3160-60, Corning, NY).

2.3. Time-Lapse Live-Cell Imaging of mRNP Dynamics in Living Cells

1. Fluorescent microscope of choice (confocal or wide-field) used at a 60× magnification (oil objective). A wide-field microscope should be equipped with a CCD camera for obtaining best results when capturing the dynamics of single mRNPs.
2. Microscope incubation chamber including temperature and CO_2 control for growing cells during imaging.

3. Glass bottomed tissue culture plates – 35 mm petri dishes with a 14 mm glass-bottomed microwell (0.16–0.19 mm thickness; Cat. No. P35G-1.5-14-C, MatTek, Ashland, MA).

4. Image analysis software such as: Imaris (Bitplane, Switzerland), Metamorph (Molecular Devices, Downingtown, PA), or ImageJ (NIH, Bethesda, MD; http://rsb.info.nih.gov/ij/).

5. Deconvolution software such as: Huygens Deconvolution Software (Scientific Volume Imaging, The Netherlands), AutoQuant (Media Cybernetics, Bethesda, MD), or DeltaVision (Applied Percision, Issaquah, WA).

2.4. RNA FISH

1. PBS solution for washing.

2. 4% paraformaldehyde in PBS (Cat. No. 19208, Electron Microscopy Science, Hatfield, PA).

3. 70% ethanol.

4. 20× Saline–sodium citrate buffer (SSC): 3 M NaCl, 0.3 M sodium citrate at pH 7 (Cat. No. 0804, Amresco).

5. Formamide (Sigma). For 40% formamide: 60 ml 4× SSC (diluted with ddH_2O) + 40 ml 100% formamide.

6. ssDNA/tRNA. Mix equal vol. of 10 mg/ml ssDNA (Cat. No. D-7656, Sigma) and 10 mg/ml tRNA (Cat. No. 109541, Roche).

7. BSA (10 mg/ml).

8. Fluorescently labeled DNA probe. Use ~10 ng probe per coverslip (from stock of 40 ng/μl). The probe against the MS2 sequence repeats is a 51 nucleotide long DNA probe with a Cy3 molecule conjugated to the 5′-end (can use different fluorophores such as Cy's, FITC, Alexa etc). Since there are many MS2 sequence repeats in each transcript, many probes will bind to each mRNA thus providing a strong FISH signal. The MS2 probe sequence: 5′-TTT CTA GGC AAT TAG GTA CCT TAG GAT CTA ATG AAC CCG GGA ATA CTG CAG-3′.

9. Solution 1: 2.5 μl probe (40 ng/μl stock, for 10 coverslips); 3.6 μl of 20× SSC; 2 μl of 5 mg/ml of ssDNA/tRNA; 23 μl of DDW; 160 μl of 100% formamide. Adjust to 200 μl with DDW.

10. Solution 2: 198 μl of DDW; 2 μl of BSA; 50 μl of 20× SSC.

3. Methods

Overview of the problematic issues that might arise when dealing with the MS2-CP-GFP system are listed in Table 1. Subheading 4 – Notes – relays important information from real-life experiments to assist the reader in transforming the steps into a successful experiment.

3.1. Construction of an Expression Vector Containing the MS2 Sequence Repeats

1. Insert your GOI into an expression vector. If you require fluorescent tagging of the protein product then insert the coding region of your gene in-frame with a vector containing a fluorescent fusion-protein of choice (see Note 2). A large variety of fluorescent proteins now exist that can be purchased from different companies. In order to obtain stable cell lines, antibiotic resistance will be required. At this point it is important to consider that the promoter driving this gene will define the levels of expression, namely, viral promoters (e.g., CMV, SV2) will result in high levels of expression, while endogenous promoters (e.g., β-actin promoter) will lead to moderate levels of expression (see Note 3).

2. Insert the MS2 sequence repeats from the pSL24MS2 plasmid into the 3′UTR of your GOI. Preferable restriction sites will be the *BamHI* at the 5′ of the MS2 sequence repeats and *BglII* at the 3′ of the MS2 sequence repeats. *BamHI/BglII* digestion will result in a 1,308 bp fragment containing the 24 MS2 sequence repeats (see Notes 1 and 4).

An adaptor with suitable restriction sites can be added to the 3′UTR sequence of the GOI if there are no suitable sites for the insertion of the MS2 sequence repeats fragment. Fig. 1 depicts the cloning steps to be taken for generating a plasmid expressing a CFP-GOI fusion protein containing 24 MS2 sequence repeats in the 3′UTR (see Note 5).

3.2. Generation of a Cell Line Stably Expressing the CFP/GOI/MS2 Construct

1. Split the cells 1 day prior to electroporation. Make a single cell suspension and plate in a 10 cm tissue culture dish in fresh medium. On the next day, confluence should reach 50–80%. Transfection efficiency is reduced if cells are too aggregated.

2. Day of transfection: Wash the cells with 1× PBS, trypsinze gently by adding 1–1.5 ml of trypsin to the cells and incubate for 1–5 min at 37°C. Add medium containing 10% FBS and

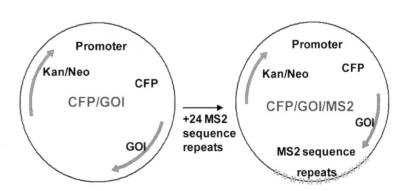

Fig. 1. Scheme of the structure of the gene construct and insertion of the 24 MS2 sequence repeats. The gene of interest (GOI) is fused to a fluorescent protein, in this case, cyan fluorescent protein (CFP). The 24 MS2 sequence repeats are cloned into the 3′UTR of the gene. Antibiotic resistance is conveyed by a resistance gene, here the kanamycin/neomycin gene for bacterial and mammalian selection, respectively.

transfer the cells to a 15 ml tube. Centrifuge for 5 min at 200 g and aspirate the medium.

3. Suspend in 1 ml cold medium plus serum.
4. Apply 200–250 μl of the cells (approx. 200,000 cells) to a sterile cuvette and add the CFP/GOI/MS2 plasmid (2–10 μg of DNA per transfection) to the cells.
5. Tap gently to mix and wait for 10 min at room temperature.
6. Electroporate using either pre-set protocols or your own settings. The following electroporation conditions have been successfully used with the BioRad Gene-Pulsar Xcell when transfecting these human cell lines: U2OS: 170 V, 950 μF; HeLa: 150 V, 500 μF; HEK-293: 300 V, 500 μF. Different protocols should be tried to check for best transfection efficiency (see Note 6).
7. Plate electroporated cells in a 10 cm plate with 10 ml fresh medium plus serum. Mix cells and medium gently, and incubate at 37°C.
8. Next day: Add appropriate antibiotics to the medium to begin selection. Change the medium and antibiotics every 3 days. Continue with the selection for about 2–3 weeks until single colonies develop.
9. Use autoclaved cloning cylinders to collect colonies. Place the cylinders on the well separated colonies. Pipette 100 μl of trypsin into each cylinder, incubate for 1–5 min, and then add 100 μl of medium to each cylinder. Gently pipette and suspend the colony, and transfer to a 24-well plate. Add 0.5 ml fresh medium plus antibiotics to each well.
10. After the colonies expand in the wells, screen the colonies for positive colonies which integrated the GOI (see Note 7). A fluorescent microscope is used for easy and rapid detection of the fluorescent protein expressed by the GOI. If the gene is cloned without a fluorescent tag then RT-PCR with specific primers to the GOI can be performed. We do not recommend RT-PCR through the MS2 region due to the repeated nature of the sequence. Another option is performing Western blotting using a specific antibody directed against the protein product expressed from the GOI coding region (see Note 8).

3.3. Screening for MS2-RNA Positive Colonies by RNA FISH

1. Grow cells from a positive clone on 18-mm round coverslips in a 12 well TC dish.
2. Wash briefly in 1× PBS and fix in 4% paraformaldehyde (PFA) in PBS 20 min.
3. Wash briefly in 1× PBS and then add 70% ethanol. Leave overnight at 4°C.
4. Next day: Rinse twice prior to hybridization with 1× PBS for 10 min (with gentle shaking).

5. Wash 10 min in 0.5% Triton X-100 in PBS. Wash for 10 min in 1× PBS.

6. Prehybridization: Wash twice for 5 min in 40% formamide.

7. During rinses prepare Solution 1 and Solution 2. This volume can be used for ten 18-mm coverslips. The fluorescent probe directly added to solution 1 (detailed above in Subheading 2.4) is directed against the MS2 sequence repeats and therefore will bind many times to each transcript. This will allow clear detection of the transcription sites and mRNAs produced from the stably integrated genes (see Note 9).

    Just before hybridization, boil solution 1 in an eppendorf tube for 5 min and cool on ice for 5 min. Add 200 µl of Solution 2 to 200 µl of boiled solution 1 and keep on ice.

8. Hybridization: place a 40 µl drop of probe solution mix in a petri dish. Gently apply the coverslip onto the drop, with fixed cells facing down. To avoid drying, place a small reservoir with hybridization solution (40% formamide) in the dish. Close the petri dish and seal with parafilm. Place the hybridization dish in a 37°C incubator and hybridize for 3 h (or overnight).

9. Next day: Half an hour before rinsing, warm up the remaining 40% formamide solution to 37°C.

10. Open hybridization chamber and transfer coverslips face up back into a 12-well dish containing the prewarmed 40% formamide. Rinse twice for 15 min at 37°C.

11. Rinse for 2× 1 h in 1× PBS at RT, with gentle shaking.

12. If necessary, perform nuclear staining for 5 min (DAPI or Hoechst).

13. Brief wash in PBS. Mount slides in mounting solution.

14. Screen for cell clones that contain the MS2 signal using a fluorescent microscope. This is seen as one strong fluorescent spot in the nucleus (= transcription site) and many distributed dots throughout the cytoplasm (= cellular mRNAs). See Fig. 2. Positive clones can be further used for the live-cell imaging assays.

3.4. Live-Cell Imaging of Cellular mRNPs

1. Split the cells to 50–80% confluence 1 day prior to transfection.

2. Day of transfection: Trypsinize cells, wash in PBS, and transfer ~2×10^5 cells suspended in 200 µl of cold medium into a 0.4 cm cuvette (as explained above in Subheading 3.2).

3. Add 1–4 µg of MS2-CP-GFP DNA plasmid and 40 µg salmon sperm carrier DNA to the cells. Tap gently to mix and wait for 10 min at RT (see Notes 6, 10 and 11).

4. Electroporate the cells.

5. Transfer the cells to the glass-bottomed tissue culture plates, and plate in the center of the coverslip. After first attachment

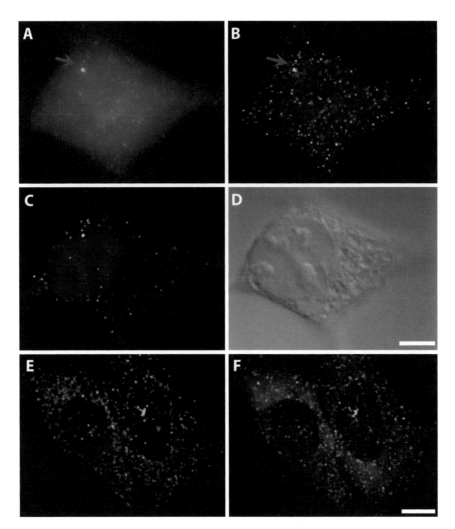

Fig. 2. Detection of MS2 mRNAs in fixed and living cells. (a) A transcription site (*red arrow*) and cellular mRNA (*green dots*) transcribed from a GOI are detected throughout a fixed cell using RNA FISH. The FISH probe is targeted to MS2 sequence repeats found in the 3′UTR of the mRNA transcribed from the GOI. The transcription site is always found in the nucleus and harbors several mRNAs simultaneously. Therefore, the fluorescent signal at the transcription site is typically stronger than the cellular mRNPs. The image depicts the complete volume of the cell while being viewed from above. (b) The cell in (a) after deconvolution: the mRNP signal was enhanced together with a reduction in the noise, resulting in the clear detection of mRNPs distributed throughout the cell volume. (c) A single plane from the deconvolved imaged 3D volume showing the transcription site in the nucleus (*red*) and the mRNAs in context of the nucleus and the cytoplasm. (d) DIC image of the cell. Bar 5 μm. Imaged by Sharon Yunger (19). (e) mRNPs imaged in a living cell using a YFP-MS2-CP fusion protein (*green*) that binds to the MS2 sequence repeats in the mRNA. Two cells are seen in this image. (f) A red cytoplasmic protein depicts the borderline between the cytoplasm and nucleus. The strong dot in the center of the nucleus (*right hand* cell) shows the transcribing gene. Bar 5 μm. Imaged by Amir Mor (20).

of the cells to the glass (~2–5 h), add 1 ml of fresh medium to the plate.

6. Once the cells have spread properly onto the glass (several hours) (see Notes 12 and 13), transfer the plate to a fluorescent microscope equipped with an incubator that provides

conditions of 37°C and 5% CO_2. Search for transfected cells containing a strong transcription site and expressing labeled mRNPs. The mRNPs are seen as many small round dots moving in the cells (see Note 9). Adjust the imaging conditions according to the level of the expressed signal versus the diffusive level of background. Avoid exposing the cells during the search to high intensity illumination in order to reduce bleaching of the fluorescent signal.

7. Determine the experimental parameters required for time-lapse imaging such as exposure time, number of fields, number of Z-slices, and the number of repeats. The number of Z-slices is dependent on several parameters such as cell shape and mobility. We suggest to start with a range of 5–15 slices with a step-width of 0.3–0.6 µm. Then adjust the parameters as needed. The light intensity and exposure time need to be low as possible to prevent bleaching of the signal but high enough to observe the signal over the background. These parameters are dependent on the light source of the microscope, but preferably should allow imaging at frequencies that are lower than 1 s.

8. Analyze the acquired time-lapse movies using image analysis software. If required, enhance the mRNP signal using a deconvolution algorithm. We find that 5–20 iterations with the Huygens Essential Deconvolution Software efficiently enhances mRNP signal in the time-lapse movies.

4. Notes

1. *Cloning strategy*: Bacteria tend to discard repeated sequences and therefore the MS2 sequence repeats are prone to shortening during transformation. One possibility is to use bacteria e.g., Stbl2 competent cells that grow at 30°C, that are not supposed to dispose of these repeats. However, we have found that careful monitoring of the number of repeats following transformation and single colony picking with regular component bacteria, allows the detection of bacterial clones that harbor the 24-sequence repeats in a constant manner. This requires the presence of flanking restriction sites on both sides of the MS2 sequence repeats in order to accurately assess the number of repeats within the vector. Due to these limitations it is useful that the MS2 sequence repeats be cloned last into the expression vector.

2. *Fluorescent marker for protein expression*: We find it helpful to insert a fluorescent fusion-protein sequence upstream of the

coding region. This serves two purposes: (a) Proof of completion of gene expression via the production of the cytoplasmic protein product. (b) Easier scanning for stable cell clones. The cyan fluorescent protein (CFP) is considered a good marker since it will not interfere with the YFP-MS2 proteins that are usually used for following the mRNA in the MS2 tagging-system. Note: throughout the chapter we refer to MS2-CP-GFP, but the fluorescent protein can be of any color, such as YFP-MS2-CP that we typically use. Tagging of the coding region of the expressed gene with GFP is not recommended since it will then require the use of a weak red fluorescent MS2-CP protein (mCherry-MS2 or RFP-MS2) for tagging of the mRNA, which is the molecule of interest in this assay.

3. *Promoter*: The choice of promoter will affect the expression levels of the mRNA, and therefore it is important to select a promoter according to the study goals. While the use of a strong promoter e.g., CMV promoter, will generate high mRNA levels, it might be better to choose an inducible promoter such as the Tet-On system, thereby allowing the control of the timing and level of mRNA expression. This issue might be vital since an overflow of labeled mRNAs in imaging experiments might affect the ability to detect the process of interest. Therefore, another option is the use of an endogenous promoter that typically results in moderate levels of mRNA expression e.g., the β-actin promoter. The use of gene-specific promoters requires the verification that the promoter is indeed active in the cell line of choice.

4. *Number of MS2 sequence repeats*: The tagging of a particular mRNA species to be followed in individual living cells is fulfilled by the insertion of DNA repeats termed "MS2 sequence repeats" into the GOI. While the insertion of one or two repeats has been useful in biochemical purification procedures, this is not sufficient for live-cell studies. Three versions containing increasing numbers of MS2 sequence repeats have been tested. While 6 or 12 MS2 sequence repeats were found to be useful in mRNA tagging in yeast cells, 24 repeats produced reasonable signal in mammalian cells (21). Since each MS2 sequence repeat will bind a dimer of MS2-CP-GFP proteins, it is preferable to choose the 24 sequence repeats version in order to obtain a high signal-to-noise ratio i.e., to detect the signal of mRNPs versus the diffusive background of the MS2-CP-GFP protein. It has been shown that an average of 30 MS2-CP-GFP proteins bind to one mRNA containing the MS2 sequence repeats (19, 21).

5. *The location of MS2 sequence repeats in the gene construct*: A typical gene construct contains the cDNA and an upstream

promoter. The MS2 sequence repeats are normally cloned after the coding region and within the 3′UTR (before the poly(A) signal), thereby not interfering with the translation of the gene. While another option is to place these repeats upstream of the coding region within the 5′UTR, this seems less preferable since the presence of the stem-loop secondary structures could interfere with ribosomal scanning prior to translation.

6. *Transfectability*: The MS2-CP-GFP protein is usually expressed by transient transfection and therefore it is important to use cells that easily transfect. Also we suggest using stable cell lines in which the GOI +MS2 sequence repeats are stably integrated. We find the U2OS and HeLa cell lines useful for this line of work, while the HEK293 cell line that is easily transfectable can be problematic due to small cell size, cell clumping, and easy detachment from the cell surface.

7. *Stable cell line generation*: In order to have isogenic levels of the expressed mRNA, the generation of stable cell lines in which the GOI (+MS2 sequence repeats) is integrated into the genome is highly recommended. Moreover, since stable integration results in the generation of tandem arrays of the integrated gene, it will be easier to detect the site of transcription within the nuclear volume. Stable integration of the expression vector containing the MS2 sequence repeats is performed using standard transfection protocols such as electroporation or calcium phosphate, together with selection for antibiotic resistance. Verification of positive cell clones is performed either on the basis of the protein expression of the fluorescent protein product, or by RNA FISH with a fluorescent probe against the MS2 sequence repeats.

8. *Tag sequences*: In order to use the protein product in biochemical assays it might be useful to consider the cloning of Flag or HA-tags at the N-terminal region of the gene.

9. *Detection of mRNA and active transcription sites*: This can be performed at first during the scanning of the stable cell clones by RNA FISH. Positive clones should portray a strong fluorescent cytoplasmic signal indicative of processive mRNA export, and another important characteristic is the detection of active nuclear transcription sites. The next step of verification would be in living cells via transient transfection of the MS2-CP-GFP protein into the stable cell line containing the GOI with 24 MS2 repeat in its 3′UTRs. This should result in the appearance of an apparent fluorescent spot in the nucleus termed the "transcription site." Smaller and mobile spots that will be detected within the nucleus and the cytoplasm are the mRNPs (Fig. 2).

10. *Fluorescent MS2-CP protein*: In order to obtain an ideal signal-background ratio, several optimization steps are required. First, as mentioned above, the MS2-CP protein should preferably be fused to GFP or YFP. The green emitting fluorophores provide strong signal and are also preferable for the photobleaching experiments. While GFP is brighter, YFP (but not GFP) can be used in conjunction with CFP in the same cell (protein product). Second, high levels of the MS2-CP-GFP protein might interfere with the detection of the single mRNPs due to masking of the specific signal by the high background. It should be noted that the MS2-CP-GFP protein usually harbors a nuclear localization sequence (NLS) in order to have the MS2-CP-GFP protein present at the site of transcription as the nascent chain is being transcribed. Expressing the MS2-CP-GFP protein under a moderate to weak promoter will result in optimal signal-to-noise ratios. Also, in some cases, high levels of MS2-CP-GFP seem to enhance the appearance of nucleoli even though there is no specific binding to that structure.

11. *MS2-CP-GFP transfection*: We find that MaxiPrep DNA usually yields the best expression levels. Using a stable cell clone is preferable since all cells express similar levels of the integrated gene products. If transient transfection is the preferred method of expression, it should be noted that different methods have different response times. If electroporation is the selected method, the MS2-CP-GFP protein will be expressed already after several hours post-transfection, in contrary to liposome or calcium phosphate based methods that require at least an overnight period of incubation.

12. *Adherence of the cells to the tissue culture plate*: It is strongly recommended to work with an adherent cell line in order to avoid problems with cell detachment during imaging or cell manipulation under the microscope. Cell lines that loosely attach to the surface will create difficulties during imaging of live cells as well as in fixed cell procedures such as FISH or immunofluorescence, since the routine washes will detach the cells from the plate/coverslip. In addition, it is preferable to work with cells that do not form cell clumps, thereby enabling the detection of single cells in the field of imaging.

13. *Cell shape*: It is advantageous to use cells with a large nucleus and a spread out cytoplasm in order to ease the detection of transcription sites and single mRNPs throughout the cell.

Table 1
We have noted a list of known problems that one may encounter during these experiments and the possible solutions for troubleshooting them

Problem	Possible cause	Suggested solution
1. No transfection or low transfection efficiency	Poor quality DNA Transfection method is not optimal for the cell line	The DNA should be purified using a high quality MaxiPrep kit (Qiagen kits are recommended for stable transfections) Use higher amounts of DNA Optimization of the transfection method and ratio of reagents to the cell line
2. The transcription site is not detected in living cells although the MS2-CP-GFP was introduced successfully into the cells	The expression vector containing the gene of interest is not properly constructed, MS2 repeat sequences were lost, or gene not expressed Overexpression of the MS2-CP-GFP protein causes a high background of the diffuse unbound protein, masking the transcription site signal	Verify the expression of the gene (mRNA) by methods such as RNA FISH or RT-PCR (with specific primers to the sequence of the gene that was cloned into the expression vector). Verify the presence of the 24 MS2 sequence repeats. Detect the presence of the sites in fixed cells, where they are easier to detect Use lower amounts of MS2-CP-GFP plasmid or examine the cells at earlier times post-transfection
3. The cells are moving in and out of focus in the Z axis during imaging	Normal mobility of the living cells	Acquire Z stacks for each time point Can also acquire microscope hardware that corrects for Z drifts during imaging
4. Cells are moving away from the field of observation during imaging	Cell movement due to phototoxicity	Lower illumination intensity and reduce exposure times
5. The movie is out of focus due to high cell movement in the XY axis	Cells are not strongly attached to the plate or a very motile cell line	Facilitate the attachment of the cells to plate by using collagen coated plates or other adhesive molecules (Cell-Tak [BD Bioscience, Cat. No. BD 354240]) or polylysine (Sigma) Change to a less motile cell line
6. Cell clumping not allowing single cell tracking and analysis	The cells tend to form aggregates	Carefully pipette the cells until achieving a single cell solution before plating Change to a non-clumping cell line

(continued)

Table 1
(continued)

Problem	Possible cause	Suggested solution
7. The images are noisy. Single mRNPs are not detected	High background of the MS2-CP-GFP protein, low mRNA expression levels, or sub-optimal imaging conditions	Reduce MS2-CP-GFP expression by allowing less time to express or by using less DNA Improve imaging conditions – longer exposure-times, more light intensity Deconvolution of the images using different software
8. MS2-CP-GFP photobleaching	High illumination intensity High exposure times	Lower light intensity during imaging Image using 2×2 binning Improve image quality using an EMCCD camera Image less Z slices per time point

Acknowledgements

The laboratory of Yaron Shav-Tal is supported by the European Research Council (ERC) Israel Science Foundation (250/06), ISF-Bikura, ICRF, GIF, BSF, DIP, Ministry of Science, and Ministry of Health. YST is the Jane Stern Lebell Family Fellow in Life Sciences at BIU.

References

1. Lippincott-Schwartz, J., Altan-Bonnet, N., and Patterson, G. H. (2003) Photobleaching and photoactivation: following protein dynamics in living cells. *Nat. Cell Biol. Suppl.*, S7–14.
2. Carmo-Fonseca, M., Platani, M., and Swedlow, J. R. (2002) Macromolecular mobility inside the cell nucleus. *Trends Cell Biol.* **12**, 491–495.
3. Shav-Tal, Y., Darzacq, X., and Singer, R. H. (2006) Gene expression within a dynamic nuclear landscape. *EMBO J.* **25**, 3469–3479.
4. Gorski, S. A., Dundr, M., and Misteli, T. (2006) The road much traveled: trafficking in the cell nucleus. *Curr. Opin. Cell Biol.* **18**, 284–290.
5. Janicki, S. M., Tsukamoto, T., Salghetti, S. E., Tansey, W. P., Sachidanandam, R., Prasanth, K. V., Ried, T., Shav-Tal, Y., Bertrand, E., Singer, R. H., and Spector, D. L. (2004) From silencing to gene expression; real-time analysis in single cells. *Cell* **116**, 683–698.
6. Bertrand, E., Chartrand, P., Schaefer, M., Shenoy, S. M., Singer, R. H., and Long, R. M. (1998) Localization of ASH1 mRNA particles in living yeast. *Mol. Cell* **2**, 437–445.
7. Shav-Tal, Y., Darzacq, X., Shenoy, S. M., Fusco, D., Janicki, S. M., Spector, D. L., and Singer, R. H. (2004) Dynamics of single mRNPs in nuclei of living cells. *Science* **304**, 1797–1800.

8. Darzacq, X., Shav-Tal, Y., de Turris, V., Brody, Y., Shenoy, S. M., Phair, R. D., and Singer, R. H. (2007) In vivo dynamics of RNA polymerase II transcription. *Nat. Struct. Mol. Biol.* **14**, 796–806.
9. Lamond, A. I., and Swedlow, J. R. (2007) RNA polymerase II transcription in living color. *Nat. Struct. Mol. Biol.* **14**, 788–790.
10. Wada, Y., Ohta, Y., Xu, M., Tsutsumi, S., Minami, T., Inoue, K., Komura, D., Kitakami, J., Oshida, N., Papantonis, A., Izumi, A., Kobayashi, M., Meguro, H., Kanki, Y., Mimura, I., Yamamoto, K., Mataki, C., Hamakubo, T., Shirahige, K., Aburatani, H., Kimura, H., Kodama, T., Cook, P. R., and Ihara, S. (2009) A wave of nascent transcription on activated human genes. *Proc. Natl. Acad. Sci. U S A* **106**, 18357–18361.
11. Singh, J., and Padgett, R. A. (2009) Rates of in situ transcription and splicing in large human genes. *Nat. Struct. Mol. Biol.* **16**, 1128–1133.
12. Golding, I., and Cox, E. C. (2004) RNA dynamics in live Escherichia coli cells. *Proc. Natl. Acad. Sci. U S A* **101**, 11310–11315.
13. Golding, I., Paulsson, J., Zawilski, S. M., and Cox, E. C. (2005) Real-time kinetics of gene activity in individual bacteria. *Cell* **123**, 1025–1036.
14. Chubb, J. R., Trcek, T., Shenoy, S. M., and Singer, R. H. (2006) Transcriptional pulsing of a developmental gene. *Curr. Biol.* **16**, 1018–1025.
15. Elowitz, M. B., Levine, A. J., Siggia, E. D., and Swain, P. S. (2002) Stochastic gene expression in a single cell. *Science* **297**, 1183–1186.
16. Ozbudak, E. M., Thattai, M., Kurtser, I., Grossman, A. D., and van Oudenaarden, A. (2002) Regulation of noise in the expression of a single gene. *Nat. Genet.* **31**, 69–73.
17. Shav-Tal, Y., and Gruenbaum, Y. (2009) Single-molecule dynamics of nuclear mRNA. *F1000 Biology Reports* **1**, 29–32.
18. Boulon, S., Basyuk, E., Blanchard, J. M., Bertrand, E., and Verheggen, C. (2002) Intranuclear RNA trafficking: insights from live cell imaging. *Biochimie* **84**, 805–813.
19. Yunger, S., Rosenfeld, L., Garini, Y., and Shav-Tal, Y. (2010) Single allele analysis of transcription kinetics in living mammalian cells. *Nat. Methods* **7**, 631–633.
20. Mor, A., Suliman, S., Ben-Yishay, R., Yunger, S., Brody, Y., and Shav-Tal, Y. (2010) Dynamics of single mRNP nucleo-cytoplasmic transport through the nuclear pore in living cells. *Nat. Cell Biol.* **12**, 543–552.
21. Fusco, D., Accornero, N., Lavoie, B., Shenoy, S. M., Blanchard, J. M., Singer, R. H., and Bertrand, E. (2003) Single mRNA molecules demonstrate probabilistic movement in living Mammalian cells. *Curr. Biol.* **13**, 161–167.

Chapter 17

Using the mRNA-MS2/MS2CP-FP System to Study mRNA Transport During *Drosophila* Oogenesis

Katsiaryna Belaya and Daniel St Johnston

Abstract

Asymmetric mRNA localisation to specific compartments of the cell is a fundamental mechanism of spatial and temporal regulation of gene expression. It is used by a variety of organisms and cell types to achieve different cellular functions. However, the mechanisms of mRNA localisation are not well understood. An important advance in this field has been the development of techniques that allow the visualisation of mRNA movements in living cells in real time. In this paper, we describe one approach to visualising mRNA localisation in vivo, in which RNAs containing MS2 binding sites are labelled by the MS2 coat protein fused to fluorescent reporters. We discuss the use of this mRNA-MS2/MS2CP-FP system to study mRNA localisation during *Drosophila* oogenesis, and provide a detailed explanation of the steps required for this approach, including the design of the mRNA-MS2 and MS2CP-FP constructs, the preparation of fly oocytes for imaging, the optimal microscope configurations for live cell imaging, and strategies for image processing and analysis.

Key words: MS2 coat protein, mRNA localization, Live cell imaging, Widefield deconvolution microscopy

1. Introduction

mRNA localisation is a widespread phenomenon that occurs in organisms as diverse as yeast and humans. It contributes to many cellular processes including the establishment of intracellular asymmetry, directed cell motility, asymmetric cell division, and possibly also synaptic plasticity (1).

The molecular mechanisms of mRNA localisation have been intensively studied in *Drosophila*, because the formation of the body axes of this organism depends on the correct localisation of several maternal mRNA species to specific regions of the oocyte. Localisation of *oskar* (*osk*) mRNA to the posterior pole of the oocyte specifies the site of the pole plasm assembly, and thus where the abdomen and germ cells develop in the embryo (2–4). *nanos* (*nos*) mRNA is recruited to the pole plasm downstream of *oskar* mRNA and functions as the posterior determinant to specify the formation of the abdomen (5). Localisation of *bicoid* (*bcd*) mRNA to the anterior pole is essential for proper patterning of the head and the thorax (6, 7). Finally, the localisation of *gurken* (*grk*) mRNA to the anterior-dorsal corner of the oocyte specifies the dorso-ventral axis of the future embryo (8).

Initially, the localisation of different mRNAs was studied using in situ hybridisation techniques. This method visualises mRNA localisation in fixed oocytes, and therefore provides a picture of the steady-state distribution of mRNA molecules, but suffers from several limitations. Firstly, the hybridisation technique is not ideal for studying the localisation of mRNA during late stages of oogenesis, since the deposition of the vitelline membrane makes the oocyte impermeable to in situ probes. Secondly, the method does not provide any information about the dynamics of mRNA movement, and this information is crucial for a complete understanding of the mechanisms of mRNA transport. A great deal of effort has therefore been directed towards developing a system that allows visualisation of mRNA movements in living oocytes.

1.1. Methods for Visualisation of mRNA Localisation In Vivo

There are currently four major methods for visualising mRNA localisation in vivo (Fig. 1). The first is the injection of the fluorescently labelled mRNA molecules into the cell (Fig. 1a). This method has been successfully used to study the movement of *bcd* and *grk* mRNAs during oogenesis (9–12). However, it has two major drawbacks. Firstly, labelling of the RNA molecule can affect its structure, and thus disrupt the natural localisation process. Second, localisation of many mRNAs ultimately depends on nuclear events such as splicing and processing, during which the RNAs start to assemble with different trans-acting factors into higher-order particles (13–15). Thus, this technique may be not applicable to the analysis of certain mRNAs (for example, *osk* mRNA).

The second method is the injection of molecular beacons (Fig. 1b, (16)). A molecular beacon is a small oligonucleotide, which is complementary to the RNA of interest. This oligonucleotide has a fluorescent molecule attached to one end and a fluorescence quencher attached to the opposite end. The secondary structure of the oligonucleotide is such that when it is not bound to its target mRNA, it is folded in a way that quenches the fluorescence. Once it hybridises to its target RNA, it unfolds, leading to the appearance of the fluorescence. The technique has been developed and first applied for the study of the *osk* mRNA localisation (16). The major advantage of this technique is that it can be used to label endogenous mRNAs. However, the sensitivity of this technique is quite low and the oligonucleotides tend to bind non-specifically to other structures.

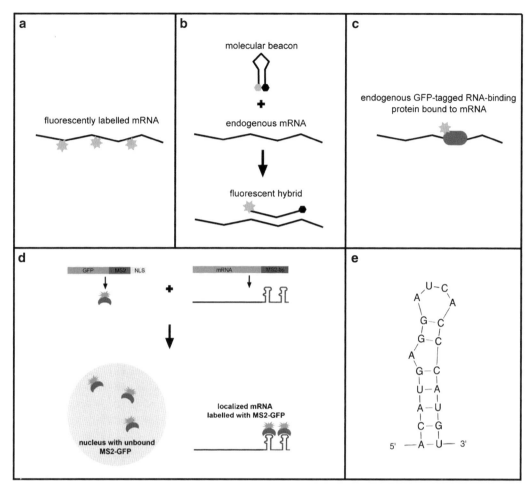

Fig. 1. Methods for visualising mRNAs in vivo. (**a**) Injection of fluorescently labelled mRNA. The mRNA is transcribed in vitro in the presence of a fluorescently labelled nucleotides. These labelled transcripts are injected into the living cells. (**b**) Injection of molecular beacons. Beacons are small oligonucleotides with a fluorophore attached to one end and a fluorescence quencher attached to the other. Upon hybridisation to the target mRNA, the beacon unfolds, thereby separating the quencher from the fluorophore which can then fluorescence. (**c**) GFP-tagging of an RNA-binding protein. An endogenous RNA-binding protein that is known to be specific for the mRNA of interest is tagged with GFP. (**d**) RNA-MS2/MS2CP-GFP tagging system. The 19-nucleotide MS2-binding sequence is inserted into the mRNA of interest. This mRNA is co-expressed with the NLS-MS2CP-GFP construct. MS2CP-GFP-NLS binds to the MS2-binding sequence and allows visualisation of the target mRNA. Unbound MS2CP-GFP-NLS is confined to the nuclei due to the presence of the NLS signal. (**e**) Secondary structure of the 19-nucleotide MS2-binding RNA sequence. The structure was predicted using the mfold program (44).

Another method of mRNA visualisation involves tagging with GFP an endogenous protein that binds to the RNA of interest (Fig. 1c). This technique has been used to study *bcd* (with Exu-GFP, (17)) and *osk* (with Stau-GFP, (18)) mRNAs. The major disadvantage of the method is that most RNA-binding proteins bind to several different RNA molecules. This method therefore requires additional controls to prove that the observed GFP-labelled particles contain the RNA of interest, and this is hard to implement.

A solution to many of the problems described above came with the advent of the mRNA-MS2/MS2CP-FP tagging system (Fig. 1d, (19)). This fourth method relies on the use of two constructs. The first construct is a fusion between a fluorescent protein (FP), the MS2 bacteriophage coat protein (MS2CP) and a nuclear localisation sequence (NLS). The second construct is the mRNA of interest containing a number of copies of the MS2CP binding site (a 19-nucleotide RNA sequence; Fig. 1e). When both of these constructs are expressed in the same cell, the NLS-MS2CP-FP protein binds to the mRNA containing the MS2CP binding sites, and the localisation of the RNA is therefore revealed by the FP fluorescence signal. The unbound NLS-MS2CP-FP protein is retained in the nucleus due to the presence of the NLS.

The MS2CP-FP protein binds mRNAs containing its binding sites (mRNA-MS2) with high specificity and high sensitivity. Since mRNA-MS2 molecules are transcribed in the cell nucleus, this technique allows one to follow endogenous mRNA molecules, which have undergone all of the required pre-mRNA processing events and have associated with the appropriate trans-acting factors. This latter property is especially important for studying such mRNAs as *osk* mRNA, whose proper localisation is known to ultimately depend on pre-mRNA splicing (13).

The technique was developed for the visualisation of Ash1 mRNA in yeast, but has since been applied to study other mRNAs in a variety of organisms (20). During *Drosophila* oogenesis, this technique has now been developed for the study of the localisation of four different mRNAs: *bcd*, *grk*, *nos* and *osk*.

1.2. Possible Applications of mRNA-MS2/MS2CP-FP Labelling System

mRNA-MS2/MS2CP-FP system can be used to study different steps of mRNA localisation process. And the use of different microscopy approaches can provide complimentary information which, taken together, may provide a full picture of mRNA transport.

Slow time-lapse imaging can provide information about the gross changes in mRNA localisation during different stages of oogenesis. For example, Weil et al. (21), used this approach to describe different steps in the localisation of *bcd* mRNA to the anterior pole of the oocyte (21). Additionally, similar low resolution imaging approaches have been used to study the effect of pharmacological disruption of the actin and microtubule cytoskeletons on the process of *bcd*, *grk* and *nos* mRNA localisation (22–24).

Imaging at high temporal and spatial resolution may allow visualisation of movement of individual mRNA particles in real time, provided that the particles are large enough to be resolved by light microscopy (see for example Fig. 2j–m). Subsequent analysis of the parameters of particle movement can reveal whether the transport is active or not, whether it is unidirectional or bidirectional, and whether the RNA switches between different

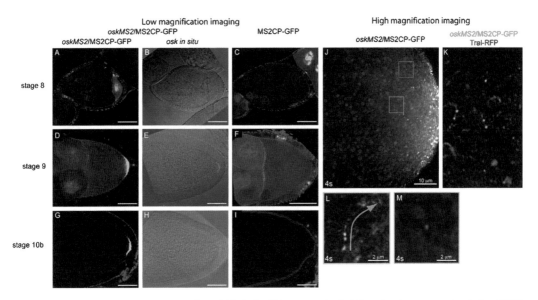

Fig. 2. Visualisation of *osk* mRNA with the RNA-MS2/MS2CP-GFP system. (**a, d, g**) Egg chambers expressing both oskMS2 and MS2CP-GFP. (**a**) oskMS2 is localised to the middle of the stage 8 oocyte. (**d, g**) oskMS2 mRNA localises to the posterior of the oocyte from stage 9 onwards. The pattern of localisation of oskMS2 mRNA is indistinguishable from that of the endogenous *osk* mRNA at all stages of oogenesis (**b, e, h**). (**c, f, i**) When MS2CP-GFP is expressed on its own, the GFP signal remains confined to the nuclei of the nurse and follicle cells. Scale bars are 50 μm. (**j**) Overlay of six frames from a high magnification time-lapse movie of an oskMS2/MS2CP-GFP expressing oocyte to show the movements of *osk* mRNA particles. Examples of individual tracks are highlighted by *coloured rectangles*, and are shown in separate *boxes* with higher magnification (**l, m**). (**k**) Co-visualisation of oskMS2/MS2CP-GFP particles and Tral-RFP in a *Drosophila* oocyte. The *red* and *green* channels were imaged sequentially. (**l**) Example of the fast-moving particle track highlighted by the *green rectangle* in (**j**). (**m**) Example of a stationary particle highlighted by the *orange rectangle* in (**j**).

modes of movement (e.g. movement and pauses). This analysis can be extended to the analysis of particle behaviour in different mutant backgrounds that disrupt mRNA localisation. This provides information about the roles of these proteins in mRNA targeting. Zimyanin et al. (18) have used this approach to characterise the process of *osk* mRNA transport in stage 9 oocytes (18).

Co-localisation studies using labelled RNA and labelled trans-acting factors have been used to provide information about whether the trans-acting factor is a component of the transport complex. This approach can also elucidate the timing of association of the trans-acting factor with the mRNA. This has been performed for the characterisation of the role of Staufen in the process of *bcd* and *osk* mRNA localisation (18, 24).

FRAP, FLIP, photoactivation and photoswitching (when MS2CP is fused to photoactivatable or photoswitchable FP) experiments are useful in elucidating how the mRNAs are maintained at their final destination (e.g. whether the molecules are anchored or continuously transported). This approach has revealed that *bcd* mRNA is continuously transported to the anterior of the oocyte during stages 10–13 of oogenesis, whereas *grk* mRNA is more stably anchored at the dorsal/anterior of the oocyte at stage 9 (21, 23).

Introduction of affinity tags into the MS2CP-FP protein may allow the biochemical purification of RNP complexes and thus provide information about the composition of the RNP transport particles. This approach has not yet been implemented in fly oocytes, but has been successfully used in other systems (25). Purified cytoplasmic fractions containing labelled mRNPs could also be added to in vitro motility assays containing preassembled cytoskeletal filaments to analyse the dynamics and directionality of mRNP particle movement in greater detail.

Finally, the mRNA-MS2/MS2CP-FP system can be used in conjunction with another RNA labelling technique, the mRNA-BoxB/λN-FP system, to co-visualise two different mRNA molecules in the same cell and thus to determine whether they share any step of the transport process (see for example (26)). In yeast, this approach has been used to show that different mRNAs co-assemble into the same localisation complex and are transported to the target localisation site together.

Thus, the mRNA-MS2/MS2CP-FP technique can be used to study the mechanisms of mRNA localisation from many different angles. In this chapter, we will describe the development and use of the mRNA-MS2/MS2CP-FP system to study of mRNA transport during *Drosophila* oogenesis.

2. Materials

2.1. Preparation of Drosophila Oocytes for Live Cell Imaging

1. *Drosophila* females and males of correct genotype, 2–3 days old.
2. Vials or bottles for growing *Drosophila*. Standard *D. melanogaster* food (27). Yeast paste for fattening flies overnight prior to dissection.
3. Colcemid (Sigma) for microtubule depolymerisation.
4. Latrunculin A (Sigma) for actin depolymerisation.
5. Dissection tools: fine tweezers (DUMOSTAR from Dumont; style No. 5 thickness) and tungsten needles.
6. Dissection microscope.
7. Voltalef PCTFE 10S oil (ARKEMA).
8. Coverslips – borosilicate cover glass thickness No. 1 (VWR).

2.2. Imaging and Image Analysis

1. A Microscope system optimised for imaging FP, for example a DeltaVision Core (Applied Precision) with the following configuration: Olympus inverted IX71 microscope, xenon or

mercury arc light source, highly sensitive Cascade II EMCCD camera (Photometrics), objectives for imaging (UPlanSApo 20× 0.75 NA air [Olympus]; UPlanSApo 100×, 1.40 NA oil [Olympus]), specifically chosen filter sets, DeltaVision soft-WoRx software for image acquisition and processing, DualView or QuadView system for simultaneous imaging of several channels (optional), TIRF module (optional).

Immersion oil kit for microscope objective (Applied Precision).

2. Deconvolution software, for example softWoRx (Applied Precision) for images obtained on a DeltaVision microscope.
3. Image analysis software, for example Metamorph (Molecular Devices).
4. Quantification software, e.g. Microsoft Excel or MatLab (MathWorks).

3. Methods

The major steps in the development of mRNA-MS2/MS2-FP labelling system for the visualisation of mRNA localisation in *Drosophila* oocytes are outlined in Fig. 3.

3.1. Cloning

3.1.1. Generation of mRNA-MS2 Construct

To create the mRNA-MS2 construct, one needs to insert multiple copies of the MS2 binding site into the mRNA of interest. The MS2 binding site is a 19nt sequence that forms a stem–loop secondary structure. Biochemical experiments have shown that a single nucleotide mutation (U at the -5 position to C) in the wild-type MS2 stem–loop sequence increases its affinity for the MS2CP protein about 50-fold (28, 29). Thus, this mutated C-variant version of MS2 stem-loop is recommended for use in creating the mRNA-MS2 construct (ACAUGAGGAUCACCCAUGU – the underlined nucleotide shows the U-5C mutation; [a construct with ten copies of MS2 binding sites is available from the St Johnston lab upon request]).

In creating this construct, it is important to consider several factors. Firstly, it is necessary to decide which form of the gene to use for cloning – the genomic sequence including all exons and introns, the cDNA with the introns spliced out, or a hybrid which includes a reporter of interest (fluorescent proteins, LacZ) fused to the mRNA localisation sequence. In deciding between these possibilities, it is essential to know the location of the mRNA localisation elements, as these sequences must be included in the construct without perturbation. Such elements are often entirely contained within the 3′UTR of the mRNA (for example *bcd* and *nos* mRNAs (30, 31)), and in these cases the use of 3′UTR alone is sufficient to create a functional localisation construct. In some

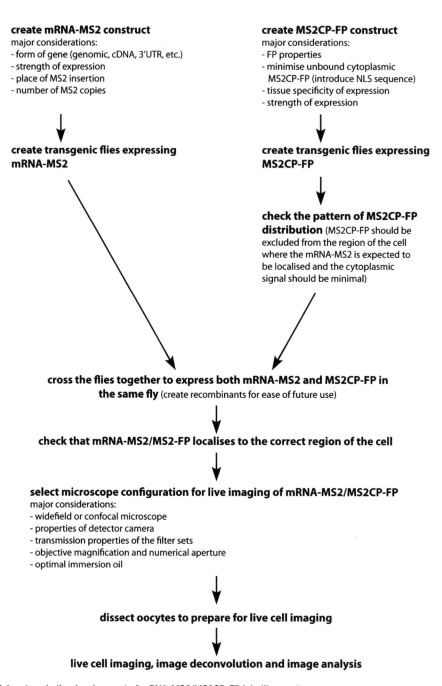

Fig. 3. Major steps in the development of mRNA-MS2/MS2CP-FP labelling system.

cases, such as *gurken* mRNA, targeting elements are located in the coding region of the mRNA, and the use of the entire cDNA is required (12, 32). Moreover, for certain mRNAs it has been shown that the nuclear events, such as splicing, are essential for proper localisation (for example, splicing of the first intron of *osk* mRNA (13)). In these cases, it is essential to use the genomic region of the gene that includes the required introns. In cases where the targeting elements have not been well described, it is advisable to use a genomic region containing all of the regulatory elements of the gene.

The expression level of the construct is important as well. For example, over-expression of certain mRNAs may lead to production of an excess of mRNA molecules, which overload the localisation pathway and prevent proper localisation (for example, overexpression of *osk* mRNA driven by the UAS promoter leads to ectopic accumulation of the mRNA in the middle of the oocyte (33)). In such cases, it is important to place the construct under the control of its endogenous promoter or a promoter that provides similar levels of expression. In cases where the endogenous promoter of the gene is used, the mRNA-MS2 construct can be inserted into the pCaSpeR vector (DGRC). pCaSpeR is a *Drosophila* P-element transformation vector, and contains a white gene selectable marker for selecting transgenic flies and Amp marker for growing the plasmid in bacteria, but lacks any regulatory sequences for construct expression (34). In cases where an exogenous promoter is to be used, the mRNA-MS2 construct can be put into such vectors as pUASp or pUMAT, which already contain the UAS and maternal α tubulin promoters respectively (35).

Secondly, it is crucial to choose an appropriate position to insert the MS2-binding sites into the mRNA of interest. Insertion of this element should not disrupt the sequences responsible for the post-transcriptional control of mRNA (mRNA transport, stability and translation). Such sequences are often located in the 5′- and 3′UTR regions of the mRNAs. In cases where the location of such sequences is unknown, it is often safe to insert MS2 binding sites immediately after the STOP codon of the coding region and before the start of the 3′UTR. Since there is unlikely to be a convenient restriction site in this position, it is usually necessary to introduce a restriction site here by site-directed mutagenesis. Another potential insertion site may be just before the poly-adenylation signal of the mRNA. Alternatively, one can perform a sequence alignment of the 3′UTRs of the orthologues from related species to identify the least conserved regions, and then insert the MS2 sites into one of these regions, as they are less likely to have important regulatory functions.

Finally, one needs to decide how many copies of MS2 binding site to introduce into the mRNA of interest. Each MS2-binding RNA motif can associate with two MS2CP proteins simultaneously. However, two MS2CP proteins (and hence two FP molecules) are probably insufficient to detect a single mRNA molecule and to distinguish it from the unbound cytoplasmic MS2CP-FP. Thus, it is advisable to introduce several tandem copies of the MS2-binding site, and mRNA reporters containing 6–24 copies of MS2 binding sites have been described. The exact number of the sites will depend on the fluorescence intensity levels required and the copy number of mRNA molecules in each localisation complex. For mammalian cells, it was found that insertion of 24 copies of the MS2 sequence is sufficient to visualise single mRNA molecules (36). Although insertion of multiple copies of MS2-binding sites will increase the fluorescence intensity of the labelled mRNA, one also needs to bear in mind that the presence of an excessive number of MS2 sites may disrupt the secondary or tertiary structure of the mRNA, destabilise the molecule, disrupt the assembly of the functional localisation complex, affect the movement of the RNP complex through an increase in the size of the transport particle, or lead to aggregation of several mRNAs together. Thus, it is advisable to insert only as many MS2 repeats as are sufficient for the detection of mRNA particles. The exact number might need to be determined empirically.

Cloning of constructs that have multiple tandem repeats of MS2-binding sequence can lead to recombination in bacteria and to loss of some repeats. It is therefore advisable to grow such plasmids in bacteria that have decreased recombination rates (for example SURE cells [Stratagene] or Stbl2 cells [Invitrogen]).

3.1.2. Generation of MS2CP-FP Construct

The MS2CP coat protein of the RNA bacteriophage MS2 has two functions: it binds to the MS2 binding site in the RNA genome of the virus and acts as a translational repressor, and it associates with other MS2CP proteins to form the icosahedral shell of the virus. It is therefore best not to use the wildtype MS2CP protein for the mRNA-MS2/MS2CP-FP labelling of mRNA, since it will multimerise in the cytoplasm of the cell to form aggregates. Instead, one should use the dlFG deletion mutant, which lacks the FG loop involved in the intersubunit interaction during capsid formation, and is therefore unable to multimerise (37). In addition, a V29I mutant of MS2CP that binds to its RNA recognition motif with higher affinity has been described (37). Thus, the best choice is to use the V29I, dlFG double mutant for the MS2CP-FP (construct available from the St Johnston lab upon request).

The MS2 needs to be fused to a fluorescent protein (FP), and experience in our and other labs has shown that fusing of GFP to the C-terminus of MS2CP does not compromise its ability to bind to the MS2 motif. A large number of fluorescent proteins are now available, and the choice of which one to use will largely depend on the intended application. Some important considerations when choosing the fluorescent protein include its brightness, its photostability and whether the protein is monomeric or forms dimers. In previous work in our lab, the constructs containing eGFP, eYFP (Venus) and Tomato have been successfully used to visualise mRNAs in *Drosophila* ovary. However, new improved versions of the fluorescent proteins have been described recently and might provide certain advantages. Additionally, photoswitchable and photoactivatable fluorescent proteins provide new opportunities for the analysis of the dynamics of mRNA localisation.

In order to increase the signal-to-noise ratio in labelling experiments, it is important to maximally reduce the levels of the unbound MS2CP-GFP in the cytoplasm. This can be achieved by fusing the MS2CP-FP to a nuclear localisation signal (for example the SV40 NLS), which will target unbound MS2CP-FP to the nucleus. We have found that one copy of the SV40 NLS is insufficient to target all of the free MS2CP-FP to the nuclei in the *Drosophila* female germ line, and it might therefore be worth including a second NLS in the construct.

Another important consideration when designing the construct is to decide which promoter to use. Different promoters will express the construct in different tissues, and the strength of expression can vary widely. The use of a ubiquitous promoter can be very convenient, since the same construct can then be used to analyse mRNA localisation in many different tissues. In some cases, however, expression in one tissue may make imaging the neighbouring tissue quite challenging. For example, use of the *hsp83* promoter to express MS2CP-FP will lead to accumulation of GFP in the nuclei of follicle cells, which makes it more difficult to image mRNA molecules in the underlying oocyte. In these cases, the use of a germline specific promoter (for example the *osk* promoter) helps overcome this issue.

The strength of MS2CP-FP expression is equally important. The amount of MS2CP-FP produced should be sufficient to occupy all of the MS2-binding sites in the mRNA-MS2 molecule and thus to ensure the RNA has maximum fluorescence intensity. On the other hand, its important to avoid over-expression of the MS2CP-FP construct, as an excess of these molecules in the cytoplasm will lead to a low signal-to-noise ratio and will make it difficult to identify the specific mRNA-MS2/MS2CP-FP complexes. For *Drosophila* egg chambers, the *osk* promoter (germline specific) and *hsp83* (ubiquitous) promoters are expressed at low but sufficient levels, and both have been successfully used to study the localisation of different mRNAs in the oocyte ((18, 21, 24, 38), our unpublished results). At the same time, our experience with using UAS-GAL4 system to express NLS-MS2CP-FP showed that this promoter drives over-expression of the protein that leads to a high background of cytoplasmic MS2CP fluorescence, aggregation of the MS2CP-FP protein in the cytoplasm, and an excessively high intensity of fluorescence in the nuclei, which makes it difficult to image the cytoplasm.

The MS2-FP construct can also incorporate other tags, if required. For example, it is possible to introduce protein tags for affinity purification into the construct, and then use these to pull down the RNP complexes for biochemical analysis.

3.1.3. Creation of Transgenic Flies Expressing mRNA-MS2 and MS2CP-FP Constructs

In order to visualise the mRNA molecules, one needs to co-express the MS2CP-FP protein and mRNA-MS2 construct in the same cell.

Once the constructs are ready they have to be injected into fly embryos in order to create transgenic fly stocks (39, 40). The flies expressing each construct on its own can be crossed together in order to express both constructs in the same fly. This will label the mRNA of interest with FP. To simplify the analysis of mRNA localisation in different genetic backgrounds, it is advisable to create recombinants that carry both transgenes on the same chromosome (41). In cases where the fluorescence level of the construct is too low, it is possible to improve the signal by increasing the number of copies of the mRNA-MS2 transgene (for example through recombination of several constructs onto the same chromosome or through mobilisation of an existing insert).

Before embarking on the analysis of mRNA transport in these flies, it is important to perform several crucial controls. For the fly stock expressing MS2CP-FP on its own, it is important to check that the fluorescence signal is restricted to the nuclei of the cells (if the NLS was included into the construct) and that the fluorescent signal outside of the nuclei is minimal (see for example Fig. 2a–i). Importantly, MS2CP-FP should not be enriched in any region of the cell where the mRNA-MS2 is expected to localise.

For the mRNA-MS2 construct it is important to confirm that the insertion of the MS2 sites has not disrupted the regulatory regions of the transcript. First of all, one should check whether the mRNA-MS2 localises to the correct region of the cell. The pattern of mRNA-MS2 distribution should be identical to that of the endogenous mRNA visualised by in situ hybridisation (see for example Fig. 2a–i). The most rigorous test of whether the RNA-MS2 is functional is to check whether this transgene can rescue the phenotype of an RNA null mutant of the gene under investigation. This experiment will only be possible if an RNA null mutant is available and if the RNA-MS2 construct was designed to include all the essential elements of the gene (both coding and regulatory).

3.2. Preparation of Samples for Live Imaging

It is important to observe the transport of mRNA molecules in conditions as close to physiological as possible.

To study the mRNA transport in *Drosophila* oocytes, one needs to have young healthy females expressing the correct constructs. Fly oogenesis is very sensitive to the outside environment and can be arrested if the conditions are unfavourable. Additionally, it has been shown that suboptimal conditions can cause the mRNAs to be targeted to stress granules (our own observations; see also ref. 42). Thus, it is important to use young healthy well-fed females for analysis.

Usually, we collect 2–3 days old females of the correct genotype and feed them overnight with yeast paste in uncrowded vials at 25°C. To increase the number of oocytes produced it is advisable to add several males to the vial. In cases where one wants to study the effects of microtubule or actin depolymerising drugs on mRNA localisation, the flies should be starved for 2–4 h before allowing to feed overnight on yeast paste containing either 100 μg/ml colcemid (Sigma) for microtubule depolymerisation or 200 μg/ml latrunculin A (Sigma) for actin depolymerisation.

To dissect the flies, a small drop of Voltalef 10S oil is placed on a thin coverslip (thickness No. 1, VWR) and one fly briefly anaesthetised with CO_2 is placed directly into this drop of oil. The ovaries are hand-dissected with sharp tweezers directly onto the coverslip, and the single ovarioles are separated using tungsten needles (making sure to remove the muscle sheath surrounding the ovarioles). One needs to be careful to cause the minimum amount of damage to the oocytes. Once dissected, the oocytes should be imaged immediately. Oocytes of young stages can survive in Voltalef oil for 1–1.5 h, and should be imaged within this time frame.

3.3. Imaging

Dissected oocytes should be imaged on an inverted microscope. The choice of the microscope system depends on the experiment. If one wants to examine the steady state distribution of the mRNA-MS2/MS2CP-FP molecules, it is possible to image on laser scanning confocal or multiphoton microscopes. These microscopes are relatively slow, but are efficient in eliminating the noise caused by the out-of-focus light. Thus, they provide images with a good signal-to-noise ratio and are ideal for imaging deep into the sample. One can therefore image throughout the entire depth of the oocyte and recreate a 3-D image of the cell. These microscopes are also good for the photobleaching/photoswitching experiments in which one examines the dynamics of pre-localised mRNA molecules.

For the experiments where one wishes to image the movement of individual mRNA molecules in real time, one needs to use microscope configurations that allow fast image acquisition rates. It has been shown that the speed of mRNA movement can reach up to 2 μm/s, and in order to record such movements, one must image at least several frames per second.

In general, confocal scanning microscopes are too slow for such applications, and the use of wide-field microscopy or spinning disk laser microscopy is recommended. It is has also been found that wide-field deconvolution microscopy combined with image deconvolution is usually superior to laser-scanning confocal microscopy for the imaging of faint signals, which is often the case when observing single mRNA molecules labelled with mRNA-MS2/MS2CP-FP approach (43). In our lab, we successfully use a DeltaVision microscope (Applied Precision; with an inverted Olympus IX71 microscope) in order to visualise movement of oskMS2/MS2CP-GFP and bcdMS2/MS2CP-GFP RNPs. Another microscope setup that can be used for these experiments is a spinning disk confocal microscope, as it also allows imaging at relatively fast frame rates. However, we have found that this system causes faster bleaching of the MS2CP-FP fluorescence.

When selecting the microscope configuration it is important to consider the detector camera, the objectives, the filter cubes, and the ability to switch rapidly between wavelengths for imaging several channels simultaneously.

In order to visualise movement of single molecules in real time, the detector camera should collect as much of the emission signal as possible. This can be achieved with one of the modern high quality cameras such as a cooled CCD (charge coupled device) or EMCCD (electron multiplying charge coupled device) camera. These cameras are characterised by high sensitivity and are capable of fast image acquisition rates.

The next consideration is the objective. In order to detect small RNA-MS2/MS2CP-FP particles, it is essential to image at the highest magnification possible. For example, fast-moving oskMS2 and bcdMS2 RNP particles can only be detected with a 100× objective. The numerical aperture of the objective is an important attribute. The higher the numerical aperture, the better the resolution that one can achieve. For our imaging, we use UPlanSApo 100×, 1.4 NA immersion oil objective (Olympus). In cases where the molecules that need to be imaged are in close proximity to the coverslip (less than 100 nm into the sample), one can image with a total internal reflection (TIRF) system using special TIRF objectives.

Selection of the immersion oil for the objective is crucial for optimal imaging. It is useful to obtain an immersion oil kit (for example a kit from Applied Precision) that contains a series of oils of different refractive indices, and use it to determine which oil is best for the sample under investigation. Comparison of different immersion oils in our lab showed that the best oil for imaging oocytes dissected in Voltalef 10S oil is that with a refractive index of 1.534. If the imaging medium were changed from Voltalef oil to a water-based medium, the optimal oil would be different.

The choice of the filter sets is very important and depends on the properties of the FP used. The excitation and emission filters should be optimised to transmit only the relevant wavelengths of light while blocking the rest. The selection is based on the excitation–emission spectra of the FP and on the transmission curves of the filter set supplied by the manufacturer. The correct choice of filter will maximise the brightness of the RNA-MS2/MS2CP-FP particle signal and reduce the background. In the experiments where imaging of several channels is required, the speed of switching between filters becomes important. Multicolour imaging can be performed with the use of either a filter wheel based system or a device that allows simultaneous imaging of several channels (for example DualView or QuadView). The filter wheel allows the use of a wide range of different filter sets, but it is only possible to image the different wavelengths sequentially, which reduces the speed of imaging. In experiments that require simultaneous image acquisition or where the rate of image acquisition is a priority, the system of choice is either the DualView or the QuadView. These devices split the emitted light into two/four channels, which are recorded simultaneously on separate halves/quadrants of the CCD camera chip. These devices can be used to co-visualise mRNA-MS2/MS2CP-FP molecules and other fluorescently-labelled proteins.

The sample can be Illuminated with either mercury or xenon arc lamps. Xenon lamps provide radiation across the visible spectrum and are not subject to the flicker characteristic of mercury lamps. Thus, they can be used to image a wide range of fluorophores, and ensure uniform illumination of the sample. At the same time, certain mercury lamps can provide a stronger intensity of illumination at discrete wavelengths and will perform better for imaging in that specific part of spectrum. Thus, the choice between lamps will depend on the fluorophores used. In our lab we have successfully used both types of lamps for imaging MS2CP-eGFP and MS2CP-eYFP proteins.

3.3.1. Imaging of Drosophila Oocytes

Prior to imaging, it is important to locate several healthy-looking oocytes of the desired stage of oogenesis using a low magnification objective (for example $20 \times NA = 0.75$ air objective, Olympus), and then record the stage coordinates of these oocytes (softWoRx software from Deltavision has a special function that automatically records the coordinates). In order to minimise bleaching of the samples, the oocytes can be located under bright field illumination to minimise the exposure of the sample to fluorescence. Once the oocytes have been located, one switches to the high-magnification objective for high-resolution fluorescence imaging.

Once at high magnification, one has to manually focus on the focal plane of the cytoplasm containing mRNA particles and then choose the best fluorescence exposure time and fluorescence intensity (see Note 1). There is always a compromise between achieving the maximum speed of image acquisition and minimising the intensity of the fluorescence illumination to avoid photobleaching and photodamage to the cell. Before acquiring images, it is important to make sure that the softWoRx software is set up correctly and records the right pixel size, objective lens type, and fluorescence filter sets. This information will be crucial at later steps during image deconvolution. It is also important to ensure that the speed of image acquisition is constant throughout the entire time-lapse movie, as this will greatly ease the subsequent data analysis (especially when estimating the speed of particle movement).

For the visualisation of oskMS2 and bcdMS2 particles on the Deltavision microscope, we were able to image with image acquisition speeds of up to three to five frames per second for periods of 1–2 min (Fig. 2j–m). After that time, the fluorescence starts to become photobleached. Due to the limit in the rate of image acquisition, we were unable to image in three dimensions. Thus, we were not able to record the full path of mRNA molecules from their site of synthesis to their final destination. To compensate for this, we acquired many two-dimensional movies at different focal planes and compared the parameters of particle movement in different regions of the oocyte. Sufficient sampling at different focal planes at different time points of oogenesis should provide a complete picture of mRNA movement during oogenesis. Future improvements in imaging technologies and the development of new fluorescent proteins may allow the imaging of the complete paths of mRNA molecules in three dimensions.

The Deltavision microscope automatically saves the acquired images in the Deltavision format and also creates a log file, which contains all data relevant to the properties of the movie created. The movie in the Deltavision format can be exported to the DeltaVision stand-alone analysis computer and can be deconvolved using the Applied Precision softWoRx software. This software uses a constrained iterative deconvolution algorithm, which efficiently reduces noise through the removal of the out-of-focus light, increasing the contrast of the resulting image. Once the deconvolution of the image is complete, the movie can be exported as a series of TIFF files, which can be used for image analysis in other software packages.

3.3.2. Image Analysis and Particle Tracking

The data from FRAP and FLIP experiments is analysed by measuring the changes in fluorescence intensities of bleached and unbleached regions of the oocyte over time. There are different software packages that allow such analyses, such as Metamorph. The resulting data can be used to estimate the dynamics of mRNA-MS2 in the analysed region of the cell. With such experiments, it is important to correct for the photobleaching due to image acquisition. This can be done by measuring the changes in fluorescence intensity in a cell that is in the same field of view, but was not subjected to bleaching (for example a follicle cell nucleus or a neighbouring oocyte).

For the analysis of high-resolution movies, it is often desirable to obtain quantitative information about the parameters of mRNA particle movement, and this requires the tracking of many individual fluorescent particles over time. It is important to track a statistically sufficient number of particle trajectories to be able to draw significant conclusions from these data. In addition, it is important to analyse a sufficient number of oocytes per genotype in order to account for the variation between oocytes. The exact number required for the analysis will depend on the quality of the data and on the type of conclusions one wishes to draw.

There are a number of different software packages available for the automatic tracking of fluorescent particles. These programs are worth investigating to see whether they are capable of detecting all the particles of interest in a particular setting, and ignore other fluorescent or autofluorescent objects (e.g. yolk and vesicles). All those we have tested on *oskar* mRNA failed to track all of the particles because of the high particle density and the difficulty in distinguishing between fast moving particles that pass close to each other.

In many cases, the fluorescence of RNA-MS2/MS2CP-GFP particles is too low, while the background fluorescence of the ooplasm too high to use these types of software. In such cases, one has to use manual tracking algorithms. One software package that can allow this is MetaMorph Premier Offline (using "Track points" application; [Molecular Devices]). This software allows manual tracking of all the particles of interest and then logs the coordinates of each data point into a separate file. These data can then be used for the determination of parameters, such as particle speed, direction of movement, and length of movement. Analysis of these data can be performed in MatLab or Microsoft Excel. When performing manual tracking, it is important to track all of the detectable particles in the cell, in order to avoid introducing of biases into the data.

4. Notes

1. In cases where no movement of RNA particles is observed, one should check whether the oocyte is still alive by acquiring images every 10 s to see whether there are any cytoplasmic movements. In young oocytes, there are noticeable short-range movements of limited amounts of cytoplasm called cytoplasmic seething. In older oocytes, the entire ooplasm moves in circular fashion around the oocyte – a phenomenon called cytoplasmic streaming. The presence of these movements indicates that the oocyte is viable.

Acknowledgements

K.B. was supported by the Darwin Trust of Edinburgh. D. St J. was supported by a Wellcome Trust Principal Research Fellowship.

References

1. St Johnston, D. (2005) Moving messages: the intracellular localization of mRNAs, *Nat Rev Mol Cell Biol 6*, 363–375.
2. Kim-Ha, J., Smith, J. L., and Macdonald, P. M. (1991) *oskar* mRNA is localized to the posterior pole of the *Drosophila* oocyte, *Cell 66*, 23–35.
3. Ephrussi, A., Dickinson, L. K., and Lehmann, R. (1991) Oskar organizes the germ plasm and directs localization of the posterior determinant *nanos*, *Cell 66*, 37–50.
4. Ephrussi, A., and Lehmann, R. (1992) Induction of germ cell formation by *oskar*, *Nature 358*, 387–392.
5. Gavis, E. R., and Lehmann, R. (1992) Localization of *nanos* RNA controls embryonic polarity, *Cell 71*, 301–313.
6. Berleth, T., Burri, M., Thoma, G., Bopp, D., Richstein, S., Frigerio, G., Noll, M., and Nusslein-Volhard, C. (1988) The role of localization of *bicoid* RNA in organizing the anterior pattern of the *Drosophila* embryo, *Embo J 7*, 1749–1756.
7. Ephrussi, A., and Johnston, D. (2004) Seeing is believing the bicoid morphogen gradient matures, *Cell 116*, 143–152.
8. Neuman-Silberberg, F. S., and Schupbach, T. (1993) The Drosophila dorsoventral patterning gene *gurken* produces a dorsally localized RNA and encodes a TGF alpha-like protein, *Cell 75*, 165–174.
9. Cha, B., Koppetsch, B., and Theurkauf, W. (2001) In vivo analysis of Drosophila bicoid mRNA localization reveals a novel microtubule-dependent axis specification pathway, *Cell 106*, 35–46.
10. MacDougall, N., Clark, A., MacDougall, E., and Davis, I. (2003) *Drosophila gurken* (TGFalpha) mRNA localizes as particles that move within the oocyte in two dynein-dependent steps, *Dev Cell 4*, 307–319.
11. Manseau, L., Calley, J., and Phan, H. (1996) Profilin is required for posterior patterning of the *Drosophila* oocyte, in *Development 122*, 2109–2116.
12. Van De Bor, V., Hartswood, E., Jones, C., Finnegan, D., and Davis, I. (2005) *gurken* and the I factor retrotransposon RNAs share common localization signals and machinery, *Dev Cell 9*, 51–62.
13. Hachet, O., and Ephrussi, A. (2004) Splicing of *oskar* RNA in the nucleus is coupled to its cytoplasmic localization, in *Nature 428*, 959–963.
14. Long, R. M., Gu, W., Meng, X., Gonsalvez, G., Singer, R. H., and Chartrand, P. (2001) An exclusively nuclear RNA-binding protein affects asymmetric localization of ASH1 mRNA and Ash1p in yeast, *J Cell Biol 153*, 307–318.
15. Oleynikov, Y., and Singer, R. H. (2003) Real-time visualization of ZBP1 association with beta-actin mRNA during transcription and localization, in *Current Biology 13*, 199–207.
16. Bratu, D. P., Cha, B. J., Mhlanga, M. M., Kramer, F. R., and Tyagi, S. (2003) Visualizing the distribution and transport of mRNAs in living cells, in *Proc Natl Acad Sci U S A 100*, 13308–13313.
17. Wang, S., and Hazelrigg, T. (1994) Implications for *bcd* mRNA localization from spatial distribution of exu protein in *Drosophila* oogenesis., in *Nature 369*, 400–403.
18. Zimyanin, V. L., Belaya, K., Pecreaux, J., Gilchrist, M. J., Clark, A., Davis, I., and St Johnston, D. (2008) In vivo imaging of *oskar* mRNA transport reveals the mechanism of posterior localization, *Cell 134*, 843–853.
19. Bertrand, E., Chartrand, P., Schaefer, M., Shenoy, S., Singer, R., and Long, R. (1998) Localization of ASH1 mRNA particles in living yeast, *Mol Cell 2*, 437–445.
20. Rodriguez, A. J., Condeelis, J., Singer, R. H., and Dictenberg, J. B. (2007) Imaging mRNA movement from transcription sites to translation sites, *Seminars in cell & developmental biology 18*, 202–208.
21. Weil, T. T., Forrest, K. M., and Gavis, E. R. (2006) Localization of *bicoid* mRNA in late oocytes is maintained by continual active transport, in *Dev Cell 11*, 251–262.
22. Forrest, K., and Gavis, E. (2003) Live imaging of endogenous RNA reveals a diffusion and entrapment mechanism for nanos mRNA localization in Drosophila, *Current Biology 13*, 1159–1168.
23. Jaramillo, A. M., Weil, T. T., Goodhouse, J., Gavis, E. R., and Schupbach, T. (2008) The dynamics of fluorescently labeled endogenous gurken mRNA in *Drosophila*, *J Cell Sci 121*, 887–894.
24. Weil, T. T., Parton, R., Davis, I., and Gavis, E. R. (2008) Changes in *bicoid* mRNA anchoring highlight conserved mechanisms during the oocyte-to-embryo transition, *Curr Biol 18*, 1055–1061.
25. Zhang, Z., and Krainer, A. R. (2007) Splicing remodels messenger ribonucleoprotein architecture via eIF4A3-dependent and -independent recruitment of exon junction complex components, *Proc Natl Acad Sci U S A 104*, 11574–11579.

26. Lange, S., Katayama, Y., Schmid, M., Burkacky, O., Brauchle, C., Lamb, D. C., and Jansen, R. P. (2008) Simultaneous transport of different localized mRNA species revealed by live-cell imaging, *Traffic (Copenhagen, Denmark) 9*, 1256–1267.
27. Sullivan, W., Ashburner, M., and Hawley, R. S. (2000) *Drosophila Protocols*, Cold Spring Harbor Laboratory Press, Cold Spring Harbor, New York.
28. Lowary, P. T., and Uhlenbeck, O. C. (1987) An RNA mutation that increases the affinity of an RNA-protein interaction, *Nucleic acids research 15*, 10483–10493.
29. Valegard, K., Murray, J. B., Stonehouse, N. J., van den Worm, S., Stockley, P. G., and Liljas, L. (1997) The three-dimensional structures of two complexes between recombinant MS2 capsids and RNA operator fragments reveal sequence-specific protein-RNA interactions, *J Mol Biol 270*, 724–738.
30. Gavis, E. R., Curtis, D., and Lehmann, R. (1996) Identification of cis-acting sequences that control *nanos* RNA localization, *Dev Biol 176*, 36–50.
31. Macdonald, P. (1990) *bicoid* mRNA localization signal: phylogenetic conservation of function and RNA secondary structure, *Development 110*, 161–171.
32. Thio, G. L., Ray, R. P., Barcelo, G., and Schupbach, T. (2000) Localization of *gurken* RNA in *Drosophila* oogenesis requires elements in the 5′ and 3′ regions of the transcript, *Dev Biol 221*, 435–446.
33. Zimyanin, V., Lowe, N., and St Johnston, D. (2007) An Oskar-dependent positive feedback loop maintains the polarity of the *Drosophila* oocyte, in *Current Biology*.
34. Pirrotta, V. (1988) Vectors for P mediated transformation in Drosophila, *Biotechnology 10*, 437–456.
35. Rorth, P. (1998) Gal4 in the *Drosophila* female germline, *Mech Dev 78*, 113–118.
36. Fusco, D., Accornero, N., Lavoie, B., Shenoy, S. M., Blanchard, J. M., Singer, R. H., and Bertrand, E. (2003) Single mRNA molecules demonstrate probabilistic movement in living Mammalian cells, in *Current Biology 13*, 161–167.
37. Peabody, D. S., and Ely, K. R. (1992) Control of translational repression by protein-protein interactions, *Nucleic Acids Research 20*, 1649–1655.
38. Wagner, C., Palacios, I., Jaeger, L., and St Johnston, D. (2001) Dimerization of the 3′ UTR of *bicoid* mRNA involves a two-step mechanism, *Journal of Molecular Biology 313*, 511–524.
39. Rubin, G. M., and Spradling, A. C. (1982) Genetic transformation of *Drosophila* with transposable element vectors, *Science 218*, 348–353.
40. Spradling, A. C., and Rubin, G. M. (1982) Transposition of cloned P elements into *Drosophila* germ line chromosomes, *Science 218*, 341–347.
41. Greenspan, R. J. (2004) *Fly Pushing: the theory and practice of Drosophila Genetics*, 2nd ed., Cold Spring Harbor Press, Cold Spring Harbor, New York.
42. Snee, M. J., and Macdonald, P. M. (2009) Dynamic organization and plasticity of sponge bodies, *Dev Dyn 238*, 918–930.
43. Swedlow, J. R., Hu, K., Andrews, P. D., Roos, D. S., and Murray, J. M. (2002) Measuring tubulin content in Toxoplasma gondii: a comparison of laser-scanning confocal and wide-field fluorescence microscopy, *Proc Natl Acad Sci U S A 99*, 2014–2019.
44. Zuker, M. (2003) Mfold web server for nucleic acid folding and hybridization prediction, *Nucleic acids research 31*, 3406–3415.

Part IV

Use of Cell Fractionation to Demonstrate the Sub-Cellular Localization of RNA

Chapter 18

Genome-Wide Analysis of RNA Extracted from Isolated Mitochondria

Erez Eliyahu, Daniel Melamed, and Yoav Arava

Abstract

Isolating mitochondria by subcellular fractionation is a well-established method for retrieving intact and functional mitochondria. This procedure has been used to identify proteins of the mitochondria and to explore import mechanisms. Using the same method, it was shown that mitochondria can be purified along with cytoplasmic ribosomes and nuclear-encoded mRNAs attached to the outer membrane. Combining this procedure with DNA microarray analysis allows for global identification of the mRNAs associated with mitochondria, and hence a better understanding of the underlying molecular mechanisms. In this chapter, we will describe a procedure for the isolation of mitochondria from yeast and RNA purification. We will then describe the process of labeling and hybridization to DNA microarrays, and comment on a few aspects of the data analysis.

Key words: DNA microarrays, Mitochondria, RNA localization

1. Introduction

A key feature of eukaryotic cells is their organization into distinct compartments having specific functions. Each cellular compartment contains a unique set of proteins that are essential for its activity. As protein synthesis occurs throughout the cytoplasm, these proteins need to be targeted to their proper compartment. This targeting may occur through the utilization of various chaperones and import receptors that bind the fully synthesized protein and escort it to the correct site (1–3). It is well established, however, that many organellar proteins are synthesized near their site of activity and not at distant sites in the cytoplasm. ER resident proteins, for example, are translated directly into the ER membrane or lumen (4). Such a localized protein synthesis minimizes the chances of function at inappropriate sites and may increase the efficiency of protein targeting. Consistent with this process, many mRNAs were found to localize near the site of action of their encoded protein. Moreover, in many cases, mRNA localization occurs prior to the initiation of protein synthesis, and mRNA mislocalization may be associated with mislocalization of its encoded protein (5).

The mitochondrion, the cell respiratory organelle, is enclosed with a double membrane. It contains one circular chromosome that contains a small number of genes with roles mainly in protein synthesis. However, the majority of mitochondrial proteins is encoded in the nucleus and must be transported into the mitochondria (6). It was suggested more than 30 years ago that many of these proteins are synthesized near the mitochondria and transported co-translationally. This was based on the observation that ribosomes associated with mRNAs encoding mitochondrial proteins accumulate on the surface of yeast mitochondria (7–10). More recent in vivo studies, in which either the import of fully synthesized proteins into mitochondria (11) or the targeting of ribosomes to the mitochondria surface was inhibited (12), further supported the notion that many mitochondrial proteins are inserted into the mitochondria while being translated. The predominance of this co-translational process became apparent following genome-wide studies in *Saccharomyces cerevisiae*, which identified the mRNA population associated with mitochondria (13–15). Whether these mRNAs approach the mitochondria prior to translation initiation or while being translated is unclear; some models suggest that targeting occurs while the protein is being synthesized, and the translated mitochondrial targeting sequence (MTS) assists in the association of the translation complex (mRNA-ribosome and nascent peptide) with the mitochondria (16). In agreement with this model, we have recently shown that either changing just two amino acids in an MTS or deleting a mitochondria import receptor leads to a profound decrease in the mitochondrial association of mRNAs (15). On the other hand, noncoding domains (most significantly 3′ UTRs) were also shown to be involved in mRNA association to the mitochondria (17, 18). The association through these domains appears to necessitate in some cases an RNA-binding protein from the PUM family (Puf3) (19). Since many PUM family proteins are known to be involved in translation inhibition (20), it is possible that the role of Puf3 is to exclude translation of its target mRNA while en route to the mitochondria. Future studies may reveal the exact mechanisms and dynamics of mRNA targeting to the mitochondria.

2. Materials

1. Galactose Growth Medium – autoclaved 1% yeast extract, 2% Bacto peptone medium supplemented with 2% filtered galactose.
2. Dithiothreitol (DTT) Buffer – freshly prepared 0.1 M Tris–HCl at pH 9.4 supplemented with 10 mM DTT.
3. Sorbitol Buffer – 1.2 M sorbitol in Tris–HCl at pH 7.4 (use filtered buffer).
4. Zymolyase – Zymolyase 20T (Seikagaku America, INC). Should be weighed and dissolved in Sorbitol buffer just before use.
5. Recovery Medium – Galactose growth medium supplemented with 1 M sorbitol.
6. Lysis Buffer – 0.6 M mannitol, 30 mM Tris–HCl at pH 7.6, 5 mM MgAc, 100 mM KCl and freshly added 0.1 mg/ml cycloheximide (CHX), 0.5 mg/ml Heparin, and 1 mM phenylmethanesulfonyl fluoride (PMSF). Filter and use ice-cold. Can be stored at 4°C for at least a month.
7. Phenol:Chloroform (5:1) at pH 4.7 (Sigma P1944).
8. ImProm-II reverse transcription system (Promega, A3802).
9. 25× Amino-allyl mix – 12.5 mM of each dATP, dGTP, dCTP, 5 mM dTTP, and 7.5 mM amino-allyl dUTP (Ambion #8439).
10. DNA clean and concentrator kit (Zymo Research, D4004).
11. Slides Blocking Buffer – 0.5% BSA (A7906 sigma), 5× saline-sodium citrate buffer (SSC), and 1% SDS.
12. Hybridization Buffer – 2× MWG buffer (Ocimum Biosolutions 1180-000010).
13. PolyA (Sigma, P9403) dissolved to 10 µg/µl with water. Keep at −20°C.

3. Methods

The following protocol is designed for global determination of mRNA association with the mitochondria of yeast *S. cerevisiae*. It entails enzymatic degradation of the cell wall and gentle breakage of the plasma membrane by moderate mechanical force, without the use of detergent. Heavy complexes, which include mitochondria and other large compartments, are precipitated at $10,000 \times g$ and RNA is extracted from this crude mitochondrial pellet. The

RNA is then labeled with red fluorescent dye and hybridized to DNA microarrays together with green-labeled reference RNA. The resulting fluorescent signals represent the relative association of mRNAs to the mitochondria. Herein we provide a detailed protocol for this procedure, which could be modified easily for the analyses of cells grown under different conditions or of deleted genes that may be involved in the targeting process.

3.1. Mitochondrial Fractionation

This part describes a procedure for the isolation of mitochondria that results in a crude mitochondrial fraction (i.e., that may include other cellular compartments) (Fig. 1). Further purification using a sucrose density gradient (e.g., according to (21)) is possible, yet it may lead to loss of mitochondria-associated mRNAs and therefore reduce the quality of the results. In some experimental settings (such as comparative analyses), the presence of other cellular parts (e.g., plasma membrane and ER) may not pose a limitation as they will be present at similar levels in all samples.

1. Grow 500 ml of cells to OD 0.8 at A_{600} in 30°C on Galactose Growth Medium (see Note 1).
2. Centrifuge cells at 3,000×g for 4 min at room temperature, resuspend cells in double distilled water, centrifuge again, and discard the supernatant.
3. Resuspend the pellet in 20 ml of DTT Buffer. Incubate at 30°C for 10 min with gentle shaking (see Note 2).
4. Centrifuge the sample at 3,000×g for 4 min at room temperature, discard the supernatant, and resuspend the pellet in 10 ml Sorbitol Buffer. Measure the OD, at A_{600}, of a 20 μl aliquot diluted with 980 μl of water.

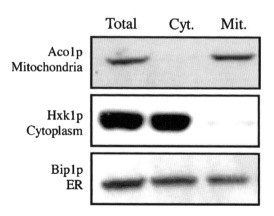

Fig. 1. Western analysis of fractionation quality. Equal fractions from the unfractionated extract (Total), cytosolic extract (Cyt), and mitochondria (Mit) were resolved by SDS-PAGE, transferred to a nitrocellulose membrane and tested with antibodies recognizing the indicated marker proteins.

5. To break the cell wall and convert cells to spheroplasts, add 36 mg of Zymolyase. Incubate cells with gentle shaking for 15 min at 30°C. Completion of cell wall hydrolysis is verified by mixing 20 μl from the sample with 980 μl of water and measuring the OD at A_{600}; a tenfold decrease in the OD, at A_{600}, compared to the previous step is expected because spheroplasts are quickly lysed in water. If not, continue the incubation until the required decrease is reached (see Note 3).

6. Centrifuge the sample at 2,000 × g for 4 min at room temperature. Wash once with 20 ml of Sorbitol Buffer, discard the supernatant, and resuspend the pellet in 100 ml Recovery Medium. Incubate the spheroplasts for 2 h at 30°C while shaking (see Note 4).

7. Add 0.1 mg/ml CHX and immediately spin the samples at 2,000 × g for 4 min at 4°C (see Note 5).

8. Wash the sample twice with cold 5 ml Lysis Buffer.

9. Resuspend the sample with 5 ml Lysis Buffer and gently break the cells with 15–20 strokes with a tight-fitting Dounce homogenizer. Move the sample to 13-ml tubes (e.g., Sarstedt D-51588).

10. Centrifuge the sample at 450 × g for 6 min at 4°C and carefully place the supernatant in a new tube. Avoid touching the pellet, which contains unbroken cells and nuclei. Set aside 25% of the supernatant, which can be used to extract RNA (by the "hot phenol" method (22)) or proteins (by adding a protein loading buffer). This will serve as an unfractionated ("Total") control.

11. Centrifuge the supernatant at 10,000 × g for 10 min at 4°C. Place the supernatant in another tube. RNA or proteins can be extracted from this sample and serve as a cytosolic control.

12. Resuspend the pellet, which includes the mitochondria, with 5 ml of cold Lysis Buffer, centrifuge at 10,000 × g for 10 min at 4°C, and remove residual cytosolic components.

13. Resuspend the pellet with 3 ml of cold Lysis Buffer. This is the crude mitochondrial fraction.

14. The quality of the fractionation procedure can be evaluated by Western analysis. Aliquots of the unfractionated (Total) sample (Subheading 3.1, step 10), cytosolic sample (Subheading 3.1, step 11), and mitochondrial sample (Subheading 3.1, step 13) are resolved by SDS-PAGE and probed with antibodies to various cellular markers. Figure 1 presents representative results of such an analysis. Two aspects should be evaluated: (1) losses during fractionation that will be apparent if the sum of signals in the Mit and Cyt fractions

is not equal to the signal in the Total sample; and (2) purity of the fractions – the mitochondrial sample should be devoid of cytosolic markers and the cytosolic fraction devoid from mitochondria markers. It should be noted that markers of some cellular organelles (most significantly the ER) usually appear in the mitochondria sample. This presence may be due to their similar mass or to their established physical connections (23–25).

3.2. RNA Extraction

1. Add one volume of 8 M *Guanidinium-HCl* to the mitochondrial fraction, mix gently, add two volumes of 100% ethanol and mix gently by inverting the tube. Incubate in –20°C for at least 2 h. (The same extraction procedure can be done for the cytosol fraction or the unfractionated sample if RNA extraction is desired).

2. Centrifuge the samples at $11,000 \times g$ for 20 min at 4°C and discard the supernatant. Wash the pellet with 5 ml of ice-cold 80% ethanol (do not resuspend) and centrifuge again at $11,000 \times g$ for 20 min at 4°C. The pellet is visible and stable at this stage due to the presence of heparin.

3. Remove the supernatant and resuspend the pellet with 400 μl RNAse-free water. Transfer the sample to a microtube, and precipitate the RNA by adding 0.1 volume of 3 M sodium acetate at pH 5.2 and two volumes of 100% ethanol. Incubate for at least 2 h at –20°C and spin at maximal speed for 20 min at 4°C. Wash with ice-cold 80% ethanol and spin again for 20 min at 4°C.

4. Resuspend the pellet with 650 μl of RNase-free water. Add an equal volume of Phenol:Chloroform (5:1) (pH 4.7) and vortex vigorously. Spin at top speed for 5 min at room temperature. Take 500 μl of the aqueous phase (upper layer) into a new microtube (see Note 6).

5. Add 350 μl of RNase-free water and 150 μl 10 M LiCl (final concentration of 1.5 M), and incubate overnight at –20°C. Thaw the sample on ice and centrifuge at top speed for 20 min at 4°C. Wash the pellet carefully (as the pellet is transparent and unstable) with 200 μl of cold 80% ethanol and resuspend in 150 μl of RNase-free water (see Note 7).

6. Precipitate the RNA with sodium acetate as in Subheading 3.2, step 3, wash with 80% ethanol, and air dry. Resuspend the pellet in RNase-free water. The sample can be stored at –80°C.

3.3. Fluorescent Labeling

The protocol below is a general one that can be used for labeling RNA obtained by various methods (e.g., "hot phenol" extraction, polysomal mRNA preparations, or co-immunoprecipitated mRNAs).

The RNA is converted to cDNA that contains an amino-allyl modified nucleotide (usually deoxyuridine) by reverse transcription reaction. This nucleotide is then coupled with a fluorescent dye. We usually label the mitochondria fraction with a Cy5 fluorescent dye and hybridize it to a DNA microarray together with a reference RNA that is labeled with Cy3 dye. The reference sample allows correcting problems in hybridization and variations in spot quality (see Note 8 for selection of reference sample). Thus, for every gene, the microarray results are best interpreted as the ratio of fluorescence signals between the sample and the reference.

1. Mix 15–50 µg RNA with 5 µg Oligo dT (T20VN (V = any nucleotide except T)). Adjust to 15.5 µl with nuclease-free water (see Note 9).

2. Denature RNA secondary structures by incubating the mixture for 10 min at 70°C and then transfer to ice for 10 min.

3. We utilize Promega's ImProm-II reverse transcription system for cDNA synthesis. Mix 6 µl 5× reaction buffer, 4 µl 25 mM $MgCl_2$, 1.2 µl 25× amino-allyl mix, 3.0 µl reverse transcriptase, and 0.3 µl nuclease-free water (total volume of 14.5 µl) and incubate at 42°C for 2 h.

4. Degrade RNA by adding 10 µl of 1 N NaOH and 10 µl of 0.5 M EDTA, and incubate at 65°C for 15 min.

5. Neutralize by adding 25 µl of 1 M HEPES (pH 7.0) and 25 µl of nuclease-free water.

6. Purify the cDNA using a DNA clean and concentrator kit (Zymo Research, D4004), and resuspend in 9 µl of nuclease-free water. The amino-allyl labeled cDNAs can be stored at −80°C for at least a month.

7. Add 1 µl of sodium bicarbonate 1 M (pH 9.0) to the cDNA and 1 µl of a fluorescent dye (see Note 10).

8. Incubate at room temperature protected from light for 1 h to allow coupling of the dye to the amino-allyl groups.

9. Purify the Cy-labeled cDNA using a DNA clean and concentrator kit (Zymo Research, D4004). Elute with nuclease-free water in a final volume of 5 µl.

3.4. Microarray Hybridization

The following hybridization procedure is for aminosilane-coated glass slides that are spotted with long (70 mer) oligonucleotides (Operon AROS for yeast). The oligos were designed to yield minimal cross-hybridization between genes. Slides are stored in the dark under desiccation and are handled with powder-free gloves. As the spotted DNA will not be visible during the process, the array boundaries should be marked with a diamond pen on the back of the slide prior to their use.

1. Place the slides at 42°C for 1 h in a coupling jar containing preheated and filtered slide-blocking buffer.

2. Carefully remove the slides from the blocking solution and wash three times with 0.1× SSC for 5 min each. Wash the slide in double distilled *water* for 30 s and spin at 500 rpm for 5 min at room temperature to dry. Slides are ready for hybridization.

3. Prepare the hybridization mixture by combining 10 μl labeled cDNA (5 μl of the mitochondria sample and 5 μl of the reference sample) with 12.5 μl of hybridization buffer and 1 μl of PolyA that was preheated at 95°C for 3 min.

4. Set the microarray slide in a hybridization chamber. Carefully pipette the hybridization mixture on the spotted area and cover with a cover slip. Incubate in a water bath at 42°C overnight in the dark.

5. Wash the slides at the end of the hybridization by three consecutive washes, each for 12 min with gentle shaking: first wash with 2× SSC and 0.2% SDS that were prewarmed to 42°C, then with 2× SSC, and finally with 0.2× SSC (see Note 11).

6. Spin-dry the slide by centrifugation at $50 \times g$ for 5 min at room temperature. Store the dried slides in a dark box until scanning.

7. Scan with a laser scanner (e.g., GenePix 4000B) with settings that yield less than 10% saturated signals. If necessary, scan at two laser intensities. Spots with signals that are too close to their local background include too few pixels or those having an irregular structure are flagged out (we use "GenePix" software from Molecular Devices to assist with the flagging procedure and the data retrieval).

3.5. Data Analysis

Hybridization procedures in which two differently labeled samples are hybridized to the same DNA microarray result in two values per gene: a value that represents its transcript levels in one sample and a second value that represent the amount of transcripts in the second sample. In the specific example described herein, one value represents the mRNAs that are associated with the mitochondria, and the other value represents presence in the cytosolic fraction. Thus, the ratio between these values is the relative association of the transcripts of a gene with the mitochondria. Changes in this ratio upon altering of growth conditions or utilizing mutants of various genes provide important information regarding the mechanisms of targeting to the mitochondria. These studies necessitate multiple microarray hybridizations that are used to obtain relative associations to the mitochondria for thousands of genes. The simplest way to identify changes in mRNAs association is by dividing the ratios obtained in one experiment by those obtained in another experiment. This will give the fold-change in mRNAs association to the mitochondria. Genes, in which the association of their mRNAs to the mitochondria was most affected (increased or decreased), can be selected for further analysis by defining a cut-off (e.g., differences having twofold changes from the mean change of all genes). An alternative and more robust way to identify affected targets is to generate a scatter plot (e.g., Fig. 2a) in which the ratios obtained from one sample are plotted against those from another sample. This plot identifies global changes occurring throughout the genome, which will be apparent from the slope of the best-fit trend line and from the distribution of spots around that line. It is possible to set standard deviation lines around the trend line and extract genes that deviate beyond these lines.

The extended experimental procedure and the fact that many mRNAs are not associated with mitochondria usually result in a low number of genes that pass the filtration criteria. This relatively low number limits statistical analyses because many genes may not pass the filtration criteria in all/most experimental

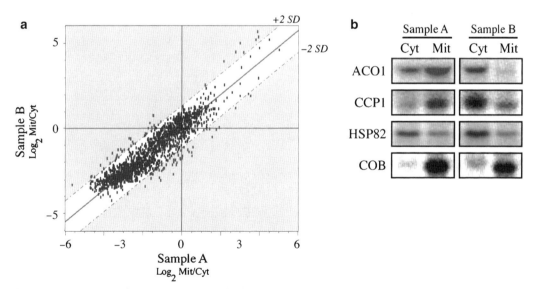

Fig. 2. Representative results of microarray analysis of two fractionated samples. Two yeast strains were subjected to the fractionation procedure. The Mit and Cyt fractions from each strain were labeled differentially and cohybridized to a DNA microarray. This resulted in a Mit/Cyt ratio for transcripts of thousands of genes from either Strain A or Strain B. (a) Scatter plot representation of microarray results from Strain A (X-axis) versus Strain B (Y-axis). Each *dot* indicates the Log_2 of the Mit/Cyt ratio of transcripts of a particular gene. The best-fit linear trend line is indicated and the ±2 standard deviation (SD) lines (*dashed lines*). Spots below the −2 SD or above the +2 SD lines (*gray areas*) indicate mRNAs that are relatively less associated or more associated with mitochondria in Strain B compared to Strain A, respectively. (b) Northern analysis for representative genes. Equal amounts of RNA from the Cyt or Mit fractions of Strain A or Strain B were subjected to Northern analyses using probes recognizing the indicated mRNAs. *ACO1* and *CCP1* mRNAs present significant change in their Cyt:Mit distribution and are likely to appear in the *gray area* in the scatter plot. The distribution of *HSP82* and *COB* does not change and is therefore likely to be within the two SD lines (*white area*).

repeats. Thus, the validity of the results for mRNAs of interest should be tested using alternative methods, such as Northern analysis (Fig. 2b), qRT-PCR, or Fluorescence In Situ Hybridization (FISH) (14). Yet, the genome-wide analysis of mitochondrial association provides important information beyond identification of a gene with altered association. It allows selection of a group of mRNAs with altered association and testing for features that are common to this group. These features may be their intracellular localization (e.g., regions of the mitochondria such as inner or outer membrane), or their molecular function (e.g., oxidative phosphorylation and mitochondria translation). Hints regarding their mechanisms of targeting can be obtained from examining physical properties of the encoded protein (e.g., the presence of a signal peptide and its charge) or mRNA features such binding motifs for RNA-binding proteins. Several tools are available for these analyses. The *Saccharomyces* genome database (SGD) Gene Ontology Term Finder, and Slim Mapper tools (http://www.yeastgenome.org/GOContents.shtml) allow rapid and reliable identification of enrichment of certain intracellular functions, locations, or molecular processes among a group of genes. Some physical properties of mitochondria targeting signals can be obtained from the Signal P server (http://www.cbs.dtu.dk/services/SignalP/) and from a recent publication (26). For identifying mRNA motifs, we usually utilize the MEME algorithm (27).

4. Notes

1. It is preferable to use galactose or a non-fermentable carbon source for mitochondrial enrichment. Based on our experience, growth in selection media rather than rich media yields similar amounts of RNA.

2. Treatment with DTT enhances cell wall hydrolysis by breaking the disulfide bonds within the cell wall proteins, thereby easing later access of the 1,3-glucanase to the glucan linkages.

3. Cells that retain their cell wall will not be broken properly in later steps, hence the efficiency of the Zymolyase step is critical. Since spheroplasts are very sensitive to hypo-osmotic conditions, they must be maintained in an isotonic solution during and after Zymolyase treatment to prevent premature lysis. Here, the non-metabolized sugar alcohol sorbitol provides the osmotic support.

4. The stress imposed by the Zymolyase treatment or by the hyper-osmotic buffer may disrupt the normal mRNA localization, and hence the recovery step is necessary.

5. Cycloheximide inhibits ribosomal translocation yet it does not lead to the disassembly of ribosomes from the mRNA. As an outcome, mRNAs that are associated with mitochondria through their ribosomes remain associated throughout the procedure. It should be noted, however, that CHX was shown to induce a differential effect on mRNA association with the mitochondria (19).

6. This step removes any residual DNA and proteins. Avoid taking any of the phenol (lower) layer as it will inhibit later enzymatic steps.

7. The LiCl precipitation is necessary to remove residual Heparin, which inhibits the fluorescent labeling reaction.

8. To analyze enrichment to the mitochondria, it is possible to use two types of references: the first is unfractionated RNA, i.e., an RNA sample that was collected just prior to the spin down of the mitochondria from the cytosolic fraction (Subheading 3.1, step 10). Since the cell lysis procedure induces a significant change in the transcriptome, this sample will represent the amount of each gene's transcripts prior to fractionation. Thus, the resulting ratio between the mitochondria and the unfractionated RNA signal will represent the relative enrichment of RNAs in the mitochondria, with high ratios indicating a relatively high enrichment. A second possible reference is RNA that was extracted from the cytosolic fraction (Subheading 3.1, step 11). This RNA sample is the complement of the mitochondrial fraction, and the resulting hybridization ratio reports directly on the distribution of mRNAs between these two populations. High ratios represent a relatively high association with the mitochondria, and low ratios represent a relatively high cytosolic association. Such a direct comparison appears to be refractory to changes in total RNA levels, yet is problematic for genes that are highly associated with one of the fractions.

9. The amounts of RNA removed from each fraction may vary without affecting the data interpretation. This is because the signal per mRNA is relative to all other mRNAs and this relativity (or ranking) will be maintained regardless of the amount of RNA taken.

10. Cy dyes are dissolved in DMSO (12 µl) and divided into aliquots of 1 µl each. Unused dyes are dried and stored at 4°C in a desiccator for later use. It is important to minimize the exposure of the dyes or labeled samples to light.

11. The cover slip should be gently removed from the slides during the first wash (using a needle can ease the process; do not let the cover slip smear on the printed area). Avoid the slides' drying during the procedure.

References

1. Girzalsky, W., Platta, H. W., and Erdmann, R. (2009) Protein transport across the peroxisomal membrane. *Biol Chem* **390**, 745–51
2. Hormann, F., Soll, J., and Bolter, B. (2007) The chloroplast protein import machinery: a review. *Methods Mol Biol* **390**, 179–93
3. Terry, L. J., Shows, E. B., and Wente, S. R. (2007) Crossing the nuclear envelope: hierarchical regulation of nucleocytoplasmic transport. *Science* **318**, 1412–6
4. Schwartz, T. U. (2007) Origins and evolution of cotranslational transport to the ER. *Adv Exp Med Biol* **607**, 52–60
5. Holt, C. E., and Bullock, S. L. (2009) Subcellular mRNA localization in animal cells and why it matters. *Science* **326**, 1212–6
6. Neupert, W. (1997) Protein import into mitochondria. *Annu Rev Biochem* **66**, 863–917
7. Kellems, R. E., and Butow, R. A. (1972) Cytoplasmic-type 80 S ribosomes associated with yeast mitochondria. I. Evidence for ribosome binding sites on yeast mitochondria. *J Biol Chem* **247**, 8043–50
8. Kellems, R. E., Allison, V. F., and Butow, R. A. (1974) Cytoplasmic type 80 S ribosomes associated with yeast mitochondria. II. Evidence for the association of cytoplasmic ribosomes with the outer mitochondrial membrane in situ. *J Biol Chem* **249**, 3297–303
9. Kellems, R. E., and Butow, R. A. (1974) Cytoplasmic type 80 S ribosomes associated with yeast mitochondria. 3. Changes in the amount of bound ribosomes in response to changes in metabolic state. *J Biol Chem* **249**, 3304–10
10. Kellems, R. E., Allison, V. F., and Butow, R. A. (1975) Cytoplasmic type 80S ribosomes associated with yeast mitochondria. IV. Attachment of ribosomes to the outer membrane of isolated mitochondria. *J Cell Biol* **65**, 1–14
11. Fujiki, M., and Verner, K. (1993) Coupling of cytosolic protein synthesis and mitochondrial protein import in yeast. Evidence for cotranslational import in vivo. *J Biol Chem* **268**, 1914–20
12. George, R., Walsh, P., Beddoe, T., and Lithgow, T. (2002) The nascent polypeptide-associated complex (NAC) promotes interaction of ribosomes with the mitochondrial surface in vivo. *FEBS Lett* **516**, 213–6
13. Marc, P., Margeot, A., Devaux, F., Blugeon, C., Corral-Debrinski, M., and Jacq, C. (2002) Genome-wide analysis of mRNAs targeted to yeast mitochondria. *EMBO Rep* **3**, 159–64
14. Garcia, M., Darzacq, X., Delaveau, T., Jourdren, L., Singer, R. H., and Jacq, C. (2007) Mitochondria-associated yeast mRNAs and the biogenesis of molecular complexes. *Mol Biol Cell* **18**, 362–8
15. Eliyahu, E., Pnueli, L., Melamed, D., Scherrer, T., Gerber, A. P., Pines, O., Rapaport, D., and Arava, Y. (2010) Tom20 mediates localization of mRNAs to mitochondria in a translation-dependent manner. *Mol Cell Biol* **30**, 284–94
16. Lithgow, T. (2000) Targeting of proteins to mitochondria. *FEBS Lett* **476**, 22–6
17. Sylvestre, J., Margeot, A., Jacq, C., Dujardin, G., and Corral-Debrinski, M. (2003) The role of the 3' untranslated region in mRNA sorting to the vicinity of mitochondria is conserved from yeast to human cells. *Mol Biol Cell* **14**, 3848–56
18. Corral-Debrinski, M., Blugeon, C., and Jacq, C. (2000) In yeast, the 3' untranslated region or the presequence of ATM1 is required for the exclusive localization of its mRNA to the vicinity of mitochondria. *Mol Cell Biol* **20**, 7881–92
19. Saint-Georges, Y., Garcia, M., Delaveau, T., Jourdren, L., Le Crom, S., Lemoine, S., Tanty, V., Devaux, F., and Jacq, C. (2008) Yeast mitochondrial biogenesis: a role for the PUF RNA-binding protein Puf3p in mRNA localization. *PLoS ONE* **3**, e2293
20. Wickens, M., Bernstein, D. S., Kimble, J., and Parker, R. (2002) A PUF family portrait: 3'UTR regulation as a way of life. *Trends Genet* **18**, 150–7
21. Meisinger, C., Sommer, T., and Pfanner, N. (2000) Purification of Saccharomcyes cerevisiae mitochondria devoid of microsomal and cytosolic contaminations. *Anal Biochem* **287**, 339–42
22. Schmitt, M. E., Brown, T. A., and Trumpower, B. L. (1990) A rapid and simple method for preparation of RNA from Saccharomyces cerevisiae. *Nucleic Acids Res* **18**, 3091–2
23. Goetz, J. G., and Nabi, I. R. (2006) Interaction of the smooth endoplasmic reticulum and mitochondria. *Biochem Soc Trans* **34**, 370–3
24. Pinton, P., Giorgi, C., Siviero, R., Zecchini, E., and Rizzuto, R. (2008) Calcium and apoptosis: ER-mitochondria Ca2+ transfer in the control of apoptosis. *Oncogene* **27**, 6407–18

25. Kornmann, B., Currie, E., Collins, S. R., Schuldiner, M., Nunnari, J., Weissman, J. S., and Walter, P. (2009) An ER-mitochondria tethering complex revealed by a synthetic biology screen. *Science* **325**, 477–81

26. Dinur-Mills, M., Tal, M., and Pines, O. (2008) Dual targeted mitochondrial proteins are characterized by lower MTS parameters and total net charge. *PLoS ONE* **3**, e2161

27. Bailey, T. L., Boden, M., Buske, F. A., Frith, M., Grant, C. E., Clementi, L., Ren, J., Li, W. W., and Noble, W. S. (2009) MEME SUITE: tools for motif discovery and searching. *Nucleic Acids Res* **37**, W202–8

Chapter 19

Analyzing mRNA Localization to the Endoplasmic Reticulum via Cell Fractionation

Sujatha Jagannathan, Christine Nwosu, and Christopher V. Nicchitta

Abstract

The partitioning of secretory and membrane protein-encoding mRNAs to the endoplasmic reticulum (ER), and their translation on ER-associated ribosomes, governs access to the secretory/exocytic pathways of the cell. As mRNAs encoding secretory and membrane proteins comprise approximately 30% of the transcriptome, the localization of mRNAs to the ER represents an extraordinarily prominent, ubiquitous, and yet poorly understood RNA localization phenomenon.

The partitioning of mRNAs to the ER is generally thought to be achieved by the signal recognition particle (SRP) pathway. In this pathway, mRNA localization to the ER is determined by the translation product – translation yields an N-terminal signal sequence or a topogenic signal that is recognized by the SRP and the resulting mRNA–ribosome–SRP complex is then recruited to the ER membrane. Recent studies have demonstrated that mRNAs can be localized to the ER via a signal sequence and/or translation-independent pathway(s) and that discrete sets of cytosolic protein-encoding mRNAs are enriched on the ER membrane, though they lack an encoded signal sequence. These key findings reopen investigations into the mechanism(s) that govern mRNA localization to the ER.

In this contribution, we describe two independent methods that can be utilized to study this important and poorly understood aspect of eukaryotic cell biology. These methods comprise two independent means of fractionating tissue culture cells to yield free/cytosolic polyribosomes and ER membrane-bound polyribosomes. Detailed methods for the fractionation and characterization of the two polyribosome pools are provided.

Key words: mRNA localization, Endoplasmic reticulum, Cytosol, Polyribosome, rRNA, mRNA

1. Introduction

The endoplasmic reticulum (ER) is the site of synthesis of secretory and membrane proteins, which comprise ca. 30% of the cell's proteome (1). A fundamental question in cell biology concerns the cellular mechanisms that compartmentalize the synthesis of secretory and membrane proteins to the ER. Pioneering in vitro studies in the 1980s by Blobel and colleagues established the signal recognition particle (SRP) pathway as (a) the mechanism of mRNA partitioning to the ER (2–5). In this model, all newly exported mRNAs initiate translation on cytosolic ribosomes. In the case of mRNAs encoding secretory or membrane proteins, translation yields the synthesis of an N-terminal signal sequence or a transmembrane domain which is recognized by the SRP, resulting in a suppression of protein synthesis. The ribosome–nascent polypeptide–SRP complex is then recruited to the ER via binding interactions with the ER-resident SRP-receptor. Upon binding of the ribosome–nascent polypeptide–SRP complex to the ER, SRP is released, translation resumes, and the growing peptide is co-translationally translocated into the ER for further processing.

While the SRP pathway has been widely accepted to be the mechanism by which mRNA are partitioned between the ER and the cytosol, several experimental observations indicate that SRP-independent mechanisms contribute to mRNA partitioning in the cell. For example, several groups have observed significant overlap in the composition of cytoplasmic and membrane-associated mRNAs (6–8). Though the Signal Hypothesis predicts that signal sequence-encoding mRNAs would be present, perhaps at low enrichment, in cytosolic polysomes, this model does not provide a mechanism for how mRNAs lacking encoded signal sequences would be partitioned to the ER. In addition, genetic ablation of components of the SRP pathway in yeast does not affect the viability of the organism (9). Instead, disabling of the SRP pathway function is compensated by an expansion of ER, indicating that the cells are able to adapt to the absence of the SRP pathway by activating or expanding compensatory pathways (9). Consistent with these findings, depletion of SRP54 (the signal peptide binding SRP subunit) by RNA interference in trypanosome is not lethal, and loss of SRP54 does not affect the processing of signal peptide containing proteins (10). In HeLa cells, RNAi knockdown of SRP54 does not have any effect on the growth or viability of the cells. Further, loss of SRP54 affects the expression of the membrane receptor DR4 but not DR5, suggesting that cells have multiple pathways to bring about mRNA partitioning between the cytosol and ER (11). In support of the existence of an alternative, SRP-independent mRNA partitioning mechanism, our group recently reported that deletion of the encoded signal sequence of Grp94, an ER-localized mRNA, or mutational loss of its translation function did not disrupt mRNA localization to the ER (12). In fact, specific subsets of mRNAs encoding cytoplasmic and nucleoplasmic proteins have been consistently observed to be enriched on the ER, even though they do not encode a signal peptide (6–8, 13). A very recent report from the Walter lab suggests a mechanism by which such noncanonical mRNA partitioning can occur; in yeast, the localization of HAC1 mRNA to the

ER is mediated by a conserved element in the 3' UTR (14). However, it is not known whether such direct mRNA localization is an exception to the common rule of SRP-dependent partitioning or if it represents a broader, primary mechanism by which all mRNAs are sorted between the cytosol and the ER. In order to study this phenomenon, it is essential to systematically analyze individual mRNAs and assess their partitioning between the cytosol and the ER.

This paper describes two methods for analyzing mRNA partitioning between the ER and the cytosol. The first method, sequential detergent extraction, takes advantage of the difference in the lipid composition of the plasma membrane and the ER membrane. Digitonin, a β-sterol binding detergent that selectively solubilizes the cholesterol-rich plasma membrane and leaves the ER and nuclear membrane intact, is used to release cytosolic polysomes. Then, the permeabilized cells are washed and the ER-bound polyribosomes are solubilized by any of a variety of detergents, including dodecylmaltoside, NP40, or an admixture of Nonidet P-40 (NP40) and sodium deoxycholate (DOC). The second method, mechanical homogenization followed by differential centrifugation, is a variation of the classical differential centrifugation method for cellular fractionation that was developed by Claude in the 1940s and later perfected by Palade and others (15, 16).

Over the past decades, the field of mRNA localization has been niched to studies in *Drosophila* embryos and budding yeast (17, 18). There is a critical need for more studies on the mechanisms of subcellular mRNA localization in higher eukaryotic cells to enable a broader understanding of this important biological phenomenon. The methods outlined in this paper provide tools needed to study mRNA partitioning between the cytosol and the ER, a ubiquitous mRNA sorting process that, conservatively, directs the subcellular localization of >30% of the transcriptome.

2. Materials

As compared to DNA, RNA is very susceptible to degradation due both to nonspecific cleavage in the presence of divalent cations, and more importantly, the near ubiquitous presence of RNAse activity. Thus, great care should be taken to avoid RNAse contamination at every step of the following experiments. The most common sources of RNAse are human skin and microbial growth in stock solutions. Hence, gloves should always be worn while handling RNA and stock solutions should be stored in small aliquots and discarded at frequent intervals.

To reduce/eliminate RNAse contamination, buffers and solutions can be treated with 0.1% (v/v) diethyl pyrocarbonate

(DEPC) (see Note 1) overnight at 37°C and then autoclaved for 15 min to remove unreacted DEPC. Buffers containing free amine groups (TRIS, HEPES, etc.) and solutions/buffers that cannot be autoclaved (sucrose, MOPS, etc.) cannot be DEPC-treated. Such solutions/buffers should be made up using molecular biology grade, RNAse-free reagents and DEPC-treated water. All glassware should be baked at 400°C for 4 h to inactivate RNAses. Sterile disposable tips and tubes are generally RNAse-free. Non-disposable plasticware (such as ultracentrifuge tubes) can be treated twice with RNAZap (Ambion) or 0.2% SDS, and rinsed thoroughly in DEPC-treated water.

2.1. Cell Culture

1. HEK293 cell line (ATCC).
2. Cell-culture medium: Dulbecco's modified Eagle's medium (DMEM; Mediatech) supplemented with 10% fetal bovine serum (FBS; Invitrogen).
3. Trypsin 0.05% in 0.53 mM ethylenediaminetetraacetate (EDTA) (Invitrogen).
4. 1× Phosphate buffered saline (PBS): 10 mM sodium phosphate dibasic (Na_2HPO_4), 2 mM potassium phosphate monobasic (KH_2PO_4), 2.7 mM potassium chloride (KCl), 137 mM sodium chloride (NaCl), 0.5 mM magnesium chloride ($MgCl_2$), 1 mM calcium chloride ($CaCl_2$). (Mediatech; Cat. no. 21-030-CV).

2.2. Fractionation by Sequential Detergent Extraction

1. 1% (w/v) digitonin (Calbiochem; Cat. no. 300410) in DMSO (freeze in 100 μl aliquots; see Note 2).
2. RNaseOut™ Recombinant Ribonuclease Inhibitor: 40 U/μl stock. Store at −20°C (Invitrogen; Cat. no. 17-0969-01).
3. Complete™ Protease Inhibitor Cocktail: Complete™ EDTA-free (Roche Molecular Biochemicals; Cat. no. 1-873-580). Make a 100× stock in DMSO and store at −20°C. Use at a final concentration of 1×.
4. Diethyl pyrocarbonate (DEPC)-treated water. Prepare as a 0.1% (v/v) solution and incubate at 37°C overnight. Autoclave for 15 min to destroy unreacted DEPC.
5. Stock solutions: 4 M potassium acetate (KOAc); 1 M potassium 4-(2-hydroxyethyl)-1-piperazineethanesulfonate (K-HEPES); 1 M magnesium acetate [$Mg(OAc)_2$]; 0.2 M ethyleneglycol bis(2-aminoethylether)-N,N,N',N'–tetraacetic acid (EGTA) at pH 8.0; 10% (v/v) Nonidet P-40 (NP-40); 10% (w/v) sodium deoxycholate (DOC); 20% (w/v) n-dodecy-β-D-maltoside (DDM).
6. Permeabilization buffer: 110 mM KOAc, 25 mM K-HEPES, pH 7.2, 2.5 mM $Mg(OAc)_2$, 1 mM EGTA, 0.03% digitonin, 1 mM DTT, 50 μg/ml cycloheximide (CHX), 1× Complete

Protease Inhibitor Cocktail, and 40 U/mL RNaseOUT™. Digitonin, DTT, CHX, Complete Protease Inhibitor Cocktail, and RNaseOUT™ must be added fresh.

7. Wash buffer: 110 mM KOAc, 25 mM K-HEPES at pH 7.2, 2.5 mM Mg(OAc)$_2$, 1 mM EGTA, 0.004% digitonin, 1 mM DTT, and 50 µg/ml CHX. Digitonin, DTT and CHX must be added fresh.

8. Lysis buffer: 400 mM KOAc, 25 mM K-HEPES at pH 7.2, 15 mM Mg(OAc)$_2$, 1 mM DTT, 50 µg/ml CHX, 1× Complete Protease Inhibitor Cocktail, and 40 U/mL RNase Out with either 2% (w/v) DDM or 1% (v/v) NP-40 and 0.5% (w/v). DTT, CHX, Complete Protease Inhibitor Cocktail and RNaseOUT™ must be added fresh.

9. Sucrose cushion: 0.5 M sucrose in lysis buffer.

2.3. Fractionation by Differential Centrifugation

1. Cell scraper (Fisher Scientific).
2. Cell homogenizer (Isobiotec), obtained from Isobiotec, Ortenauer Strasse 13, 69126 Heidelberg, Germany.
3. 3 ml Luer-Lok™ syringes (BD Pharmingen).
4. Stock solutions: 200 mM KCl, 150 mM MgCl$_2$, 1 M Tris–HCl at pH 7.4, 2 M sucrose, 100 mM dithiothreitol (DTT; freeze in 100 µl aliquots), 10 mg/ml CHX (freeze in 100 µl aliquots).
5. Hypotonic lysis buffer (HLB): 10 mM KCl, 7.5 mM MgCl$_2$, 50 mM Tris–HCl, pH 7.4, 1 mM DTT, 50 µg/ml CHX, 1× Complete Protease Inhibitor Cocktail, 40 U/ml RNAseOUT™. Add CHX, DTT, Complete Protease Inhibitor Cocktail, and RNAseOUT™ just prior to using the solution.
6. Polycarbonate ultracentrifuge tubes, 11 × 34 mm (Beckman; Cat. No. 343778).

2.4. Immunoflourescence Microscopy

1. Coverslips.
2. Fixation buffer: 4% paraformaldehyde in 1× PBS.
3. Acetone.
4. Blocking solution: 1% (w/v) BSA and 0.2% (v/v) Triton X-100 in 1× PBS.
5. Primary antibody diluted in 1% (w/v) BSA and 0.05% (v/v) Triton X-100 in 1× PBS.
6. Fluor-conjugated secondary antibody in 1% (w/v) BSA and 0.05% (v/v) Triton X-100 in 1× PBS.
7. DAPI (DNA stain) diluted in 1% (w/v) BSA and 0.05% (v/v) Triton X-100 in 1× PBS.
8. Antifade mounting medium.
9. Nail polish – coverslip sealant.

2.5. Polysome Analysis

1. 15% sucrose in lysis buffer.
2. 40% sucrose in lysis buffer.
3. Polyallomer centrifuge tubes, 14×89 mm (Beckman; cat # 331372).
4. Teledyne/Isco gradient fractionator with a continuous UV flow cell.

2.6. RNA and Protein Extraction

1. TRIzol® Reagent (Invitrogen; see Note 3).
2. Chloroform.
3. Isopropanol.
4. 0.3 M guanidine chloride in 95% isopropanol.
5. Ethanol.
6. 75% (v/v) ethanol.
7. Nuclease-free water (to resuspend RNA).
8. Protein sample buffer: 0.5 M unbuffered Tris and 5% SDS.

2.7. Denaturing Formaldehyde Agarose Gel Electrophoresis and Northern Blotting

1. 10× MOPS: 0.2 M 3-(N-morpholino)propanesulfonic acid (MOPS), 80 mM sodium acetate (NaOAc), 10 mM EDTA at pH 7.4. This buffer is light sensitive and should be stored in an amber bottle. The color of this solution slowly changes to orange with time. This does not affect its activity.
2. Agarose (electrophoresis grade).
3. 37.5% (w/v) formaldehyde.
4. Formamide (deionized).
5. RNA tracking dye (Ambion).
6. SyBR safe RNA dye (Invitrogen) (optional).
7. DNA/RNA gel electrophoresis apparatus.
8. 20× SSC: 3 M NaCl and 0.3 M NaOAc at pH 7.0.
9. 10 N Sodium hydroxide (NaOH).
10. Northern transfer buffer: 5× SSC and 10 mM NaOH.
11. Methylene blue stain: 0.02% (w/v) methylene blue in 0.2 M NaOAc at pH 5.2.
12. Hybond XL membrane (Amersham Pharmacia Biotech).
13. Whatman 3MM filter paper or equivalent.
14. Paper towels.
15. Stratalinker UV Crosslinker (Stratagene).
16. T4 Polynucleotide kinase (PNK) (New England Biolabs).
17. 100 µM oligonucleotide directed against the RNA sequence to be detected.
18. γ-[^{32}P]-dATP at 6,000 Ci/mmol; end-labeling grade (MP Biomedicals).

19. Sephadex G-25 quick spin column or equivalent (Roche).
20. Scintillation fluid, scintillation vials, and liquid scintillation spectrometer.
21. ExpressHyb hybridization solution (Clontech).
22. Hybridization oven and glass tubes.
23. Low-stringency wash buffer: 0.5× SSC and 0.1% (w/v) SDS in deionized water.
24. High-stringency wash buffer: 0.1× SSC and 0.1% (w/v) SDS in deionized water.
25. Phosphorimager plates, cassettes, and scanner (Typhoon 9400; GE Healthcare).

2.8. SDS Polyacrylamide Gel Electrophoresis and Western Blotting

1. 12.5% denaturing polyacrylamide gel, 0.75 mm thick.
2. Gel electrophoresis system (Bio-Rad).
3. 5× gel loading dye: 0.2 M Tris–HCl at pH 6.8, 10% (v/v) glycerol, 10% (w/v) SDS, 0.05% (w/v) bromophenol blue, 10 mM β-mercaptoethanol (BME). Add β-BME to the sample buffer just prior to use.
4. 5× SDS running buffer: 250 mM Tris–HCl, 2 M glycine, and 1% (w/v) SDS.
5. CAPS transfer buffer: 50 mM 3-[cyclohexylamino]-1-propane sulfonic acid (Sigma C2632) at pH 11, 0.075% SDS, and 20% (v/v) methanol.
6. Nitrocellulose membrane (Bio-Rad).
7. Ponceau stain: 0.1% (w/v) Ponceau S and 5% (v/v) acetic acid.
8. 1× PBS-T: 1× PBS and 0.2% Tween-20.
9. 5% milk in 1× PBS-T.
10. Primary and secondary antibodies.
11. Enhanced chemiluminescence (ECL) reagents (Denville Scientific).
12. X-ray film and cassette (Denville Scientific).
13. Automated X-ray film processor.

3. Methods

3.1. Methods for Cell Fractionation

Cell fractionation has been used for several decades to analyze the molecular composition and functionalities of the organelles/metabolic compartments of eukaryotic cells. Here we describe cell fractionation methods developed for the study of the molecular mechanism(s) that govern mRNA partitioning between the cytosol and the ER. The experimental objective of these methods is to separate, to high enrichment, cytosolic polysomes and membrane-bound polysomes. This chapter describes two independent methods of cellular fractionation namely, (1) sequential detergent extraction, and (2) mechanical fractionation followed by differential centrifugation.

3.1.1. Fractionation by Sequential Detergent Extraction

This method takes advantage of the relatively high cholesterol content of the plasma membrane, as compared to other cellular membranes. Digitonin is a β-sterol binding detergent that selectively solubilizes the plasma membrane, leaving the ER and nuclear membranes intact. Hence, sequential treatment with digitonin followed by a more lytic detergent, such as an NP-40/DOC cocktail, yields cytosolic- and membrane-bound polysome fractions, respectively (schematically illustrated in Fig. 1a). The various steps of the sequential detergent extraction procedure have been validated in HEK293 cells by immunofluorescence microscopy, where it can be seen that disruption of the plasma membrane with digitonin results in the release of (depolymerized) tubulin, without affecting the ER, the actin cytoskeleton, or the intermediate filament network (Fig. 1b). Following addition of the ER lysis buffer, the ER fraction is recovered in a soluble fraction and the nuclei, actin cytoskeleton, and intermediate filament network remain (Fig. 1b). Companion immunoblot analyses of marker protein distributions show that the cytosolic proteins GAPDH and tubulin are present in the cytosol fraction, as expected, and the ER-membrane proteins, TRAPα and ER-lumenal protein, GRP94 are present in the ER fraction (Fig. 2a). The detergent-insoluble material consists primarily of nuclear and cytoskeletal elements, as evidenced by the marker proteins histone H3 and actin, respectively (Fig. 2a). Similarly, Northern blot analysis of the mRNA composition of the cytosol and membrane fractions show that the cytosol fraction is enriched for mRNAs encoding histone (H3F3A) and GAPDH, whereas the membrane fraction is enriched in mRNAs encoding ER resident proteins, such as GRP94 and calreticulin (Fig. 2b).

The method described below is for cells grown in monolayer. However, the protocol can be easily adapted for nonadherent cells by performing permeabilization, wash and lysis in suspension, and pelleting cells at $3,000 \times g$ for 5 min between the different steps. The volumes of reagents mentioned in the following protocol are scaled to extract polysomes from ten million cells.

1. Seed HEK293T cells in a T75 flask to be 80–90% confluent on the day of the experiment.
2. Aspirate the media and wash the cells once with 10 ml of 1× PBS (room temperature).
3. Treat the cells with 10 ml of ice-cold PBS (1×) containing 50 μg/ml CHX for 10 min on ice (see Notes 4 and 5). Perform all remaining steps on ice using ice-cold reagents.

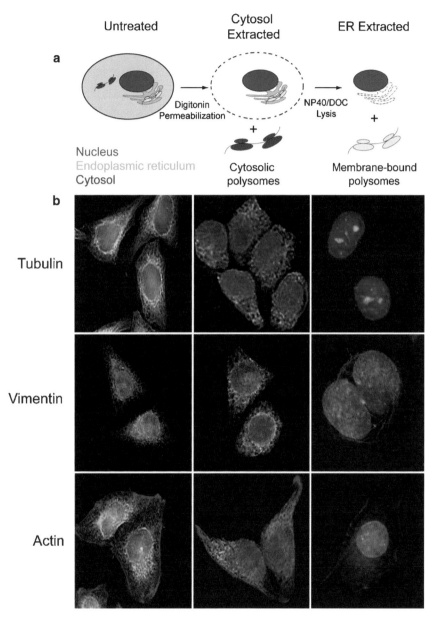

Fig. 1. Characterization of the detergent fractionation method: immunofluorescence microscopy in HeLa cells. (a) Schematic representation of the effect of the permeabilization and lysis on cells. Upon digitonin treatment, the plasma membrane is solubilized, allowing the recovery of cytosolic polysomes (shown in *red*). During the subsequent lysis with NP40/DOC mixture, the ER is solubilized and membrane-bound polysomes (shown in *green*) recovered, leaving the nucleus intact (shown in *blue*). (b) HeLa cells were immunostained during the three stages of detergent extraction (untreated, permeabilized and lysed) using antibodies against TRAPα (*green*), tubulin (*red*), vimentin (*red*), and actin (*red*). Nucleus is stained with DAPI (*blue*).

4. Add 1 ml of permeabilization buffer to the cells, taking care not to dislodge the cells (see Notes 6 and 7) and incubate for 5 min. Tilt the flask to drain the soluble material (cytosol fraction) and collect the cytosol in a pre-cooled microcentrifuge tube.

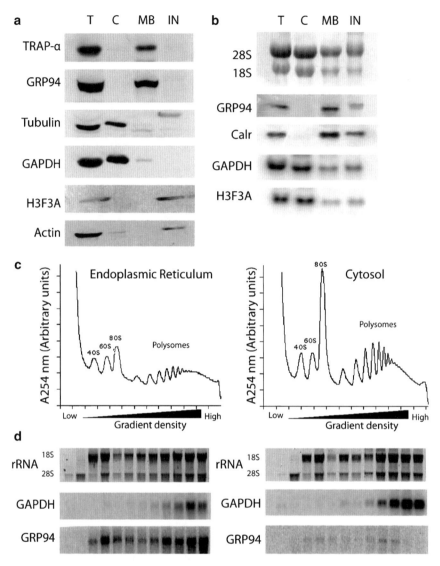

Fig. 2. Validation of the detergent fractionation method via Western and Northern blot analysis. (a) Protein from the total (T), cytosol (C), membrane-bound (M), and insoluble (IN) fractions of HEK293 cells were extracted using the TRIzol® Reagent and equivalent amounts of protein were resolved on a 10% SDS polyacrylamide gel. The proteins were transferred to a nitrocellulose membrane and the analyzed for the presence of ER resident (TRAPα and GRP94), cytosolic (tubulin and GAPDH), and nuclear (Histone H3F3A) proteins by western blotting. Primary antibodies were used at a dilution of 1:3,000 and HRP-conjugated secondary antibodies were used at 1:5,000. (b) RNA from the T, C, M, and IN fractions derived from the sequential detergent fractionation procedure were resolved on a denaturing agarose gel and transferred to a nylon membrane. The ribosomal RNA were stained using methylene blue and the profile was documented. The membrane was then probed for membrane-bound mRNAs (GRP94 and calreticulin) and cytosolic mRNAs (GAPDH and Histone H3F3A) using γ-[32P] labeled antisense oligonucleotides. Following overnight exposure on phosphorimager plates, images were collected using Typhoon 9400 and image size/contrast adjusted using Adobe Photoshop v7. 0. (c) Cytosol and membrane-bound (endoplasmic reticulum) polysomes were resolved on 15–50% sucrose gradients. The position of the ribosomes in the gradient was assessed by UV spectrometry (A_{254} nm). (d) RNA extracted from the sucrose gradient fractions were assessed by denaturing agarose gel electrophoresis followed by Northern blot for GRP94 and GAPDH. The RNA integrity is demonstrated by methylene blue staining of rRNA.

5. Wash cells gently with 1 ml of wash buffer and combine the wash with the cytosol fraction.

6. Treat the cells with 1 ml of lysis buffer for 5 min. Drain and collect the soluble material (membrane fraction).

7. Clarify both the cytosolic and membrane fractions at $7,500 \times g$ for 10 min to remove cell debris. Transfer the supernatants to clean, prechilled microcentrifuge tubes.

8. The various steps of this process can be visualized by immunofluorescence microscopy by staining for TRAPα (ER), tubulin (cytosol), vimentin (intermediate filaments), and actin (cytoskeleton) (see Subheading 3.2; Fig. 1b).

9. To analyze the RNA/protein content in the cytosol and membrane fractions (Fig. 2b), directly extract using 1 ml TRIzol® Reagent per 0.25 ml of sample (see Subheading 3.4). Samples in TRIzol® can be frozen at −70°C for storage prior to processing.

10. Alternately, the polysome profiles of the cytosolic and membrane-bound fractions can be analyzed by layering 1 ml of the cytosolic and membrane fractions on 15–40% linear sucrose gradients and subjecting them to velocity sedimentation (see Subheading 3.3). The ribosome composition in the gradient fractions can be analyzed by generating an A_{254} nm trace manually or by using an automated gradient analyzer (see Subheading 3.3; Fig. 2c). The gradient fractions can then be extracted using TRIzol® reagent and assessed for RNA quality and composition by denaturing formaldehyde agarose gel and Northern blotting (see Subheading 3.5; Fig. 2d).

11. If the polyribosomes in the cytosol and membrane fractions need to be recovered for downstream applications, layer the clarified lysate over 1/3 volume of 0.5 M sucrose in the same buffer as that of the sample. Centrifuge at $100,000 \times g$ for 40 min in a Beckman TLA 100.2 rotor at 4°C. Ribosome pellets will appear clear and glassy.

3.1.2. Fractionation by Differential Centrifugation

A cell cracker, or more technically, a ball-bearing homogenizer, is a precision device that efficiently and reproducibly disrupts cell structure while maintaining organelle integrity. The following protocol (schematically outlined in Fig. 3a) allows for efficient recovery of membrane-bound and cytosolic polysomes. By immunoblot analysis of marker protein distribution, we identified conditions that allowed separation of the nucleus from the ER (as evidenced by the relative absence of the nuclear marker histone H3 in the ER fraction; Fig. 3b, lane 2) and the ER-bound polysomes from free-polysomes (Fig. 3b, lanes 2–4). Using this method, ER membranes (indicated by the ER membrane marker, TRAPα) were recovered in a $15,000 \times g$ spin

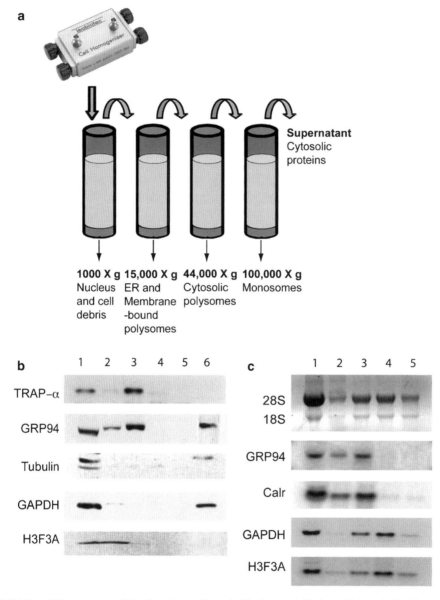

Fig. 3. Validation of the mechanical fractionation method via Western and Northern blot analysis. (a) A schematic representation of the mechanical fractionation method of cellular fractionation. (b) Protein extracted from the fractions Total (1), pellets from centrifugation at 500 × g (2), 15,000 × g (3), 44,000 × g (4), 100,000 × g (5) and that from the supernatant from the 100,000 × g spin (6) were analyzed for ER, cytosol, and nuclear proteins by Western blot as described in Fig. 1a. (c) RNA extracted from the fractions 1–6 were analyzed by Northern blot as described in Fig. 1b.

(Fig. 3a) as were, appropriately, ER-bound polysomes (Fig. 3b). The 44,000 × g spin, which should theoretically sediment free polysomes (based on K-factor calculations) yields a fraction enriched for cytosolic polysomes (Fig. 3b). The supernatant from the 100,000 × g spin (which theoretically sediments monosomes and individual ribosomes) contains the cytosolic proteins such as histone H3F3A and GAPDH, as expected, and also

GRP94 (ER lumen protein), which is partially released during cell homogenization.

1. Aspirate media from cells and wash the cells once with 10 ml PBS (1×) at room temperature.
2. Add 6 ml of ice-cold 1× PBS to each well and scrape cells with a cell scraper. All subsequent steps were performed on ice, using ice-cold solutions.
3. Pellet cells at 1,000×g for 4 min at 4°C. (Extract cell pellets from two wells of the 6-well plate with 1 ml of the TRIzol® Reagent; see Note 8).
4. Resuspend pellets from the remaining four wells in 4 ml of ice-cold HLB.
5. Let cells swell on ice for 10 min.
6. Rinse the cell cracker (precooled and kept on ice) with cold HLB using 5-ml Luer-Lok™ syringes (see Note 9). Fill a precooled 5-ml Luer-Lok™ syringe with the cell suspension, avoiding air bubbles. Pass the cell suspension through precooled Cell Cracker with the 12 μm clearance ball bearing 12 times (six passes on each syringe). Pool the lysates and adjust the volume to 4 ml if needed.
7. Adjust homogenate to 250 mM sucrose and 4 mM $MgCl_2$ using 2 M sucrose and 150 mM $MgCl_2$.
8. Centrifuge at 1,000×g for 5 min at 4°C to pellet unbroken cells and nuclei (solubilize the pellets in TRIzol® Reagent; see Note 8).
9. Centrifuge the supernatant from step 8 at 15,000×g for 15 min at 4°C (solubilize the pellets in TRIzol® Reagent).
10. Centrifuge the supernatant from step 9 at 44,000×g for 15 min at 4°C (solubilize the pellets in TRIzol® Reagent).
11. Centrifuge the supernatant from step 10 at 100,000×g for 1 h at 4°C (solubilize the pellets in TRIzol® Reagent).
12. Add 1 ml TRIzol® Reagent to 250 μl of the supernatant from 100,000×g spin (see Note 10).
13. Proceed to TRIzol® extraction (Subheading 3.5) or freeze the samples at −80°C until RNA/Protein extraction.

3.2. Immunofluorescence Microscopy

1. Plate cells onto glass coverslips 12–24 h prior to analysis.
2. Rinse the cells in cold 1× PBS.
3. Fix the cells in ice-cold 4% paraformaldehyde in 1× PBS for 15 min, on ice.
4. Rinse thrice with cold 1× PBS for 5 min each.
5. Permeabilize the fixed cells in either cold PBS (1×) with 1% (v/v) Triton X-100 for 10 min on ice, or cold acetone, with incubation at −20°C for 10 min.

6. Rinse thrice with cold 1× PBS for 5 min each.
7. Block using the blocking solution for 1 h at room temperature (RT), or overnight at 4°C.
8. Add primary antibody solution (sufficient to cover the cell layer) and incubate at RT for 1 h. Alternatively, invert the coverslip onto a 50 μl drop of the primary antibody solution, on parafilm, and incubate for 1 h at RT.
9. Rinse thrice with 1× PBS for 5 min each at RT.
10. Repeat step 8 with secondary antibody solution and DAPI solution, and incubate for 1 h at RT, in the dark.
11. Rinse thrice with 1× PBS for 5 min each at RT.
12. Mount the cover slip on a slide using antifade mounting medium.
13. Seal the slides with nail polish.

3.3. Polysome Gradient Analysis

Pour 15–40% linear sucrose gradient as follows:

1. Add 5 ml of 15% sucrose in a Beckman centrifuge tube (SW40).
2. Underlay 5 ml of 40% sucrose solution slowly so that the interface is not disturbed (see Note 11).
3. Cover the tube with parafilm and slowly tip the tube so that it is laying on its side. Support the tube on both sides with microcentrifuge tube racks to prevent rolling.
4. Let the gradient form over 2 h. A gradient maker such as SG 15 Gradient Maker (GE Healthcare) can also be used to generate a linear gradient.
5. Carefully tip the tube back up and place it on ice for at least 30 min. If needed, the gradient can be stably stored on ice overnight.
6. Overlay the sample on the sucrose gradient and centrifuge at $45,000 \times g$ for 3 h.
7. Fractionate the gradient, either via an automated gradient fractionator (i.e., Teledyne/ISCO) or manually; in the absence of a gradient fractionator the centrifuge tubes can be manually punctured, fractions collected, and the UV absorbance (254 nm) measured for each fraction to generate the polyribosome trace.
8. The gradient fractions can further be extracted using TRIzol® reagent and the RNA and protein composition can be assessed by Northern and Western blotting, respectively.

3.4. RNA and Protein Extraction Using the TRIzol® Reagent

TRIzol® Reagent can be used to sequentially extract RNA and protein from the same biological sample.

1. Incubate the TRIzol®-treated samples at room temperature for 10 min to allow dissociation of nucleoprotein complexes (see Note 8).

2. Spin the TRIzol®-treated samples at $10,000 \times g$ for 10 min to remove insoluble material (optional).

3. Add 200 µl of chloroform per ml of TRIzol(R) and vortex for 15 s.

4. Incubate at room temperature for 3 min to allow phase separation.

5. Spin at maximum speed for 15 min on a table-top centrifuge at 4°C.

6. RNA isolation: Carefully transfer 0.6 ml of the aqueous phase (containing the RNA) to a clean tube (see Note 12) and add 0.5 ml of isopropanol. Mix well and incubate at room temperature for 10 min to precipitate RNA (see Note 13). Spin at $13,000 \times g$ for 10 min to pellet the RNA. Wash the RNA pellet with 1 ml of 75% ethanol. Air dry the pellet for 2–3 min (do not allow the pellet to over dry) and re-suspend the RNA in appropriate volume of DEPC-treated water (see Note 14). The RNA sample can be stored in nuclease-free water at –80°C for 1–2 years, or in 70% ethanol at –80°C, indefinitely.

7. Protein isolation: Remove any remaining aqueous phase and as much of the interphase as possible without removing the organic phase which contains the protein. Add 0.3 ml of ethanol to 0.6 ml of the organic phase to precipitate any DNA. Incubate for 5 min at room temperature and spin at $3,000 \times g$ for 5 min to remove the DNA. The DNA pellet will be barely visible and very soft. Take care to not disturb it while removing the supernatant. To precipitate the protein, add 0.75 ml of isopropanol per 0.4 ml of the organic phase–ethanol mixture and incubate for 10 min at room temperature to precipitate the protein. Spin at $13,000 \times g$ for 10 min to pellet the protein. Wash the pellet for 20 min with 1 ml of 0.4 M guanidine hydrochloride in 95% isopropanol at room temperature. Repeat this step for a total of three washes. Wash using 1 ml of ethanol to remove the salt and air-dry the pellet for 2–3 min. Resuspend the protein pellet in protein sample buffer. If the pellet does not go into solution upon incubation at room temperature for 15–20 min, heating the sample at 65°C for 15 min with occasional vortexing will assist solubilization (see Note 15).

8. RNA and protein quantification: Measure the RNA concentration by UV absorption at 260 nm. The yield of RNA from ten million cells is typically about 50–100 µg, and can be used in a variety of downstream applications including, but not limited to Northern blotting, RT-PCR, and cDNA microarrays (see Note 16). The protein concentration is best measured by standard assays such as the BCA assay (Pierce). Absorption at 280 nm can be used to get a rough estimate of the protein

concentration, but these measurements are often dubious due to possible nucleic acid contributions to the absorbance readings.

3.5. Denaturing Formaldehyde Agarose Gel Electrophoresis and Northern Blotting

1. Rinse the gel tray, comb, and the electrophoresis apparatus with 0.2% SDS and DEPC-treated water.

2. Formaldehyde agarose gel: Combine 1 g of agarose in 82 ml of DEPC-treated water and dissolve by heating. Add 10 ml of 10× MOPS buffer to the melted agarose. When the solution cools to about 60°C, add 8 ml of formaldehyde in a fume hood (see Notes 17 and 18).

3. Formamide/formaldehyde sample buffer: Mix 200 µl of formamide, 70 µl of formaldehyde, 30 µl of 10× MOPS, and 27 µl tracking dye (Ambion). Add 24 µl of sample buffer to 8 µl of RNA sample and heat the samples for 10 min at 65°C. Cool the samples to room temperature and load.

4. Run the gel in 1× MOPS buffer at 120 V for 2 h (see Note 19).

5. Set up the Northern transfer as follows: Cut one piece of Hybond™ membrane and six pieces of Whatman 3MM filter paper to the size of the gel. Soak the gel and the membrane in Northern transfer buffer for 5–10 min. Assemble the Northern transfer by sandwiching the gel and the membrane between six filter paper squares (three on each side) and place this assembly on a stack of paper towels. Gently roll a glass pipette over the assembled transfer stack to eliminate any trapped air. Wet two long pieces of the filter paper precut to the width of the gel and place one end on top of the transfer stack and dip the other end in transfer buffer. This will serve as a wick to facilitate downward capillary transfer (see Note 20). Place a plastic tray on top of the transfer stack to limit buffer evaporation. Transfer for 2–12 h.

6. Crosslink RNA to the membrane using the auto crosslink option setting on a Stratagene Stratalinker. After this point, the reagents do not need to be RNAse-free.

7. Stain the membrane with methylene blue stain and record the rRNA banding pattern.

8. Destain the membrane in water and prehybridize the membrane using ExpressHyb™ hybridization solution for 30 min. The hybridization should be carried out at 42°C when using oligo probes, and at 50°C when using random probes (see Note 21).

9. Radiolabeled probes: oligonucleotide probes can be end-labeled with γ-[^{32}P] ATP using T4 polynucleotide kinase

(New England Biolabs). Alternately, probes can be made by random priming of a fragment of the target gene using α-[^{32}P] CTP, with a first strand cDNA synthesis kit (Ambion). Unincorporated α-[^{32}P] ATP is removed using a G-25 Sephadex quick spin column.

10. Quantify the radiolabeled probe using a liquid scintillation counter and add 1×10^7 cpm [^{32}P]-labeled DNA probe to the prehybridized membrane. It is important to denature the probes generated by random priming by boiling at 95°C and rapidly cooling on ice before adding to the blot.

11. Hybridize the probe for 8 h to overnight.

12. Rinse the membrane in low-stringency wash buffer for 30 min, followed by high-stringency wash buffer for 30 min.

13. Place the blot on prewet filter paper, cover with Saran Wrap and expose to a phosphorimager screen for 3 h or longer and scan (see Note 22).

14. Radiolabeled probes can be stripped off a membrane by washing in 95°C 0.5% SDS solution (see Note 23).

15. The membrane can be reprobed using a different probe by repeating steps 8–14.

3.6. SDS Polyacrylamide Gel Electrophoresis and Western Blotting

1. Dilute the protein samples 1:4 in 4× sample buffer. Boil the samples at 95°C for 5 min. Cool and load 20 µg of protein per lane on a 10% SDS polyacrylamide gel.

2. Run the gel at a constant voltage of 120 V until the dye front reaches the end of the gel, in 1× SDS running buffer.

3. Soak the gel in CAPS transfer buffer.

4. Cut nitrocellulose membrane to the dimensions of the gel and soak in CAPS buffer.

5. Semi-dry Western transfer: assemble the transfer by sandwiching the membrane and the gel between four pieces of prewet Whatman 3MM filter paper (cut to the dimensions of the gel), two pieces per side.

6. Perform Western transfer at a constant current of 100 mA for 30 min.

7. Ponceau stain the membrane to check for efficient transfer.

8. Destain the membrane in water and block in 5% milk in 1× PBS-T for 1 h at room temperature (see Note 24).

9. Wash the membrane in 1× PBS-T once for 15 min and twice for 5 min.

10. Add the primary antibody diluted to the appropriate final concentration in 2% milk in 1× PBS-T to the blot and incubate for 1 h at room temperature (see Note 25).

11. Repeat step 10.

12. Add the HRP-conjugated secondary antibody in 2% milk in 1× PBS-T and incubate for 1 h at room temperature (see Notes 24 and 25).
13. Repeat step 10.
14. Develop the blot using ECL reagents.

4. Notes

1. DEPC is a suspected carcinogen. Avoid inhalation and skin contact and always handle in a fume hood. However, after autoclaving, DEPC-containing solutions are no longer reactive or hazardous, though they have a slightly sweet odor.
2. Digitonin stock solutions are unstable to long-term storage. We recommend making up small volumes of stock solution and discarding at monthly intervals.
3. TRIzol® Reagent contains phenol and should be used with caution. Standard safety procedures should be followed to dispose phenol-containing solutions.
4. HEK293T cells are weakly adherent. Care should be taken to avoid dislodging them, e.g. pipetting reagents gently on to the sides of the tissue culture vessel. Coating the surface of tissue culture plasticware or coverslips with poly-d-lysine or collagen will enhance cell attachment. For fractionation experiments, collagen coating is preferred.
5. The 10-min incubation of cells on ice at this point is important to enable microtubule depolymerization. If cells are not incubated on ice for a sufficiently long period, tubulin will be primarily present in the insoluble fraction, rather than the cytosol.
6. This procedure can be easily scaled up or down by correspondingly changing the volume of reagents used. The main factor to consider is that the amount of the reagents used should be sufficient to cover the entire surface of the cell monolayer.
7. For adherent cells such as HEK293T, it is preferred to perform the permeabilization/cytosol release step on the monolayer. When lifted by trypsinization, the recovery of the cytosol fraction from adherent cells can be variable and should be thoroughly evaluated prior to experimentation.
8. It may be necessary to pass the TRIzol®-ized sample through 27½ gauge needle to completely shear the DNA. If shearing the DNA by this method, extreme care should be taken to avoid direct contact with TRIzol® Reagent.

9. Luer-Lok™ syringes are necessary to avoid sample loss while using the cell cracker; standard syringes tend to detach due to the high pressure generated when cells pass through.

10. Alternately, precipitate the protein from the supernatant fraction using trichloroacetic acid method and RNA using lithium chloride method. Please refer to Maniatis et al. (19) for detailed protocols.

11. Keep the tip of the needle only slightly under the interface so that when the needle is retracted, it does not leave a track of 40% sucrose through the 15% layer.

12. Be careful to not touch the interphase with the micropippetor tip as this reduces the quality of the RNA preparation. It would be best to try not to recover the last 5% of the aqueous phase.

13. If working with small quantities of RNA, precipitate at −20°C with and include yeast tRNA or glycogen as a carrier to aid precipitation.

14. Even trace amounts of DEPC can inactivate most polymerases and other enzymes. Hence, use commercial nuclease-free water if RNA will be used for molecular biology applications downstream.

15. While there is anecdotal suggestions that the protein pellet from TRIzol® extraction is difficult to get into solution, we have found that our sample buffer efficiently solubilizes the pellet and works well in the standard Laemmli buffer system. We have confirmed by SDS-PAGE that the profile of proteins from HEK293T extracted by TRIzol®, or directly extracted in sample buffer, are identical.

16. For applications that require the RNA sample to be strictly free of DNA, it may be necessary to include a step of DNAse-treatment using RNAse-free DNAse such as Turbo-DNAse from Ambion.

17. Formaldehyde is highly toxic. All waste generated (including the gel) should be disposed following standard safety procedures.

18. If required, 2 μl of SyBr safe dye can be added to the gel solution to enable visualization of the RNA after electrophoresis.

19. The 1× MOPS buffer can be reused several times unless RNAse contamination is suspected.

20. Make certain that there is no buffer short circuit between the wick and the paper towels. If required, place pieces of parafilm around the edge of the gel to prevent the wick from touching the paper towels. This is to ensure that the buffer transfer occurs only through the gel.

21. This is only a general rule of thumb and works in our hands for most of the probes we use in the lab. The stringency of the hybridization depends on the temperature of hybridization. Hence it may be required to identify an optimal hybridization temperature for a given probe depending on its length, Tm, etc. to ensure specificity and sensitivity.
22. Drying of the membrane hybridized with the probe fixes the probe to the membrane and make it very difficult to strip. Hence care should be taken to avoid membrane drying if it is to be reprobed. In the event of the membrane drying, it could be placed in −20°C for an extended period, to allow the radiolabel to decay before reprobing for another RNA.
23. Efficient stripping may require multiple washes in 95°C 0.5% SDS. Stripping of the probes may need to be checked by exposing the blot overnight to a phosphorimager screen.
24. This step can be done at 4°C overnight.
25. If you reuse antibody solutions, omit sodium azide in the HRP-conjugated secondary antibody stock solution. Sodium azide inhibits the activity of the HRP enzyme.

Acknowledgements

The authors thank Angela Jockheck-Clark for critical reading of the manuscript. We thank Tianli Zheng and other present and past members of the Nicchitta laboratory for their contribution to the development of the techniques described in this article. We also wish to thank Mayya Shveygert and the Gromeier lab for providing GAPDH and H3 antibodies. This work was supported by NIH grant GM-077382 to CVN.

References

1. Stevens, T. J., and Arkin, I. T. (2000) Do more complex organisms have a greater proportion of membrane proteins in their genomes?, *Proteins 39*, 417–420.
2. Walter, P., Ibrahimi, I., and Blobel, G. (1981) Translocation of proteins across the endoplasmic reticulum. I. Signal recognition protein (SRP) binds to in-vitro-assembled polysomes synthesizing secretory protein, *J Cell Biol 91*, 545–550.
3. Walter, P., and Blobel, G. (1981) Translocation of proteins across the endoplasmic reticulum. II. Signal recognition protein (SRP) mediates the selective binding to microsomal membranes of in-vitro-assembled polysomes synthesizing secretory protein, *J Cell Biol 91*, 551–556.
4. Walter, P., and Blobel, G. (1981) Translocation of proteins across the endoplasmic reticulum III. Signal recognition protein (SRP) causes signal sequence-dependent and site-specific arrest of chain elongation that is released by microsomal membranes, *J Cell Biol 91*, 557–561.
5. Lingappa, V. R., and Blobel, G. (1980) Early events in the biosynthesis of secretory and membrane proteins: the signal hypothesis, *Recent Prog Horm Res 36*, 451–475.
6. Lerner, R. S., Seiser, R. M., Zheng, T., Lager, P. J., Reedy, M. C., Keene, J. D., and Nicchitta,

C. V. (2003) Partitioning and translation of mRNAs encoding soluble proteins on membrane-bound ribosomes, *RNA 9*, 1123–1137.

7. Diehn, M., Eisen, M. B., Botstein, D., and Brown, P. O. (2000) Large-scale identification of secreted and membrane-associated gene products using DNA microarrays, *Nat Genet 25*, 58–62.

8. Mueckler, M. M., and Pitot, H. C. (1981) Structure and function of rat liver polysome populations. I. Complexity, frequency distribution, and degree of uniqueness of free and membrane-bound polysomal polyadenylate-containing RNA populations, *J Cell Biol 90*, 495–506.

9. Mutka, S. C., and Walter, P. (2001) Multifaceted physiological response allows yeast to adapt to the loss of the signal recognition particle-dependent protein-targeting pathway, *Mol Biol Cell 12*, 577–588.

10. Liu, L., Liang, X. H., Uliel, S., Unger, R., Ullu, E., and Michaeli, S. (2002) RNA interference of signal peptide-binding protein SRP54 elicits deleterious effects and protein sorting defects in trypanosomes, *J Biol Chem 277*, 47348–47357.

11. Ren, Y. G., Wagner, K. W., Knee, D. A., Aza-Blanc, P., Nasoff, M., and Deveraux, Q. L. (2004) Differential regulation of the TRAIL death receptors DR4 and DR5 by the signal recognition particle, *Mol Biol Cell 15*, 5064–5074.

12. Pyhtila, B., Zheng, T., Lager, P. J., Keene, J. D., Reedy, M. C., and Nicchitta, C. V. (2008) Signal sequence- and translation-independent mRNA localization to the endoplasmic reticulum, *RNA 14*, 445–453.

13. Diehn, M., Bhattacharya, R., Botstein, D., and Brown, P. O. (2006) Genome-scale identification of membrane-associated human mRNAs, *PLoS Genet 2*, e11.

14. Aragon, T., van Anken, E., Pincus, D., Serafimova, I. M., Korennykh, A. V., Rubio, C. A., and Walter, P. (2008) Messenger RNA targeting to endoplasmic reticulum stress signalling sites, *Nature*.

15. Claude, A. (1946) Fractionation of mammalian liver cells by differential centrifugation: II. Experimental procedures and results, *Journal of Experimental Medicine 84*, 61–89.

16. Duve, C. (1975) Exploring cells with a centrifuge, *Science 189*, 186–194.

17. Palacios, I. M., and St Johnston, D. (2001) Getting the message across: the intracellular localization of mRNAs in higher eukaryotes, *Annu Rev Cell Dev Biol 17*, 569–614.

18. Paquin, N., and Chartrand, P. (2008) Local regulation of mRNA translation: new insights from the bud, *Trends Cell Biol 18*, 105–111.

19. Sambrook J, F. E. F., and Maniatis T (2001) *Molecular Cloning: A laboratory manual*, Cold Spring Harbor Press, Cold Spring Harbor, New York, USA.

Chapter 20

Isolation of mRNAs Encoding Peroxisomal Proteins from Yeast Using a Combined Cell Fractionation and Affinity Purification Procedure

Gadi Zipor, Cecile Brocard, and Jeffrey E. Gerst

Abstract

Targeted mRNA localization to distinct subcellular sites occurs throughout the eukaryotes and presumably allows for the localized translation of proteins near their site of function. Specific mRNAs have been localized in cells using a variety of reliable methods, such as fluorescence in situ hybridization with labeled RNA probes, mRNA tagging using RNA aptamers and fluorescent proteins that recognize these aptamers, and quenched fluorescent RNA probes that become activated upon binding to mRNAs. However, fluorescence-based RNA localization studies can be strengthened when coupled with cell fractionation and membrane isolation techniques in order to identify mRNAs associated with specific organelles or other subcellular structures. Here we describe a novel method to isolate mRNAs associated with peroxisomes in the yeast, *Saccharomyces cerevisiae*. This method employs a combination of density gradient centrifugation and affinity purification to yield a highly enriched peroxisome fraction suitable for RNA isolation and reverse transcription-polymerase chain reaction detection of mRNAs bound to peroxisome membranes. The method is presented for the analysis of peroxisome-associated mRNAs; however it is applicable to studies on other subcellular compartments.

Key words: mRNA, Peroxisome, Peroxins, Affinity purification, Hemagglutinin epitope, Reverse transcription-polymerase chain reaction, Yeast

1. Introduction

A number of studies over the past two decades have demonstrated that mRNAs can undergo targeting to specific subcellular sites in eukaryotic cells in order to regulate cell division, cell polarity and differentiation, and body plan development in metazoans (1–3). Thus, mRNA targeting is now considered to be a widespread phenomenon that occurs in unicellular organisms, animal and plant tissues, and in developing embryos from a variety of animal phyla (2–4). The transport and localization of mRNAs within the cell can be achieved by different mechanisms, such as local RNA synthesis, local protection from degradation, diffusion and anchoring, or active transport by molecular motors (5).

Because of the importance of mRNA localization in the correct placement of proteins within cells, there is a need for specialized tools to monitor mRNA trafficking and localization in individual cells using fluorescence microscopy. The most commonly used technique is fluorescence in situ hybridization (FISH); a method that uses short DNA or RNA probes, that either bear fluorophores or can be detected using fluorescent-labeled antibodies, to hybridize to denatured RNA within cells. The advantage of FISH is that it can detect the endogenously expressed mRNAs, however, this method necessitates the use of fixed material and does not allow for the monitoring mRNAs in vivo. A tool that is commonly used for the localization of mRNAs in vivo is the bacteriophage MS2 coat protein (MS2-CP) that can bind to a short RNA stem-loop sequence (the MS2 aptamer) inserted into an mRNA of interest. When fused with a fluorescent protein, such as green fluorescent protein (GFP) or red fluorescent protein (RFP), the MS2-CP reporter can visualize mRNAs tagged with the MS2 loop sequences (6). A more recent application of this method is m-TAG (7, 8), a gene-tagging procedure which allows the sustained visualization of endogenously expressed mRNAs in living yeast. This procedure uses a PCR-based strategy for insertion of the MS2 loops into any gene of interest in the genome by homologous recombination. Upon co-expression of MS2-CP fused with GFP(3×), it is possible to examine the localization of endogenous mRNAs in vivo. This technique has been successfully used to localize specific mRNAs (many for the first time) to the endoplasmic reticulum (ER), mitochondria, peroxisomes, and to the bud tip (7). While the advantage of FISH and m-TAG is that both methods can efficiently localize endogenous mRNAs, their weakness is that by focusing on specific genes they are less applicable for rapid large-scale studies aimed at visualizing all mRNAs that localize to a given intracellular structure.

In contrast, a method that enables mRNA localization on a larger scale is subcellular fractionation, which yields purified organelles/membranes that can be assayed by reverse transcription-polymerase chain reaction (RT-PCR) with specific oligonucleotides or using DNA microarrays to detect mRNAs in the purified fractions. For example, microarray analyses on preparations derived from both yeast and mammalian cells revealed that thousands of mRNAs, including those coding for cytosolic proteins, are enriched in ER membranes (9, 10). Likewise, others have shown that mRNAs encoding mitochondrial proteins of prokaryotic origin preferentially localize to the vicinity of mitochondria (11) and that >500 mRNAs associate with mitochondria in yeast (12). In the procedure outlined below, we describe a method to obtain a highly enriched peroxisome fraction, using subcellular fractionation followed by affinity purification, and identify the associated mRNAs using RT-PCR. Application of this procedure has successfully demonstrated that mRNAs encoding certain peroxisomal proteins localize to peroxisome membranes (13).

Peroxisomes are single membrane-bound organelles containing diverse enzymes related to lipid metabolism and are found exclusively in eukaryotes (14–17). In yeast, peroxisome function is essential for the catabolism of fatty acids, which makes this organism useful for the study of peroxisome biogenesis (18). While many studies have explored the mechanism by which proteins are imported into the peroxisome matrix (19), less is known of how peroxisome membrane proteins and peroxins, the proteins involved in peroxisome biogenesis (20), are imported. The targeted transport of mRNA to the peroxisome membrane may confer localized synthesis near the import machinery and, thus, facilitate protein translocation into the organelle.

1.1. Overview of the mRNA Isolation Method Using Cell Fractionation and Affinity Purification

The principles of the peroxisome purification method are illustrated in Fig. 1. The basic procedure includes the separation of organelles using isopycnic centrifugation on a Nycodenz gradient, followed by affinity purification. Peroxisomes are specifically purified from the gradient fraction using an epitope-tagged anchor protein. Briefly, an HA epitope-encoding sequence was appended at the 3′end of the *PEX30* locus in wild type yeast cells to allow for endogenous expression of the HA-tagged Pex30

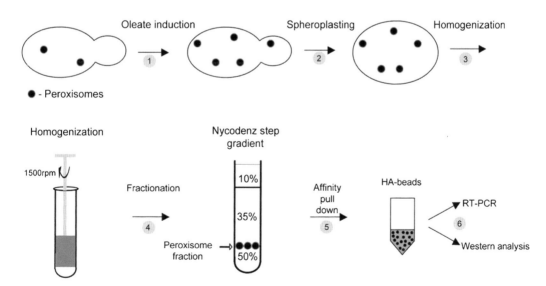

Fig. 1. Isolation of mRNAs encoding peroxisomal proteins from yeast using a combined cell fractionation and affinity purification procedure. An illustration of the different stages of the protocol: (1) Grow Pex30-HA tagged yeast cells in peroxisome induction medium (containing oleic acid). (2) Yeast lytic enzyme is used to spheroplast the cells. (3) Spheroplasts are broken and homogenized using Teflon dounce-type homogenizer. (4) Fractions are separated by ultracentrifugation on a Nycodenz density step gradient. (5) The peroxisome fraction is then affinity purified using anti-HA beads. (6) The different fractions are analyzed using both western analysis and RT-PCR (to identify bound mRNAs).

peroxisomal membrane protein. Peroxisomes were then purified from the enriched gradient fraction using agarose beads conjugated with anti-HA antibodies. After purification, RNAs were isolated from the purified peroxisomes and assayed by RT-PCR using oligonucleotide pairs to specific genes.

In yeast, peroxisomes are the only site for the β-oxidation of fatty acids. Therefore peroxisome function is required for cells to grow on medium containing fatty acids as the sole carbon source. First (see Subheading 3, step 1), yeast cells expressing HA-Pex30 are pre-grown to log phase on glucose-containing medium, followed by overnight culture in oleate-containing medium to induce peroxisome proliferation. Second (step 2), cells are harvested, washed, and treated with Zymolase (yeast lytic enzyme) to remove the cell wall and create spheroplasts. Next (step 3), the spheroplasts are lysed and homogenized using a Dounce homogenizer. The lysed spheroplasts are then centrifuged and the post-nuclear supernatant (PNS) obtained is loaded on top of a Nycodenz step density gradient (step 4) and further centrifuged to yield an enriched peroxisome fraction that is separated from the cytosol and other membrane-containing fractions, such as the mitochondria. Next (step 5), the peroxisome fraction is further purified using agarose beads conjugated to anti-HA antibodies. Both total RNA and protein are then extracted (step 6) from a sample of the PNS, the membrane/mitochondrial fraction, and the purified peroxisome fraction. The purity of the peroxisome fraction is verified by western analysis using specific antibodies against ER, Golgi, plasma membrane, cytosolic, and peroxisomal proteins, while mRNAs present in the peroxisome fraction are examined using either semi-quantitative or real-time PCR. See ref. 13 for examples of western and PCR analysis.

2. Materials

2.1. Cell Culture and Harvesting

1. SC medium: 1 l liquid medium is composed of 7 g of Synthetic Dry Mix [mix composed of: 294 g ammonium sulfate; 30 g dibasic potassium phosphate; 0.3 g each of arginine, cysteine, and proline; 0.45 g each of isoleucine, lysine, and tyrosine; 0.75 g each of glutamic acid, phenylalanine, and serine; 1.0 g each of aspartate, threonine, and valine; and 100 g Yeast nitrogen base lacking ammonium sulfate and amino acids (Becton, Dickinson and Co., cat. no. 233520)], 10 ml amino acid stock mix [200 ml of 100× synthetic complete medium composed of: 0.4 g each of adenine, uracil, tryptophan, histidine, methionine, and leucine in 150 ml DDW; acidify with 3 ml concentrated HCl; fill to 200 ml with DDW], 5 g or 20 g glucose (as required), 0.35 ml 10 N NaOH; and DDW

to 1 l (without diethylpyrocarbonate; DEPC); final pH = 6–7. For further details on preparation, see recipe in Haim-Vilmovsky and Gerst (8).

2. Oleate mix: 10% (vol/vol) oleic acid (Merck, cat. no. k3711571), 1% (vol/vol) Tween 80 (Sigma, cat. no. p8074), 17.7% (vol/vol) 1 N NaOH. Boil for 2 h in a water bath and again for 1 h before each use (to emulsify and sterilize).

3. Oleate induction medium: 0.3% (wt/vol) Yeast extract (Becton, Dickinson and Co., cat. no. 212750), 0.5% (wt/vol) Peptone (Becton, Dickinson and Co., cat. no. 211677), 0.5% (wt/vol) KH_2PO_4, and adjust solution to pH 6 with 1 M K_2HPO_4; autoclave for 30 min and add 2% oleate mix after cooling.

4. Ultra-pure water or DEPC-treated DDW.

5. Tris–sulfate 1 M, pH 9.4.

6. Sorbitol/Potassium phosphate buffer: 1.2 M Sorbitol and 20 mM Potassium phosphate buffer at pH 6.

7. Dithiothreitol (DTT).

2.2. Spheroplasting

1. Yeast lytic enzyme (Zymolase); 77,200 U/g (ICN Biomedicals, cat. no.152270).

2. 2% (wt/vol) Sodium dodecyl sulfate (SDS) in DDW.

2.3. Homogenization

1. PEG/Sucrose Buffer (PSB): 5 mM 2-(N-morpholino)ethanesulfonic acid (Mes), 1 mM KCl, 0.5 mM EDTA (adjusted to pH 6.0 with KOH), 12% (wt/vol) PEG 1500, and 160 mM Sucrose. Add freshly made protease inhibitors and RNAsin [Final concentrations: 2 μg/ml aprotinin, 1 μg/ml pepstatin, 0.5 μg/ml leupeptin, 1 mM phenylmethanesulphonylfluoride (PMSF), and 40 U/ml RNAsin (Promega, cat. no. W251B)].

2. Teflon dounce-type homogenizer (7 ml volume).

2.4. Nycodenz Layering and Density Gradient Centrifugation

1. Nycodenz gradient: 35%, and 50% (wt/wt) Nycodenz (AG Axis shield PoC, cat. no. 1002424), refraction index: 1.4070, and 1.4280, respectively) in PSB. Before use add freshly made protease inhibitors + RNAsin (see Subheading 2.3, item 1) to the Nycodenz solutions.

2. Ultracentrifuge, SW41-type rotor.

2.5. Affinity Purification

1. Pull-Down Buffer (PDB): 50 mM Tris/HCl at pH 7.5, 1 mM DTT, 12% (wt/vol) PEG 1500, 160 mM sucrose, and freshly prepared protease inhibitors and RNAsin (see Subheading 2.3, item 1).

2. Ultracentrifuge, SW60-type rotor.

3. Monoclonal Anti-HA Agarose beads (Sigma, cat. no. A2095).

2.6. RNA Purification and RT-PCR

1. MasterPure™ yeast RNA purification kit (Epicentre Biotechnologies, cat. no. MPY03100).
2. Molony murine leukemia virus reverse transcriptase and RNase H minus (M-MLV H(−)RT; Promega, cat. no. M368B).

2.7. Western Blotting

1. 8% SDS–PAGE gel (mini).
2. Protein sample buffer (1×): 80 mM Tris–HCl at pH 6.8, 10% (vol/vol) glycerol, 2% (wt/vol) SDS, and 0.25% (wt/vol) bromophenol blue.
3. 1% SDS.

3. Methods

3.1. Cell Culture and Harvesting

1. Inoculate an overnight culture of yeast in 10 ml SC medium containing 2% (wt/vol) glucose.
2. Next day, dilute cells to OD = 0.2 at A_{600} in 1 l SC medium containing 0.5% (wt/vol) glucose and grow cells to OD = ~2 at A_{600}.
3. Spin down cells and inoculate to OD = 1 at A_{600} in 1 l Oleate induction medium (harvest 500 ml of the SC medium OD = ~2 at A_{600}, and resuspend with 1 l oleate medium).
4. Incubate cells overnight at 30°C with shaking at ~200 rpm.
5. Harvest cells by centrifugation at 3,400 × g for 5 min using a Sorvall-type centrifuge with a rotor for use with large volumes (see Note 1).
6. Wash 2× with DDW.
7. Resuspend cell pellet with 0.1 M Tris/Sulfate at pH 9.4 and 10 mM DTT; Use 1 ml buffer for each 0.5 g of cells (i.e. 6 ml buffer for 3 g of cells).
8. Incubate for 15 min at 30°C with gentle shaking.
9. Spin down for 3 min at 1,400 × g at room temperature.
10. Wash in SB; use 1 ml SB per gram of cells.

3.2. Spheroplasting

1. Resuspend cell pellet at 0.15 g cells/ml SB containing Yeast lytic enzyme (10 mg/g cells).
2. Incubate with gentle shaking at 30°C for 15–30 min.
3. Microscopy: Verify spheroplast formation using 2% (wt/vol) SDS buffer (mix 2 µl of cells with 2 µl of 2% SDS buffer on a microscope slide); spheroplasts lyse upon contact with SDS-containing buffer and appear as dark spots under phase contrast, as opposed to intact cells which still bear their cell wall.

4. Harvest spheroplasts using a Sorvall centrifuge with an SS34-type rotor at $1,500 \times g$ for 5 min at 4°C (see Note 2).

5. Wash 2× in SB using gentle resuspension.

3.3. Homogenization

1. Gently resuspend spheroplasts in 1.5 ml/g cells with PEG/Sucrose Buffer (PSB) containing protease inhibitors + RNAsin. For example, for 3 g cells, add 4.5 ml PSB buffer (see Note 3).

2. Homogenize using a motor-driven Potter-Elvehjem small clearance teflon pestle in a glass tube homogenizer at 1,500 rpm (use 20 slow strokes) on ice (see Note 4).

3. Harvest nuclei/unbroken cells and cell debris using a Sorvall centrifuge with an SS34-type rotor for 5 min at 4°C at $1,400 \times g$ (see Note 5). Remove soluble fraction (i.e. post-nuclear supernatant/PNS) to ice; save ~30 μl for western analysis (see Note 6).

4. Carefully add 50% Nycodenz (wt/wt in PSB; refraction index = 1.4280) to PNS to reach a final concentration of 10% (vol/vol) Nycodenz (see Note 7).

5. Load the PNS (containing 10% Nycodenz) on top of a Nycodenz step gradient (see below) (see Note 8).

3.4. Nycodenz Layering and Density Gradient Centrifugation

1. Layer the bottom of a 13-ml Beckman centrifuge tube (Polyallomer; cat. no. 331372 for use with a SW41-type rotor) with 2.5 ml of 50% (wt/wt) Nycodenz in PSB.

2. Next gently layer 6.5 ml of 35% (wt/wt) Nycodenz in PSB on top of the 50% Nycodenz bottom layer. Use care not to disturb the layer.

3. Finally, add 2.5 ml of the PNS in 10% (wt/wt) Nycodenz. Fill up the tubes with PNS containing 10% Nycodenz, if necessary.

4. Centrifuge in an ultracentrifuge using an SW41-type rotor at $\sim 80,000 \times g$ for 1 h at 4°C.

5. Collect the visible cloudy peroxisome and membrane/mitochondria fractions (located just above the 50%:35% and 35%:10% interfaces, respectively) using a 2–5 ml syringe (needle gauge = 21 G, length = 0.8×40 mm). Usually ~1–2 ml should be collected per fraction (see Note 9).

3.5. Affinity Purification

1. Carefully dilute the peroxisome fraction with three volumes of PDB (containing proteases and RNasin). Add to a 3-ml Beckmann centrifuge tube (Polyallomer; cat. no. 355636).

2. Pellet the peroxisomes using an ultracentrifuge with an SW60-type rotor. Spin at $\sim 80,000 \times g$ for 20 min at 4°C.

3. Decant supernatant and gently resuspend peroxisomes in 400 μl of PDB (containing proteases and RNasin).

4. Divide the peroxisome fraction into two equal aliquots (~200 μl each) and incubate each suspension with 25 μl anti-HA antibody cross-linked beads, pre-washed 3× in PDB as follows: add 50 μl aliquots of the 50% bead slurry to separate tubes before washing each bed with PDB; and add the peroxisome suspension directly to the washed beads.

5. Incubate overnight at 4°C with rotation.

6. Wash 3× for 30 min each with 200 μl PDB.

7. Decant supernatant without removing beads.

3.6. RT-PCR and Western Analysis

To one aliquot perform RNA purification, reverse transcription, and PCR (steps 1–3). To the other aliquot perform western analysis (step 4).

1. RNA purification:

   We employ the MasterPure™ yeast RNA purification kit:
   Add 300 μl of extraction reagent (containing Proteinase K) to the beads and boil for 15 min. Follow mRNA extraction instructions according to the protocol of the manufacturer (including the optional DNase treatment step).

   Typically 500–1,000 ng of total RNA is obtained from the peroxisome fraction originating from 3 g of cells.

   In parallel, use 200 μl of the PNS and 10%/35% fractions for mRNA purification: dilute the fractions 1:1 with DDW, add 300 μl extraction reagent (containing Proteinase K), and follow the mRNA extraction instructions.

2. Reverse transcription (M-MLV H(−)RT):

   Use 250–500 ng RNA from each fraction for each RT reaction. Use random hexamers (Fermentas, cat. no. S0142) and follow the manufacturer instructions, e.g. we use 10 min RT and 42°C for 50 min.

   As a control, perform the same procedure without adding reverse transcriptase.

3. PCR:

   Semiquantitative PCR (between 18 and 24 cycles for detection) should be performed with gene-specific primers (~20 bp; 10 pmol each per reaction) complimentary to a region of 300–500 nucleotides of the ORF. The PCR reaction is performed in a 20 μl reaction volume under standard conditions recommended by the manufacturer (i.e. we use 2× Taq Master mix purple; Lambda Biotech, cat. no. D123P; final (1×) composition: 1.5 mM $MgCl_2$, 200 μM dNTPs, and 2.5 U/25 μl of *taq* polymerase). Use 1 μl of the reverse transcriptase reaction (e.g. cDNA) per PCR reaction.

PCR should be performed using the following conditions:

Melting temp – 95°C

Annealing temp – 55°C

Elongation temp – 72°C

Number of cycles typically employed for detection: 18, 21, and 24 cycles.

Load 20 μl of the PCR reactions onto a 1% (wt/vol) agarose gel in TAE buffer (1×); electrophorese at 120 V for ~20 min. For an example of the expected results, see Fig. 3 in Zipor et al. (13) (see Note 10).

4. Western analysis:

To the other aliquot of beads add 90 μl elution buffer (consisting of: 1× sample buffer containing 2% SDS and freshly added 10 μl of 1 M DTT), boil for 10 min and collect the soup; add and load aliquots of 25–30 μl onto a 8% SDS–PAGE mini-gel. Load PNS and 10%/35% fractions samples (e.g. Load 2–3 μl and ~6 μl of the PNS and 10%/35% fractions, respectively). Assess the relative level of purity of the peroxisome fraction (relative to samples of the PNS and 10%/35% fractions) using specific antibodies against proteins of the peroxisome, Golgi, mitochondria, plasma membrane, cytoplasm, etc.

4. Notes

1. Oleic acid-containing medium (oleate induction medium) is somewhat viscous and can make it more difficult to pellet the yeast cells, thus, the RPM used to pellet (Subheading 3.1, step 5) is higher than that typically used. Make sure that the size of the yeast pellet (Subheading 3.1, step 5) is sufficient for the procedure (obtain at least 3 ml of cells – i.e. ~3 g of cells). If not, elevate the speed and time of centrifugation or increase the culture volume.

2. Following zymolase treatment (Subheading 3.2), it is essential that all steps are performed on ice.

3. It is very important to use RNase-free instruments and solutions. The process contains many steps and the amount mRNA obtained is small, so one should be careful with the handling of the material.

4. Make sure that most of the cells are broken during the homogenization step (Subheading 3.3, step 2). If not, add several additional strokes and verify that the teflon pestle tightly fits the glass homogenizer.

5. It is very important to properly remove the nuclei and cell debris after homogenization in order to avoid contamination

of the PNS. If required, the PNS can be centrifuged in a new clean tube for another 15 min at 4°C using an SS34-type rotor at $1,400 \times g$.

6. To monitor the degree of purification, we recommend removing aliquots at each step of purification; determine protein concentration and immediately add SDS-containing sample buffer and freeze until needed for analysis by western blotting.

7. By no means should the PNS be centrifuged at high speed and separated into supernatant and pellets before the dilution with Nycodenz (and loading onto the gradient). High-speed centrifugation and pelleting in the absence of dilution can lead to contacts between organelles that remain following ultracentrifugation and, thus, lead to contamination of the peroxisome fraction.

8. The objective of isopycnic centrifugation is to obtain an enriched peroxisome fraction that is separate from other organelles (i.e. mitochondria) and membranes, and has been established here for the use with Nycodenz only.

9. The enriched peroxisome fraction collected after gradient centrifugation can be kept at 4°C for up to 24 h without noticeable protein degradation.

10. The PCR reactions (Subheading 3.6, step 3) should show differences in the mRNA content between the various fractions. If no products are observed, one can increase the number cycles. Always include positive and negative controls for the PCR reactions. One can employ yeast genomic DNA as a template for the positive controls.

Acknowledgements

This work was supported by grants to J.E.G. from the Minerva Foundation, Germany and Josef Cohn-Minerva Center for Biomembrane Research at the Weizmann Institute of Science, Israel. J.E.G. holds the Besen-Bender Chair in Microbiology and Parasitology. C.B. is supported by the Elise-Richter Program of the Austrian Science Fund (FWF).

References

1. Du, T.G., Schmid, M., and Jansen, R.P. (2007) Why cells move messages: the biological functions of mRNA localization. *Semin. Cell Dev. Biol.* **18**, 171–177.
2. Bashirullah, A., Cooperstock, R.L., and Lipshitz, H.D. (1998) RNA localization in development. *Annu. Rev. Biochem.* **67**, 335–394.
3. Darzacq, X., Powrie, E., Gu, W., Singer, R.H., and Zenklusen D. RNA asymmetric distribution and daughter/mother differentiation in yeast. *Curr. Opin. Microbiol.* **6**, 614–620.
4. Crofts, A.J., Washida, H., Okita, T.W., Ogawa, M., Kumamaru, T., and Satoh, H. (2004) Targeting of proteins to endoplasmic

reticulum-derived compartments in plants. The importance of RNA localization. *Plant Physiol.* **136**, 3414–3419.
5. St Johnston D. (2005) Moving messages: the intracellular localization of mRNAs. *Nat. Rev. Mol. Cell. Biol.* **6**, 363–375.
6. Beach, D.L., Salmon, E.D., and Bloom, K. (1999) Localization and anchoring of mRNA in budding yeast. *Curr. Biol.* **9**, 569–578.
7. Haim, L., Zipor, G., Aronov, S., and Gerst, J.E. (2007) A genomic integration method to visualize localization of endogenous mRNAs in living yeast. *Nat. Methods* **4**, 409–412.
8. Haim-Vilmovsky L and Gerst, J.E. (2009) m-TAG: a PCR-based genomic integration method to visualize the localization of specific endogenous mRNAs in vivo in yeast. *Nat. Protoc.* **4**, 1274–1284.
9. Lerner, R.S., Seiser, R.M., Zheng, T., *et al.* (2003) Partitioning and translation of mRNAs encoding soluble proteins on membrane-bound ribosomes. *RNA* **9**, 1123–1137.
10. Aronov, S., Gelin-Licht, R., Zipor, G., Haim, L., Safran, E., and Gerst J.E. (2007) mRNAs encoding polarity and exocytosis factors are cotransported with the cortical endoplasmic reticulum to the incipient bud in *Saccharomyces cerevisiae*. *Mol. Cell. Biol.* **27**, 3441–3455.
11. Sylvestre, J., Vialette, S., Corral-Debrinski, M., and Jacq, C. (2003) Long mRNAs coding for yeast mitochondrial proteins of prokaryotic origin preferentially localize to the vicinity of mitochondria. *Genome Biol.* **4**, R44.
12. Garcia, M., Darzacq, X., Devaux, F., Singer, R.H., and Jacq C. (2007) Yeast mitochondrial transcriptomics. *Methods Mol. Biol.* **372**, 505–528.
13. Zipor, G., Haim-Vilmovsky, L, Gelin-Licht, R., Brocard, C., and Gerst J.E. (2009) Localization of mRNAs coding for peroxisomal proteins in the yeast, *Saccharomyces cerevisiae*. *Proc. Natl. Acad. Sci. USA*, **106**, 19848–19853.
14. Antonenkov, V.D., Sormunen, R.T., and Hiltunen J.K. (2004) The behavior of peroxisomes in vitro: mammalian peroxisomes are osmotically sensitive particles. *Am. J. Physiol. Cell Physiol.* **287**, C1623–1635.
15. Subramani S. (1993) Protein import into peroxisomes and biogenesis of the organelle. *Annu. Rev. Cell Biol.* **9**, 445–478.
16. Wanders, R.J. and Waterham H.R. (2006) Biochemistry of Mammalian peroxisomes revisited. *Annu. Rev. Biochem.* **75**, 295–332.
17. Hayashi, M. and Nishimura, M. (20030 Entering a new era of research on plant peroxisomes. *Curr. Opin. Plant Biol.* **6**, 577–582.
18. Erdmann, R., Veenhuis, M., Mertens, D., and Kunau, W.H. (1989) Isolation of peroxisome-deficient mutants of *Saccharomyces cerevisiae*. *Proc. Natl. Acad. Sci. USA* **86**, 5419–5423.
19. Girzalsky, W., Platta, H.W., and Erdmann R. (2009) Protein transport across the peroxisomal membrane. *Biol. Chem.* **390**, 745–751.
20. Distel, B., Erdmann, R., Gould, S., *et al.* (1996) A unified nomenclature for peroxisome biogenesis factors. *J. Cell Biol.* **135**:1–3.

Chapter 21

Profiling Axonal mRNA Transport

Dianna E. Willis and Jeffery L. Twiss

Abstract

The importance of mRNA localization and localized protein synthesis to spatially modulate protein levels in distinct subcellular domains has increasingly been recognized in recent years. Axonal and dendritic processes of neurons represent separate functional domains of the cell that have shown the capacity to autonomously respond to extracellular stimuli through localized protein synthesis. With the vast distance often separating distal axons and dendrites from the neuronal cell body, these processes have provided an appealing and useful model system to study the mechanisms that drive mRNA localization and regulate localized mRNA translation. Here, we discuss the methodologies that have been used to isolate neuronal processes to purity, and provide an in-depth method for using a modified Boyden chamber to isolate axons from adult dorsal root ganglion neurons for analyses of axonal mRNA content. We further show how this method can be utilized to identify specific mRNAs whose transport into axons is altered in response to extracellular stimuli, providing a means to begin to understand how axonal protein synthesis contributes to the proper function of the neuron.

Key words: mRNA localization, RNA transport, Axonal protein synthesis, Local translation

1. Introduction

The ability to amplify very small quantities of nucleic acids has provided new possibilities to query the RNA profiles of subcellular compartments, at least when the compartment or region can be isolated to purity (1). Neurons have provided a particularly appealing experimental model for analyses of RNA localization, since primary neurons can extend processes for hundreds of microns to a few centimeters in culture. The possibility that proteins can be synthesized in the efferent processes of neurons, referred to as dendrites, was suggested by ultrastructural studies showing that ribosomes concentrate adjacent to postsynaptic regions of these cells (2). Subsequent studies looking at this localized protein synthesis have supported the early hypotheses that the proteins generated from dendritically localized mRNAs play a role in synaptic plasticity (3). Synaptic plasticity is thought to underlie learning and memory, so stimuli regulating mRNA transport into and translation within dendrites could have profound effects on brain function. Protein synthesis in the axonal compartment has been more closely linked to growth of this process, both during development and after injury (4, 5). As mentioned in other chapters herein and in recent reviews (6, 7), subcellular localization of mRNAs is a common feature of polarized eukaryotic cells, and neurons provide a convenient experimental system for profiling the populations of localized mRNAs and for determining how mRNA transport is regulated. In this chapter, we will detail methodology that we have developed for analyses of axonal mRNA content, both for profiling the populations of mRNAs that localize and for quantifying stimulus-dependent transport of mRNAs into the axonal compartment.

Investigators have used several different approaches to isolate subcellular domains of neurons. From in vivo preparations, density gradient ultracentrifugation has been used to fractionate growth cones and synaptic structures (synaptosomes and postsynaptic densities) from cell bodies and non-neuronal cells (8–10). The growth cone is the terminal portion of the growing axon where new membrane and cytoskeleton is laid down for extension of this process; actively growing dendrites also have growth cones, but dendrites only extend for a few millimeters while axons can grow for hundreds of millimeters to many centimeters. The synaptosome preparations contain both presynaptic (axonal) and postsynaptic (dendritic) components, while the postsynaptic density is only the dendritic. Though mRNAs have been detected and protein synthesis has been demonstrated using the synaptosome-type preparations (10–13), contamination from other cellular elements, particularly glial cells, is a major concern with this approach. Proteomics analyses of synaptosome and postsynaptic density preparations clearly show that a small complement of glial proteins cofractionate with these structures (14, 15). Thus, most analyses of localized protein synthesis and mRNA transport have used cultured neurons.

In vitro, neurons isolated from the central and peripheral nervous systems (CNS and PNS, respectively) will extend processes, axons and dendrites, for several hundred microns away from the cell body. For rodent preparations, culture of CNS neurons is largely restricted to embryonic and the early postnatal period. In contrast, PNS neurons can be cultured from adult animals. With the large geographic separation of the ends of the neuronal processes from the cell body, one simply needs to determine a means to compartmentalize the growing processes away from the cell body and any non-neuronal cells that are included in the preparation.

Since neurons are postmitotic, the non-neuronal cells (typically astrocytes in CNS cultures and Schwann cells in PNS cultures) can be minimized by inclusion of mitotic inhibitors. However, in our hands this requires several days to effectively kill off these glial cells and the glia are known to exert varying trophic effects on neurons (16–18).

Several methods have been used to physically separate neuronal processes of PNS and CNS neurons from their cell bodies in culture (Fig. 1). In the following section, we detail use of the "modified Boyden" chamber approach for isolation of axonal processes from dissociated cultures of adult DRG neurons. Torre and Steward initially used this approach to show glycosylation of dendritically synthesized proteins in cultures of rodent hippocampal neurons (24). In their method, *Matrigel*™ was used in the lower compartment, providing a substrate for dendrites to grow into and facilitating separation of the processes. While this was effective for these glycosylation studies, the proteins within the Matrigel are too concentrated to allow for reasonable fractionation of any metabolically labeled proteins from lysates. The high protein content of Matrigel does not allow for electrophoretic methods for detection and, in our hands, some lots of matrigel have generated nonspecific amplifications by reverse transcriptase-coupled polymerase chain reaction (RT-PCR). In searching for a more versatile method, we found that DRG axons would traverse pores of the laminin-coated membrane and adhere to the undersurface to provide an effective separation of cell bodies and distal processes (25). Other groups have used an analogous method to purify axonal processes from developing DRG neurons (26), dendrites from hippocampal neurons (27), and pseudopodia from fibroblasts (28). In the sections below, we outline the use of this culture method to profile axonal mRNA levels. The reader should take note that the methods for RNA analysis provided here could readily be adapted to the other culture methods outlined in Fig. 1, particularly the microfluidics approach that Taylor et al. have recently used for profiling RNA transport in cortical neurons (29).

2. Materials

2.1. DRG Cell Culture and Axonal Treatment

1. 8 μm polyethylene terephthalate (PET) tissue culture inserts for 6-well plates (Falcon #353093; Franklin Lakes, NJ) – these membranes are translucent so they can also be used for imaging.
2. 6-Well tissue culture plates, deep-welled (Falcon #353046).
3. 6-Well tissue culture plates, shallow-welled (Greiner Bio-One #657160; Monroe, NC).

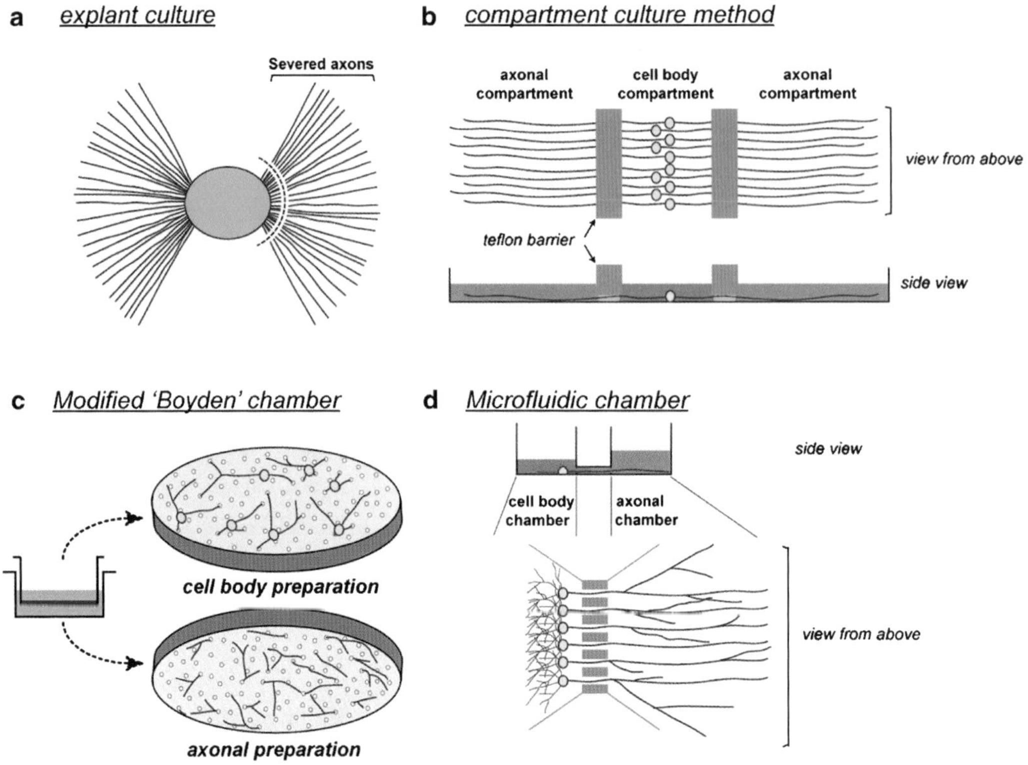

a *explant culture*

b *compartment culture method*

c *Modified 'Boyden' chamber*

d *Microfluidic chamber*

Fig. 1. Culture methods for isolation of neuronal processes. A variety of methods have been developed to isolate axons and dendrites from cell bodies and non-neuronal cells. (**a**) The simplest approach is to physically dissect away the neurites as Olink-Coux and Hollenbeck did for explant cultures of rodent PNS ganglia (19). The much larger size of some invertebrate neurons allows for such dissections even in dissociated cultures (20). (**b**) The compartmented or "Campenot" culture approach has been used for sympathetic and developing DRG cultures (21). Vogelaar et al. recently modified this chamber method for explant DRG cultures from adult rodents (22). This is an appealing approach since the Teflon barrier not only blocks passage of glial cells, but also provides physically separate media compartment where processes can be treated directly with trophic/tropic and pharmacologic agents (23). The separate media compartments also facilitate preparation of axonal vs. cell body lysates. A drawback for this approach is the distance that processes must extend to reach the "axonal compartment," since the barrier is typically 1 mm thick and cell body compartment can add a few mm more depending on where the neuron initially adheres. Consequently, these cultures are maintained for several days to weeks. (**c**) Several groups, including the authors, have used a modified Boyden chamber to isolate axonal or dendritic processes (24–27). The porous membrane used in these preparations is only a few microns thick (8 μm for Falcon PET membrane with 8 μm diameter pores), so axons can traverse the membrane pores to reach the undersurface within hours of plating. This short distance allows isolation of dendrites from neurons that are more polarized than the DRGs and sympathetic neurons (24, 27). Neuronal processes are isolated from cell bodies and nonneuronal cells by physically scraping, which requires care to reach rigorous RT-PCR based standards of purity that we and others have typically used (see Subheading 3.1.3). Notably the Macara lab has used this approach to isolate pseudopodia from cultured fibroblasts (28). (**d**) Taylor et al. developed a microfluidic device for isolation of axonal processes (29, 30). Neuronal processes extend through grooves in this device to reach a separate chamber, and similar to the Campenot chamber cultures, media of the axonal and cell body compartments are separate providing a means to uniquely stimulate the neuronal processes (31). In contrast to the Campenot chambers, the cell body and axonal compartments can be separated by only a few hundred microns in these microfluidic devices. Beyond the methods outlined in this schematic, a wafer chip based method was very recently published for generating directed growth of neuronal processes and also allows for axonal isolation and fluorescence-based imaging (32). Additionally, a pure axoplasm preparation from peripheral nerve preparations was recently published that holds promise for analysis of axonal RNA transport and localized protein synthesis in vivo (33).

4. Cell culture tested poly-L-lysine, mol. wt 70,000–150,000 (Sigma #P4707).

5. Mouse laminin (Millipore #CC095; Billerica, MA).

6. DMEB/F12 cell culture media – DMEM and Ham's F-12 1:1 mix supplemented with 0.1% cell culture tested BSA, fraction V.

7. Complete culture media – DMEM and Ham's F-12 1:1 mix, with N1 medium supplement to 1× (Sigma #N6530; St. Louis, MO) and 10% horse serum. This is made up in 50 ml batches and kept for up to 1 month at 4°C.

8. Phosphate buffered saline at pH 7.4 (PBS).

9. Collagenase, type XI (Sigma #C9697); stock solution of 500,000 units/ml in PBS is stored at –80°C in 100 μl aliquots.

10. 5 3/4″ glass pipettes – Fire-polished to decrease cell damage during trituration (see Note 1).

11. Cytosine arabinoside (Sigma #C1768).

12. 5,6-Dichlorobenzimidazole riboside (DRB; Sigma #D1916).

13. 20 μm diameter carboxylated polystyrene microparticles (Polyscience, Inc. #24811-2; Warington, PA).

14. Mouse 2.5S Nerve Growth Factor (NGF; Harlan; Indianapolis, IN); 50 μg per experiment.

15. *PolyLink* protein kit for carboxylated microparticles (Polysciences, Inc. #24350-1).

2.2. RNA and Protein Isolation, Quantification and Normalization

1. *RNaqueous-micro* kit (Ambion #AM1931; Austin, TX).

2. *RiboGreen* RNA quantification reagent and kit (Invitrogen #R11490; Carlsbad, CA).

3. *NanoOrange* protein quantitation kit (Invitrogen #N6666).

4. *VersaFluor* fluorometer (BioRad #170-2402; Hercules, CA) with RNA and protein analyses filter sets (excitation 485–495 nm/emission 515–525 nm and excitation 470–490 nm/emission 585–595 nm, respectively).

5. Microcon centrifugal filter, 10 kilodalton (kDa) cutoff (Millipore).

2.3. Reverse Transcription and Real-Time PCR

1. *iScript* cDNA synthesis kit (BioRad #170-8891).

2. *HotStarTaq* 2× Master Mix kit (Qiagen #203445-14; Valencia, CA).

3. 2× *SybrGreen* Master Mix kit (Qiagen #204145-14).

4. Real-time primers for each transcript of interest optimized for comparative cycle threshold (C_t) method (see Note 2).

5. *7900HT* real-time amplification and sequence detection system (Applied Biosystems; Foster City, CA).
   6. *Biomek 2000* liquid handling robotic system (Beckman Coulter; Brea, CA).

3. Methods

3.1. Culture Method to Isolate Neuronal Processes

Use of the modified Boyden chamber for culturing DRG neurons from adult rats to isolate axonal processes to purity is detailed here. Since the DRG neurons only extend axonal processes in culture (25), this system provides a pure preparation of axons. Neurons with full dendritic–axonal polarity (e.g., cortical or motor neurons) would generate a mixed preparation of axons and dendrites. However, with a single axonal process and many dendrites for each of these more polarized neurons, Poon et al. have shown that the majority of the processes reaching the lower membrane surface are dendritic when culturing hippocampal neurons on these porous membranes (27).

3.1.1. Preparation of Tissue Culture Inserts

1. The day before culturing, place 8 μm PET tissue culture inserts into deep-welled 6-well plates (see Note 3). Add approximately 3 ml of poly-L-lysine (50 μg/ml) into each insert *(see Note 4)*.
2. Leave poly-L-lysine on plates for 30–60 min.
3. Remove poly-L-lysine and air-dry the inserts and plates in the hood for at least 30 min. Inspect to ensure that the plates are dry. If not, continue drying until no visible liquid remains.
4. Rinse the plates one time with sterile tissue culture grade dH_2O. Aspirate the dH_2O taking care to not disturb (i.e., puncture or tear) the membrane and air-dry for a minimum of 15 min.
5. Cover the plates and inserts with 3.0 ml of laminin (5 μg/ml in PBS) (see Note 5).
6. Incubate overnight at 4°C with gentle rocking to ensure complete coverage of the surface of the membranes.

3.1.2. DRG Culture

1. Remove the L4 and L5 DRGs from the animals and place all of the isolated DRGs into one well of a 12-well plate containing 1 ml of DMEB/F12 + 1× N1 supplement *(see Note 6)*.
2. After all of the DRGs have been removed, wash them briefly by moving them from well to well through six wells containing 0.5 ml of DMEM/F12 + 1× penicillin/streptomycin (see Note 7).
3. After the penicillin/streptomycin rinse, put the DRGs into a well containing 0.5 ml of complete culture media, add collagenase

to a final of 5,000 units/ml and incubate at 37°C for 20 min.

4. After 20 min, triturate the DRG cell suspension 15–20 times by gently pipetting up and down using a fire-polished pipette to break apart the ganglia. Return to incubator for 5 min (see Note 1).

5. Triturate again and remove DRGs from plate and put into 15 ml conical tube. Add 8.5 ml of DMEB/F12 media.

6. Pellet the cells at $160 \times g$ for 5 min in a swinging bucket rotor at room temperature.

7. Aspirate media. Add 1 ml of DMEB/F12 media to the pellet and triturate 15–20 times with a fire polished Pasteur pipette until fully dissociated.

8. Add 8 ml of DMEB/F12 to the dissociated pellet, invert to mix, then centrifuge at $160 \times g$ for 5 min.

9. Repeat steps 7 and 8 two additional times for a total of three washes.

10. After the final wash, resuspend the DRGs in complete culture medium plus 10 µM cytosine arabinoside and 80 µM 5,6-dichlorobenzimidazole riboside. Use 2 ml of media per each filter. Add 2 ml of cells + media to the top of each filter. Add 2 ml of media containing no cells into the bottom of each well (for a total of 4 ml of media/well).

11. Culture overnight at 37°C and 5% CO_2.

3.1.3. Isolation of Axonal Compartment

1. Rinse both the upper and lower surface of the tissue culture inserts with 1× PBS to remove media and any nonadherent cell material.

2. Gently scrape the upper surface of each insert with a cotton-tipped applicator. Use four applicators per insert, turning the insert 90° after each swab to ensure that the entire surface of the insert is swabbed and that no areas are missed. It is imperative to remove all of the cell body side from the inserts used for axonal RNA preps without pressing so hard that material is pressed from the top surface through the filter (see Note 8).

3. After shearing, rinse the upper surface of the inserts very briefly with 1× PBS. This will remove any cell body material that may have remained. Do this gently as the axons are now sheared from their cell bodies and will not tolerate being treated forcefully.

4. Remove the insert from its housing with a scalpel. Do not cut too close to the housing as this area is very hard to get clean on the upper surface and tends to be the biggest source of contamination (see Note 9).

5. Place the excised insert axonal side down into a 35-mm dish containing 0.5 ml of RNA lysis buffer from the RNaqueous-micro kit.

6. Continue with all six inserts, combining them into a single 35-mm dish.

7. Place on a rocker at 4°C for 20 min.

3.1.4. RNA Isolation

1. Remove lysates from the 35-mm dish and place into 1.5-ml tubes (see Note 10).

2. Vortex for 30 s to ensure lysis.

3. Add one-half volume (~250 µl) 100% ethanol to the lysate and vortex briefly.

4. Load lysate/ethanol onto an *RNaqueous-micro* column (150 µl at time).

5. Centrifuge at max speed for 10 s. Repeat this several times to filter all of the axonal lysate.

6. Follow the *RNaqueous-micro* kit instructions to wash the columns.

7. After the final wash, transfer the columns to a new, RNase-free microcentrifuge tube and elute the RNA with the *RNaqueous-micro* elution solution (a total of 15 µl elution volume).

3.1.5. RNA Quantification and Reverse Transcription

1. Quantify 1.0 µl of each RNA sample using the RiboGreen RNA quantification kit with a high-range working solution in a total volume of 200 µl (see Note 11).

2. Read the RNA-RiboGreen fluorescence on a fluorometer such as a BioRad VersaFluor Fluorometer per quantification kit manufacturer's protocol.

3. Use approximately 50 ng of total RNA for the reverse transcriptase (RT) reaction using the *iScript* cDNA synthesis kit (see Note 12).

4. Dilute the RT product 1:5 with RNase-free water and use 1 µl for standard PCR to test for axonal purity. All axonal preparations must first be assessed for purity by standard RT-PCR for axonal, somatic and nonneuronal transcripts (see Note 13).

3.2. Approach for Quantitative Analyses of Axonal mRNA Transport

Effects of neurotrophins and other growth regulating stimuli on general axonal and dendritic transport have long been known. However, it has only been in recent years that effects on RNA transport have been documented. Extracellular stimuli that modulate synaptic activity and axonal growth have been shown to trigger changes in transport of mRNAs from the cell body into neuronal processes (5). In cultures of developing CNS cortical neurons, neurotrophins were early on shown to modulate transport of axonal RNAs using *SYTO®* dyes (34). Bassell's group went on to show that β-actin is among the mRNAs whose axonal transport is increased by neurotrophin 3 application to cortical neurons (35). The majority of studies of mRNA transport have been limited to single transcripts. We have used the culture method outlined above to broadly screen for ligand-dependent alterations in axonal RNA levels and the signaling mechanisms regulating these changes (36, 37). The general scheme for this quantitative method is outlined below (see also Fig. 2). Although we cannot entirely exclude the possibility that alterations in axonal mRNA stability contribute to the changes in axonal mRNA levels that we detect using this method, analysis of cell body mRNA levels in parallel experiments show an opposite trend in response to environmental stimuli compared to those seen in the axon (37). This strongly suggests that DRG sensory neurons alter mRNA delivery into the axons.

3.2.1. Exposure of Axons to Neurotrophins

1. Place 12.5 mg (approximately 0.5 ml) of 20 μm carboxylated polystyrene microparticles into a 1.5-ml centrifuge tube (see Note 14).

2. Pellet at $1,000 \times g$ for 5 min and resuspend in 0.4 ml *PolyLink Coupling Buffer*.

3. Pellet at $1,000 \times g$ for 5 min and resuspend in 0.17 ml *PolyLink Coupling Buffer*.

4. Add 20 μl of freshly prepared 200 mg/ml EDAC solution to the resuspended microparticles and mix by gently vortexing.

5. Add 50 μg of NGF or BSA (i.e., control) dropwise to the microparticles suspension, mixing gently after each drop (see Note 15).

6. Allow the protein to couple to the microparticles for 60 min at room temperature with end over end mixing.

7. Pellet at $1,000 \times g$ for 10 min and wash the microparticles two times with 0.4 ml *PolyLink Coupling Buffer*. Save the wash buffer after removal, as this will be used to calculate coupling efficiency to determine dosage of ligand on microparticles.

8. Quantify the amount of protein in the wash buffer by fluorometry using *NanoOrange* reagent to determine the amount of bound protein. Generally, 90–95% of the protein is bound to the microparticles. This calculation is based on the mass of protein added to coupling mixture less the mass of protein in the wash from #7.

9. Add the bound microparticles to 12 ml of complete culture media and add 2 ml per well to a shallow 6-well plate; the

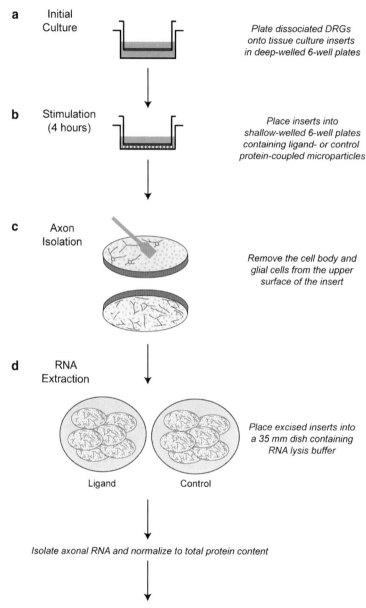

Fig. 2. Method for analysis of stimulus-dependent axonal mRNA transport. (**a**) Using a modified Boyden chamber, dissociated neurons are plated onto tissue culture inserts in 6-well plates with deep wells and allowed to grow until axons have extended through the pores. For injury-conditioned, adult DRG neurons, generally an overnight culture of 14–16 h is sufficient to allow significant growth through the porous membrane. For cultures of neurons that have not been conditioned by injury, or "naive" neurons, this initial culture period can be extended so that more robust axonal growth is seen along the undersurface. However, the smaller diameter pore membranes than 8 μm should be considered for longer term cultures. The deep wells are critical so that the bottom of the insert does not rest upon the bottom of the wells and hinder axonal growth onto the under surface of the insert. (**b**) The tissue culture inserts are then transferred to a new 6-well plate with shallow wells with ligand- or control protein-coupled microparticles placed at the bottom of each well. The shallow depth of the wells allows the bottom surface of the insert to rest on the surface of the well, facilitating contact between the axons and the microparticles. (**c**) After stimulation (typically, 4 h has been sufficient to see alterations in axonal mRNA transport in adult DRG neurons), the top surfaces of the inserts are scraped to remove the cell body and non-neuronal cells.

density of microparticles can be varied based on the coverage desired and the protein concentration on the microparticles.

10. Remove the tissue culture inserts from the original 6-well plate after overnight culture, remove the media from the top of the filter and place each insert into the new, shallow 6-well plate containing the bound microparticles. Add 2 ml of complete culture media to the top of the inserts and return the plate to the incubator.

11. Incubate at 37°C, 5% CO_2 for 4 h. The time for exposure can be varied to account for transit of mRNAs. For example, if one wants to evaluate transport of newly transcribed mRNAs, then the exposure time should be extended to allow for transcriptional activation and DRB should be excluded from the culture preparations.

3.2.2. Isolation of Axonal Compartment

1. Rinse both the upper and lower surface of the tissue culture inserts with 1× PBS to remove media, any nonadherent cell material and any microparticles that might be stuck to the under surface.

2. Gently scrape the upper surface of each insert with a cotton-tipped applicator, rinse, excise the membrane, and place in RNA lysis buffer as outlined in Subheading 3.1.3, steps 2–7.

3.2.3. RNA Isolation

The RNA isolation for this approach follows the approach outlined in Subheading 3.1.4 except that the flow through from step 5 is saved for analyses of protein content in Subheading 3.2.4 below. This will be used to normalize axonal mass between the control and ligand stimulated axonal RNA preparation samples.

3.2.4. Quantification of Protein for Normalization

1. Place the 750 μl of flow-through sample generated in step 6 above into a Microcon centrifugal filter with molecular weight cutoff of 10 kDa.

2. Following the manufacturers instructions, centrifuge the filter to initially concentrate the samples.

3. Resuspend the concentrated protein in 0.5 ml 1× PBS and repeat step 2.

4. Resuspend the concentrated protein in 0.5 ml 1× PBS (see Note 16).

Fig. 2. (continued) Various methods for removing the material on the upper surface of the filter have been used. Most commonly, we used cotton-tipped applicators from a scientific supply vendors. However, we have found that over-the-counter cotton-tipped swabs that can be autoclaved to ensure sterility provide a more consistent result. (**d**) After removal of the material on the top surface, the membrane insert is removed from its housing with a scalpel and all inserts treated with the same stimulus are combined into a 35-mm dish containing RNA lysis buffer. The isolated axonal RNA is quantified, normalized to total axonal protein content, reverse transcribed and checked for purity. Validated preparations are used for real-time PCR.

5. Quantify the protein in each sample by fluorometry using the *NanoOrange* protein quantification kit.

6. Determine the normalization ratio between the NGF and BSA treated samples.

3.2.5. RNA Quantification and Reverse Transcription

1. Quantify 1.0 μl of each RNA sample using the RiboGreen RNA quantification kit with a high-range working solution in a total volume of 200 μl as outlined in Subheading 3.1.5, steps 1 and 2. This is done as a confirmation for presence of RNA in the axonal samples and provides a confirmation of normalization; that is, the 4-h stimulations used here do not seem to change overall axonal RNA content, but rather change the populations of mRNAs in transport.

2. Plan to use approximately 50 ng of total RNA for the RT reaction; however, the exact levels are normalized to the protein content of the lysates determined in Subheading 3.2.4 above. Using the iScript cDNA synthesis kit for RT, dilute the RT reactions by 1:5 with RNase-free water, and then test for purity as outlined in Subheading 3.1.5, steps 3 and 4 and Note 13.

3.2.6. Real-Time PCR

The methods outlined below are based on comparative threshold (C_t) method for real-time analysis with *SyberGreen* reagent labeled PCR products. We have optimized this for our system with careful choice of controls for normalization as detailed below. This approach could also be adopted for multiplex analyses with other detection methodologies:

1. Primers used for C_t method real-time analysis with *SyberGreen* reagent are first run on a validation experiment to ensure that the efficiencies of the target and endogenous control amplifications are approximately equal (see Note 2).

2. A robotic system (Biomek 2000) is used to standardize pipetting of samples and reagents into the 384-well plates. Automatic outlier calculations are performed using the ABI *Prism SDS 2.3* software (see Note 17).

3. Quantify the relative levels of the individual transcripts by normalizing to an endogenous control. The C_t for each transcript is determined using the automatic C_t algorithm of the *SDS 2.3* software to calculate the optimal baseline range and threshold values (see Note 18).

4. Calculate individual ΔC_t values by subtracting the control RNA C_t value from the individual transcript C_t values (see Note 19).

5. Determine the $\Delta\Delta C_t$ value by subtracting the BSA calibrator ΔC_t value from the NGF ΔC_t value.

6. Express the fold difference as $2^{-\Delta\Delta C_t} \pm S$, where S is the standard deviation of the $\Delta\Delta C_t$ values.

3.3. Concluding Remarks

We consider the analysis method detailed above as a first step in assessing which mRNAs are present in the axon and how the transport of these mRNAs may be altered by extracellular signals. Subsequent techniques are then needed to more fully understand this regulated and dynamic process. For example, quantitative fluorescence in situ hybridization (FISH) can be used to confirm the results obtained from this method, as well as provide information on the spatial distribution of the mRNA along the axon shaft and in the growth cone, particularly relative to the source of the stimulus (37). Similarly, diffusion limited fluorescent proteins, such as myristoylated GFP (38, 39), can be used to generate reporter constructs useful for the identification of motifs that mediate axonal localization as well as the signaling mechanisms that drive alterations in axonal mRNA localization and localized translation.

Although we have outlined methods for profiling and quantitative analyses of RNA transport in neuronal processes, the general methods provided here can be adopted for nonneuronal cells if the desired subcellular compartment can be isolated to purity. Mili et al. have used this general scheme for profiling mRNAs in pseudopodia of fibroblasts stimulated with fibronectin vs. lysophosphatidic acid (28). Likewise, subcellular fractionation of synaptoneurosomes from postmortem human brain was recently used to profile potential alterations in mRNA transport in Alzheimer's disease (12). A key issue for success in these approaches is finding means to confirm the purity of the preparations; that is, validating the absence of cell body restricted RNAs. The universal expression of localizing β-actin and nonlocalizing γ-actin mRNAs is likely the best starting point with additional control transcripts based on those discussed and referenced above.

4. Notes

1. Glass Pasteur pipettes are fire-polished to remove the sharp edge at the opening and to reduce the diameter of the opening. Once cooled, these fire-polished pipettes are used to mechanically dissociate the DRG tissue by gently pipetting up and down 15–20 times (i.e., triturate) until the DRG suspension is homogenous.

2. Real-time primers for use in comparative C_t experiments must first be tested in validation experiments to confirm their suitability for relative quantitation. In order to be used, the absolute value of the slope of the log input amount vs. ΔC_t must be less than 0.1 over a minimum of 5 log units. In addition,

the dissociation curves must show single melting peaks and there must be no nonspecific product formation with the no template controls. For a more complete explanation on comparative C_t validation experiments, a good resource is the *Guide to Performing Relative Quantitation of Gene Expression Using Real-Time Quantitative PCR*, which can be found at *appliedbiosystems.com*.

3. All manipulations for the culture methods described here should be performed in a laminar flow hood ("biosafety cabinet") and all materials appropriately sterilized.

4. It takes a minimum of 2.5 ml of liquid to ensure that the bottom and top of the insert are fully covered. It is critical that the entire bottom surface of the insert be covered in order to facilitate axon growth. Also note that we typically use membranes with 8 μm diameter pores for the adult rat DRG neurons and the methods here describe that. Smaller pore diameters are preferred for longer cultures or cultures from adult mouse DRGs; we have used 3 μm diameter pores for both of these. Although our experience with embryonic preparations is limited, other groups have used membranes with 1 μm diameter pores for embryonic/early postnatal DRGs (26).

5. Higher concentrations of laminin will diminish the membrane pore size and this becomes very problematic for membranes with smaller diameter pores; the reader may consider a lower laminin concentration for membranes with 3 or 1 μm diameter pores.

6. Based on adult L4-5 DRG cultures, two ganglia per well produces optimal cell density for axons to penetrate the membrane pores and yields a robust preparation of axonal processes. Higher densities tend to decrease penetration and lower densities do not provide sufficient axon mass for subsequent manipulations.

7. This is the only antibiotic that the cells will be exposed to. Use a pair of micro-scissors to move each DRG from well to well during the rinse. This will ensure that the perineurium has been nicked to facilitate penetration of the collagenase in subsequent steps.

8. We have experienced variations in sources for cotton-tipped applicators that require reoptimizing this shearing approach. We have found that the most consistent source is the "cotton-tipped swabs" available from local drug or department stores.

9. Leave an edge of about 3 mm of filter attached to the side to ensure that any cell body material remaining stays behind.

In practice, using a #11 scalpel blade fitted to a surgical handle provides a 3 mm edge based on the thickness of the handle.

10. Note that all subsequent steps below require RNAse free solutions and RNAse free plasticware.

11. We prefer to use fluorometry to quantitate the small mass of RNA that this approach yields. The standard curve required for quantification by fluorometry gives confidence that the absorption values generated by the samples are within the dynamic range of the instrument. However, one can use other approaches that do not require the use of standard curve (e.g., nano-format UV spectrometry).

12. From our experience with adult rat DRG neurons, L4-5 DRGs from three animals (i.e., 12 ganglia) generate 25–100 ng of axonal RNA. With amplification, this is more than sufficient for generating probes for cDNA arrays (37) and quantitative RT-PCR approaches to gain a relative comparison of localized mRNA levels (see below) (36). Although analyses of axonal proteins requires more input since there are no means to exponentially amplify signals like there is for mRNA (i.e., with PCR), one can scale up and generate sufficient levels of protein for immunoblotting, immunoprecipitations, and even co-immunoprecipitation with the DRG cultures (25, 36, 40–42). Other methods for isolating axonal processes have similarly been used for such immunoprecipitations (21).

13. Axonal preparations from adult DRGs show robust amplification of β-actin mRNA, but are devoid of γ-actin mRNA (which remains confined to the cell body) and the glial-specific mRNAs MAP2 and GFAP (36, 37). Additional purity markers, such as mRNAs encoding histones and transcription factors, have proven useful in other culture methods using different neuronal cell types (27, 29, 31).

14. In this approach, we limit the neurotrophin exposure to the axonal compartment by immobilizing the ligands onto polystyrene microparticles that are of too large diameter to pass through the membrane pores (20 μm diameter particles vs. 8 μm diameter pores).

15. Although the method described here is for NGF, a member of the neurotrophin family, it is easily adapted for other polypeptide ligands (e.g., semaphorins (37)). Care should be taken to ensure desired level of receptor occupancy. Proteoglycans can also be used, but this requires a passive adsorption process rather than covalent linkage used here. However, a more diffuse localized gradient of ligand in the media could be generated by use of passively adsorbed ligands

as these will be more easily leached from the particles. Additionally, avidin-conjugated microparticles could effectively be used to increase affinity for any biotinylated agents.

16. These two spins, each concentrating the samples 20-fold and effectively exchanging 95% of the RNA lysis buffer for PBS, makes the flow-through sample compatible with *NanoOrange* reagent for protein quantitation.

17. Normalized RT reactions from pure axonal preparations treated with NGF or BSA are diluted 1:5 with RNase-free water and 1 μl assayed in quadruplicate from a minimum of three independent experiments. 96-Well formats can also be used but they do not allow for the same level of throughput that is needed to simultaneously anlayze many samples and controls in quadruplicate.

18. Controlling for the variability in the number of axons that traverse through the pores of the tissue culture inserts (or into an isolated chamber as in other methods) presents a unique challenge in analyzing changes in axonal mRNA transport. We have controlled for axon number based on the total protein content of the isolated axons and have used this axonal protein amount to normalize the axonal load between the NGF- and BSA-treated samples (see Subheading 3.2.4). We have also used a comparative threshold method to provide an additional control for axonal content as well as an internal control for reverse transcription efficiency. We prefer this to an absolute determination of localized RNA levels, since the 12S RNA that we use for normalization also gives an independent assessment of RT efficiency. When dealing with minute RNA quantities, such as here, efficiency of RT reaction can fall dramatically as the RNA template levels decrease. *iScript* has behaved well for us down to 15 ng of total RNA, but below this the efficiency quickly falls.

19. Since transport of some mRNAs that are classically considered as "housekeeping genes" is robustly regulated by ligands (e.g., β-actin mRNA), care must be taken in choice of RNAs used for the C_t calculation. We have used the 12S mitochondrial ribosomal RNA as mitochondria are frequent in axons and the mitochondrial genome is transcribed as a single polycistronic RNA (36).

Acknowledgements

The methods presented here were developed using funds from the National Institutes of Health (R01-NS041596 and R01-NS049041 to JLT and K99-NR010797 to DEW).

References

1. Martin, K.C. and A. Ephrussi. (2009). mRNA localization: gene expression in the spatial dimension. Cell 136: 719–30.
2. Steward, O. and W.B. Levy. (1982). Preferential localization of polyribosomes under the base of dendritic spines in granule cells of the dentate gyrus. J Neurosci 2: 284–91.
3. Bramham, C.R. and D.G. Wells. (2007). Dendritic mRNA: transport, translation and function. Nat Rev Neurosci 8: 776–89.
4. Lin, A.C. and C.E. Holt. (2008). Function and regulation of local axonal translation. Curr Opin Neurobiol 18: 60–68.
5. Yoo, S., E.A. van Niekerk, T.T. Merianda, and J.L. Twiss. (2009). Dynamics of axonal mRNA transport and implications for peripheral nerve regeneration. Exp Neurol 223: 19–27.
6. Holt, C.E. and S.L. Bullock. (2009). Subcellular mRNA localization in animal cells and why it matters. Science 326: 1212–6.
7. Willis, D.E. and J.L. Twiss. (2010). Regulation of protein levels in subcellular domains through mRNA transport and localized translation. Mol Cell Proteomics in press 9: 952–62.
8. Ellis, L., F. Katz, and K.H. Pfenninger. (1985). Nerve growth cones isolated from fetal rat brain. J. Neurosci. 5: 1393–1401.
9. Witzmann, F.A., R.J. Arnold, F. Bai, P. Hrncirova, M.W. Kimpel, Y.S. Mechref, W.J. McBride, M.V. Novotny, N.M. Pedrick, H.N. Ringham, and J.R. Simon. (2005). A proteomic survey of rat cerebral cortical synaptosomes. Proteomics 5: 2177–201.
10. Jimenez, C.R., M. Eyman, Z.S. Lavina, A. Gioio, K.W. Li, R.C. van der Schors, W.P. Geraerts, A. Giuditta, B.B. Kaplan, and J. van Minnen. (2002). Protein synthesis in synaptosomes: a proteomics analysis. J Neurochem 81: 735–44.
11. Matsumoto, M., M. Setou, and K. Inokuchi. (2007). Transcriptome analysis reveals the population of dendritic RNAs and their redistribution by neural activity. Neurosci Res 57: 411–23.
12. Williams, C., R. Mehrian Shai, Y. Wu, Y.H. Hsu, T. Sitzer, B. Spann, C. McCleary, Y. Mo, and C.A. Miller. (2009). Transcriptome analysis of synaptoneurosomes identifies neuroplasticity genes overexpressed in incipient Alzheimer's disease. PLoS One 4: e4936.
13. Weiler, I.J. and W.T. Greenough. (1993). Metabotropic glutamate receptors trigger postsynaptic protein synthesis. Proc. Natl. Acad. Sci., USA 90: 7168–7171.
14. Klemmer, P., A.B. Smit, and K.W. Li. (2009). Proteomics analysis of immuno-precipitated synaptic protein complexes. J Proteomics 72: 82–90.
15. Schrimpf, S.P., V. Meskenaite, E. Brunner, D. Rutishauser, P. Walther, J. Eng, R. Aebersold, and P. Sonderegger. (2005). Proteomic analysis of synaptosomes using isotope-coded affinity tags and mass spectrometry. Proteomics 5: 2531–41.
16. Cotrina, M.L., J.H. Lin, J.C. Lopez-Garcia, C.C. Naus, and M. Nedergaard. (2000). ATP-mediated glia signaling. J Neurosci 20: 2835–44.
17. Slezak, M. and F.W. Pfrieger. (2003). New roles for astrocytes: regulation of CNS synaptogenesis. Trends Neurosci 26: 531–5.
18. Shea, T.B. (1994). Toxic and trophic effects of glial-derived factors on neuronal cultures. Neuroreport 5: 797–800.
19. Olink-Coux, M. and P.J. Hollenbeck. (1996). Localization and Active Transport of mRNA in Axons of Sympathetic Neurons in Culture. Journal of Neuroscience 16: 1346–1358.
20. Moccia, R., D. Chen, V. Lyles, E. Kapuya, Y. E, S. Kalachikov, C.M. Spahn, J. Frank, E.R. Kandel, M. Barad, and K.C. Martin. (2003). An unbiased cDNA library prepared from isolated Aplysia sensory neuron processes is enriched for cytoskeletal and translational mRNAs. J Neurosci 23: 9409–17.
21. Eng, H., K. Lund, and R.B. Campenot. (1999). Synthesis of beta-tubulin, actin, and other proteins in axons of sympathetic neurons in compartmented cultures. J Neurosci 19: 1–9.
22. Vogelaar, C.F., N.M. Gervasi, L.F. Gumy, D.J. Story, R. Raha-Chowdhury, K.M. Leung, C.E. Holt, and J.W. Fawcett. (2009). Axonal mRNAs: characterisation and role in the growth and regeneration of dorsal root ganglion axons and growth cones. Mol Cell Neurosci 42: 102–15.
23. Hillefors, M., A. Gioio, M. Mameza, and B. Kaplan. (2007). Axon viability and mitochondrial function are dependent on local protein synthesis in sympathetic neurons. Cell Mol Neurobiol 27: 701–16.

24. Torre, E.R. and O. Steward. (1996). Protein synthesis within dendrites: glycosylation of newly synthesized proteins in dendrites of hippocampal neurons in culture. J Neurosci 16: 5967–78.
25. Zheng, J.-Q., T. Kelly, B. Chang, S. Ryazantsev, A. Rajasekaran, K. Martin, and J. Twiss. (2001). A functional role for intra-axonal protein synthesis during axonal regeneration from adult sensory neurons. J Neurosci 21: 9291–9303.
26. Wu, K.Y., U. Hengst, L.J. Cox, E.Z. Macosko, A. Jeromin, E.R. Urquhart, and S.R. Jaffrey. (2005). Local translation of RhoA regulates growth cone collapse. Nature 436: 1020–4.
27. Poon, M.M., S.H. Choi, C.A. Jamieson, D.H. Geschwind, and K.C. Martin. (2006). Identification of process-localized mRNAs from cultured rodent hippocampal neurons. J Neurosci 26: 13390–9.
28. Mili, S., K. Moissoglu, and I.G. Macara. (2008). Genome-wide screen reveals APC-associated RNAs enriched in cell protrusions. Nature 453: 115–9.
29. Taylor, A.M., N.C. Berchtold, V.M. Perreau, C.H. Tu, N. Li Jeon, and C.W. Cotman. (2009). Axonal mRNA in uninjured and regenerating cortical mammalian axons. J Neurosci 29: 4697–707.
30. Taylor, A.M., M. Blurton-Jones, S.W. Rhee, D.H. Cribbs, C.W. Cotman, and N.L. Jeon. (2005). A microfluidic culture platform for CNS axonal injury, regeneration and transport. Nat Methods 2: 599–605.
31. Cox, L.J., U. Hengst, N.G. Gurskaya, K.A. Lukyanov, and S.R. Jaffrey. (2008). Intra-axonal translation and retrograde trafficking of CREB promotes neuronal survival. Nat Cell Biol 10: 149–59.
32. Wu, H.I., G.H. Cheng, Y.Y. Wong, C.M. Lin, W. Fang, W.Y. Chow, and Y.C. Chang. A lab-on-a-chip platform for studying the subcellular functional proteome of neuronal axons. Lab Chip 10: 647–53.
33. Rishal, I., I. Michaelevski, M. Rozenbaum, V. Shinder, K.F. Medzihradszky, A.L. Burlingame, and M. Fainzilber. (2009). Axoplasm isolation from peripheral nerve. Dev Neurobiol 70: 126–33.
34. Knowles, R.B. and K.S. Kosik. (1997). Neurotrophin-3 signals redistribute RNA in neurons. Proceedings of the National Academy of Sciences of the United States of America 94: 14804–8.
35. Zhang, H.L., R.H. Singer, and G.J. Bassell. (1999). Neurotrophin regulation of beta-actin mRNA and protein localization within growth cones. J Cell Biol 147: 59–70.
36. Willis, D., K.W. Li, J.Q. Zheng, J.H. Chang, A. Smit, T. Kelly, T.T. Merianda, J. Sylvester, J. van Minnen, and J.L. Twiss. (2005). Differential transport and local translation of cytoskeletal, injury-response, and neurodegeneration protein mRNAs in axons. J Neurosci 25: 778–91.
37. Willis, D.E., E.A. van Niekerk, Y. Sasaki, M. Mesngon, T.T. Merianda, G.G. Williams, M. Kendall, D.S. Smith, G.J. Bassell, and J.L. Twiss. (2007). Extracellular stimuli specifically regulate localized levels of individual neuronal mRNAs. J Cell Biol 178: 965–80.
38. Aakalu, G., W.B. Smith, N. Nguyen, C. Jiang, and E.M. Schuman. (2001). Dynamic visualization of local protein synthesis in hippocampal neurons. Neuron 30: 489–502.
39. Yudin, D., S. Hanz, S. Yoo, E. Iavnilovitch, D. Willis, Y. Segal-Ruder, D. Vuppalanchi, K. Ben-Yaakov, M. Hieda, Y. Yoneda, J. Twiss, and M. Fainzilber. (2008). Localized regulation of axonal RanGTPase controls retrograde injury signaling in peripheral nerve. Neuron 59: 241–52.
40. Hanz, S., E. Perlson, D. Willis, J.Q. Zheng, R. Massarwa, J.J. Huerta, M. Koltzenburg, M. Kohler, J. van-Minnen, J.L. Twiss, and M. Fainzilber. (2003). Axoplasmic importins enable retrograde injury signaling in lesioned nerve. Neuron 40: 1095–104.
41. van Niekerk, E.A., D.E. Willis, J.H. Chang, K. Reumann, T. Heise, and J.L. Twiss. (2007). Sumoylation in axons triggers retrograde transport of the RNA-binding protein La. Proc Natl Acad Sci U S A 104: 12913–8.
42. Wang, W., E. van Niekerk, D.E. Willis, and J.L. Twiss. (2007). RNA transport and localized protein synthesis in neurological disorders and neural repair. Develop Neurobiol 67: 1166–82.

Chapter 22

RNA Purification from Tumor Cell Protrusions Using Porous Polycarbonate Filters

Jay Shankar and Ivan R. Nabi

Abstract

Actin-rich cellular protrusions or pseudopodia form via local actin filament polymerization and branching and represent a variety of polarized cellular domains including lamellipodia, filipodia, and neuronal growth cones. RNA localization and local protein translation in these domains are important for various cellular processes. RNA transport and local synthesis have been implicated in cell migration and tumor cell metastasis as well as in neuronal plasticity in neurons. Characterization of the mRNAs present in these domains is key to understanding the functional role of mRNA translocation and local protein translation in cellular processes. We describe here a method to segregate pseudopodia of metastatic cancer cells from the cell body using porous polycarbonate filters. This approach enables the purification and identification of RNAs and proteins in these protrusive cellular domains.

Key words: Pseudopodia, RNA localization, Cancer metastasis, Microarray, Confocal microscopy

1. Introduction

mRNA localization has been observed in diverse cells and organisms ranging from yeast, protozoa, and plants, to various animal cell types such as fibroblasts, nerve cells, and oocytes (1). Actin-dependent cellular protrusion is the fundamental mechanism by which cells move over a substrate (2). Membrane protrusions are formed via local actin filament polymerization and branching and stabilized by de novo establishment of focal adhesions at the leading edge (3). Actin-rich protrusions, or pseudopodia, represent polarized cellular domains that include, but are not limited to, lamellipodia and filipodia in fibroblasts and dendrites and growth cones in neurons (4). mRNA targeting to actin-rich cellular protrusions has long been described, indeed, β-actin was the first mRNA shown to be polarized to fibroblast lamellipodia (5).

Interestingly, polarized distribution of β-actin was shown to be inversely correlated with metastatic ability of cancer cells (6). In cancer cells, pseudopodial domains are enriched for select mRNAs and the protein translation machinery suggesting that they are active sites of de novo protein synthesis (7, 8). In neurons, mRNA localization and local protein synthesis have been hypothesized to contribute to synaptic plasticity (9).

Critical to understanding the functional role of local protein translation in cellular processes is the characterization of those mRNAs present in these domains. This is limited by the ability to obtain sufficient quantities of pseudopodial domains cleanly segregated from the rest of the cell. For instance, earlier studies identified mRNAs present in mechanically dissected and aspirated individual dendrites (10). Porous polycarbonate filters of varying pore size are available (0.4 μm up to 8 μm). The smaller pore size, restricting cell passage across the filter, has been extensively used to selectively access, for instance, the basolateral surface of polarized epithelial cells. Larger pore sizes, that allow cell passage through the pores, have been used to study cell migration and invasion. We determined that use of an intermediate pore size, 1 μm, selectively allowed passage of actin-rich pseudopodia of the MSV-MDCK-INV cancer cell line but not the nucleus or a cellular organelle, mitochondria (11). Similarly, pseudopodia of some metastatic cancer cell lines such as MDA-231, MDA-435, HT1080, DU145, U87, and U251 selectively pass through 1 μm pore filters that restrict passage of nuclei and mitochondria (12) (Fig. 1). Use of larger 3 μm pores allows passage of some nuclei and mitochondria over a period of 24 h enabling potential contamination of the isolated pseudopodial fraction (Fig. 1). However, not all cells, even cancer cells, project pseudopodia through 1 μm pores (Fig. 2). For instance, studies of fibroblast protrusions have used 3 μm pore filters, chemotactic gradients and shorter times of cell plating to ensure that only pseudopodia pass through the pores (13, 14).

Various studies have used different pore size filters to characterize mRNA present in actin-rich protrusions including neurons (15), fibroblasts (14), and cancer cells (7, 12). Comparison of the cohorts of mRNAs present in protrusions of neurons and cancer cells identified similar enrichment for mRNAs coding for proteins involved in protein translation, however, network analysis also identified significant differences between these mRNA cohorts (7, 15). Microarray analysis of mRNAs of pseudopodia of MSV-MDCK-INV cancer cells led to the determination that Rho/ROCK signaling was crucial for polarized localization of mRNAs to these pseudopodial domains (7), as reported for β-actin mRNA (16). In fibroblasts, studies of pseudopodial mRNAs led to the identification of a novel microtubule-dependent localizing mechanism for mRNAs that interact with the adenomatous polyposis coli (APC) tumor suppressor in RNA granules (14). A recent

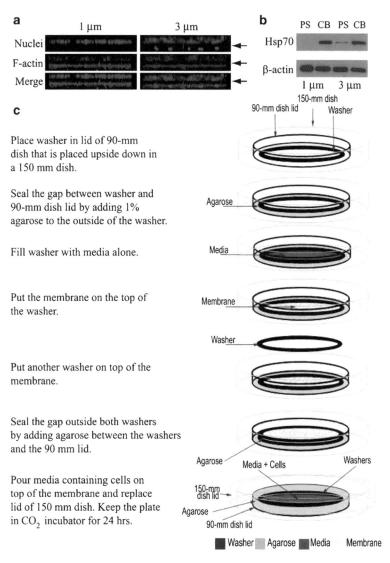

Fig. 1. Purity of pseudopodial fraction of cells plated on 1 and 3-μm pore-size filters: (a) Metastatic breast MDA-231 cells were plated overnight on 1 and 3 μm pore filters and F-actin labeled with rhodamine phalloidin (F-actin; *red*) and nuclei with Hoechst (*blue*). Confocal Z-sections were acquired and *arrows* mark the position of the filter. Nuclei are not seen on the underside of the 1 μm filter, but are present on the underside of the 3 μm filter. (b) To further assess the purity of the pseudopod fraction, cell body (CB) and pseudopod (PS) fractions were purified by scraping the top and bottom of 100 μm diameter 1 μm pore filters incubated with cells in a filter assembly, and analyzed by Western blot for β-actin and mitochondrial HSP70 (mHSP70). mHSP70 was excluded from the pseudopod fraction of 1 μm pore filters, but was present in the pseudopod fraction of 3 μm pore filters. (c) Schematic diagram showing the steps required to assemble the large filter units to harvest pseudopodia for RNA and protein isolation.

combined microarray and proteomics study led to the identification of pseudopod-enriched proteins that regulate epithelial–mesenchymal transition or EMT (12). Use of filters of defined pore size to further characterize the mRNA complement of various cell types can therefore lead to a better understanding of the cellular specificity of mRNAs localized to actin-rich protrusions as

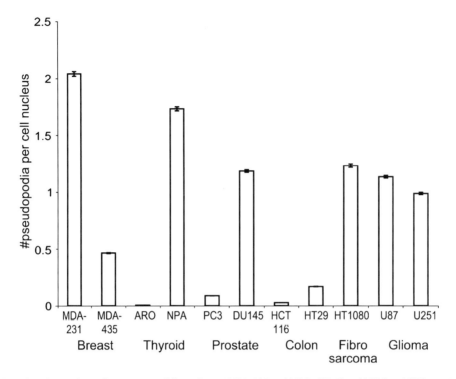

Fig. 2. Pseudopod count for various cancer cell lines: Breast MDA-231 and MDA-435, thyroid ARO and NPA, prostate PC3 and DU145, colon HCT116 and HT29, fibrosarcoma HT1080, and glioma U87 and U251 cancer cells were plated on 1 μm filter inserts overnight, fixed and F-actin labeled with rhodamine phalloidin and nuclei with Hoechst. Confocal X–Y planes at the level of the pseudopodia below the filter and at the level of the nuclei above the filter were taken from the same field. The number of pseudopodia per nuclei was counted using ImagePro software from multiple fields for each cell line (mean ± SEM).

well as the regulation and functional significance of their local translation. In this chapter, we describe the protocol for the harvesting, storage, and handling of RNA and protein from the pseudopodia of the metastatic tumor cell lines grown on 1 μm pore filters.

2. Materials

2.1. Cell Culture

1. Roswell Park Memorial Institute 1640 medium (RPMI, 1640) supplemented with 5% (v/v) L-glutamine, 5% (v/v) penicillin–streptomycin solution, 25 mM HEPES, and 10% (v/v) FBS.

2. Trypsin solution (0.25%) and sterile phosphate buffered saline (PBS, pH 7.4).

3. 100 mm diameter 1 μm pore filters (SPI Supplies, West Chester, PA), 76.2–88.9 mm custom-made washers (Boulons Plus, Anjou, QC, Canada), 150 and 100-mm Falcon Petri

dish, 2% agarose (w/v), pair of tweezers. 24-mm Falcon transparent PET membrane filters (polyethylene terephthalate) (1, 3 μm pore size), 6-well plates (VWR International, Canada). Commercial inserts (both Falcon and Corning Transwells) and filters from S&P are available in various pore sizes. Washers can be obtained from any local metal washer company; however a custom order will likely be required.

2.2. Harvesting Pseudopodia

1. Autoclaved PBS (1×), plastic tray (18×16×5″) filled with ice, metal sheet to be placed on the ice-tray (12×10″), piece of parafilm (6×4″), non-sterile 150-mm dish for rinsing the filters, single-edge industrial 0.22-mm razor blade (VWR International, ON, Canada), sterile 1.5-mL labeled microfuge tubes, 200 μL autoclaved tips (for protein), 200 μL filter tips (for RNA), 70% ethanol (v/v), 14.3 M β-mercaptoethanol (β-ME), QIAshredder homogenizer (Qiagen, Germany), and total RNA isolation kit (Qiagen, Germany).

2. M2 lysis buffer: 20 mM Tris–HCl at pH 7.6, 0.5% NP-40 (v/v), 250 mM NaCl, 3 mM EGTA, and 3 mM EDTA. To 1 mL of lysis buffer, add fresh 2 μL of 1 M DTT (dithiothreitol), 5 μL of PMSF (phenylmethylsulfonyl fluoride), 2 μL of 4 mg/mL leupeptin, 5 μL of 0.5 M NaF, 5 μL of 0.5 M Na_3VO_4 (see Note 1). For protein estimation use Bradford reagent (Bio-Rad, Hercules, CA).

3. RNA buffers: The buffers are supplied along with the total RNA isolation kit (Qiagen, Germany).

2.3. SDS-Polyacrylamide Gel Electrophoresis (SDS-PAGE) and Western Blot

1. Separating buffer (5×): 1.5 M Tris–HCl at pH 8.8. Keep at room temperature.

2. Stacking buffer (5×): 0.5 M Tris–HCl at pH 6.8. Keep at room temperature.

3. 30% Acrylamide/bis solution (29.2:0.8 w/v acryl:bisacryl) and N,N,N,N-tetramethyl-ethylenediamine (TEMED, Bio-Rad, Hercules, CA).

4. 10% (w/v) SDS solution (see Note 2).

5. 10% (w/v) ammonium persulfate solution made fresh.

6. Running buffer (10×): 2.5 M Tris-base, 19.2 M Glycine, 10% (w/v) SDS at pH 8.3. To make 1× take 100 mL of stock solution and add 900 mL of ddW.

7. Sample loading buffer 3× stock: 1 M Tris–HCl at pH 6.8 (2.5 mL of 1 M Tris–HCl), 20% SDS (w/v, 3 mL 20% SDS), Glycerol (3 mL), β-mercaptoethanol (0.3 mL), and bromophenol blue 0.6 mg. Total volume: 10 mL (store at 4°C).

8. Prestained molecular weight markers (Fermentas, MD, USA).

9. Semi-dry transfer buffer (10×): 48 mM Tris, 39 mM glycine, (20% methanol) at pH 9.2. Dissolve 58.2 g Tris and 29.3 g

glycine (and 3.75 g SDS or 37.5 mL of 10% SDS) in 1 L of dd H_2O to make 10×. For 1×, to 100 mL of 10× add 200 mL of methanol and make the volume up to 1,000 mL (see Note 3).

10. Hybond-C Extra nitrocellulose membrane from Amersham Biosceinces (GE Lifesciences, UK) and 3MM chromatography paper from Whatman (Maidstone, UK).

11. Phosphate buffer saline with Tween (PBS-T): Prepare 10× stock with 137 mM NaCl, 2.7 mM KCl, 100 mM Na_2HPO_4, and 2 mM KH_2PO_4. Dissolve 80 g NaCl, 2 g KCl, 20.8 g Na_2HPO_4, and 2 g KH_2PO_4, at pH 7.4, add ddH_2O to make volume to 1 L. For 1× of PBS-T, to 100 mL of 10× of PBS add 0.5% (v/v) of Tween-20 and make the volume up to 1 L.

12. Blocking buffer: 5% (w/v) nonfat dry milk in PBS-T.

13. Primary antibody dilution buffer: PBS-T, Primary antibody: mouse mitochondrial-Hsp70 (mHSP70) antibody from Ab-Biochem and mouse anti-β-actin from Sigma.

14. Secondary antibody: Antimouse IgG conjugated to horse radish peroxidase (Cedarlanelabs, ON, Canada).

15. Enhanced chemiluminescent (ECL) reagents from Millipore, MA, USA and BioFlex scientific imaging film (8 × 10″, Clonex Corporation, ON, Canada).

2.4. Immunofluorescence Labeling and Confocal Microscopy

1. 24-mm Falcon transparent PET membrane filters (polyethylene terephthalate) (1 μm, 3 μm pore size), 6-well plates (VWR International, ON, Canada).

2. Microscope cover slips (22 × 40 × 0.15 mm) from Fisher, Pittsburgh, PA and slides (3 × 1 × 1 mm) (VWR International, ON, Canada).

3. PBS-CM buffer: To 1× PBS buffer add 1 mM $MgCl_2$ and 0.1 mM $CaCl_2$.

4. Paraformaldehyde (Sigma): Prepare 3% (w/v) solution in PBS, freeze in 10 mL aliquots at −20°C, and thaw and warm to 37°C a fresh aliquot prior to each experiment (see Note 4).

5. Permeabilization solution: 0.1% (v/v) Triton X-100 in PBS-CM.

6. Antibody dilution buffer: 1% (w/v) BSA in PBS-CM.

7. For nuclear stain dissolve 200 nM Hoechst (Bisbenzimide; p-(5-(5-(4-Methyl-1-piperazinyl)-1H-2-benzimidazolyl)-1H-2-benzimidazolyl)phenol trihydrochloride) in water and use at 1:100 dil. in blocking buffer.

8. For actin labeling use Alexa Fluor-568 phalloidin (1:500 dil. Molecular Probes, Eugene, OR) in blocking buffer. Store at −20°C in dark.

9. Mounting medium: Airvol (Air Products, Inc.).

3. Methods

An important consideration is whether the cell line of interest projects pseudopodia through filter pores. This can be easily tested by confocal microscopy of Hoechst and phalloidin labeled cells grown on commercially available cell culture inserts (Subheading 3.1). While our preference, due to minimal contamination of the pseudopodial fraction with cell body components (Fig. 1a, b), is to use 1 µm pore filters, for some cell lines it may not be possible to obtain sufficient pseudopodial material using 1 µm pores. However, use of larger pore sizes can lead to cell body contamination of the pseudopod fraction (Fig. 1a, b). Commercial filters in supports are available in various pore sizes and should be used to test pseudopodial protrusion through the filters prior to undertaking the purification. In all cases, it is critical to verify pseudopodial fraction purity by Western blot. It is important to note that identification of proteins highly enriched in the pseudopod fraction has proven difficult, likely because not all pseudopodia pass through the filter pores to the bottom of the filter and therefore remain in the cell body fraction. Indeed, β-actin is only slightly enriched in the pseudopod fraction by Western blot (Fig. 1a). Exclusion of mitochondrial proteins from the pseudopod fraction has proven, in our experience, to be a reliable marker of pseudopod purity (7, 8, 11, 12).

3.1. Confocal Microscopy to Assess Pseudopodial Passage Through Filter Pores

1. Place 1 and 3 µm pore size 24-mm commercial filter inserts into 6-well plates. Add 2 mL of appropriate media to the bottom of filter and then add 2 mL of media containing 1×10^6 cells on top of the insert. Grow cells for 24 h (or for shorter times) in a humidified CO_2 incubator.

2. After 24 h quickly remove the media from bottom and top of the insert, wash once with PBS and then add prewarmed 3% paraformaldehyde to the top and bottom of the containing the insert. Incubate for 15 min in a fume hood.

3. After incubation quickly remove paraformaldehyde and add 2 mL of PBS to each well (top and bottom). Wash rapidly thrice with PBS-CM in fume hood and dispose off used paraformaldehyde and first three PBS washes in appropriate chemical waste. Wash four times with PBS-CM on the bench with 10 min incubation between washes.

4. After washing, cells grown on cell culture inserts are permeabilized by incubation in PBS-CM containing 0.1% Tween-20 for 15 min at room temperature, and then rinsed thrice with PBS-CM with 5 min incubation between washes.

5. After permeabilization, cell culture inserts are incubated in blocking buffer for 45 min. After blocking, filters from the

cell culture inserts are carefully cut out of the plastic support with a razor blade, placed on parafilm with bottom (pseudopod) surface up and the filter is cut in half (or quarters if so desired). Each filter section is washed by placing it top up in a nonsterile 6-well plate containing PBS-CM. Once the filter has been removed from the support it is critical to always maintain the same orientation of the filter (top or bottom up) during washings. It is also important that the filter does not float on top, but be immersed in the washing solution (see Note 5).

6. Place drops (150 µL) of staining solution containing Hoechst and Alexa-fluor phalloidin on a piece of parafilm placed on damp Whatman paper in a 100 mm dish (humidified chamber). Remove excess PBS-CM by dabbing the filters on a tissue paper and place one filter section top (cell body) up and one bottom (pseudopod) up on separate drops of staining solution and then add 150 µL of staining solution containing Hoechst and Alexa-fluor phalloidin on top of the filter. Replace the top of the 100 mm dish to keep humid and incubate for 30–45 min at room temperature.

7. After incubation, wash the membrane with blocking buffer thrice with 10 min incubations between the washes. Dry the filter by dabbing with a tissue paper and mount (top up if labeled top up and bottom up if labeled bottom up) onto clean glass-slides with a drop of mounting media on both top and bottom of the filter. Cover with a clean 22-mm cover slip. Confocal z sections are as shown in Fig. 1b with nuclei in blue and actin in red. Some nuclei from cells plated on 3 µm pore-size filters have passed through the filter while the 1 µm pores completely prevent nuclear passage through the filter pores. This suggests that the PS fraction obtained on 1 µm pore size filters is a pure fraction (see Note 6).

8. The average number of pseudopodia per cell protruding through the filter is quantified from X–Y confocal images taken of cell nuclei above the filter and actin-rich pseudopodia below the filter. Multiple images of actin-labeled pseudopodia and cell nuclei are segmented and counted and dividing the number of pseudopodia by the number of nuclei provides the average number of pseudopodia per cell (Fig. 2).

3.2. Plating Cells to Harvest Pseudopodia

1. Cells (i.e., MDA-231 breast cancer cells) are plated in RPM1-1640 media and after reaching 80% confluence after 2 days the cells are removed from the dish with trypsin/EDTA.

2. Prior to trypsinizing the cells, melt 1% agarose and place the sterile lid of a 100-mm dish in the center of a sterile 150-mm dish. Carefully take a sterile washer with sterile tweezers and place the washer in the lid of the 100-mm dish. Seal the gap between the outside of the lid and the washer with 1% agarose. Place approximately 20 mL of RPMI-1640 media within

the washer and then carefully place the membrane on top. Place a second sterile washer on top of the membrane and seal the outside of the two washers with 1% agarose. Seed approximately 1×10^7 cells on the top of the filter in 20 mL RPMI media. Replace the lid of the 150-mm dish and place for 24-h in CO_2 incubator. Figure 1c shows a schematic diagram of the filter assembly process.

3.3. Harvesting Pseudopodia for Protein Isolation

When harvesting cell body and pseudopodial fractions it is important that all the reagents should be well prepared in advance except for the lysis buffer. To avoid contamination of the fractions, use different razor blades to scrape the cell body and pseudopodia from the top and bottom of the filters, respectively:

1. Put four 150-mm dishes containing 1× PBS, the tray containing ice with the metal sheet on it, filter tips, pipette, parafilm, razor blades, labeled microfuge tubes, and lysis buffer in the cold room for an hour before isolation of PS and CB fractions. Cover with parafilm the metal sheet in the ice-tray. Keep filter paper near the ice tray (see Note 7).

2. Carefully transport the 150-mm petridish containing the filter assembly with cells to the cold room. Carefully remove the agarose sealing around the two washers, use forceps to remove the membrane with cells and wash it four times with PBS by dipping sequentially in PBS containing petri dishes. While washing the filters, always keep the top of the membrane (containing the cell body) facing towards you.

3. Remove the membrane and dry it by holding the membrane vertically on top of the filter paper. Now, place the membrane with the top (cell body side) facing down and bottom (pseudopod side) facing up on the parafilm covered metal sheet. With the razor-blade, scrape first the bottom of the membrane containing the PS fraction. Drip approximately 20 μL of lysis buffer along the edge of the blade to collect the PS fraction into a sterile microfuge tube.

4. Flip the membrane, gently peeling it off the parafilm, and use a separate razor blade to scrape the top of the filter again dripping approximately 20 μL of lysis buffer along the edge of the blade to collect the CB fraction into a sterile microfuge tube.

5. Rock the tubes containing PS and CB fractions gently for 1-h in the cold room then centrifuge the microfuge tubes for 30 min at 4°C at $16,100 \times g$. Collect the PS and CB fractions in prelabeled tubes and store at –80°C.

6. To determine protein concentrations, add 2 μL of CB and PS fraction in a microfuge and add 788 μL of ddw to make the volume up to 800 μL, then add 200 μL Bradford reagent and incubate at room temperature for 5 min. Assay absorbance

at 595 nm and compare to a standard curve made using known protein concentrations of bovine serum albumin (see Note 8).

3.4. Harvesting Pseudopodia for RNA Isolation

1. Follow Subheading 3.3 until step 3 except use sterile PBS and sterile 200 μL filter tips. Drip approximately 20 μL of RLT plus buffer (supplied with the Total RNA kit) containing β-mercaptoethanol along the edge of the blade into a sterile microfuge for both fractions. Make up the volume up to 600 μL for both fractions with the same buffer and vortex to mix (see Note 9).

2. Carefully pipet the lysate using 1 mL filter tips directly into a QIAshredder spin column (supplied with the kit) placed in a 2 mL collection tube (supplied with the kit) without touching the edge of the column, and centrifuge for 2 min at $16,100 \times g$.

3. Transfer the lysate to a gDNA Eliminator spin column (supplied with the kit) placed in a 2-mL collection tube (supplied with the kit). Centrifuge for 1 min at $9,300 \times g$. Discard the column and save the flow through.

4. Add one volume (600 μL) of 70% ethanol to the flow through and mix well by pipetting.

5. Add 700 μL Buffer RW1 (supplied with the kit) to the RNeasy spin column. Close the lid gently, and centrifuge for 1 min at $9,300 \times g$ to wash the spin column membrane. Discard the flow-through.

6. Add 500 μL Buffer RPE (supplied with the kit) to the RNeasy spin column. Close the lid gently, and centrifuge for 1 min at $9,300 \times g$ to wash the spin column membrane.

7. Place the RNeasy spin column in a new 2-mL collection tube (supplied with the kit) and discard the old collection tube with the flowthrough. Centrifuge at $16,100 \times g$ for 2 min.

8. Place the RNeasy spin column in a new 1.5-mL sterile collection tube. Add 30 μL of RNase-free water directly to the spin column membrane. Close the lid gently, and centrifuge for 1 min at $9,300 \times g$ to elute the RNA.

9. Quantify the RNA concentration using NanoDrop ND-1000 (*NanoDrop*, Fisher Thermo, Wilmington, DE, USA). An A_{260} reading of 1.0 is equivalent to about 40 μg/mL of RNA and the OD at A_{260} nm is used to determine the RNA concentration in a solution. Ratio of OD A_{260}/A_{280} and OD A_{260}/A_{230} should be around 2.0 (see Note 10).

3.5. SDS-PAGE

1. Prepare a 1.5-mm thick 12% resolving or separating gel by mixing 5.0 mL of 5× separating buffer, with 8 mL of acrylamide/bis solution, 6.6 mL of water, 200 μL of 10% ammonium

persulfate solution and 10% SDS, and 20 μL of TEMED. Pour the gel and overlay with water. The gel should polymerize in about 30 min.

2. Prepare the 5% stacking gel by mixing 1.25 mL of 5× stacking buffer with 1.7 mL of acrylamide/bis solution, 6.8 mL of water, 100 μL of 10% ammonium persulfate solution and 10% SDS, and 10 μL of TEMED. Remove the water on top of the separating gel and then pour the stacking gel on top of the separating gel and insert the 10-well comb.

3. Prepare 1,000 mL of running buffer. Carefully remove the comb and place the polymerized gel into the running gel unit and pour running buffer into both lower and upper chamber.

4. In the meantime, boil samples in 1× loading buffer for 5 min and centrifuge at $9,300 \times g$ for 10 min. Load the supernatant of centrifuged samples (40–50 μg of protein) along with protein markers and run the gel at 80 V till the dye front enters the separating gel and then run at 100 V.

3.6. Western Blotting to Determine the Purity of PS and CB Fractions

1. SDS-PAGE separated samples are transferred to nitrocellulose membranes electrophoretically using a semi-dry transfer apparatus (Transblot, SD transfer cell, Bio-Rad). Remove the gel from the running unit and wash gel in distilled water. A sheet of nitrocellulose membrane slightly larger than the size of the separating gel is laid in the tray containing transfer buffer to allow the membrane to wet by capillary action for 15–30 s.

2. Wet extra thick blot paper, Protean XL size (18.5 × 19 cm, Bio-Rad, Hercules, CA), 3MM Whatman paper cut to the size of the nitrocellulose membrane in the transfer buffer. Completely saturate the filter paper by soaking in transfer buffer.

3. Place a presoaked sheet of extra thick filter paper and Whatman paper onto the platinum anode. Roll a pipet over the surface of the filter paper to exclude all air bubbles. Place pre-soaked membrane on the top of filter paper and Whatman paper. Carefully place the equilibrated gel on top of the transfer membrane, aligning the gel on the center of the membrane. Place the other sheet of pre-soaked filter paper on top of the gel, carefully removing air bubbles between the gel and filter paper using a pipet.

4. Carefully place the cathode onto the stack. Press to engage the latches with the guide posts without disturbing the filter paper stack. Place the safety cover on the unit. Plug the unit into the power supply. Run transfer at 25 V for 45 min (see Note 9).

5. Once the transfer is complete, carefully remove the nitrocellulose membrane, and discard gel and Whatman paper. Stain the membrane with Ponceau S (Sigma, St. Louis, MO) to verify protein transfer and mark the position of the lanes for the CB and

PS fractions. Wash the Ponceau stain from the membrane by rinsing the membrane in PBS-T for 5 min. The nitrocellulose membrane is then incubated in 50 mL of blocking buffer for 1 h at room temperature on a rocker.

6. The blocking buffer is discarded and the membrane is quickly rinsed with PBS-T. Mouse mitochondrial HSP70 antibody diluted 1:1,000 in PBS-T is then added to the membrane for 1 h at room temperature or overnight at 4°C while rocking.

7. Primary antibody is removed and the membrane is washed thrice for 15 min each with PBS-T and twice with PBS for 5 min each before adding secondary antibody. Add secondary antibody at 1:1,000 dilution and incubate for 1 h at room temperature on a rocker. After incubation, wash the membrane thrice for 15 min each with PBS-T and twice for 5 min each with PBS. In the meantime, 2 mL aliquots of each portion of the ECL reagent are warmed separately to room temperature in the dark, mixed, and then immediately added onto the top of the blot.

8. The blot is removed from the ECL reagents, blotted with tissue paper, and then placed between the leaves of a plastic sheet protector in the X-ray film cassette. Place on top of the membrane a similar sized piece of X-ray film with its top-left corner cut to identify the orientation of the membrane. After exposing film for a minute, film is developed in an automatic processor. Re-exposure for different times might be required to obtain optimal band density.

9. The same membrane is then probed for β-actin. For this, the membrane is incubated with blocking buffer overnight. After overnight incubation, membrane is washed with PBS-T and then 1:20,000 dil. primary antibody (anti-β-actin) is added to the membrane for 1 h at room temperature on a rocker. Primary antibody is removed and then membrane is washed thrice for 15 min each with PBS-T and twice with PBS for 5 min each before adding secondary antibody. Add secondary antibody at 1:10,000 dil. for 1 h at room temperature on a rocker; then membrane is washed thrice for 15 min each with PBS-T and twice with PBS for 5 min. Process the film as described above. The processed film will look as shown in Fig. 1a.

4. Notes

1. Make and store aliquots of PMSF, leupeptin, and Na_3VO_4 at −20°C, DTT at 4°C and NaF at room temperature.

2. Wear mask and gloves while preparing the SDS solution as SDS is a well known skin irritant.

3. Semi-dry transfer buffer should not be reused as the methanol in solution evaporates resulting in improper transfer.

4. Care should be taken while making PFA solution. As a formaldehyde releasing agent, paraformaldehyde is a suspected carcinogen. Always wear mask, gloves, and labcoat while making PFA solution and it should be made and aliquoted in fume hood.

5. For immunofluorescence labeling, the filters should always be kept moist. Do not let the filters dry as it will affect subsequent labeling.

6. Care should be taken to avoid the air bubbles in the mounting medium. Carefully with the help of needle place cover slip on the top of the filter with the mounting medium.

7. All the solutions and the tips for harvesting the pseudopodia should be kept at 4°C for at least 1 h before starting the isolation process. Agarose that sticks to the filter should be carefully removed without causing damage to the membrane. Be sure to maintain and remember the orientation of the filter.

8. From one filter we get around 50–60 μg protein in PS fractions and around 200–250 μg protein from the CB fraction. For a typical experiment such as Western blot, we generally harvest pseudopodia from four 100-mm filters. Yield might vary between cell lines depending on the extent of pseudopodial protrusion through the filter (see Fig. 2).

9. Add 10 μL of β-ME per 1 mL of Buffer RLT plus. Dispense in a fume hood and wear appropriate protective clothing. Buffer RLT plus is stable at room temperature for 1 month after addition of β-ME. All steps are performed at room temperature.

10. From one filter we get around 4–5 μg of RNA in PS fractions and around 20–25 μg of RNA from the CB fractions. We generally isolate RNA four to five filters per cell line to get sufficient amount of RNA for real-time PCR or microarray analysis.

Acknowledgements

The authors would like to thank Bharat Joshi for his contributions to the development of the pseudopod purification approach and for a critical reading of the text. This work was supported by grants from the Cancer Research Society Inc.

References

1. Jansen, R. P. (2001) mRNA localization: message on the move. *Nat. Rev. Mol. Cell Biol.* **2**, 247–56.
2. Lauffenberger, D. A., and Horwitz, A. F. (1996) Cell migration: A physically integrated molecular process. *Cell.* **84**, 359–369.
3. Pollard, T. D., and Borisy, G. G. (2003) Cellular Motility Driven by Assembly and Disassembly of Actin Filaments. *Cell.* **112**, 453–465.
4. Nabi, I. R. (1999) The polarization of the motile cell. *J. Cell Sci.* **112** 1803–1811.
5. Lawrence, J., and Singer, R. (1986) Intracellular localization of messenger RNAs for cytoskeletal proteins. *Cell.* **45**, 407–415.
6. Shestakova, E. A., Wyckoff, J., Jones, J., Singer, R. H., and Condeelis, J. (1999) Correlation of beta-actin messenger RNA localization with metastatic potential in rat adenocarcinoma cell lines. *Cancer Res.* **59**, 1202–5.
7. Stuart, H., Jia, Z., Messenberg, A., Joshi, B., Underhill, T. M., Moukhles, H., and Nabi, I. R. (2008) RhoA/ROCK signaling regulates the delivery and dynamics of a cohort of mRNAs in tumor cell protrusions. *J. Biol. Chem.* **283**, 34785–34795.
8. Jia, Z., Barbier, L., Stuart, H., Amraei, M., Pelech, S., Dennis, J. W., Metalnikov, P., O'Donnell, P., and Nabi, I. R. (2005) Tumor cell pseudopodial protrusions. Localized signaling domains coordinating cytoskeleton remodeling, cell adhesion, glycolysis, RNA translocation, and protein translation. *J. Biol. Chem.* **280**, 30564–73.
9. Wang, D. O., Kim, S. M., Zhao, Y., Hwang, H., Miura, S. K., Sossin, W. S., and Martin, K. C. (2009) Synapse- and stimulus-specific local translation during long-term neuronal plasticity. *Science.* **324**, 1536–40.
10. Miyashiro, K., Dichter, M., and Eberwine, J. (1994) On the nature and differential distribution of mRNAs in hippocampal neurites: implications for neuronal functioning. *Proc. Natl. Acad. Sci. USA.* **91**, 10800–10804.
11. Nguyen, T. N., Wang, H. J., Zalzal, S., Nanci, A., and Nabi, I. R. (2000) Purification and characterization of beta-actin-rich tumor cell pseudopodia: role of glycolysis. *Exp. Cell Res.* **258**, 171–83.
12. Shankar, J., Messenberg, A., Chan, J., Underhill, T. M., Foster, L. J., and Nabi, I. R. (2010) Pseudopodial Actin Dynamics Control Epithelial-Mesenchymal Transition in Metastatic Cancer Cells. *Cancer Res.* **Epub**, 0008-5472.CAN-09-4439.
13. Cho, S. Y., and Klemke, R. L. (2002) Purification of pseudopodia from polarized cells reveals redistribution and activation of Rac through assembly of a CAS/Crk scaffold. *J. Cell Biol.* **156**, 725–736.
14. Mili, S., Moissoglu, K., and Macara, I. G. (2008) Genome-wide screen reveals APC-associated RNAs enriched in cell protrusions. *Nature (Lond.).* **453**, 115–119.
15. Poon, M. M., Choi, S. H., Jamieson, C. A., Geschwind, D. H., and Martin, K. C. (2006) Identification of process-localized mRNAs from cultured rodent hippocampal neurons. *J. Neurosci.* **26**, 13390–9.
16. Latham, J., Vaughan M., Yu, E. H. S., Tullio, A. N., Adelstein, R. S., and Singer, R. H. (2001) A Rho-dependent signaling pathway operating through myosin localizes [beta]-actin mRNA in fibroblasts. *Curr. Biol.* **11**, 1010–1016.

Part V

Affinity Purification of mRNAs and the Identification of *Trans*-Acting Factors

Chapter 23

RNA-Binding Protein Immunopurification-Microarray (RIP-Chip) Analysis to Profile Localized RNAs

Alessia Galgano and André P. Gerber

Abstract

Post-transcriptional gene regulation is largely mediated by RNA-binding proteins (RBPs) that modulate mRNA expression at multiple levels, from RNA processing to translation, localization, and degradation. Thereby, the genome-wide identification of mRNAs regulated by RBPs is crucial to uncover post-transcriptional gene regulatory networks. In this chapter, we provide a detailed protocol for one of the techniques that has been developed to systematically examine RNA targets for RBPs. This technique involves the purification of endogenously formed RBP–mRNA complexes with specific antibodies from cellular extracts, followed by the identification of associated RNAs using DNA microarrays. Such RNA-binding protein immunopurification-microarray profiling, also called RIP-Chip, has also been applied to identify mRNAs that are transported to distinct subcellular compartments by RNP–motor complexes. The application and further development of this method could provide global insights into the subcellular architecture of the RBP–RNA network, and how it is restructured upon changing environmental conditions, during development, and possibly in disease.

Key words: Post-transcriptional gene regulation, RNA-binding protein, Immunopurification, Microarray, RIP-Chip

1. Introduction

Gene expression is controlled at multiple levels to ensure the coordinated synthesis, assembly, and localization of the cells' macromolecular components (1, 2). Besides transcriptional control, which is mediated by chromatin modifiers and transcription factors, it is now becoming increasingly recognized that post-transcriptional events have a substantial impact upon gene expression and alterations thereof can be the cause or consequence of human disease (3, 4). On the one hand, post-transcriptional regulation is achieved by hundreds of RNA-binding proteins (RBPs) that control every aspect of an RNA's life from RNA maturation, localization, translation, and decay (5, 6). On the other hand, a vast number of noncoding RNAs such as microRNAs (miRNAs) affect mRNA fate at translation and/or decay, or silence chromatin (7). Thus, the hundreds of known RBPs and noncoding RNAs (ncRNAs) suggest elaborate post-transcriptional regulatory networks that are comparable and linked to other cellular networks such as transcriptional circuits (8).

The application of genome-wide analysis tools such as DNA microarrays allows for the systematic exploration of the post-transcriptional system (9). For instance, it is now feasible to broadly map all of the RNAs that are associated with a particular RBPs using RNA-binding protein immunoprecipitation-microarray (RIP-Chip) profiling (Fig. 1). This method involves the isolation

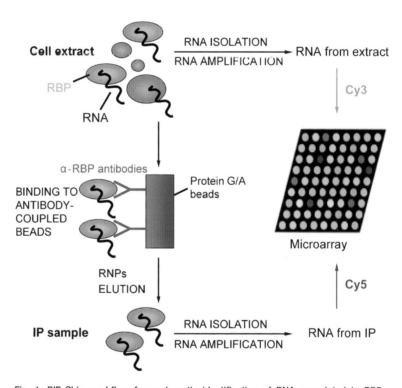

Fig. 1. RIP-Chip workflow for systematic identification of RNA associated to RBPs. Ribonucleoprotein (RNP) complexes are captured from cell-free extracts with specific antibodies coupled to either protein G or protein A sepharose beads – depending on the antibody used – and then eluted with SDS–EDTA. To control for nonspecifically enriched RNAs, the same procedure is performed with beads that are not coupled with immunoprecipitating antibody (mock samples). RNA is isolated from extracts (input) and from the immunopurified (IP) samples, amplified, and labeled with Cy3 and Cy5 fluorescent dyes, respectively. The labeled amplified RNA probes from total RNA and IPed RNA are mixed and competitively hybridized to human cDNA microarrays. In this assay, the ratio of the two RNA populations at a given array element reflects the enrichment of the respective mRNA by the affinity purification.

of RNP complexes from cell extracts with an antibody selectively recognizing an epitope of a constituent protein. The purified RNPs are then dissociated into proteins and RNAs, and the identities of RNAs are determined on a global-scale with DNA microarrays or by sequencing. Likewise, protein components can be identified with mass-spectrometry or by immunoblot analysis to detect particular proteins.

RIP-Chip was first established in Jack Keene's lab to study RNAs associated with RBPs in human cancer cells using antibodies against endogenous proteins (10). Diverse groups have then further developed and modified the RIP-Chip protocol, such as the affinity isolation of recombinantly or endogenously expressed epitope-tagged RBPs. RIP-Chip has been employed in diverse species including yeast (11–14), flies (15), worms (16), plants (17), and in various cell lines (18–23) (for a more comprehensive list of references for RIP-Chip experiments performed in various tissues and species, see ref. 24). In this regard, RIP-Chip has also been applied to identify the messages that are actively transported by RNP–motor complexes to particular subcellular regions (25–27). For instance, in the yeast *Saccharomyces cerevisiae*, mRNAs that are asymmetrically distributed between mother and daughter cells were revealed upon affinity-purification of the RNA transport components She2p, She3p, and the myosin motor protein Myo4, followed by DNA microarray analysis combined with a secondary GFP-based RNA reporter assays to visualize RNAs in living cells (25, 26). In addition to the known bud-localized *ASH1* and *IST2* messages, this analysis revealed 22 additional polarized mRNAs, suggesting the existence of widespread cytoplasmic mRNA localization in yeast (26). Besides its application to specific RNP complexes, RIP-Chip has also been implemented for the isolation of broad-specificity RBPs, such as poly(A)-binding protein, or entire ribosomes via affinity-tagged ribosomal proteins, to reveal the messages that are expressed within particular tissues or cell types, or to reveal changes of the translatome upon different stress conditions (16, 28–31). Finally, mRNAs targeted by miRNAs have been identified by the biochemical isolation of miRNA induced silencing complexes (miRISC), which are comprised of Argonaute and other proteins as well as miRNAs and target mRNAs (32–38).

A major advantage of RIP is that it is a straightforward protocol that allows for the concomitant identification of RNA and protein components of RNPs. Drawbacks of RIP are concerns that during the procedure certain RNAs or proteins may fall-off and others associate with RNP complexes (39). However, how often this occurs has not been conclusively answered and may depend on the RBP under investigation. For the isolation of unstable RNP complexes, modifications of the RIP procedure have been proposed to better preserve RNP complex stability and integrity (40). RNPs have also been cross-linked either by UV or with chemicals prior to RIP-Chip (41). A more elaborate but very specific UV-light cross-linking-immunoprecipitation (CLIP) protocol has been developed by Ule, Darnell and colleagues, which allows for the fairly precise mapping of RNA–protein or RNA–RNA interactions sites on transcripts (42–47). The drawbacks of this method are, however, a complicated experimental set-up, and it hardly allows for the analysis of co-purifying proteins. In addition, UV irradiation may chemically and physically alter the RNP structures and cause some sequence bias due to the unequal photo reactivity between bases and amino acids (43, 48).

In conclusion, the application of RIP or CLIP to analyze RNP-bound messages has provided great insight into the infrastructure of coordinated eukaryotic post-transcriptional gene expression, leading to the finding that specific RBPs may bind and thus coordinate groups of mRNAs coding for functionally related proteins organized into the so called "post-transcriptional operons" or "RNA regulons" (49, 50). Furthermore, these methods combined with high throughput RNA localization assays also provide a valuable tool to systematically dissect RNP complexes involved in RNA transport. The implementation of high-throughput sequencing (HITS) to identify bound RNAs will further increase the reproducibility and the sensitivity of these assays, possibly enabling to quantify the number of message partitioned into distinct subcellular compartments (47, 51, 52).

In the following, we present a detailed RIP-Chip protocol, which is based on our experiences with human PUF-family RNA binding proteins PUM1 and PUM2 (23). The protocol involves two major steps which can be further divided. In the first step, we describe how RNPs are captured from cellular extracts using antibodies specific for the RBP of interest, and how RNPs are eluted from the beads by treatment with SDS–EDTA. In the second part, which involves dual-color DNA microarrays, we describe how RNAs are isolated from cell extracts and from isolated RNP complexes, amplified, labeled with fluorescent dyes, and competitively hybridized to DNA arrays (Fig. 1).

2. Materials

The reagents described below have been used to purify RPBs from cultured human cells, but they can be adapted to cells derived from other organisms as well. Working with RNA implies that tips, tubes, and solutions must be RNase-free. Therefore, work surfaces, pipetters, and equipment can be treated with RNase Zap (Ambion, cat. # 9780). Ultrapure distilled DNase and RNase-free water (Gibco, Invitrogen, cat. # 10977-03) should be used to prepare buffers and solutions, which should be passed through a 0.2-μm filter. Filtered or autoclaved salt solutions can be stored at 4°C for up to 6 months, unless otherwise indicated.

2.1. Cell Culture and Lysis

1. Cells of interest for RIP experiments.
2. Cell culture media. For HeLaS3 cells and diverse cell lines, Dulbecco's Modified Eagle's Medium (DMEM,) supplemented with 10% FBS and antibiotics (1% penicillin/streptomycin).
3. 150-mm Culture dishes.
4. Cell scrapers.
5. 1× Phosphate-buffered saline (PBS).
6. 0.5 M 4-(2-hydroxyethyl)-1-piperazineethanesulfonic acid (HEPES), pH 7. Dissolve HEPES in water and adjust pH with 1 M KOH. Autoclave.
7. IGEPAL CA-630 (Sigma, cat. # I8896). Severe eye and skin irritant: wear protective glasses and gloves.
8. Dithiothreitol (DTT). Prepare 1 M aliquots in water. Store at −20°C and avoid repeated freezing and thawing.
9. Heparin sodium salt (Sigma, cat. # H3400). Prepare 20 mg/mL aliquots in water. Store at −20°C and avoid repeated freezing and thawing.
10. RNase OUT (Invitrogen, cat. # 10777-019). Store at −20°C.
11. SuperaseIN (Ambion, cat. # 2696). Store at −20°C.
12. Complete ethylenediamine tetraacetic acid (EDTA)-free tablets, proteinase inhibitor (Roche, cat. #11-873-580-001). Store at 4°C.
13. Polysome lysis buffer (PLB): 10 mM HEPES at pH 7.0, 100 mM KCl, 5 mM $MgCl_2$, 25 mM EDTA at pH 8, and 0.5% IGEPAL. Store at 4°C for up to 6 months, or prepare a 10× PLB solution to be stored at room temperature for up to 6 months. Prior to use, add the following components per 50 mL of PLB buffer: one tablet of complete proteinase inhibitor, 2 mM DTT, 50 U/mL RNase OUT, 50 U/mL SuperaseIN, and heparin 0.2 mg/mL. This complete PLB can be stored in aliquots at −80°C for up to 6 months.
14. Protein quantification assay (i.e. Bradford method, Bio-Rad Protein Assay, cat. # 1-800-424-6723). Store at 4°C.

2.2. Antibody Coupling to Protein A or Protein G Sepharose Beads

1. Antibody to RNA binding protein of interest.
2. Protein G Sepharose 4 Fast Flow beads (Amersham, cat. # 17-0648-01) or Protein A Sepharose CL-4B (Amersham, cat. # 17-0780-01). Store at 4°C.

3. Bovine serum albumin (BSA) which is protease, DNase, and RNase-free (Equitech Bio, Cat. # BAH67-0050; Roche, cat. # 8100350). Store at 4°C.

4. 10% Sodium azide (Fluka, cat. # 71290). Highly toxic: handle with gloves under a fumehood.

5. NT2 buffer: 50 mM Tris–HCl at pH 7.5, 150 mM NaCl, 1 mM $MgCl_2$, and 0.05% IGEPAL. Store at 4°C for up to 6 months.

6. NT2-coupling buffer: NT2 buffer supplemented with 5% BSA, 0.02% sodium azide, and 0.02 mg/mL heparin. Prepare fresh before use.

2.3. Immunopurification of mRNPs

1. NT2-RIP buffer: NT2 buffer supplemented with 50 U/mL RNase OUT, 50 U/mL SuperaseIN, 2 mM DTT, 30 mM EDTA at pH 8, and heparin 0.02 mg/mL. Prepare fresh before use (see Note 1).

2. SDS–EDTA elution buffer: 50 mM Tris–HCl at pH 8, 100 mM NaCl, 10 mM EDTA, and 1% (w/v) SDS. Prepare fresh before use.

2.4. RNA Isolation, RNA Amplification, and Labeling

1. *mir*VanaPARIS kit (Ambion, cat. # 1556). Store at 4°C.

2. Amino Allyl MessageAmp II aRNA kit (Ambion, cat. # AM1753). Store tubes and columns at room temperature and the reagents at −20°C.

3. Amersham CyDye (Cy3/Cy5) Post-Labeling Reactive dye Pack (GE Healthcare, cat. # RPN 5661). Store at −20°C.

2.5. cDNA Microarray Analysis

1. Human cDNA microarrays (Stanford Functional Genomics Facility, SFGF, cat. # sfgf 001).

2. 22 × 60 mm regular thin microscope slide cover slip (Menzel-Glaser, cat. # BB022060A1).

3. Hybridization chamber (Corning, cat. # 2551).

4. 20× SSC (3 M NaCl and 300 mM sodium citrate at pH 7.0). Store at room temperature.

5. 10% (w/v) sodium dodecyl sulfate (SDS). Store at room temperature.

6. Pre-hybridization solution (5× SSC, 0.1% [w/v] SDS, and 0.1 mg/mL BSA). Prepare fresh before use.

7. Post-processing washing solution: 0.1× SCC. Filter sterilized. Prepare fresh before use.

8. Poly(A) RNA (Invitrogen, cat. # POLYA.GF). Dissolve in water at 10 mg/mL. Store aliquots at −20°C.

9. Yeast tRNA (Invitrogen, cat. # 1541-011). Dissolve in water 10 mg/mL (shaking/vortexing for 60 min). Store aliquots at −20°C.

10. Human Cot-1 DNA (Invitrogen, cat. # 15279-011). Store at −20°C.

11. Three array wash solutions: 2× SSC, 0.1% SDS; 1× SSC; and 0.2× SSC. Prepare solutions fresh before use and filter.

3. Methods

3.1. Preparation of Whole-Cell RNP Lysates-Timing ~2 h

For practical reasons, the cell-free extract can be prepared several days in advance. Cell-free extract should be frozen in liquid nitrogen and stored in aliquots at −80°C.

1. Use 100-mm or 150-mm dishes to grow the desired cells in culture (see Note 2).

2. Wash the cells twice with 5–8 mL of ice-cold PBS, harvest them in 3–5 mL PBS with a cell scraper and transfer to a 15-mL centrifuge tube with a pipette.

3. Pellet the cells by centrifugation at $3,000 \times g$ for 5 min at 4°C. Aspirate PBS from cell pellet, snap-freeze cells in liquid nitrogen and store them at −80°C until use. Alternatively, you can proceed directly and prepare the cell extract for IP (see Note 3).

4. Thaw the cells on ice.

5. Estimate the pellet volume and add approximately 1.5 volumes of PLB (see Note 4). The cells are lysed by pipetting several times (via aspiration and expulsion) until the lysate looks uniform without visible cell clumps. Transfer the lysate into a 2-mL tube and spin the suspension in microcentrifuge at $14,000 \times g$ for 10 min at 4°C. Transfer the lysate into a new tube and repeat the centrifugation step twice (see Note 5).

6. Determine protein concentration of the cell-free extract with a colorimetric assay [i.e. Bradford method with bovine serum albumin (BSA) as reference standard].

7. *Optional.* Resuspend the pellet in one volume of PLB, repeat pipetting and centrifugation and combine with the supernatant from step 7 (above) (see Note 6).

8. Prepare 20 mg aliquots of extracts (or the amount of protein required per RIP), snap-freeze in liquid nitrogen and store at −80°C.

9. Save at least 2–5% of cell extract for total RNA isolation (see Subheading 3.4).

3.2. Antibody Coupling to Protein A or Protein G Sepharose Beads-Timing 1–12 h

We recommend protein G sepharose 4 Fast Flow beads (Amersham) to couple mouse or goat antibodies and protein A sepharose CL-4B (Amersham) for rabbit antibodies.

1. Protein A sepharose CL-4B beads are supplied dried and have to be swollen for at least 12 h. Therefore, 0.5 g of protein A beads are resuspended in 5 mL of NT2-coupling buffer in a 15-mL tube and further incubated on a rotating device overnight. Store swollen beads at 4°C up to 6 months.

2. Protein G sepharose 4 Fast Flow beads are supplied in 20% ethanol. Remove the buffer/ethanol from beads by centrifugation at $10,000 \times g$ for 3 min at 4°C.

3. The beads are equilibrated three times with at least eight bed volumes of NT2-coupling buffer for 10 min at room temperature and collect beads by centrifugation as described in step 2. The beads are now ready for coupling with antibodies.

4. Usually, we use 100 μL of 50% (v/v) slurry beads for each antibody coupling. Add 20–50 μg of antibody to the beads and incubate overnight (~12 h) on a rotating device at 4°C or at least 1 h at room temperature (see Note 7).

5. Spin down the coupled beads (3 min, $10,000 \times g$, 4°C) and remove the supernatant with a pipette. Wash the beads with 1 mL of NT2-coupling buffer. Beads can be stored for several months at 4°C (see Note 8).

3.3. Immunopurification of mRNPs-Timing ~1–8 h

1. Thaw the cell extract on ice. Centrifuge at $14,000 \times g$ for 10 min at 4°C. Transfer the supernatant to a new 1.5-mL tube placed on ice.

2. Equilibrate the antibody-coupled beads three times with 1 mL of NT2 buffer at room temperature by flicking the tube several times with a finger, and spinning at $10,000 \times g$ for 3 min at room temperature. Resuspend the equilibrated beads in 1 mL of NT2 buffer and transfer to a 15-mL conical bottom tube. Pellet the beads by centrifugation at $2,000 \times g$ for 2 min at 4°C and repeat the washing once. Keep the beads on ice.

3. Likewise, equilibrate an equal amount of blocked but uncoupled (without antibody) beads to perform a mock RIP to assess unspecific binding of RNAs to the beads during the procedure. Alternatively, control RIPs can be performed with equal amounts of a suitable isotype-matched antibody (unrelated antibody of animal origin and immunoglobulin type as the test antibody).

4. Resuspend the antibody-coated beads in 4.5 mL of NT2-RIP buffer (corresponding to ~9× volumes of extract) and combine with 500 μL of extract (see Note 9). Tumble the tube end-over-end on a rotating device at 4°C.

5. Collect analytical samples of the extract (~50 μg protein) and of the supernatants obtained by pelleting the beads at $2,000 \times g$ for 2 min at 4°C after 2, 4, and 6 h, to monitor the recovery of the target RBP by immunoblot analysis, and to visualize RNA integrity with an agarose gel (Fig. 2) (see Note 10).

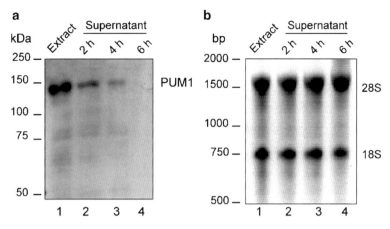

Fig. 2. Monitoring RBP and total RNA stability during RIP. (**a**) Immunoblot analysis following immunoprecipitation of PUM1 (127 kDa) with anti-PUM1 antibody. Lane 1: cell extract (25 μg); lanes 2–4: supernatants (25 μg) after 2, 4, and 6 h-incubation of extracts with PUM1 antibody-coupled protein G sepharose beads. (**b**) 1% non-denaturating agarose gel visualizing RNA stained with ethidium bromide. Lane 1: total RNA from cell extract (1 μg); lanes 2–4: total RNA from supernatants (1 μg) after 2, 4, and 6 h immunopurification. Ribosomal RNAs (rRNAs) appear as sharp bands on the stained gel indicating the integrity of total RNA. 28S rRNA band shows an intensity approximately twice that of the 18S rRNA band. The real size of the rRNAs species (28S = 5 kb; 18S = 1.9 kb) can only be seen with a denaturing formaldehyde agarose gel.

6. Collect the beads after the appropriate time of incubation to recover at least 80% of the RNPs from the extracts (for PUM proteins: 6 h at 4°C). Collect the beads by centrifugation (step 4) and save the supernatant.

7. Add 1 mL (10–20 bed volumes) of ice-cold NT2 buffer to the beads and transfer them to a 1.5-mL conical microtube. Wash the beads four times with 1.5 mL of NT2 buffer by vigorously shaking of the tube, and pellet the beads at 10,000 × g for 3 min at 4°C (see Note 11). Save 50 μL of analytical samples from each washing step.

8. To release the RNP complexes from the beads, resuspend them in 200–500 μL of SDS–EDTA elution buffer and incubate for 10 min at 65°C, mixing every 2 min by inverting the tube (see Note 12).

9. Pellet the beads at 10,000 × g for 3 min at room temperature.

10. Save the supernatant as first elution in a new 2-mL tube and keep it at room temperature.

11. Repeat steps 8–10.

12. Save analytical samples (5%) from both elutions and from the beads to control for release from antigen from the beads by SDS–PAGE and immunoblot analysis.

3.4. RNA Isolation and Synthesis of Labeled Antisense Amplified RNA-Timing ~20 h Including Incubation Time

1. Isolate RNA from at least 10 µL of cell extract and from 500 µL of immunopurified sample with the *mir*Vana PARIS Kit (Ambion) according to the manufacturer's instructions. Alternatively, RNA can be isolated with Trizol reagent (Invitrogen), or by phenol–chloroform–isoamyl alcohol extraction followed by precipitation with isopropanol (see Note 13). We usually obtained about 2–3 µg of total RNA from extracts and 100–200 ng of RNA from immunopurified fractions.

2. Resuspend RNA in 50 µL of RNase-free water.

3. Quantify RNA with an UV spectrophotometer. We regularly use a NanoDrop device (Witeg AG) and continuously monitor the absorbance from 220 to 350 nm of a 1.5 µL RNA sample. An A_{260} reading of 1.0 is equivalent to ~40 µg/mL single-stranded RNA. The A_{260}/A_{280} ratio is used to assess the quality of the RNA: A ratio of 1.8–2.1 is indicative for RNA of good quality. Lower ratios indicate the presence of protein, phenol, or other contaminants that absorb at or near 280 nm. Chloroform extraction and ethanol precipitation of the RNA sample helps to remove such contaminants. The A_{260}/A_{230} ratio of absorbance provides another indication for nucleic acid purity, and should be in the range of 2.0–2.2. Lower ratios indicate contaminants such as EDTA or carbohydrates which absorb near 230 nm. However, the A_{260}/A_{230} ratio is often very low (0.5–1) in less concentrated (5–50 ng/µL) RNA samples from IP eluates, but we did not observe negative effects in downstream applications.

4. About 0.5–1 µg of RNA isolated from extract and from the supernatants is run on a 1% agarose gel to control for RNA degradation during incubation with antibody-coupled beads (Fig. 2). Of note for the quantification of low amounts of RNA (i.e. such as from IP samples), the RNA Nano or Pico Chip Kit (Agilent) can be used with an Agilent 2100 BioAnalyzer. For total RNA, net peaks representing 18S and 28S ribosomal RNA should be visible, and the 28S/18S ratio should range between 1.5 and 2.1. Ratios lower than 1.2 indicate partial degradation of the RNA sample. Messenger RNAs are between 250 and 5,500 nucleotides (nts) long, with most of them between 1,000 and 1,500 nts.

5. Amplify 50–100 ng of RNAs isolated from cell extracts and from the IP fractions in the presence of aminoallyl-UTP with the Amino Allyl MessageAmp II aRNA kit (Ambion). Thereby, the poly-adenylated RNA is reverse-transcribed with T7-oligo(dT) primers to produce cDNA containing a T7 promoter sequence, which is followed by transcription with T7 RNA polymerase in the presence of aminoallyl-UTP next to the four natural NTPs to produce amino-allyl modified

amplified RNA (aRNA). The concentration of the aRNA is determined with a NanoDrop spectrophotometer (see Note 14). The aRNA can be stored at −20°C for several weeks; for longer periods store at −80°C.

6. 8 μg aRNA is fluorescently labeled with NHS-monoester Cy3 or Cy5 dyes.

7. Combine Cy3-labeled probe representing aRNA from input with Cy5-labeled aRNA probe from the IP sample for competitive hybridization on DNA microarrays (see Note 15).

3.5. cDNA Microarray Analysis-Timing ~20 h, Including Incubation Time

The following protocol has been applied for analysis with human cDNA microarrays containing ~45,000 arrayed features spotted on amino-silane slides (Corning).

1. Post-process microarray: Mark region of spots with a diamond scraper on the back of the slide. Cross-link spotted cDNAs with 65 mJ of UV irradiation. Block slides for 1 h at 42°C in pre-hybridization solution. Wash the slides twice in 400 mL of 0.1× SSC for 5 min, dunk in 400 mL of ultrapure water for 30 s, and dry them by centrifugation at 550 rpm for 5 min. The slides can be stored under vacuum in a desiccator at room temperature for several weeks.

2. Prepare hybridization solution: Add 6.8 μL of 20× SCC, 0.8 μL of 1 M HEPES at pH 7, 1.2 μL of 10% SDS, 2 μL of 10 mg/mL poly(A) RNA, 1.6 μL of 10 mg/mL yeast tRNA, and 2 μL of 10 mg/mL Cot-1 DNA to 25.6 μL of the combined Cy3- and Cy5-labeled aRNA probes.

3. Denature probes for hybridization for 2 min at 100°C. Let the probes cool down to RT and centrifuge at maximum speed for 5 min at room temperature.

4. Loading of the array: Remove dust from the array and the cover slip, and place the array into the hybridization chamber. Carefully pipette the denatured probe drop-wise in the center of the array and apply a 22 × 60 mm regular thin microscope slide cover slip (avoid bubbles) (see Note 16). Add three times 10 μL on both the short sides of the array and fill the liquid reservoir with 3× SCC.

5. Assemble the hybridization chamber and place in a water bath at 65°C for 16 h.

6. Wash arrays in 400 mL of three subsequent wash buffers made of 2× SSC, 0.1% SDS (bath 1), 1× SSC (bath 2), and 0.2× SSC (bath 3). The first wash is performed at 65°C for 5 min (1 min by manual shaking followed by 4 min on a shaker), and the following washes are preformed at room temperature for 5 min each. Dry the arrays by centrifugation at 550 rpm for 5 min at room temperature. Keep arrays in the dark and scan immediately.

7. Scan the arrays with a fluorescence scanner like the Axon Instruments Scanner 4200A (Molecular Devices). Adjust PMTs levels for the red (Cy5) and the green (Cy3) channel that less than 1% of arrayed features reach saturation.

8. Data are collected with GENEPIX 5.1 (Molecular Devices). Data is exported into a database for normalization and further analysis. For instance, arrays can be exported into Acuity 4.0 (Molecular Devices), which runs on SQL set-up on local desktop computers. Our arrays have been deposited and normalized computationally by the Stanford Microarray Database (SMD) (53).

9. Data filtering: Stringent filtering for irregular spots and signals with low intensity reduces the number of false-positives, but increases the number of false-negatives. For PUM IPs, the data were filtered for signal over background greater than 1.5 in the channel measuring aRNA from extract, and only features that met this criterion in >50% of the arrays were included for further analysis.

10. To identify transcripts that are specifically enriched with an RBP compared to the control, we usually perform unpaired two class Significance Analysis of Microarrays (SAM) on median centered arrays and determined false discovery rates (FDRs) for each arrayed element (54) (see Note 17).

4. Notes

1. Special care should be taken for the purification of labile RNP complexes. In this case, total salt concentration should not exceed physiological conditions (~125 mM). RNA–protein interactions may also be magnesium dependent (e.g. ribosomes) and hence, the entire RIP (including PLB) should be performed with 2.5 mM $MgCl_2$ instead of EDTA. Of note, heparin and proteases/RNase inhibitors may bind to some RBPs and interfere with RNA binding.

2. We typically use two 150-mm dishes containing 90% confluent HeLa S3 cells (~4×10^7 cells) for each RIP, yielding approximately 20 mg protein extract. However, well-expressed RBPs have been immunopurified from considerably less cells ($2-20 \times 10^6$ cells collected from two 100-mm dishes, 2–5 mg of total protein) (40). Therefore, the number of cells required for each immunopurification (IP) varies from cell type to cell type and depends on the expression of the RBPs.

3. Samples of cells can be collected and stored at −80°C until enough cell mass is obtained for RIP.

4. Extracts prepared from HeLa cells collected from ten 150-mm dishes were highly concentrated (30–40 mg/mL of protein). We used 20 mg (protein) of extract per RIP experiment. If less extract is required, the cell pellet should be resuspended in an adjusted volume of lysis buffer that yields to a protein concentration of at least 2–5 mg/mL. Since the volume of the extract further determines the volumes of buffer in later steps, we recommend that extract volume does not exceed 1.5 mL so that RIP can be performed in 15-mL conical bottom tubes.

5. The centrifugation steps are crucial to clear the extract from any cell debris and clouds of membranes that could stick to the antibody-coupled beads added later on, and interfere with RBP binding. Remove the lipid layer with a pipette tip and then transfer the cleared supernatant into a new tube.

6. The first lysis produced extracts containing 20–50 mg/mL of protein. Second lysis increases the amount of total protein, but it also dilutes the sample (step 6).

7. Check the binding capacity of antibodies to the beads with a titration experiment. The antibodies should be added in excess to saturate the beads and minimize background binding during RIP. We used 20 μg of PUM1 and 50 μg of PUM2 antibodies, as recommended in the antibody data sheet.

8. The washing removes unbound antibodies and RNases. Therefore, we prefer to bind antibodies to beads before addition to extracts.

9. A 1:10-fold dilution of the extract minimizes the likelihood for exchange of proteins and mRNAs during the mRNP isolations, and decreases background. Usually, we use 300–800 μL of extract and the final volume for RIP ranges from 3 to 8 mL.

10. Immunopurification should be performed as quickly as possible. Long incubation times increase the possibility for exchange of proteins and mRNAs and may lead to the disruption of mRNP complexes. For instance, the use of magnetic Dynabeads may shorten incubation times to less than 30 min (55). RNP cross-linking agents, such as formaldehyde, may prevent such rearrangements, but they substantially increase the background and may interfere with subsequent mRNA detection methods (19, 41).

11. Thorough washing is critical to reduce background and to give optimal results. On the one hand, the stringency is increased by adding salt (200–400 mM KCl), detergents (SDS), or 1–3 M urea to the wash buffer. However, it is important to first determine whether these conditions do not interfere with the binding of the antibody to the target protein

and/or with RBP–mRNA interactions. On the other hand, weak antibody–protein interactions require less stringent conditions to prevent dissociation of antigens during washing steps, e.g. wash the beads in buffer containing $MgCl_2$ by inverting the tube five times at room temperature, and collect beads at $5,000 \times g$ for 10–20 s.

12. Other protocols used a 1:1 mixture of NT2 buffer with (1) TE-SDS buffer (2×: 20 mM Tris–HCl [pH 7.5], 2 mM EDTA, and 2% SDS), (2) proteinase K buffer (2×, 200 mM Tris–HCl [pH 7.5], 20 mM EDTA, 100 mM NaCl, and 2% SDS supplemented with 30 μg proteinase K), or (3) 100 μL NT2 buffer containing 30 μg proteinase K (Ambion, cat. # 2546). In all cases, reactions are incubated for 30 min at 55°C, and occasionally mixed by flicking the tube with a finger (40, 56). In case of inefficient release of the RNP complexes, the elution may be repeated with buffer containing proteinase K. Therefore, beads should be stored at −80°C after elution.

13. Residual heparin inhibits RT, which is used in the next steps. Heparin can removed by precipitation of the RNA with 1.5 M LiCl (overnight at −20°C) followed by washing the pellet twice with 70% ethanol. Moreover, the addition of 20 μg of glycogen as a carrier for RNA precipitation increases the recovery of RNA. We recommend the use of a colored carrier such as GlycoBlue (Ambion), which makes the salt–RNA pellet better visible. RNA pellets from precipitations easily detach from the centrifuge tube and care should be taken during washing (remove ethanol with a pipette instead of decanting is recommended).

14. The yields of aRNA varies between different tissues or cell-types and critically depends on the amount and the quality of the input poly(A) RNA. We usually obtain around 60 μg of aRNA starting with 500 ng of total RNA isolated from HeLa cell extract. The amount of aRNA obtained from IP samples varies considerably: We obtained between 10 and 40 μg aRNA from PUM RIPs and less than 10 μg from mock eluates (23). To get best amplification results, it is recommended to use a thermal cycler with adjusted lid temperature or an incubation oven. Otherwise, better turn-off the lid-heating of the thermal cycler. No condensation water should be formed in the tube during the reaction, especially for the *in vitro* transcription, where as little as 1–2 μL of condensate throws off the concentrations of the nucleotides and magnesium and reduces the yield of aRNA.

15. Usually, we competitively hybridize aRNA amplified from the input (total RNA from extract) versus aRNA from IP samples. In this way, the input aRNA serves as a common reference for

both aRNA obtained from IPs of RBPs, and from control (mock) IPs. This approach corrects for highly expressed mRNAs, and integrates internal standard/control for the array hybridization. Alternatively, and in particular if there are high background signals in the mock IP sample, the RNA IPed from the RBP is directly compared to the RNA IPed from the control sample. This procedure is advantages to detect weaker enrichments of RNAs; however, the lack of a reference RNA may increase the variability between experiments.

16. The use of lifter slips (Erie Scientific, cat. # M5516) instead of cover slips is recommended especially with oligo microarrays. With lifter slips, the volume of the hybridization solution is increased and thicker hybridization chambers are required (e.g. Monterey Industries).

17. Cyber-T, which is based on a general Bayesian statistical framework provides an alternative to SAM (57). Both SAM and Cyber-T implement probabilistic methods that are developed to manage large data sets with at least 4–5 replicates, thus allowing the identification of even small subsets of genes that are differentially expressed.

References

1. Maniatis, T., and Reed, R. (2002) An extensive network of coupling among gene expression machines, *Nature 416*, 499–506.
2. Orphanides, G., and Reinberg, D. (2002) A unified theory of gene expression, *Cell 108*, 439–451.
3. Tazi, J., Bakkour, N., and Stamm, S. (2009) Alternative splicing and disease, *Biochim Biophys Acta 1792*, 14–26.
4. Cooper, T. A., Wan, L., and Dreyfuss, G. (2009) RNA and disease, *Cell 136*, 777–793.
5. Moore, M. J. (2005) From birth to death: the complex lives of eukaryotic mRNAs, *Science 309*, 1514–1518.
6. Anantharaman, V., Koonin, E. V., and Aravind, L. (2002) Comparative genomics and evolution of proteins involved in RNA metabolism, *Nucleic Acids Res 30*, 1427–1464.
7. Filipowicz, W., Bhattacharyya, S. N., and Sonenberg, N. (2008) Mechanisms of post–transcriptional regulation by microRNAs: are the answers in sight?, *Nat Rev Genet 9*, 102–114.
8. Kanitz, A., and Gerber, A. P. (2010) Circuitry of mRNA regulation, *Wiley Interdisp Rev Syst Biol Med 2*, 245–51.
9. Halbeisen, R. E., Galgano, A., Scherrer, T., and Gerber, A. P. (2008) Post-transcriptional gene regulation: from genome-wide studies to principles, *Cell Mol Life Sci 65*, 798–813.
10. Tenenbaum, S. A., Carson, C. C., Lager, P. J., and Keene, J. D. (2000) Identifying mRNA subsets in messenger ribonucleoprotein complexes by using cDNA arrays, *Proc Natl Acad Sci USA 97*, 14085–14090.
11. Hieronymus, H., and Silver, P. A. (2003) Genome-wide analysis of RNA-protein interactions illustrates specificity of the mRNA export machinery, *Nat Genet 33*, 155–161.
12. Inada, M., and Guthrie, C. (2004) Identification of Lhp1p-associated RNAs by microarray analysis in Saccharomyces cerevisiae reveals association with coding and noncoding RNAs, *Proc Natl Acad Sci USA 101*, 434–439.
13. Gerber, A. P., Herschlag, D., and Brown, P. O. (2004) Extensive association of functionally and cytotopically related mRNAs with Puf family RNA-binding proteins in yeast, *PLoS Biol 2*, E79.
14. Hogan, D. J., Riordan, D. P., Gerber, A. P., Herschlag, D., and Brown, P. O. (2008) Diverse RNA-binding proteins interact with functionally related sets of RNAs, suggesting an extensive regulatory system, *PLoS Biol 6*, e255.
15. Gerber, A. P., Luschnig, S., Krasnow, M. A., Brown, P. O., and Herschlag, D. (2006) Genome-wide identification of mRNAs associated with the translational regulator PUMILIO

15. in Drosophila melanogaster, *Proc Natl Acad Sci USA* **103**, 4487–4492.
16. Roy, P. J., Stuart, J. M., Lund, J., and Kim, S. K. (2002) Chromosomal clustering of muscle-expressed genes in Caenorhabditis elegans, *Nature* **418**, 975–979.
17. Schmitz-Linneweber, C., Williams-Carrier, R., and Barkan, A. (2005) RNA immunoprecipitation and microarray analysis show a chloroplast Pentatricopeptide repeat protein to be associated with the 5' region of mRNAs whose translation it activates, *Plant Cell* **17**, 2791–2804.
18. Lopez de Silanes, I., Zhan, M., Lal, A., Yang, X., and Gorospe, M. (2004) Identification of a target RNA motif for RNA-binding protein HuR, *Proc Natl Acad Sci USA* **101**, 2987–2992.
19. Penalva, L. O., Tenenbaum, S. A., and Keene, J. D. (2004) Gene expression analysis of messenger RNP complexes, *Methods Mol Biol* **257**, 125–134.
20. Lopez de Silanes, I., Galban, S., Martindale, J. L., Yang, X., Mazan-Mamczarz, K., Indig, F. E., Falco, G., Zhan, M., and Gorospe, M. (2005) Identification and functional outcome of mRNAs associated with RNA-binding protein TIA-1, *Mol Cell Biol* **25**, 9520–9531.
21. Townley-Tilson, W. H., Pendergrass, S. A., Marzluff, W. F., and Whitfield, M. L. (2006) Genome-wide analysis of mRNAs bound to the histone stem-loop binding protein, *RNA* **12**, 1853–1867.
22. Morris, A. R., Mukherjee, N., and Keene, J. D. (2008) Ribonomic analysis of human Pum1 reveals cis-trans conservation across species despite evolution of diverse mRNA target sets, *Mol Cell Biol* **28**, 4093–4103.
23. Galgano, A., Forrer, M., Jaskiewicz, L., Kanitz, A., Zavolan, M., and Gerber, A. P. (2008) Comparative analysis of mRNA targets for human PUF-family proteins suggests extensive interaction with the miRNA regulatory system, *PLoS One* **3**, e3164.
24. Morris, A. R., Mukherjee, N., and Keene, J. D. (2009) Systematic analysis of posttranscriptional gene expression, *WIREs Syst Biol Med*, DOI: 10.1002/wsbm.54
25. Takizawa, P. A., DeRisi, J. L., Wilhelm, J. E., and Vale, R. D. (2000) Plasma membrane compartmentalization in yeast by messenger RNA transport and a septin diffusion barrier, *Science* **290**, 341–344.
26. Shepard, K. A., Gerber, A. P., Jambhekar, A., Takizawa, P. A., Brown, P. O., Herschlag, D., DeRisi, J. L., and Vale, R. D. (2003) Widespread cytoplasmic mRNA transport in yeast: identification of 22 bud-localized transcripts using DNA microarray analysis, *Proc Natl Acad Sci USA* **100**, 11429–11434.
27. Elson, S. L., Noble, S. M., Solis, N. V., Filler, S. G., and Johnson, A. D. (2009) An RNA transport system in Candida albicans regulates hyphal morphology and invasive growth, *PLoS Genet* **5**, e1000664.
28. Kunitomo, H., Uesugi, H., Kohara, Y., and Iino, Y. (2005) Identification of ciliated sensory neuron-expressed genes in Caenorhabditis elegans using targeted pull-down of poly(A) tails, *Genome Biol* **6**, R17.
29. Penalva, L. O., Burdick, M. D., Lin, S. M., Sutterluety, H., and Keene, J. D. (2004) RNA-binding proteins to assess gene expression states of co-cultivated cells in response to tumor cells, *Mol Cancer* **3**, 24.
30. Heiman, M., Schaefer, A., Gong, S., Peterson, J. D., Day, M., Ramsey, K. E., Suarez-Farinas, M., Schwarz, C., Stephan, D. A., Surmeier, D. J., Greengard, P., and Heintz, N. (2008) A translational profiling approach for the molecular characterization of CNS cell types, *Cell* **135**, 738–748.
31. Halbeisen, R. E., and Gerber, A. P. (2009) Stress-Dependent Coordination of Transcriptome and Translatome in Yeast, *PLoS Biol* **7**, e105.
32. Karginov, F. V., Conaco, C., Xuan, Z., Schmidt, B. H., Parker, J. S., Mandel, G., and Hannon, G. J. (2007) A biochemical approach to identifying microRNA targets, *Proc Natl Acad Sci USA* **104**, 19291–19296.
33. Beitzinger, M., Peters, L., Zhu, J. Y., Kremmer, E., and Meister, G. (2007) Identification of human microRNA targets from isolated argonaute protein complexes, *RNA Biol* **4**, 76–84.
34. Hendrickson, D. G., Hogan, D. J., Herschlag, D., Ferrell, J. E., and Brown, P. O. (2008) Systematic identification of mRNAs recruited to argonaute 2 by specific microRNAs and corresponding changes in transcript abundance, *PLoS One* **3**, e2126.
35. Azuma-Mukai, A., Oguri, H., Mituyama, T., Qian, Z. R., Asai, K., Siomi, H., and Siomi, M. C. (2008) Characterization of endogenous human Argonautes and their miRNA partners in RNA silencing, *Proc Natl Acad Sci USA* **105**, 7964–7969.
36. Landthaler, M., Gaidatzis, D., Rothballer, A., Chen, P. Y., Soll, S. J., Dinic, L., Ojo, T., Hafner, M., Zavolan, M., and Tuschl, T. (2008) Molecular characterization of human Argonaute-containing ribonucleoprotein complexes and their bound target mRNAs, *RNA* **14**, 2580–2596.
37. Zhang, L., Hammell, M., Kudlow, B. A., Ambros, V., and Han, M. (2009) Systematic analysis of dynamic miRNA-target interactions during C. elegans development, *Development* **136**, 3043–3055.

38. Hendrickson, D. G., Hogan, D. J., McCullough, H. L., Myers, J. W., Herschlag, D., Ferrell, J. E., and Brown, P. O. (2009) Concordant regulation of translation and mRNA abundance for hundreds of targets of a human microRNA, *PLoS Biol 7*, e1000238.
39. Mili, S., and Steitz, J. A. (2004) Evidence for reassociation of RNA-binding proteins after cell lysis: implications for the interpretation of immunoprecipitation analyses, *RNA 10*, 1692–1694.
40. Baroni, T. E., Chittur, S. V., George, A. D., and Tenenbaum, S. A. (2008) Advances in RIP-chip analysis : RNA-binding protein immunoprecipitation-microarray profiling, *Methods Mol Biol 419*, 93–108.
41. San Paolo, S., Vanacova, S., Schenk, L., Scherrer, T., Blank, D., Keller, W., and Gerber, A. P. (2009) Distinct roles of non-canonical poly(A) polymerases in RNA metabolism, *PLoS Genet 5*, e1000555.
42. Ule, J., Jensen, K. B., Ruggiu, M., Mele, A., Ule, A., and Darnell, R. B. (2003) CLIP identifies Nova-regulated RNA networks in the brain, *Science 302*, 1212–1215.
43. Ule, J., Jensen, K., Mele, A., and Darnell, R. B. (2005) CLIP: a method for identifying protein-RNA interaction sites in living cells, *Methods 37*, 376–386.
44. Jensen, K. B., and Darnell, R. B. (2008) CLIP: crosslinking and immunoprecipitation of in vivo RNA targets of RNA-binding proteins, *Methods Mol Biol 488*, 85–98.
45. Yeo, G. W., Coufal, N. G., Liang, T. Y., Peng, G. E., Fu, X. D., and Gage, F. H. (2009) An RNA code for the FOX2 splicing regulator revealed by mapping RNA-protein interactions in stem cells, *Nat Struct Mol Biol 16*, 130–137.
46. Wang, Z., Tollervey, J., Briese, M., Turner, D., and Ule, J. (2009) CLIP: construction of cDNA libraries for high-throughput sequencing from RNAs cross-linked to proteins in vivo, *Methods 48*, 287–293.
47. Chi, S. W., Zang, J. B., Mele, A., and Darnell, R. B. (2009) Argonaute HITS-CLIP decodes microRNA-mRNA interaction maps, *Nature 460*, 479–486.
48. Gaillard, H., and Aguilera, A. (2008) A novel class of mRNA-containing cytoplasmic granules are produced in response to UV-irradiation, *Mol Biol Cell 19*, 4980–4992.
49. Keene, J. D., and Tenenbaum, S. A. (2002) Eukaryotic mRNPs may represent posttranscriptional operons, *Mol Cell 9*, 1161–1167.
50. Keene, J. D. (2007) Biological clocks and the coordination theory of RNA operons and regulons, *Cold Spring Harb Symp Quant Biol 72*, 157–165.
51. Licatalosi, D. D., Mele, A., Fak, J. J., Ule, J., Kayikci, M., Chi, S. W., Clark, T. A., Schweitzer, A. C., Blume, J. E., Wang, X., Darnell, J. C., and Darnell, R. B. (2008) HITS-CLIP yields genome-wide insights into brain alternative RNA processing, *Nature 456*, 464–469.
52. Fox, S., Filichkin, S., and Mockler, T. C. (2009) Applications of Ultra-high-Throughput Sequencing, *Methods Mol Biol 553*, 79–108.
53. Ball, C. A., Awad, I. A., Demeter, J., Gollub, J., Hebert, J. M., Hernandez-Boussard, T., Jin, H., Matese, J. C., Nitzberg, M., Wymore, F., Zachariah, Z. K., Brown, P. O., and Sherlock, G. (2005) The Stanford Microarray Database accommodates additional microarray platforms and data formats, *Nucleic Acids Res 33*, D580–582.
54. Tusher, V. G., Tibshirani, R., and Chu, G. (2001) Significance analysis of microarrays applied to the ionizing radiation response, *Proc Natl Acad Sci USA 98*, 5116–5121.
55. Oeffinger, M., Wei, K. E., Rogers, R., DeGrasse, J. A., Chait, B. T., Aitchison, J. D., and Rout, M. P. (2007) Comprehensive analysis of diverse ribonucleoprotein complexes, *Nat Methods 4*, 951–956.
56. Keene, J. D., Komisarow, J. M., and Friedersdorf, M. B. (2006) RIP-Chip: the isolation and identification of mRNAs, microRNAs and protein components of ribonucleoprotein complexes from cell extracts, *Nat Protoc 1*, 302–307.
57. Baldi, P., and Long, A. D. (2001) A Bayesian framework for the analysis of microarray expression data: regularized t-test and statistical inferences of gene changes, *Bioinformatics 17*, 509-519.

Chapter 24

RaPID: An Aptamer-Based mRNA Affinity Purification Technique for the Identification of RNA and Protein Factors Present in Ribonucleoprotein Complexes

Boris Slobodin and Jeffrey E. Gerst

Abstract

RNA metabolism involves regulatory processes, such as transcription, splicing, nuclear export, transport and localization, association with sites of RNA modification, silencing and decay, and necessitates a wide variety of diverse RNA-interacting proteins. These interactions can be direct via RNA-binding proteins (RBPs) or indirect via other proteins and RNAs that form ribonucleoprotein complexes that together control RNA fate. While pull-down methods for the isolation of known RBPs are commonly used, strategies have also been described for the direct isolation of messenger RNAs (mRNAs) and their associated factors. The latter techniques allow for the identification of interacting proteins and RNAs, but most suffer from problems of low sensitivity and high background. Here we describe a simple and highly effective method for RNA purification and identification (RaPID) that allows for the isolation of specific mRNAs of interest from yeast and mammalian cells, and subsequent analysis of the associated proteins and RNAs using mass spectrometry and reverse transcription-PCR, respectively. This method employs the MS2 coat RBP fused to both GFP and streptavidin-binding protein to precipitate MS2 aptamer-tagged RNAs using immobilized streptavidin.

Key words: RNA aptamer, Affinity purification, MS2 coat protein, Streptavidin-binding protein, mRNA localization, RNA–protein complexes, RNA-interacting proteins, Yeast, Mammalian cells

1. Introduction

1.1. mRNA Metabolism

Much evidence has shown that messenger RNAs (mRNAs) have far more complex lives than just bridging between genome-encoded information and proteins. Already during transcription, which is a highly regulated process (1), regulatory factors have been reported to bind mRNAs and control their fate (2, 3). Newly transcribed messages are further subjected to editing and/or splicing events, reviewed in refs. 4 and 5, respectively. After the export from the nucleus, messages are packed into mRNA–protein (mRNP) complexes or particles that are trafficked to distinct locales within the cell, as shown for messages encoding for mitochondrial (6), peroxisomal (7), and ER-localized (8) proteins, as well as factors related to cell fate (9) and polarity (10) determination. These localization events might depend upon the association with cytoskeleton (11–14), cell membranes, reviewed in refs. 15, 16 or even specific transport vesicles (17, 18). During transport, mRNAs are usually associated with proteins that prevent translation (19, 20). Then, when the mRNP granule reaches its ultimate destination, it undergoes changes that allow for anchoring of the granule and either translation, via the attenuation of translational inhibition, or storage (20, 21).

Association with stress granules and P-bodies comprises an additional level of mRNA regulation. Transcripts may associate with stress granules upon stress conditions (22) and, upon the persistence of these conditions, translocate to P-bodies for degradation (23). To add even further complexity, mRNAs may shuttle between stress granules, P-bodies, and polysomes, reviewed in ref. 24, as a part of cellular response to changing demands. Short noncoding RNAs are another class of biologically active players that control mRNA expression. Data regarding these factors has grown dramatically over the past decade and has revealed regulation at multiple levels, such as stability, translation, reviewed in ref. 25 and epigenetic silencing, reviewed in ref. 26. Taking into account that both translation and decay are tightly regulated processes, it becomes clear that throughout its life, from synthesis to decay, a single mRNA molecule is regulated at multiple levels, both by proteins and other RNAs.

1.2. Strategies Used to Isolate Specific mRNAs of Interest

Despite this level of understanding, our knowledge of regulatory proteins and RNAs that control mRNA fate is far from complete. Approaches toward identifying these factors have been employed and consist mainly of two strategies: (1) isolation of a protein known to reside in an mRNP granule and identification of additional bound proteins or RNAs; or (2) isolation of the mRNA of interest and identification of the associated factors. The first strategy has revealed the protein (27–30) and RNA (31–33) composition of several mRNP complexes. The second strategy involving the direct isolation of messages of interest has proven more difficult to achieve. Chemically modified messages have been expressed in cells (34) or incubated with crude cellular extracts (35), and subjected to pull-down procedures and subsequent analysis of the protein content. Although practical, this approach is problematic, since it is hard to ensure the correct localization and regulation of chemically modified transcripts on one hand and avoid nonspecific association with proteins, which could happen during cell breakage (36), on the other hand. To this end, approaches that involve the tagging and pull-down of exogenously expressed transcripts have been used. Small RNA tags, called aptamers (e.g., MS2, S1, D8, etc.), are added to the RNA sequence and allow for mRNA purification using molecules that recognize the aptamer. An aptamer called StreptoTag, fused to RNA sequences derived either from the U1 snRNA or MS2 replicase mRNA, was used to isolate the U1A spliceosomal protein and MS2 bacteriophage coat protein (MS2-CP), respectively, from yeast cell extracts (37). This method was enhanced (38) and used as an mRNA isolation protocol (39). RNA aptamers S1 and D8, which have affinity towards Sephadex and streptavidin, respectively, were artificially selected (40). The S1 aptamer was used to isolate the *PRP1* RNA subunit of yeast RNase P (41) and to identify the association of FXR1 and AGO2 RBPs with AU-rich elements found in the 3 UTRs of certain transcripts in mammalian cells (42). A method, called Ribotrap, used the MS2 aptamer to isolate *ASH1* mRNA and demonstrate a known interaction between the message and the She2 RBP and She3 adaptor protein in budding yeast (43).

1.3. The RaPID Procedure for Identifying Interacting Proteins and RNAs

Here we detail RaPID (*R*NA-binding *p*rotein purification and *id*entification) (44), a highly sensitive procedure that allows for the isolation of specific mRNAs of interest from eukaryotic cells (detailed here for yeast and mammalian cells). RaPID utilizes a novel MS2-CP-GFP-SBP fusion protein that: (1) interacts with an MS2 aptamer-tagged mRNA of interest via its amino terminal MS2-CP moiety; (2) allows for both visualization of mRNA localization in vivo and for Western analysis using anti-GFP antibodies via its GFP reporter moiety; and (3) interacts with high affinity with streptavidin-conjugated matrices via its carboxyl terminal *s*treptavidin *b*inding *p*rotein, or SBP, tag (45). Due to high affinity interactions between MS2-CP and the MS2 aptamer ($K_d = 3$ nM) (46) and between the SBP tag and streptavidin-conjugated media ($K_d = 2.5$ nM) (45), the MS aptamer-tagged mRNA is tightly tethered to the immobilized streptavidin and allows for stringent washing steps that greatly improve the signal-to-noise ratio of the eluted factors.

First, MS2 aptamer-tagged mRNAs are expressed within cells, either exogenously [i.e., expression of the tagged message is driven by a constitutive (e.g., cytomegalovirus) or gene-specific (e.g., β-Actin) promoter] or endogenously (i.e., the MS2 aptamer can be integrated into the genome of yeast) (47). The latter method is preferable, since it utilizes the native level of transcription and lowers the risk of artifacts due to gene over-expression. In contrast, MS2-CP-GFP-SBP expression must be tightly regulated due to its tendency to oligomerize. Although different mutants of the MS2-CP with reduced tendency to oligomerize were discovered (48), we observed that the best way to avoid it is to minimize its expression within individual cells. Therefore, we control the expression of MS2-CP-GFP-SBP under *MET25* promoter in yeast and the T-REx™ system (Invitrogen) in mammalian cells, and calibrated the level of induction in both systems. Following co-expression of the aptamer-tagged message and MS2-CP-GFP-SBP, localization of the mRNP granule is examined in vivo using fluorescence microscopy. This is important in order to determine whether to proceed with the experiment. If the observed localization is correct, the cells are then harvested and cross-linked with formaldehyde to preserve RNA–protein interactions. Following lysis, biotinylated moieties intrinsic to the cell are blocked with free avidin to prevent affinity purification with streptavidin-conjugated media. The mRNA::MS2-CP-GFP-SBP complexes are then pulled-down using streptavidin-conjugated beads, washed thoroughly, and eluted by competition with free biotin, which exhibits a much higher affinity ($K_d = 10^{-15}$ M) to streptavidin. The crosslinks of the eluted material are then reversed, and both protein and RNA fractions can be isolated and subjected to further analyzes. The schematic illustration of the RaPID procedure is presented in Fig. 1.

RaPID was shown to be highly effective for identifying both proteins and RNAs that interact with specific transcripts of interest (44). Although originally designed to identify factors that localize mRNAs in cells, it should be able to identify RNA–protein and RNA–RNA interactions important for other aspects of mRNA metabolism.

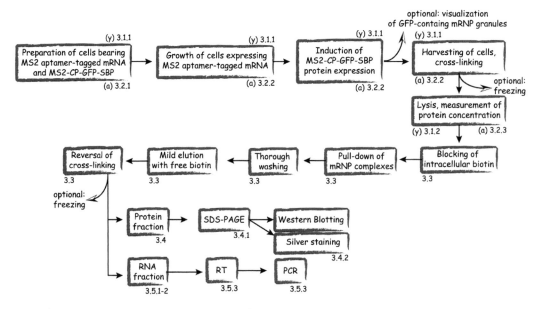

Fig. 1. Schematic flowchart of the RaPID procedure. Each major step is presented as an independent box. Numbers adjacent to the boxes indicate sections where the detailed procedure of the particular step can be found. Label (a) indicates the procedure specific for animal cells, while (y) indicates the procedure specific for yeast cells.

2. Materials

2.1. Culturing and Lysis of Yeast Cells

2.1.1. Preparation and Growth of Yeast Cells

1. Tris–ethylenediaminetetraacetic acid (EDTA) buffer (TE): 10 mM Tris–HCl and 1 mM EDTA in double distilled (or fully deionized) water (DDW), pH 7.5.
2. Lithium acetate (LiOAc) 1 M, prepared in DDW and titrated to pH 7.5 with 2 M acetic acid.
3. LiOAc 0.1 M: Mix 0.1× volumes of 1 M LiOAc with 0.9× volumes of TE buffer.
4. 50% (wt/vol) polyethylene glycol (PEG) 3350: dissolve 250 g of PEG 3350 in 300 ml of TE buffer while stirring and warming to 50°C, fill to 500 ml, filter sterilize.
5. 45% (wt/vol) PEG in TE buffer (PEG 45%): Mix 0.9× volumes of 50% PEG 3350 with 0.1× volumes of 1 M LiOAc.
6. Salmon sperm DNA (ssDNA) 5.0 mg/ml: Sheared organically extracted (i.e., phenol::chloroform (1:1), chloroform), and denatured ssDNA prepared according to standard procedures.
7. DNA for yeast transformation (~1 μg/transformation each): (a) MS2-CP-GFP-SBP plasmid (available from authors upon request). See ref. 44 for a description of the plasmid. (b) Plasmid expressing the MS2 aptamer-tagged mRNA of interest.
8. Synthetic growth medium (SC) lacking essential amino acids to allow for the selective growth of yeast. See Chapter 15 or 20 for further details on the preparation of yeast growth media.
9. Formaldehyde 37% (wt/vol), stabilized with 10–15% (vol/vol) methanol. Caution: Formaldehyde is a harmful compound and should be used with caution in a hood while wearing necessary protective wear.
10. 1 M glycine solution at pH 7.0.

2.1.2. Yeast Cell Lysis

1. Yeast lysis buffer: 20 mM Tris–HCl at pH 7.5, 150 mM NaCl, 0.5% (vol/vol) Triton X-100, 1.8 mM $MgCl_2$; prepared using RNase-free water.
2. Complete yeast lysis buffer: supplement Yeast lysis buffer with the following agents (to be freshly added before use): 10 μg/ml aprotinin (Sigma, Cat. A1153), 1 mM phenylmethanesulfonylfluoride (PMSF, Sigma, Cat. P7626), 10 μg/ml Pepstatin A (Sigma, Cat. P4265), 10 μg/ml Leupeptin (Sigma, Cat. L0649), 10 μg/ml Soybean trypsin inhibitor (Sigma, Cat. T9003), 1 mM dithiothreitol (DTT), and 80 U/ml RNAsin® ribonuclease inhibitor (Promega, Cat. N2515).

3. Glass beads, 0.5 mm in diameter (Biospec products, Cat. 11079-105).
4. BCA protein assay kit (Pierce, Cat. 23225).

2.2. Culturing and Lysis of Mammalian Cells

2.2.1. Preparation of Cell Line Stably Expressing the Inducible MS2-CP-GFP-SBP

1. Zeocin™ (InvivoGen, Cat. ant-zn-1).
2. Manually cut and autoclaved Whatmann (3 mm thickness) paper disks, 5–6 mm diameter.
3. Transfection reagent, allowing for an efficient transfection of the particular cells used. For HEK293, we usually use a calcium phosphate-based transfection protocol prepared manually according to standard protocols; alternatively, jetPEI™ transfection reagent (Polyplus transfection, Cat. 101-10) can be used according to manufacturer's instructions.
4. DNA for transfection: pcDNA4/TO-MS2-CP-GFP-SBP plasmid (see ref. 44; available from authors upon request), 10 µg/100 mm dish.
5. The medium best suitable for the growth of the particular cell line should be used. For HEK293 cells, we use DMEM medium (Gibco, Cat. 41965) supplemented with 10% (vol/vol) fetal bovine serum and 5% (vol/vol) Pen-Strep-Ampho solution (Biological Industries, Israel. Cats. 04-001-1A and 03-033-1B, respectively).
6. 0.05% Trypsin–EDTA solution (Biological Industries, Israel. Cat. 03-053-1B).

2.2.2. Growth of the Animal Cells in Culture

1. Growth medium, as listed in Subheading 2.2.1, item 5.
2. DNA for transfection: An expression vector encoding the MS2 aptamer-tagged mRNA of interest. If the cells do not stably express the inducible MS2-CP-GFP-SBP, they should also be transfected with pcDNA4/TO-MS2-CP-GFP-SBP plasmid (44) as well. We usually do not transfect more than 10–12 µg of total DNA per 10 cm dish.
3. Tetracycline (100 µg/ml stock solution; filter sterilized). Store at 4°C for up to 1 month.
4. Formaldehyde solution as detailed in Subheading 2.1.1, item 9.
5. Glycine solution as detailed in Subheading 2.1.1, item 10.

2.2.3. Lysis of Mammalian Cells

1. Mammalian lysis buffer: 50 mM Tris–HCl at pH 7.5, 150 mM NaCl, 1% (v/v) Triton X-100, and 1.8 mM $MgCl_2$; prepared in RNase-free water.
2. Complete mammalian lysis buffer: supplement the mammalian lysis buffer with Protease inhibitor cocktail for mammalian cells and tissue extracts (Sigma, Cat. P8340), 1 mM DTT, and 80 U/ml RNAsin (as in Subheading 2.1.2) before use.
3. Polystyrene round-bottom 5-ml tubes (sterile).

4. Ultrasonic cell disruptor equipped with microprobe to fit microfuge tubes (e.g., Microson XL2007).
5. BCA protein assay kit (Pierce, Cat. 23225).

2.3. RaPID for the Pull-Down of mRNA–Protein Complexes

1. Avidin solution, 1 mg/ml prepared in phosphate-buffered saline (PBS); store at 4°C (Sigma, Cat. A9275).
2. Streptavidin-conjugated Sepharose™ beads: (GE Healthcare, Cat. 17-5113-01).
3. Sterile 15-ml polypropylene centrifuge tubes.
4. Yeast tRNA, 10 mg/ml solution (Sigma, Cat. R8508).
5. BSA 4% solution, prepared from BSA fraction V (Sigma, Cat. A7906).
6. Washing buffer: 20 mM Tris–HCl at pH 7.5, 300 mM NaCl, and 0.5% (vol/vol) NP-40.
7. Biotin (Sigma, Cat. B4501), 0.2 M stock solution in dimethylsulfoxide (DMSO); store at 4°C for up to 2 weeks. For elution, prepare a 6 mM solution in PBS prewarmed to 37°C. This solution should be prepared directly before use and be well mixed.
8. 2× Cross-link reversal buffer: 100 mM Tris–HCl at pH 7.0, 10 mM EDTA, 20 mM DTT, and 2% (wt/v) sodium dodecylsulfate (SDS) (32).
9. 5× Protein sample buffer: 400 mM Tris–HCl at pH 6.8, 50% (v/v) glycerol, 10% (wt/v) SDS, and 0.5% (wt/v) Bromophenol blue. Store at room temperature. Before use, add fresh 0.5 M DTT or 5% (v/v) β-mercaptoethanol. Warning! DTT and β-mercaptoethanol may be harmful upon inhalation or skin contact; use hood while preparing buffers containing these agents.

2.4. Analysis of Precipitated Protein

1. Standard equipment for sodium dodecylsulfate-polyacrylamide gel electrophoresis (SDS-PAGE) experiments.
2. Sterile box capable to accommodate acrylamide gel.
3. Sterile DDW.
4. Silver Stain Kit (Pierce, Cat. 24612).
5. Sterile surgical blades for the excision of protein bands from gel.

2.5. Analysis of Precipitated RNA

1. MPC protein precipitation reagent (a component of the MasterPure™ Yeast RNA purification kit; Epicentre Biotechnologies, Cat. MPY03100).
2. Glycogen 20 mg/ml solution (Fermentas, Cat. R0561).
3. 3 M NaOAc solution, pH 5.2 (Fermentas, Cat. R1181).
4. Ultra-pure water (Biological Industries, Israel. Cat. 01-866-1B).

5. RQ1 RNase-Free DNase (Promega, Cat. M6101).
6. M-MLV Reverse Transcriptase (Promega, Cat. M1701).
7. Random hexamer primers, 100 μm solution (Fermentas, Cat. SO142).
8. dNTPs mixture solution 5 mM, made from 100 mM stocks of dATP, dCTP, dGTP, and dTTP (Fermentas Cat. R0141, R0151, R0161 and R0171, respectively) using Ultra-pure water. Store at −20°C.

3. Methods

For a step-by-step overview of the RaPID procedure, see Fig. 1.

3.1. Yeast

3.1.1. Growth of Yeast Cultures

1. Transformation of yeast is done using a standard LiOAc-based procedure (for details and instructions see Chapter 15). Each strain should be transformed with plasmids expressing the MS2 aptamer-tagged mRNA of interest and MS2-CP-GFP-SBP, and grown on selective media.

2. Inoculate a single colony picked from the transformation dish into a 50-ml test tube with 7–8 ml of selective medium and grow in a shaking incubator at 26°C for 6–10 h.

3. Measure absorbance of the culture at A_{600} and transfer an amount of the grown culture into a 2-l Erlenmeyer flask containing 400 ml of the selective growth medium. It is advisable to calculate the initial amount of the culture in a way that following overnight incubation, the culture will not exceed O.D. = 1.0 at A_{600}. Incubate overnight in an incubator with shaking at 26°C. For additional information, see Note 1.

4. Measure absorbance of the culture at A_{600} and, if needed, keep growing the culture up to O.D. = 1.0 at A_{600}. Collect the cells by centrifugation using a SLA3000-type rotor at $1,100 \times g$ for 5 min in a Sorvall centrifuge and discard the growth medium. Resuspend the cell pellet in 10 ml of a fresh growth medium lacking methionine.

5. Add the cells to 200 ml of selective growth medium lacking methionine in an Erlenmeyer flask and grow with shaking at 26°C for 60–75 min to induce the expression of the MS2-CP-GFP-SBP. The precise incubation time should be determined experimentally. Visualize the expression of the MS2-CP-GFP-SBP and presence of the mRNP granules by fluorescent microscopy using excitation at ~488 nm and filter allowing for GFP visualization. For additional information, see Note 2 and Table 1.

Table 1
Problems associated with the GFP signal

Description	Potential cause	Suggested solution
GFP signal is undetectable in the mammalian cells after tetracycline treatment	(a) pcDNA4/TO-MS2-CP-GFP-SBP did not respond to the tetracycline induction	• Prepare a fresh tetracycline solution • Increase the final concentration of the tetracycline added to the medium and/or exposure time • Use transient transfection of the pcDNA4/TO-MS2-CP-GFP-SBP plasmid
The mRNA granules are undetectable in yeast cells following methionine starvation	(a) Note that most, but not necessarily all transcripts form detectable mRNP granules using this method. If not, the problem is probably due to low levels of tagged mRNA expression	• Use MS2-CP-GFP(×3), which is capable to localizing endogenously expressed transcripts in yeast (46) • Use either a stronger promoter, multicopy expression plasmids, or both for expression of the tagged mRNA

6. Collect the cells by centrifugation using a Sorvall centrifuge (as in step 4) and discard the growth medium. Add 10 ml of PBS, resuspend the cells, and transfer them to a 50-ml test tube. Centrifuge at $960 \times g$ for 3 min and discard the supernatant. Add 10 ml of a freshly prepared PBS-0.01% formaldehyde solution, resuspend the cells by pipetting (DO NOT VORTEX) and incubate at room temperature with slow shaking for 10 min.

7. To terminate the cross-linking reaction, add glycine solution to a final concentration of 0.125 M (32) and incubate for an additional 2 min. Centrifuge the cells (as in step 6), discard the supernatant, and wash once with PBS.

8. Discard the PBS and quickly freeze the cell pellet using liquid nitrogen. For storage, transfer the cells to a −80°C freezer. The cells may be stored for prolonged periods under these conditions.

3.1.2. Yeast Cell Lysis

IMPORTANT: The steps described in this subheading should be done on ice or at 4°C.

1. Place the frozen cell pellet on ice and thaw by adding 4 ml of complete yeast lysis buffer, with occasional vortexing.

2. When a homogenized suspension is achieved, transfer aliquots of 0.5 ml of the cell suspension to microfuge tubes prefilled with ~0.5 ml (vol) of glass beads. Vortex using an IKA/Vibrax shaker located in the cold room (4°C) at maximum

speed for total of 40 min. Turn off the shaker every 10 min and let stand for 5 min to prevent excessive heating.

3. Centrifuge microfuge tubes in a table-top centrifuge pre-cooled to 4°C at $960 \times g$ for 2 min. Transfer the supernatant into fresh microfuge tubes and centrifuge again at $15,300 \times g$ for 10 min.

4. Collect the supernatants from each set of samples into a 14-ml test tube. Measure the protein concentration in a small aliquot (see Note 3). We do not advise to freeze the protein extract since salting out of proteins upon high concentration might occur.

5. Proceed to the RaPID pull-down step (see Subheading 3.3).

3.2. Mammalian Cells

3.2.1. Preparation of Cell Lines Stably Expressing Inducible MS2-CP-GFP-SBP

1. While isolation of an mRNA of interest is possible after transient transfection of the MS2-CP-GFP-SBP construct into mammalian cells, it is advisable to use a stable cell line that expresses MS2-CP-GFP-SBP in an inducible fashion. This alone will save a considerable amount of DNA, transfection reagents, and will allow for the controlled induction of MS2-CP-GFP-SBP expression, thus, avoiding protein aggregation and/or nonspecific RNA–protein interactions resulting from over-expression.

2. Since MS2-CP-GFP-SBP was originally constructed using the pcDNA4/TO vector (Invitrogen), it should be integrated into a cell line that stably expresses the tetracycline repressor (TR). Such cell lines can be purchased from Invitrogen or manually prepared by stable integration of the pcDNA6/TR vector into the cell genome. To do so, refer to the instructions that appear at the Invitrogen website.

3. Seed the cells stably expressing the TR on a dish and transfect them with the pcDNA4/TO-MS2-CP-GFP-SBP plasmid using a transfection protocol suitable for the cell line. After 24 h, change the medium and add Zeocin at the minimal effective concentration (see Note 4). Change the medium every 2 days to discard dead cells and grow in presence of the antibiotic until resistant colonies appear.

4. Transfer the colonies into separate wells of a 24-well tissue culture plate using sterile Whatman disks soaked in the 0.05% trypsin solution and propagate in the presence of Zeocin until the cells cover the bottom of the well. Then, discard the Whatman disks and transfer the cells into a larger dish. Grow until a sufficient amount of the cells that are resistant to antibiotic is achieved.

5. Test the clones for the inducible expression of MS2-CP-GFP-SBP protein. Seed the cells into a 6-well tissue culture plate and add different concentrations of tetracycline to the medium

(see Note 5). Incubate for 10–12 h, collect the cells, and analyze the total cellular extract using Western blotting with anti-GFP antibodies. If the expression of MS2-CP-GFP-SBP (molecular mass = ~48 kDa) is responsive to tetracycline addition and undetected when tetracycline is absent, this clone is suitable for further applications.

6. Freeze clones for future use.

3.2.2. Growth and Transfection of Mammalian Cells in Culture

This protocol is written for HEK293-TRex cells with pcDNA4/TO-MS2-CP-GFP-SBP stably integrated into their genome; however, it should fit nearly any cell line that can be prepared as described in Subheading 3.2.1.

1. Split growing cells and seed them on 100-mm dishes at ~30–40% confluence. For calculation of the number of cells necessary, see Note 1.

2. After the cells attach to the surface and reach ~60–70% confluency (usually within 18–24 h following seeding), transfect them with a MS2 aptamer-tagged mRNA of interest (8–12 μg of plasmid DNA/100-mm dish) using standard calcium phosphate-based protocol or other techniques that allow for high efficiency of transfection.

3. If using the calcium phosphate-based protocol, change the medium 6–8 h after addition of the transfection solution with fresh prewarmed culture medium. Add tetracycline (use a final concentration of 100 ng/ml or at any other concentration that yields a moderate level of protein expression, see Note 5) to induce the expression of the MS2-CP-GFP-SBP. Incubate for 12–16 h in a controlled temperature and CO_2 incubator. Following incubation, the GFP signal should become visible upon visualization with fluorescent microscopy, as described in Subheading 3.1.1, step 5. If the signal is undetectable, see Table 1.

4. Place the dishes on ice and wash once with an ice-cold PBS solution. Be careful not to detach the cells from the dish surface. Carefully aspirate the PBS solution.

5. Add 2–3 ml of the ice-cold PBS to each dish and collect the cells using a cell scraper. Transfer the cells into 15-ml tubes and centrifuge at $250 \times g$ for 2 min. Carefully aspirate the supernatant.

6. Add 10 ml of freshly prepared PBS-0.01% (vol/vol) formaldehyde solution (42), dissolve the cell pellet by pipetting, and incubate with slow shaking for 10 min at room temperature.

7. To terminate the cross-linking reaction, add glycine to a final concentration of 0.125 M (32) and incubate for an additional 2 min. Centrifuge the cells as in step 5 and aspirate the supernatant.

8. Wash once with cold PBS, centrifuge as in step 5. Aspirate the supernatant.

9. Quickly freeze the cell pellet in liquid nitrogen and store at −80°C or proceed to the lysis step.

3.2.3. Lysis of Mammalian Cells

1. Add 1 ml of ice-cold complete mammalian lysis buffer per 2.5×10^6 of frozen cells (but not less than 1 ml overall) and thaw on ice for 5 min with occasional vortexing.

2. Transfer the resuspended cells to 5 ml glass tubes and sonicate on ice for three rounds using a sonicator; each round lasting 15 s at 7–9 W, with 2 min pauses on ice in between rounds (32). Be careful not to overheat samples.

3. Transfer the sonicated cells to microfuge tubes and pellet debris by centrifugation at $15,300 \times g$ for 15 min at 4°C. Remove the supernatant to a fresh tube. Proceed to protein concentration measurement (see Note 3) and further to the RaPID pull-down step (see Subheading 3.3). We do not recommend freezing the extract at this point.

3.3. RaPID Pull-Down of mRNA–Protein Complexes

1. Transfer the protein extract destined for RaPID to a fresh 15-ml tube, add the avidin solution (10 μg of avidin per 1 mg of protein extract) (40) and incubate at 4°C for 1 h with constant rotation/shaking. In contrast to intracellular biotin and biotinylated moieties, MS2-CP-GFP-SBP does not possess any affinity toward avidin (44). Thus, avidin can be used to block undesired biotin–streptavidin interactions during the process.

2. In parallel, aliquot the streptavidin-conjugated beads (use 5 μl of the slurry per 1 mg of added protein extract, but not less than 30 μl overall) to fresh tubes (see Note 6); wash the beads twice with 1–2 ml of ice-cold PBS and once in lysis buffer; centrifuge $660 \times g$ for 2 min and discard the supernatant between washes. Fill equal volumes of lysis buffer and BSA solution (the total volume should be not less than 1 ml), add tRNA (0.1 mg per 100 μl of beads) and incubate at 4°C with constant rotation for 1 h. Centrifuge the tubes as described above and wash twice with 1 ml of lysis buffer.

3. Discard the buffer and add the avidin-blocked total cell extract from step 1 to the beads along with 0.1 mg of yeast tRNA and incubate at 4°C for 2–15 h with constant rotation. Addition of the tRNA to the pull-down reaction reduces nonspecific interactions.

4. After the pull-down reaction, centrifuge the tubes at $660 \times g$ for 2 min and remove the supernatant. Wash the beads three times with 2 ml of lysis buffer and twice with washing buffer. All steps should be performed at 4°C, each step lasting for 10 min and with rotation.

5. Wash the beads with 2 ml of ice-cold PBS, centrifuge 660 ×g for 2 min, and remove the excess buffer. For elution of the mRNA–protein complexes from the beads, add 80–150 µl of the biotin elution solution to the beads and incubate for 1 h at 4°C with rotation.

6. Centrifuge at 660 ×g for 2 min, transfer the eluate into a fresh microfuge tube, centrifuge once again, and transfer the eluate into a new tube to assure that no beads were carried over. It is advisable to use gel-loading tips or syringes in order to avoid transfer of beads. Separate the amounts destined for the isolation of RNA and protein into different tubes. See Note 7.

7. To reverse the cross-linking in the fraction destined for RNA isolation, add an equal volume of the cross-link reversal buffer to the eluate and incubate at 70°C for 1–2 h (32). Spin the solution and either freeze at −20°C or proceed directly to RNA isolation (see Subheading 3.5.1).

8. To reverse the cross-linking in the fraction destined for protein analysis, add an appropriate volume of the 5× protein sample buffer to reach the 1× concentration and incubate at 70°C for 1–2 h. Spin and either freeze at −20°C or proceed directly to SDS-PAGE separation (see Subheading 3.4.1).

3.4. Analysis of Precipitated Protein

3.4.1. SDS-PAGE Separation

1. The volume of the eluate taken for separation on an SDS-PAGE gel should fit the capacity of the wells inside the gel. Medium-sized homemade thick gels (20 cm × 15 cm, BioRad) can usually accommodate up to 250 µl per lane. We suggest using gels of this size or larger when identifying unknown proteins since the separation ability of such a gel is higher than that of standard mini-gels (7 cm × 8 cm, BioRad).

2. If the gel separation and subsequent Western blotting are employed for identification of expected proteins (e.g., using antibodies), it is more convenient to use mini-gels. To do this, the loading volume should not exceed ~60 µl per well. If the eluate volume is larger, it should be reduced using one of the following methods: (a) using a SpeedVac: Care should be taken to avoid over-drying when reducing the volume. Do not heat the sample during the incubation in the SpeedVac; (b) employing standard protein precipitation techniques using trichloroacetic acid or acetone (49), followed by resuspension in a set volume of 1× sample buffer. In all cases, the aliquots of eluate should be boiled for 5 min following these procedures and immediately prior to gel loading and electrophoresis.

3. Load the samples onto SDS-PAGE gels. We found that 9% gels gave the best separation over a wide range of proteins; however, this might vary depending upon the size of the bound proteins. Alternatively, a gradient gel might be employed.

Run the gel until satisfactory separation of the proteins is achieved and proceed to silver staining (see Subheading 3.4.2) or Western blotting using standard techniques.

3.4.2. Silver Staining of Gels

1. Open the gel box, remove the upper (stacking) gel and transfer the lower (separating) gel into a sterile container. We use 15-cm round sterile dishes, but any box capable to accommodate the gel will fit. Use gloves to avoid contamination. Tip: Pouring a small volume of sterile DDW onto the gel before picking it up will help avoid any tearing.

2. Cover the gel with sterile DDW and incubate for 5 min while slowly rotating.

3. Silver stain the gel. We use the Silver Stain Kit (Pierce), which is fully compatible with subsequent mass-spectrometry analysis. When following the manufacturer's instructions, this kit provides fast and sensitive staining with a reasonable signal-to-noise ratio.

4. Analyze the staining results. If specific bands are detected, excise them with a sterile blade and send for identification. Solutions for potential problems at this step can be found in Table 2.

Table 2
Problems associated with silver staining

Description	Potential cause	Suggested solution
Too many protein bands appearing in the gel following staining	(a) Nonspecific protein binding to the beads	• Wash the beads thoroughly following the pull-down. Increase the concentration of NaCl in the washing buffer up to 500 mM • Increase the concentration of avidin (see Subheading 3.3, step 1) twofold
	(b) MS2-CP-GFP-SBP associates nonspecifically with proteins due to high levels of expression	• Reduce the concentration of formaldehyde or the length of the cross-linking process • Reduce the level of MS2-CP-GFP-SBP expression using a shorter induction time
No specific bands were identified in the gel after staining	(a) Interactions between the RNA and protein were disrupted	• Increase cross-linking either by adding a higher concentration of formaldehyde, prolonging the time of cross-linking, or both • Lyse the cells using a more gentle method, such as grinding liquid N_2 frozen yeast cells, see ref. 50

3.5. Analysis of Precipitated RNA

3.5.1. Isolation of RNA from the Eluate

This step may be performed by using kit-derived reagents, as described here, or, alternatively, by standard phenol–chloroform extraction procedures (49).

1. Thaw (or place) the eluate fraction for RNA isolation on ice. Vortex and spin using a precooled centrifuge at ≥10,000×g for 1 min.
2. Add 175 μl of MPC protein precipitation reagent to each 300 μl of eluate and vortex vigorously for 10 s.
3. Centrifuge for 10 min at 4°C at ≥10,000×g.
4. Carefully transfer the supernatant into a new microfuge tube and centrifuge again for 5 min at 4°C at ≥10,000×g. Transfer the supernatant into a fresh tube. This step is optional and ensures that the supernatant is devoid of pelleted protein aggregates.
5. Add glycogen (80 μg/ml final conc.) and NaOAc (0.3 M final conc.), vortex thoroughly for 20 s. Add an equal volume of isopropanol, mix by inverting the tubes ~20 times, and incubate overnight at –20°C.
6. Centrifuge the tubes as in step 3. The pellet containing the RNA should be visible. Carefully discard the supernatant.
7. Wash the pellet by adding 0.5 ml of 70% ice-cold ethanol, vortex, and centrifuge for 5 min at 4°C at ≥10,000×g. Carefully discard the supernatant and spin again. Aspirate the excess volume of ethanol using a gel-loading tip.
8. Heat the tube in a dry bath at 50°C for 10 min with an open cap. This step is required to assure that all the ethanol will evaporate. However, do not over-dry, since this will make the pellet hard to dissolve.
9. Dissolve the pellet in 30 μl of Ultra-pure water. Vortex for 10 s, spin, and heat at 50°C for 5–10 min to achieve complete solubility. Store the dissolved RNA at –20°C or directly proceed to the DNase treatment step. RNA may be stored at –20°C for several weeks or even months. However, for longer storage, use a –80°C freezer.

3.5.2. DNase Treatment of the Isolated RNA

1. If frozen, thaw the dissolved RNA on ice. Add 3.7 μl of 5× DNase reaction buffer and 3 μl of DNase, and incubate for 1.5–2 h at 37°C. We use the DNase reaction buffer at a final concentration of 0.5× in order to avoid excessive amounts of $MgSO_4$ that may negatively affect the PCR. For further information, please refer to the manufacturer's instructions.
2. To stop the reaction and inactivate the DNase enzyme, add 4 μl of the Stop solution and incubate at 65°C for 10 min. Spin and freeze, or proceed to the reverse transcription step.

3.5.3. Reverse Transcription and Polymerase Chain Reaction (RT-PCR)

1. Aliquot 10 μl of the eluted and DNase-treated RNA to a fresh tube, add 1.3 μl of the random hexamer mixture and fill with Ultra-pure water to final volume of 15.4 μl. Vortex and spin.
2. Incubate at 70°C for 5 min and immediately remove to ice.
3. Prepare a mix including 5 μl of 5× RT reaction buffer, 2.6 μl of a 5 mM stock of dNTPs, and 1 μl of RT enzyme per tube. Aliquot into the tubes containing the RNA annealed to the hexamer primers. Mix by vortexing and spin.
4. Incubate at 25°C for 10 min, at 45°C for 25 min, at 42°C for 25 min, and at 70°C for 15 min in a step-wise manner, preferably using a PCR machine to ensure accurate temperature and incubation time. The tubes may then be stored at −20°C.
5. Prepare the PCR reactions as following: Add 7.5 μl of PCR mix, 0.75 μl of the 5 μM stock of each primer, 1 μl of the RT product, and 5.5 μl of Ultra-pure water per tube. Run the PCR and analyze the products on 1% agarose gels using standard agarose gel electrophoresis. See Table 3 for possible solutions to problems linked to poor RNA recovery.

Table 3
Low yields of MS2-tagged mRNA isolation

Description	Potential cause	Suggested solution
The aptamer-tagged mRNA is undetectable in the RNA fraction of the eluate	(a) The aptamer-tagged mRNA did not express	• Check the expression of the aptamer-tagged mRNA in the total RNA sample • If working with the mammalian cell cultures, check the transfection efficiency. Use more DNA for transfection if necessary
	(b) The aptamer-tagged mRNA was degraded at its distal 5′ and 3′ ends due to its instability	• Use primers that delimit a short and central part of the mRNA • Add more RNase inhibitors to the lysis buffer and be punctilious to perform all the steps of the procedure at 4°C • Shorten the time of the pull-down to 2 h
	(c) The aptamer-tagged mRNA was completely degraded due to RNAse contamination	• Check the lab equipment for RNase contamination • Prepare all buffers in DEPC-treated DDW or, preferably, in commercially available DDW having no RNase activity

4. Notes

1. The amount of the protein extract needed for the procedure varies depending on the desirable downstream application (detection of RNA, protein, or both). For example, we used yeast cultures of up to 1 l to obtain 80–100 mg of crude extract for the identification of novel RNA-interacting proteins in yeast; however larger culture volumes might be needed to help isolate poorly expressed target proteins. We have successfully used 5–15 mg of total protein extract for the identification of known proteins in Westerns and 2–10 mg of total protein extract for the analysis of RNA by RT-PCR. Thus, the culture volume of yeast or the number of dishes containing animal cells can vary, depending upon the desired downstream application, as well as the efficiency of both the lysis and transfection steps.

2. An ideal level of induction allows for clear visualization of the RNP granules and a relatively weak background of GFP fluorescence. Prolonged induction (i.e., >90 min in our yeast strains) can lead to excessive expression of the MS2-CP-GFP-SBP and, possibly, to a higher level of background of precipitated proteins using RaPID.

3. The concentration of total cellular protein should be accurately measured in order to provide an equal amount of cell extract for each RaPID reaction and allow for reproducibility. Taking an equal amount of cells for each reaction does not necessarily mean an equal amount of the extracted protein, thus, measurement of the protein concentration before performing the pull-down reaction is advisable. Since the lysates used for this procedure are usually concentrated, we dilute them 1:10 in order to accurately determine the protein concentration.

4. The minimal effective concentration is the lowest concentration of antibiotics that causes the death of cells that do not express the resistance gene encoded by the transfected plasmid. If the sensitivity of a particular cell line to Zeocin is unknown, it might be revealed by testing the viability of the untransfected cells grown in presence of different concentrations of the antibiotic.

5. Our clones (derived from HEK293 cells) demonstrated an induction of MS2-CP-GFP-SBP protein expression upon exposure to as little as 50 ng/ml of tetracycline. However, different clones might behave slightly differently. In any case, one should keep in mind that low-moderate levels of MS2-CP-GFP-SBP expression are required in order to avoid aggregation.

6. We used standard 1.7-ml conical microfuge tubes when working with small volumes of input or 15-ml sterile polypropylene conical centrifuge tubes when working with larger volumes.

7. It is also possible to isolate the RNA from the entire volume of the eluate, as detailed in Subheading 3.5.1, and to perform protein extraction from the leftover pellet (after RNA extraction) using acetone (49). However, when the identification of RNA-interacting proteins is the main goal, we prefer to separate the proteins directly from the eluate to avoid the loss of low molecular mass proteins during extraction.

Acknowledgements

We thank Pravinkumar Purushothaman and Johannes Koch for the critical reading of this manuscript. This work was supported by a grant to J.E.G. from the Yeda CEO Fund, Weizmann Institute of Science, Israel, and Minerva Foundation, Germany. J.E.G. holds the Besen-Brender Chair of Microbiology and Parasitology, Weizmann Institute of Science, Israel.

References

1. Hager, G. L., McNally, J. G., and Misteli, T. (2009) Transcription dynamics. *Mol. Cell* 35, 741–753.
2. Pan, F., Huttelmaier, S., Singer, R. H., and Gu, W. (2007) ZBP2 facilitates binding of ZBP1 to beta-actin mRNA during transcription. *Mol. Cell. Biol.* 27, 8340–8351.
3. Shen, Z., Paquin, N., Forget, A., and Chartrand, P. (2009) Nuclear shuttling of She2p couples ASH1 mRNA localization to its translational repression by recruiting Loc1p and Puf6p. *Mol. Biol. Cell* 20, 2265–2275.
4. Skarda, J., Amariglio, N., and Rechavi, G. (2009) RNA editing in human cancer: review. *Apmis.* 117, 551–557.
5. Toor, N., Keating, K. S., and Pyle, A. M. (2009) Structural insights into RNA splicing. *Curr. Opin. Struct. Biol.* 19, 260–266.
6. Sylvestre, J., Vialette, S., Corral Debrinski, M., and Jacq, C. (2003) Long mRNAs coding for yeast mitochondrial proteins of prokaryotic origin preferentially localize to the vicinity of mitochondria. *Genome Biol.* 4, R44.
7. Zipor, G., Haim-Vilmovsky, L., Gelin-Licht, R., Gadir, N., Brocard, C., and Gerst, J. E. (2009) Localization of mRNAs coding for peroxisomal proteins in the yeast, Saccharomyces cerevisiae. *Proc. Natl. Acad. Sci. USA* 106, 19848–19853.
8. Lerner, R. S., Seiser, R. M., Zheng, T., Lager, P. J., Reedy, M. C., Keene, J. D., and Nicchitta, C. V. (2003) Partitioning and translation of mRNAs encoding soluble proteins on membrane-bound ribosomes. *RNA* 9, 1123–1137.
9. Bertrand, E., Chartrand, P., Schaefer, M., Shenoy, S. M., Singer, R. H., and Long, R. M. (1998) Localization of ASH1 mRNA particles in living yeast. *Mol. Cell* 2, 437–445.
10. Aronov, S., Gelin-Licht, R., Zipor, G., Haim, L., Safran, E., and Gerst, J. E. (2007) mRNAs encoding polarity and exocytosis factors are cotransported with the cortical endoplasmic reticulum to the incipient bud in Saccharomyces cerevisiae. *Mol. Cell. Biol.* 27, 3441–3455.
11. Bohl, F., Kruse, C., Frank, A., Ferring, D., and Jansen, R. P. (2000) She2p, a novel RNA-binding protein tethers ASH1 mRNA to the Myo4p myosin motor via She3p. *EMBO J.* 19, 5514–5524.
12. Farina, K. L., Huttelmaier, S., Musunuru, K., Darnell, R., and Singer, R. H. (2003) Two ZBP1 KH domains facilitate beta-actin mRNA localization, granule formation, and cytoskeletal attachment. *J. Cell. Biol.* 160, 77–87.
13. Takizawa, P. A., and Vale, R. D. (2000) The myosin motor, Myo4p, binds Ash1 mRNA via the adapter protein, She3p. *Proc. Natl. Acad. Sci. USA* 97, 5273–5278.

14. Tekotte, H., and Davis, I. (2002) Intracellular mRNA localization: motors move messages. *Trends Genet.* **18**, 636–642.
15. Cohen, R. S. (2005) The role of membranes and membrane trafficking in RNA localization. *Biol. Cell* **97**, 5–18.
16. Gerst, J. E. (2008) Message on the web: mRNA and ER co-trafficking. *Trends Cell Biol.* **18**, 68–76.
17. Trautwein, M., Dengjel, J., Schirle, M., and Spang, A. (2004) Arf1p provides an unexpected link between COPI vesicles and mRNA in Saccharomyces cerevisiae. *Mol. Biol. Cell* **15**, 5021–5037.
18. Bi, J., Tsai, N. P., Lu, H. Y., Loh, H. H., and Wei, L. N. (2007) Copb1-facilitated axonal transport and translation of kappa opioid-receptor mRNA. *Proc. Natl. Acad. Sci. USA* **104**, 13810–13815.
19. Gu, W., Deng, Y., Zenklusen, D., and Singer, R. H. (2004) A new yeast PUF family protein, Puf6p, represses ASH1 mRNA translation and is required for its localization. *Genes Dev.* **18**, 1452–1465.
20. Huttelmaier, S., Zenklusen, D., Lederer, M., Dictenberg, J., Lorenz, M., Meng, X., Bassell, G. J., Condeelis, J., and Singer, R. H. (2005) Spatial regulation of beta-actin translation by Src-dependent phosphorylation of ZBP1. *Nature* **438**, 512–515.
21. Paquin, N., Menade, M., Poirier, G., Donato, D., Drouet, E., and Chartrand, P. (2007) Local activation of yeast ASH1 mRNA translation through phosphorylation of Khd1p by the casein kinase Yck1p. *Mol. Cell* **26**, 795–809.
22. Kedersha, N., and Anderson, P. (2002) Stress granules: sites of mRNA triage that regulate mRNA stability and translatability. *Biochem. Soc. Trans.* **30**, 963–969.
23. Kedersha, N., Stoecklin, G., Ayodele, M., Yacono, P., Lykke-Andersen, J., Fritzler, M. J., Scheuner, D., Kaufman, R. J., Golan, D. E., and Anderson, P. (2005) Stress granules and processing bodies are dynamically linked sites of mRNP remodeling. *J. Cell. Biol.* **169**, 871–884.
24. Balagopal, V., and Parker, R. (2009) Polysomes, P bodies and stress granules: states and fates of eukaryotic mRNAs. *Curr. Opin. Cell Biol.* **21**, 403–408.
25. Bartel, D. P. (2009) MicroRNAs: target recognition and regulatory functions. *Cell* **136**, 215–233.
26. Wassenegger, M. (2005) The role of the RNAi machinery in heterochromatin formation. *Cell* **122**, 13–16.
27. Jonson, L., Vikesaa, J., Krogh, A., Nielsen, L. K., Hansen, T., Borup, R., Johnsen, A. H., Christiansen, J., and Nielsen, F. C. (2007) Molecular composition of IMP1 ribonucleoprotein granules. *Mol. Cell. Proteomics* **6**, 798–811.
28. Villace, P., Marion, R. M., and Ortin, J. (2004) The composition of Staufen-containing RNA granules from human cells indicates their role in the regulated transport and translation of messenger RNAs. *Nucleic Acids Res.* **32**, 2411–2420.
29. Oeffinger, M., Wei, K. E., Rogers, R., DeGrasse, J. A., Chait, B. T., Aitchison, J. D., and Rout, M. P. (2007) Comprehensive analysis of diverse ribonucleoprotein complexes. *Nat. Methods* **4**, 951–956.
30. Hock, J., Weinmann, L., Ender, C., Rudel, S., Kremmer, E., Raabe, M., Urlaub, H., and Meister, G. (2007) Proteomic and functional analysis of Argonaute-containing mRNA-protein complexes in human cells. *EMBO Rep.* **8**, 1052–1060.
31. Gilbert, C., and Svejstrup, J. Q. (2006) RNA immunoprecipitation for determining RNA-protein associations in vivo. *Curr. Protoc. Mol. Biol.* Chapter **27**, Unit 27.4.
32. Niranjanakumari, S., Lasda, E., Brazas, R., and Garcia-Blanco, M. A. (2002) Reversible cross-linking combined with immunoprecipitation to study RNA-protein interactions in vivo. *Methods* **26**, 182–190.
33. Keene, J. D., Komisarow, J. M., and Friedersdorf, M. B. (2006) RIP-Chip: the isolation and identification of mRNAs, microRNAs and protein components of ribonucleoprotein complexes from cell extracts. *Nature Protoc.* **1**, 302–307.
34. Zielinski, J., Kilk, K., Peritz, T., Kannanayakal, T., Miyashiro, K. Y., Eiriksdottir, E., Jochems, J., Langel, U., and Eberwine, J. (2006) In vivo identification of ribonucleoprotein-RNA interactions. *Proc. Natl. Acad. Sci. USA* **103**, 1557–1562.
35. Gerber, A. P., Luschnig, S., Krasnow, M. A., Brown, P. O., and Herschlag, D. (2006) Genome-wide identification of mRNAs associated with the translational regulator PUMILIO in Drosophila melanogaster. *Proc. Natl. Acad. Sci. USA* **103**, 4487–4492.
36. Mili, S., and Steitz, J. A. (2004) Evidence for reassociation of RNA-binding proteins after cell lysis: implications for the interpretation of immunoprecipitation analyses. *RNA* **10**, 1692–1694.
37. Bachler, M., Schroeder, R., and von Ahsen, U. (1999) StreptoTag: a novel method for the isolation of RNA-binding proteins. *RNA* **5**, 1509–1516.
38. Dangerfield, J. A., Windbichler, N., Salmons, B., Gunzburg, W. H., and Schroder, R. (2006) Enhancement of the StreptoTag method for isolation of endogenously expressed proteins

with complex RNA binding targets. *Electrophoresis* 27, 1874–1877.

39. Windbichler, N., and Schroeder, R. (2006) Isolation of specific RNA-binding proteins using the streptomycin-binding RNA aptamer. *Nature Protoc.* 1, 637–640.

40. Srisawat, C., and Engelke, D. R. (2002) RNA affinity tags for purification of RNAs and ribonucleoprotein complexes. *Methods* 26, 156–161.

41. Srisawat, C., and Engelke, D. R. (2001) Streptavidin aptamers: affinity tags for the study of RNAs and ribonucleoproteins. *RNA* 7, 632–641.

42. Vasudevan, S., and Steitz, J. A. (2007) AU-rich-element-mediated upregulation of translation by FXR1 and Argonaute 2. *Cell* 128, 1105–1118.

43. Beach, D. L., and Keene, J. D. (2008) Ribotrap: targeted purification of RNA-specific RNPs from cell lysates through immunoaffinity precipitation to identify regulatory proteins and RNAs. *Methods Mol. Biol.* 419, 69–91.

44. Slobodin, B., and Gerst, J. E. (2010) A novel mRNA affinity purification technique for the identification of interacting proteins and transcripts in ribonucleoprotein complexes. *RNA* 16, 2277–90.

45. Keefe, A. D., Wilson, D. S., Seelig, B., and Szostak, J. W. (2001) One-step purification of recombinant proteins using a nanomolar-affinity streptavidin-binding peptide, the SBP-Tag. *Protein Expr. Purif.* 23, 440–446.

46. Lim, F., and Peabody, D. S. (1994) Mutations that increase the affinity of a translational repressor for RNA. *Nucleic Acids Res.* 22, 3748–3752.

47. Haim, L., Zipor, G., Aronov, S., and Gerst, J. E. (2007) A genomic integration method to visualize localization of endogenous mRNAs in living yeast. *Nat. Methods* 4, 409–412.

48. Peabody, D. S., and Ely, K. R. (1992) Control of translational repression by protein-protein interactions. *Nucleic Acids Res.* 20, 1649–1655.

49. Sambrook, J., and Russell, D. W. (2001) Molecular Cloning: A Laboratory manual, Third Edition. *Cold Spring Harbor Laboratory Press, New York.*

50. Lopez de Heredia, M., and Jansen, R. P. (2004) RNA integrity as a quality indicator during the first steps of RNP purifications: a comparison of yeast lysis methods. *BMC Biochem.* 5, 14.

Chapter 25

RIP: An mRNA Localization Technique

Sabarinath Jayaseelan, Francis Doyle, Salvatore Currenti, and Scott A. Tenenbaum

Abstract

A detailed understanding of post-transcriptional gene expression is necessary to correlate the different elements involved in the many levels of RNA–protein interactions that are needed to coordinate the cellular biomolecular machinery. The profile of mRNA, a major component of this machinery, can be examined after isolation from specific RNA-binding proteins (RBPs). RIP-Chip or *ribonomic* profiling is a versatile *in vivo* technique that has been widely used to study post-transcriptional gene regulation and the localization of mRNA. Here we elaborately detail the methodology for *m*RNA isolation using *R*BP *i*mmuno*p*recipitation (RIP) as a primary approach. Specific antibodies are used to target RBPs, which are then used to capture the associated mRNA.

Key words: Ribonomics, mRNA localization, RNA-binding protein, RIP-Chip, RNA coprecipitation

1. Introduction

The past decade has seen an explosion of research in the field of post-transcriptional gene regulation and has begun to compete with transcriptional regulation, which had been the traditional focus of gene expression. The significance of post-transcriptional gene regulation became evident when it was shown that the transcriptome was frequently not directly correlated with its corresponding proteome (1–3). This is a result of the significant role post-transcriptional/translational processes play in the overall regulation of protein expression. Technological advances in the genome-wide and targeted gene expression analysis have also enhanced our understanding of several cellular systems and the significance post-transcriptional mechanisms play in them (4–8).

Immunoprecipitation (IP) of RNA-binding proteins (RBPs) followed by microarray analysis popularly known as

RIP-Chip (3, 9, 10), can be a powerful *in vivo*, high-throughput technique for identifying specific associated RBP targets from cell extracts. This method compliments other RNA localization techniques that can have some limitations and is widely used to isolate and identify mRNA and miRNA targets (5, 9–16). This chapter will focus on providing methodological detail on how endogenous ribonucleoprotein (RNP) complexes can be isolated using RIP and then the co-purified RNA can be extracted from these complexes. Magnetic beads coated with protein A or G are incubated to bind with a specific antibody which can target an RNP complex. This method can also target recombinant RBPs in a robust *in vitro* version of this method (5).

2. Materials

2.1. Preparation of Cell Lysate

1. Polysome Lysis Buffer (1× PLB): 100 mM KCl, 5 mM $MgCl_2$, 10 mM HEPES at pH 7.00, 0.5% Nonidet P-40 (also known as Igepal CA-630; Sigma Cat. No. I8896). A 10× stock buffer is prepared prior to time of use, and may be stored at room temperature. 1 mM dithiothreitol (DTT), 200 units/ml RNase OUT (Invitrogen Cat. No. 10777-019), one Complete Mini, EDTA-free Protease Inhibitor Tablet (Roche Cat. No. 11836170001) are added at time of use. RNase free water (Ambion Cat. No. 9932) is used to raise the extraction buffer final volume to 7 ml.

2. The protease inhibitor cocktail tablet allows inhibition of serine and cysteine proteases during extractions from human cells. When preparing larger volumes of 1× PLB for cellular extraction, add additional protease inhibitor cocktail tablets as necessary.

2.2. Buffers Required for Immunoprecipitation

1. NT-2 buffer: 150 mM Tris–HCl at pH 7.0, 100 mM Tris-HCl at pH 8.0, 750 mM NaCl, 5 mM $MgCl_2$, and 0.25% Nonidet P-40 (Igepal CA-630). Typically, a 5× concentration of NT-2 stock buffer is prepared. A 1× concentration of NT-2 buffer is utilized for the duration of the RIP protocol. NT-2 buffer is stored between 2 and 8°C.

2. NET-2 buffer (binding buffer for RNP to antibody): 1× NT-2 buffer supplemented with 20 mM EDTA at pH 8.0 (Ambion Cat. No. AM9260G), 1 mM DTT, 200 units/ml RNase OUT.

3. Proteinase K digestion buffer: 1× NT-2 buffer supplemented with 1% sodium dodecyl sulfate (SDS) and 1.2 mg/ml Proteinase K (Ambion Cat. No. AM2546).

4. 1× PLB preparation: 1 mM DTT, 200 units/ml RNase OUT, one Complete Mini, EDTA-free Protease Inhibitor Tablet, 0.7 ml 10× PLB stock, and raise to 7 ml with RNase free water.

2.3. Immunoprecipitation Reaction

1. Dynabeads Protein A (Invitrogen Cat. No. 100-02D) and Dynabeads Protein G (Invitrogen Cat. No. 100-04D).

2. 5 μg of the antibody to the RNA binding protein of interest.

3. For RNA precipitation: (1) acid phenol–chloroform at pH 4.5, with isoamyl alcohol (IAA), (Ambion Cat. No. 9722); (2) chloroform (Fisher Scientific Cat. No. BP1145-1) and 5 M ammonium acetate (Ambion Cat. No. 9070G); (3) 7.5 M lithium chloride (Ambion Cat. No. 9480); (4) 5 mg/ml glycogen (Ambion Cat. No. 9570); (5) 100% ethanol; (6) 80% ethanol.

4. A magnetic rack (Invitrogen Cat. No. 123-21D or Millipore Cat. No. 20-400) is used to separate Dynabeads during immunoprecipitation. A Labquake™ tube shaker/rotator (Krackeler Cat. No. 16-4002110) is used for mixing during antibody-bead incubation, and following the addition of cell lysate (4°C overnight incubation).

5. Buffer NT-2 is used to wash the beads during immunoprecipitation, and a Chemical Duty Pump (Millipore Cat. No. WP6111560) is used to aspirate supernatant/waste.

2.4. SDS–PAGE and Western Blotting for Proteins of Interest

1. Loading buffer: 2× Laemmli buffer (Bio-Rad Cat. No. 161-0737) and 2-mercaptoethanol electrophoresis grade (Fisher Scientific Cat. No. BP176-100).

2. SDS–PAGE buffer: 10× stock, 10× Tris/Glycine/SDS (TGS) (Bio-Rad Cat. No. 161-0772).

3. Gels: 4–20% Tris–HCl (Bio-Rad Cat. No. 161-1159) and 7.5% Tris–HCl (Bio-Rad Cat. No. 161-1172).

4. SDS–PAGE ladder: Precision Plus Protein WesternC Standards (Bio-Rad Cat. No. 161-0376).

5. Transfer buffer: 100 ml of 10× Tris/Glycine (TG) (Bio-Rad Cat. No. 161-0771), 200 ml of methanol (Fisher Scientific Cat. No. A452-4), and 700 ml of deionized water. Keep transfer buffer cold at 4°C prior to use.

6. Membrane: Immobilon-P^{SQ} transfer membrane and PVDF 0.2 μm (Millipore Cat. No. ISEQ07850).

7. Blocking buffer: 5% non-fat milk in TBST (1 l of 10× TBST: 1.37 M NaCl, 250 mM Tris–HCl pH 8.0, 0.5% Tween-20, and titrate HCl to pH of 7.5).

8. Ponceau red: Stock consists of 2% Ponceau-S in 30% trichloroacetic acid and 30% sulfosalicylic acid. Dilute the Ponceau-S stock 1:10.

9. Primary antibody incubation buffer: Starting Block T20 (Pierce Cat. No 37543), and dilute the antibody of interest to a concentration of 1:1,000–1:2,000.

10. Secondary antibody incubation buffer: Starting Block T20 (Pierce Cat. No 37543), and dilute the appropriate secondary antibody-horseradish peroxidase (HRP) conjugate 1:10,000. The same dilution applies to Precision Protein StrepTactin-HRP Conjugate (Bio-Rad Cat. No. 161-0380), and is added along with the secondary antibody of interest. The Precision Protein StrepTactin-HRP Conjugate allows for the resolution of all reference bands with sizes ranging from 10 to 250 kDa.

2.5. ONE-HOUR Complete IP-Western Kit (Rapid Method Used to Identify Precipitated Protein of Interest with High Sensitivity and Minimized Background in Comparison to Conventional Western Blotting)

1. ONE-HOUR IP-Western Kit (for Rabbit primary antibody, Genscript Cat. No. L00231), ONE-HOUR IP-Western Kit (for Mouse primary antibody, Genscript Cat. No. L00232), ONE-HOUR IP-Western Kit (for Goat primary antibody, Genscript Cat. No. L00233).

2. Follow ONE-HOUR Complete IP-Western Kit Technical Manual No. 0218, Version 06192009.

3. Methods

The RIP method described here has been successfully tested on mammalian cells including stem cells, for the isolation of RNA integrally associated with RBPs. Iron oxide magnetic beads linked to specific antibodies through protein A or G were used to target the specific RBPs. A general outline of the RIP procedure is shown in Fig. 1.

3.1. Preparation of Cell Lysate

1. Cell type chosen for study may be dependent on the biological question asked in the experiment as well as the presence of the protein of interest. Typically 5×10^6 to 1×10^7 cells are used to make lysate enough for a single RIP. This may vary based on factors specific to any given experiment.

2. One volume of 1× PLB buffer is added to an equal volume of the frozen cell pellet and kept in ice for thawing. Adding the PLB buffer prior to thawing inhibits any protease activity and improves the quality of the lysate (see Notes 1–3).

3. After complete thaw, vortex the mixture vigorously (see Note 3). Split them into aliquots and store them at −80°C till further use.

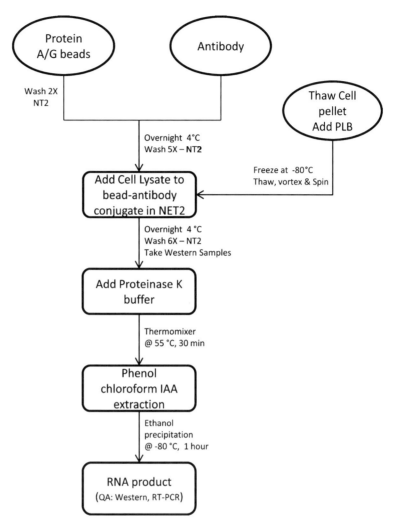

Fig. 1. Simplified process flow for RIP.

3.2. Immobilization of Antibodies on Magnetic beads

1. Prepare a 1:1 stock mixture of protein A and protein G coated magnetic beads and store at 4°C (see Notes 4 and 5).

2. As the beads tend to aggregate and settle down, vortex vigorously and take 75 µl of the protein A/G slurry in a 1.5-ml microfuge tube. Wash twice with 0.5 ml of NT-2 buffer (see Note 6). The buffer is removed by placing the vial against a magnet and using a vacuum aspiration system to remove the clear liquid (see Note 7). There will be at least two tubes for each experiment – one negative control (i.e., T7 antibody, pre-immune serum, etc.) and one experimental condition (antibody to RBP of interest). A positive control can also be included (U1-70k, PABP, etc.)

3. Resuspend the beads in 100 µl of NT-2 (optional: NT-2 with 5% bovine serum albumin) and add 5 µg antibody.

4. Vortex and then tumble the vials in rotator for a minimum of 1 h at room temperature. Preferably, the incubation should be done at 4°C overnight for maximal binding.

3.3. mRNA Localization Using RIP

1. Centrifuge the bead/antibody slurry for 15 s at $5,000 \times g$. Aspirate supernatant keeping the vial against the magnet. Add 1 ml of NT-2, mix, spin at $5,000 \times g$ for 15 s. Aspirate supernatant. Repeat this step five more times for a total of six washes.

2. During the previous step, thaw the cell lysate on ice, vortex vigorously, and centrifuge at $20,000 \times g$ for 10 min at 4°C and then keep in ice.

3. Resuspend the beads in 900 µl of NET-2 buffer. Add 100 µl of lysate (clear upper layer) to each IP mixture (see Note 8). Invert to mix and give it a quick spin. Keeping the sample on magnet, remove 100 µl and place it into a new tube. This is called the "total." Tumble the IP tubes at 4°C overnight (see Notes 9 and 10).

4. Wash six times with 1 ml of ice cold NT-2. If high background is observed, wash four times with NT-2 followed by two washes with NT-2 having 0.5–3 M urea (10, 17). Before you spin down the beads for the sixth wash, take out 100 µl of the bead slurry (out of 1 ml total volume) for IP Western (see Notes 11 and 12). Also take out 10 µl from "total," add 10 µl of 2× SDS–PAGE buffer and save it in −20°C for Western blotting (see Notes 13–25). An example western analysis is shown in Fig. 2.

3.4. RNA Purification

1. Resuspend the beads in 150 µl of Proteinase K buffer. To each "total" tube add 36 µl of NT-2 + 15 µl of 10% SDS + 9 µl of Proteinase K. Keep all the IP tubes and total tubes at 55°C for 30 min (use Eppendorf Thermomixer© R dry block heating and cooling shaker) (while incubating, spin down the aliquot taken for IP Western, remove supernatant and resuspend the beads in 30 µl of 1× SDS–PAGE buffer, store at −20°C).

2. After 30 min, give the tubes a quick spin to bring everything down, add an equal volume of (150 µl) buffer-saturated acid phenol–chloroform (pH 4.5) with isoamyl alcohol, vortex to mix and centrifuge at $20,000 \times g$ for 10 min. Remove aqueous (upper) layer carefully without disturbing the beads and proteins settled at the interface, and place in a new tube. Add 150 µl of chloroform, vortex, and spin.

3. To the supernatant add 50 µl 5 M ammonium acetate, 15 µl of 7.5 M LiCl, 5 µl of 5 mg/ml glycogen, and 1 ml of cold 100% ethanol. Keep in −80°C for at least half hour.

4. Spin at $20,000 \times g$ for 30 min at 4°C. Decant ethanol, leaving the RNA pellet. Wash once with 80% ethanol, spin again

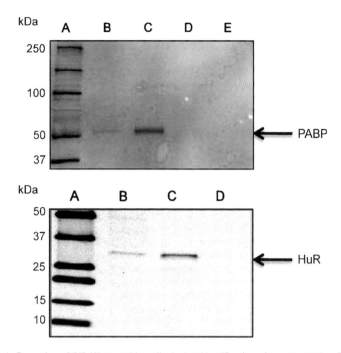

Fig. 2. Examples of RIP-Western blots displaying identification of target proteins. Each blot was resolved using Genscript ONE-HOUR IP-Western kits. Shown in the *top panel*, is a RIP from human HeLa cervical cancer cell line, using Poly (A)-binding protein (PABP) antibody (Millipore Cat. No. 05-847). Lanes within this image (*on top*) are: (A) Precision Plus WesternC Protein Standard; (B) total obtained during RIP; (C) PABP protein RIP; (D) RIP using Normal Mouse IgG (Millipore Cat. No. 12-371); (E) RIP using Anti-T7-Tag® Monoclonal antibody (Novagen Cat. No. 69522). Shown in the *bottom panel*, is a RIP from the GM12878 lymphoblastoid cell line using embryonic lethal, abnormal vision, Drosophila-like 1 protein (HuR), antibody (Santa Cruz Cat. No. sc-5261). Lanes are (image in the *bottom panel*): (A) Precision Plus WesternC Protein Standard; (B) total RNA obtained during RIP; (C) HuR protein RIP; (D) Anti-T7-Tag® Monoclonal antibody RIP.

for 30 min. Dry in Vacufuge™ concentrator (5 min at RT) and resuspend the pellet in 10 ml of RNase-free water or any other buffer of choice. The quality of RNA can be validated using RT-PCR as shown in Fig. 3. The PCR was performed using gene specific primers for GAPDH and β-actin. GAPDH mRNA does not bind to HuR, however, it does have a poly(A) tail and should bind to PABP. Therefore we expect to see enrichment for GAPDH in the PABP IP, but not in the HuR IP. Beta-actin is a known HuR target and is also polyadenylated, therefore enrichment is expected in both IPs.

5. RNA yield obtained can be expected to vary based on multiple factors including the proportion of transcribed RNA that interacts with the studied protein and the expression level of that protein in the given cells. Amount of starting material required is dependent on the specific microarray type to be used and its associated protocol. Determination of sufficient

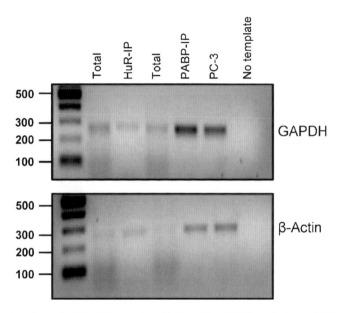

Fig. 3. The figure depicts a RIP experiment followed by RT-PCR readout on a 1.5% agarose gel of immunoprecipitations targeting HuR and PABP in K562 cell line. The PCR was performed using gene specific primers for GAPDH (*top*) and β-actin (*bottom*). mRNA from prostate carcinoma (PC-3) cells were used as the positive control. The last lane shows RT-PCR done with no template.

quantity to meet this need can be assessed with a NanoDrop spectrophotometer and confirmed as part of the quality control process at a microarray core facility.

3.5. Sample Preparation for Microarray Application

The additional use of cDNA microarrays for the analysis of RNA isolated in the described manner has been termed RIP-Chip. Various types of arrays are available with differences in both manufacturing technology and probe profile. The choice of a particular microarray platform will affect further preparation protocols and amount of starting material required. Various fee for service facilities can perform these steps as well as process the microarrays (some facilities optionally include initial bioinformatics analysis as well).

3.6. Microarray Bioinformatic Analysis

The bioinformatic portion of RIP-Chip begins with the processing of the microarray raw data (i.e., probe intensity values) to produce either a set of genes or alternatively, a list of transcriptional fragments if Tiling-arrays or deep-sequencing is used. Gene level arrays contain sets of probes designed against sequence from portions of exons in well-annotated genes. Tiling-arrays contain probes for all genomic sequence in a range of interest, regardless of annotation or expected transcriptional activity.

3.6.1. Gene Array Processing

There are a number of options in manufacturer and model of gene level microarrays as well as tools for their analysis. We frequently use "Human Gene 1.0 ST Array" from Affymetrix and Agilent's

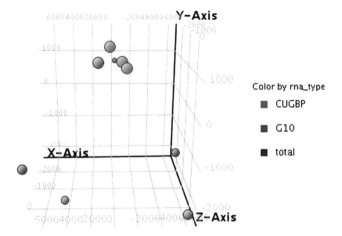

Fig. 4. Example of principal component analysis of samples from a CUGBP (CELF1) RIP. PCA provides an indication of similarity amongst samples. Note the CUGBP replicates cluster better as an enriched subset than either total transcribed RNA or the G10 controls (which are expected to primarily consist of random background signal).

GeneSpring GX10 software. A typical approach for analyzing these arrays is to gather a minimum of triplicate sets for each of total input RNA, treatment (i.e., using antibody for RBP of interest) RIP RNA, and negative control RIP RNA (i.e., antibody for the G10 protein from T7 bacteriophage). Set members are subjected to basic quality control, which involves comparing correlation values and principal component analysis (Fig. 4) among replicates.

The iterative PLIER16 algorithm is then used to gather expression results in each group, which are then filtered for the 20th to 100th percentile. PLIER16 is preferred as it has been shown to be the most consistent with RT-PCR results when preprocessing Affymetrix microarray data (18). When compared with multiple popular algorithms, PLIER16 has also shown superior performance in reducing false positives when analyzing poorly performing probe sets (19).

A list is generated using genes that show a minimum twofold increase of measured expression in treatment versus input sets. A separate list is created for genes that show a similar level of increase in the negative control versus input sets. This second list is then subtracted from the first to remove RNAs that appear to have a general propensity for inclusion by the RIP process (perhaps due to an affinity for the beads).

A one-way analysis of variance (ANOVA) is performed on this RNA subset. Since microarray expression comparison can be viewed as essentially a set of many experiments, each with an associated null hypothesis a multiple testing correction is applied to the ANOVA p-values (for instance, the Benjamini–Hochberg method (20)). The final set of genes is then filtered based on a desired confidence level, such as 95% ($p = 0.05$).

3.6.2. Tiling-Array Processing

Whereas gene level arrays consist of probe sets engineered to test various segments of exons in well-annotated transcripts, Tiling-arrays offer an unbiased interrogation of the genome by providing probes against all portions of a targeted sequence examined by the array. RIP combined with Tiling-arrays allows the potential to examine RBPs acting on introns and noncoding RNA outside the boundaries of traditionally annotated mRNA transcripts.

Affymetrix GeneChip ENCODE 2.0R Arrays are typically used for Tiling-based experiments. These arrays target the ENCODE sequence, accounting for 1% of the human genome (21). Analysis can be performed using Affymetrix's Tiling-Analysis Software (TAS version 1.1.02) in combination with their Integrated Genome Browser (IGB).

Typically, replicate groups similar to those in the gene level array experiments are produced. The TAS software has less advanced quality control analysis, but does allow for basic sample to sample comparisons (Fig. 5). Quantile normalization with linear scaling is applied to control (total input) and treatment groups (RIP). A two-sample analysis is performed by TAS using the Wilcoxon signed rank test and a 68-base bandwidth. Transcriptional fragments (Fig. 6) are generated for each of the RBP and negative

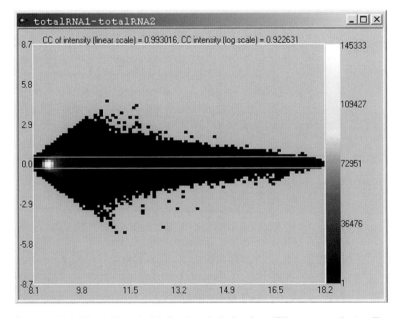

Fig. 5. An MvA (Bland-Altman) plot showing similarity of two Tiling array replicates. The plot's x-axis is the average intensity of a given probe (across replicates) and the y-axis is its difference in value. Because of the number of probes, these plots are also color coded for density. Here, we can see that the bulk of probes are in the same intensity range with minimal disparity across samples.

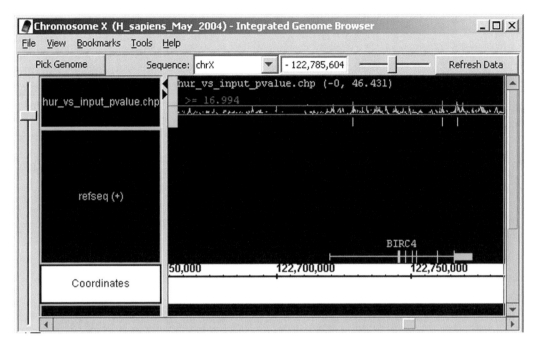

Fig. 6. Excerpt from an Integrated Genome Browser window showing HuR RIP based transfrags in two introns and the 3′ UTR of BIRC4 (XIAP), a gene that has been shown to be stabilized by HuR (25). The intron hits are interesting as HuR's (ELAVL1) drosophila homolog has been shown to affect alternative splicing by binding to introns in pre-mRNA (26).

RIP groups by including probes that meet a cut off p-value and account for a minimum run of 30 bases with a maximum gap of 60 bases. Sets are compared and fragments from the negative control RIP that overlap any RBP RIP fragments are removed from the RBP set. At the time of writing, TAS does not have an inherent means of applying a multiple testing correction.

4. Notes

1. When preparing the 1× PLB buffer, it is important to prepare a fresh buffer to ensure proper cell lysis. The main determinant of the duration of stability of the 1× PLB buffer is the activity of the protease inhibitor tablet. A solution containing a protease inhibitor tablet(s) is stable for 1–2 weeks when stored at 2–8°C, or a maximum of 12 weeks at −15 to −25°C. In most cases, the entire volume of 1× PLB is used while preparing cellular lysate.

2. One protease inhibitor tablet is sufficient for the inhibition of proteolytic activity in 10 ml of extraction buffer (1× PLB). In instances where high proteolytic activity may be present (usually assumed to be the case in unknown), one tablet is used per 7 ml of extraction buffer.

3. Cell pellets utilized for lysate preparation are free of media and have been washed with phosphate buffered saline (PBS) prior to freezing at −80°C. During cell lysate preparations, the cell pellets are allowed to thaw minimally on ice, followed by the addition of an equal volume of 1× PLB. The cell pellet-PLB mixture is occasionally vortexed to promote thawing and subsequently commence lysis. Once the cells appear fully thawed, vortex vigorously to ensure proper cell lysis, and aliquot and store at −80°C accordingly. Poor vortexing may result in low total RNA yields when conducting immunoprecipitations.

4. For general precautions on working with RNA, all instruments, items, tips, and cell lysates should be DNase-free and RNase-free. Gloves, benches, and pipettes may be cleansed using RNase Zap (Ambion Cat. No 9780) or RNase Away (Molecular BioProducts Cat. No 7001), followed by air-drying.

5. The type of beads utilized depends on immunoglobulin isotype and species. It is important to consult the bead manufacturers binding chart to determine the optimal choice of beads for the particular isotype and species of antibody being used. In most instances, mouse monoclonal antibodies have strong affinity for Dynabeads Protein G and rabbit polyclonal antibodies have strong affinity for both Dynabeads Protein G and Dynabeads Protein A.

6. When washing beads in NT-2 after cell lysate incubation (rotating at 4°C), be sure to vortex samples vigorously to eliminate potential background. This may be visualized by obtaining a Western Blot positive result for a "negative" control precipitation, due to poor vortexing.

7. While aspirating supernatants, be sure to change tips between samples and "negative" controls to avoid carryover contamination. Also, when aspirating, place the tip opposite of the bead aggregate to prevent disruption and potential contamination among immunoprecipitation samples.

8. When adding cell lysate (following a 10 min, 20,000 × g spin at 4°C) to immunoprecipitation samples, it is important to leave the resulting cellular pellet undisturbed. Only use the lysate ("upper") layer. If there is any lysate remaining, it may be saved for later use. When reusing cell lysate, it is important to allow thawing on ice followed by vigorous vortexing, and a 10 min, 20,000 × g spin at 4°C (used Eppendorf centrifuge model 5417R).

9. During immunoprecipitation, it is essential not to "pool" IP samples to prevent contaminating viable samples. Each individual precipitation must remain separate.

10. Dynabeads Protein A and Dynabeads Protein G can be mixed (Protein A + Protein G stock), washed with NT-2, and stored at 4°C until ready to use.

11. It is important to obtain an aliquot during the sixth wash (prior to Proteinase K digestion), to test the efficiency of immunoprecipitation by Western Blotting. Proteins are eluted off Dynabeads using 1× SDS–PAGE loading buffer followed by heating at 95°C. The beads can be centrifuged down at 8,000 × g for 2 min (once allowed to briefly cool), and supernatant directly applied to an SDS–PAGE. When testing an antibody for the first time, it is important to verify that the protein of interest is being precipitated during immunoprecipitation. In some instances, it may be necessary to try several antibodies before selecting an antibody capable of precipitating the target RBP of interest.

12. Millipore commercially supplies EZ-Magna RIP™ (Millipore Cat. No. 17-701), a universal RIP kit, which enables immunoprecipitation using antibodies directed against RBP's to be conducted. The kit supplies control antibodies and primers allowing for RT-PCR validation of precipitated mRNA.

13. When using Immobilon-P^{SQ} transfer membrane, PVDF 0.2 μm, always remember to soak the membrane for 10 s in methanol prior to transfer, followed by rinsing in transfer buffer to ensure proper protein transfer. When using nitrocellulose membranes, do not soak in methanol prior to use. It is beneficial to soak nitrocellulose membranes in transfer buffer briefly before preparing the gel sandwich. PVDF membranes have high protein binding capacity, physical strength, and stability (22).

14. In some instances, especially when transferring multiple blots using the same power supply, it is appropriate to extend the time of transfer.

15. When attempting to resolve high molecular weight proteins (>100 kDa), reducing the methanol percentage to 10% will allow large proteins to travel easier.

16. When resolving low molecular weight proteins (<100 kDa), keep the methanol concentration at 20%.

17. A 4–20% Tris–HCl gel resolves proteins of 10–250 kDa, and is usually utilized when conducting an SDS–PAGE electrophoresis. In some instances, it is beneficial to use a 7.5% Tris–HCl gel to resolve proteins of high molecular weight (>100 kDa). A 7.5% Tris–HCl gel is typically used to resolve a protein(s) of molecular weight(s) between 25 and 200 kDa.

18. Always remember to use a positive control lysate (usually the lysate used for immunoprecipitation) to demonstrate protocol efficiency and verify the antibody is capable of identifying

target protein, which may not be present in experimental (immunoprecipitation) samples. If desired, you may load a control with a known molecular weight (i.e., β-actin, GAPDH, Tubulin) to verify that lanes in the gel have been evenly loaded with sample and it is also beneficial when comparing protein expression levels between samples.

19. After creating the gel sandwich prior to transfer, it is important to remove possible air bubbles from within. This can be done by rolling a cylindrical tube over the sandwich. Air bubbles tend to disrupt protein transfer and will minimize transfer efficiency.

20. When conducting protein transfer, it is important to create a cold environment (it increases the efficiency of protein transfer). Ice chips can be used around the outside of the gel transfer unit or the transfer can be performed in a refrigerator to keep the transfer at 4°C.

21. To verify protein transfer (optional), Ponceau Red is added to the membrane and incubated while rocking for 5 min and wash with deionized water until the protein bands are well defined. A digital image may be obtained and used for comparison after antibody detection.

22. In some instances, primary antibodies will produce stronger signal when diluted in blocking buffer (5% non-fat milk in TBST), rather than TBST alone (23, 24). This is not recommended in cases when high background may be an issue.

23. For some antibodies, signal is stronger when primary antibody incubation is conducted at room temperature rather than 4°C. However, it is recommended to incubate the primary antibody at 4°C to minimize nonspecific binding and reduce possible bacterial contamination and subsequent protein destruction. Again, this preference varies among antibodies.

24. Membranes may be stored in a 1× TBST buffer, supplemented with 0.2% sodium azide (NaN_3) or may be dried, covered in filter paper, and stored for later use.

25. The Genscript ONE-HOUR Complete IP-Western Kit is often used to resolve target protein following protein transfer. The kit provides high sensitivity, low background, rapid protein identification, and no secondary antibody is required.

Acknowledgments

We wish to thank the members of the Tenenbaum Lab and specifically Arthur Beauregard for helpful editorial comments. This work was supported in part by NIH/NHGRI Grant 5U01HG004571 to SAT.

References

1. Futcher, B., Latter, G. I., Monardo, P., McLaughlin, C. S., and Garrels, J. I. (1999) A sampling of the yeast proteome. *Mol Cell Biol.* **19**, 7357–7368.

2. Gygi, S. P., Rochon, Y., Franza, B. R., and Aebersold, R. (1999) Correlation between protein and mRNA abundance in yeast. *Mol Cell Biol.* **19**, 1720–1730.

3. Tenenbaum, S. A., Carson, C. C., Lager, P. J., and Keene, J. D. (2000) Identifying mRNA subsets in messenger ribonucleoprotein complexes by using cDNA arrays. *Proc Natl Acad Sci U S A.* **97**, 14085–14090.

4. Perou, C. M., Jeffrey, S. S., van de Rijn, M., Rees, C. A., Eisen, M. B., Ross, D. T., Pergamenschikov, A., Williams, C. F., Zhu, S. X., Lee, J. C., Lashkari, D., Shalon, D., Brown, P. O., and Botstein, D. (1999) Distinctive gene expression patterns in human mammary epithelial cells and breast cancers. *Proc Natl Acad Sci U S A.* **96**, 9212–9217.

5. Townley-Tilson, W. H., Pendergrass, S. A., Marzluff, W. F., and Whitfield, M. L. (2006) Genome-wide analysis of mRNAs bound to the histone stem-loop binding protein. *RNA.* **12**, 1853–1867.

6. Sanchez-Diaz, P., and Penalva, L. O. (2006) Post-transcription meets post-genomic: the saga of RNA binding proteins in a new era. *RNA Biol.* **3**, 101–109.

7. Lockhart, D. J., and Winzeler, E. A. (2000) Genomics, gene expression and DNA arrays. *Nature.* **405**, 827–836.

8. Mata, J., Marguerat, S., and Bahler, J. (2005) Post-transcriptional control of gene expression: a genome-wide perspective. *Trends Biochem Sci.* **30**, 506–514.

9. Tenenbaum, S. A., Lager, P. J., Carson, C. C., and Keene, J. D. (2002) Ribonomics: identifying mRNA subsets in mRNP complexes using antibodies to RNA-binding proteins and genomic arrays. *Methods.* **26**, 191–198.

10. Keene, J. D., Komisarow, J. M., and Friedersdorf, M. B. (2006) RIP-Chip: the isolation and identification of mRNAs, microRNAs and protein components of ribonucleoprotein complexes from cell extracts. *Nat Protoc.* **1**, 302–307.

11. Brown, V., Jin, P., Ceman, S., Darnell, J. C., O'Donnell, W. T., Tenenbaum, S. A., Jin, X., Feng, Y., Wilkinson, K. D., Keene, J. D., Darnell, R. B., and Warren, S. T. (2001) Microarray identification of FMRP-associated brain mRNAs and altered mRNA translational profiles in fragile X syndrome. *Cell.* **107**, 477–487.

12. Darnell, J. C., Jensen, K. B., Jin, P., Brown, V., Warren, S. T., and Darnell, R. B. (2001) Fragile X mental retardation protein targets G quartet mRNAs important for neuronal function. *Cell.* **107**, 489–499.

13. Wang, W. X., Wilfred, B. R., Hu, Y., Stromberg, A. J., and Nelson, P. T. (2010) Anti-Argonaute RIP-Chip shows that miRNA transfections alter global patterns of mRNA recruitment to microribonucleoprotein complexes. *RNA.* **16**, 394–404.

14. Gerber, A. P., Herschlag, D., and Brown, P. O. (2004) Extensive Association of Functionally and Cytotopically Related mRNAs with Puf Family RNA-Binding Proteins in Yeast. *PLoS Biol.* **2**, E79.

15. Lopez de Silanes, I., Zhan, M., Lal, A., Yang, X., and Gorospe, M. (2004) Identification of a target RNA motif for RNA-binding protein HuR. *Proc Natl Acad Sci U S A.* **101**, 2987–2992.

16. Mazan-Mamczarz, K., Hagner, P. R., Corl, S., Srikantan, S., Wood, W. H., Becker, K. G., Gorospe, M., Keene, J. D., Levenson, A. S., and Gartenhaus, R. B. (2008) Post-transcriptional gene regulation by HuR promotes a more tumorigenic phenotype. *Oncogene.* **27**, 6151–6163.

17. Baroni, T. E., Chittur, S. V., George, A. D., and Tenenbaum, S. A. (2008) Advances in RIP-chip analysis : RNA-binding protein immunoprecipitation-microarray profiling. *Methods Mol Biol.* **419**, 93–108.

18. Gyorffy, B., Molnar, B., Lage, H., Szallasi, Z., and Eklund, A. C. (2009) Evaluation of microarray preprocessing algorithms based on concordance with RT-PCR in clinical samples. *PLoS One.* **4**, e5645.

19. Seo, J., and Hoffman, E. P. (2006) Probe set algorithms: is there a rational best bet?. *BMC Bioinformatics.* **7**, 395.

20. Benjamini, Y., and Hochberg, Y. (1995) Controlling the false discovery rate: A practical and powerful approach to multiple testing. *J. Roy. Statist. Soc. Ser.* **57**, 289–300.

21. Birney, E., Stamatoyannopoulos, J. A., Dutta, A., Guigo, R., Gingeras, T. R., Margulies, E. H., Weng, Z., Snyder, M., Dermitzakis, E. T., Thurman, R. E., Kuehn, M. S., Taylor, C. M., Neph, S., Koch, C. M., Asthana, S., Malhotra, A., Adzhubei, I., Greenbaum, J. A., Andrews, R. M., Flicek, P., Boyle, P. J., Cao, H., Carter, N. P., Clelland, G. K., Davis, S., Day, N., Dhami, P., Dillon, S. C., Dorschner, M. O., Fiegler, H., Giresi, P. G., Goldy, J., Hawrylycz, M., Haydock, A., Humbert, R., James, K. D.,

Johnson, B. E., Johnson, E. M., Frum, T. T., Rosenzweig, E. R., Karnani, N., Lee, K., Lefebvre, G. C., Navas, P. A., Neri, F., Parker, S. C., Sabo, P. J., Sandstrom, R., Shafer, A., Vetrie, D., Weaver, M., Wilcox, S., Yu, M., Collins, F. S., Dekker, J., Lieb, J. D., Tullius, T. D., Crawford, G. E., Sunyaev, S., Noble, W. S., Dunham, I., Denoeud, F., Reymond, A., Kapranov, P., Rozowsky, J., Zheng, D., Castelo, R., Frankish, A., Harrow, J., Ghosh, S., Sandelin, A., Hofacker, I. L., Baertsch, R., Keefe, D., Dike, S., Cheng, J., Hirsch, H. A., Sekinger, E. A., Lagarde, J., Abril, J. F., Shahab, A., Flamm, C., Fried, C., Hackermuller, J., Hertel, J., Lindemeyer, M., Missal, K., Tanzer, A., Washietl, S., Korbel, J., Emanuelsson, O., Pedersen, J. S., Holroyd, N., Taylor, R., Swarbreck, D., Matthews, N., Dickson, M. C., Thomas, D. J., Weirauch, M. T., Gilbert, J., Drenkow, J., Bell, I., Zhao, X., Srinivasan, K. G., Sung, W. K., Ooi, H. S., Chiu, K. P., Foissac, S., Alioto, T., Brent, M., Pachter, L., Tress, M. L., Valencia, A., Choo, S. W., Choo, C. Y., Ucla, C., Manzano, C., Wyss, C., Cheung, E., Clark, T. G., Brown, J. B., Ganesh, M., Patel, S., Tammana, H., Chrast, J., Henrichsen, C. N., Kai, C., Kawai, J., Nagalakshmi, U., Wu, J., Lian, Z., Lian, J., Newburger, P., Zhang, X., Bickel, P., Mattick, J. S., Carninci, P., Hayashizaki, Y., Weissman, S., Hubbard, T., Myers, R. M., Rogers, J., Stadler, P. F., Lowe, T. M., Wei, C. L., Ruan, Y., Struhl, K., Gerstein, M., Antonarakis, S. E., Fu, Y., Green, E. D., Karaoz, U., Siepel, A., Taylor, J., Liefer, L. A., Wetterstrand, K. A., Good, P. J., Feingold, E. A., Guyer, M. S., Cooper, G. M., Asimenos, G., Dewey, C. N., Hou, M., Nikolaev, S., Montoya-Burgos, J. I., Loytynoja, A., Whelan, S., Pardi, F., Massingham, T., Huang, H., Zhang, N. R., Holmes, I., Mullikin, J. C., Ureta-Vidal, A., Paten, B., Seringhaus, M., Church, D., Rosenbloom, K., Kent, W. J., Stone, E. A., Batzoglou, S., Goldman, N., Hardison, R. C., Haussler, D., Miller, W., Sidow, A., Trinklein, N. D., Zhang, Z. D., Barrera, L., Stuart, R., King, D. C., Ameur, A., Enroth, S., Bieda, M. C., Kim, J., Bhinge, A. A., Jiang, N., Liu, J., Yao, F., Vega, V. B., Lee, C. W., Ng, P., Yang, A., Moqtaderi, Z., Zhu, Z., Xu, X., Squazzo, S., Oberley, M. J., Inman, D., Singer, M. A., Richmond, T. A., Munn, K. J., Rada-Iglesias, A., Wallerman, O., Komorowski, J., Fowler, J. C., Couttet, P., Bruce, A. W., Dovey, O. M., Ellis, P. D., Langford, C. F., Nix, D. A., Euskirchen, G., Hartman, S., Urban, A. E., Kraus, P., Van Calcar, S., Heintzman, N., Kim, T. H., Wang, K., Qu, C., Hon, G., Luna, R., Glass, C. K., Rosenfeld, M. G., Aldred, S. F., Cooper, S. J., Halees, A., Lin, J. M., Shulha, H. P., Xu, M., Haidar, J. N., Yu, Y., Iyer, V. R., Green, R. D., Wadelius, C., Farnham, P. J., Ren, B., Harte, R. A., Hinrichs, A. S., Trumbower, H., Clawson, H., Hillman-Jackson, J., Zweig, A. S., Smith, K., Thakkapallayil, A., Barber, G., Kuhn, R. M., Karolchik, D., Armengol, L., Bird, C. P., de Bakker, P. I., Kern, A. D., Lopez-Bigas, N., Martin, J. D., Stranger, B. E., Woodroffe, A., Davydov, E., Dimas, A., Eyras, E., Hallgrimsdottir, I. B., Huppert, J., Zody, M. C., Abecasis, G. R., Estivill, X., Bouffard, G. G., Guan, X., Hansen, N. F., Idol, J. R., Maduro, V. V., Maskeri, B., McDowell, J. C., Park, M., Thomas, P. J., Young, A. C., Blakesley, R. W., Muzny, D. M., Sodergren, E., Wheeler, D. A., Worley, K. C., Jiang, H., Weinstock, G. M., Gibbs, R. A., Graves, T., Fulton, R., Mardis, E. R., Wilson, R. K., Clamp, M., Cuff, J., Gnerre, S., Jaffe, D. B., Chang, J. L., Lindblad-Toh, K., Lander, E. S., Koriabine, M., Nefedov, M., Osoegawa, K., Yoshinaga, Y., Zhu, B., and de Jong, P. J. (2007) Identification and analysis of functional elements in 1% of the human genome by the ENCODE pilot project. *Nature.* **447**, 799–816.

22. Kurien, B. T., and Scofield, R. H. (2006) Western blotting. *Methods.* **38**, 283–293.

23. Lasne, F. (2001) Double-blotting: a solution to the problem of non-specific binding of secondary antibodies in immunoblotting procedures. *J Immunol Methods.* **253**, 125–131.

24. Macphee, D. J. (2009) Methodological considerations for improving Western blot analysis. *J Pharmacol Toxicol Methods.*

25. Zhang, X., Zou, T., Rao, J. N., Liu, L., Xiao, L., Wang, P. Y., Cui, Y. H., Gorospe, M., and Wang, J. Y. (2009) Stabilization of XIAP mRNA through the RNA binding protein HuR regulated by cellular polyamines. *Nucleic Acids Res.* **37**, 7623–7637.

26. Lisbin, M. J., Qiu, J., and White, K. (2001) The neuron-specific RNA-binding protein ELAV regulates neuroglian alternative splicing in neurons and binds directly to its pre-mRNA. *Genes Dev.* **15**, 2546–2561.

Chapter 26

The Dual Use of RNA Aptamer Sequences for Affinity Purification and Localization Studies of RNAs and RNA–Protein Complexes

Scott C. Walker, Paul D. Good, Theresa A. Gipson, and David R. Engelke

Abstract

RNA affinity tags (aptamers) have emerged as useful tools for the isolation of RNAs and ribonucleoprotein complexes from cell extracts. The streptavidin binding RNA aptamer binds with high affinity and is quickly and cleanly eluted with biotin under mild conditions that retain intact complexes. We describe the use of the streptavidin binding aptamer as a tool for purification and discuss strategies towards the design and production of tagged RNAs with a focus on structured target RNAs. The aptamer site can be further exploited as a unique region for the hybridization of oligonucleotide probes and localization by fluorescent in situ hybridization (FISH). The aptamer insertion will allow the localization of a population of RNA species (such as mutants) to be viewed specifically, while in the presence of the wild type RNA. We describe the production of labeled oligonucleotide probes and the preparation of yeast cells for the localization of RNAs by FISH.

Key words: RNA, RNP isolation, Aptamer, FISH, Ribonucleoprotein, Streptavidin

1. Introduction

Aptamers are nucleic acid sequences that have been selected from a large pool of random sequences to exhibit a particular desired property, usually tight binding to a specific target molecule. The desired properties of each aptamer are obtained through multiple rounds of selection by utilizing the *s*ystematic *e*volution of *l*igands by *ex*ponential enrichment (SELEX) method (1, 2). Both DNA and RNA aptamers have been isolated with many uses including a wide range of therapeutics, biosensors, chiral reagents, and affinity tags (3).

The more recently developed RNA affinity tags complement the more widely used protein based affinity tags (4) and these aptamers have become useful research tools allowing the specific purification of RNA targets and their associated complexes from cellular extracts.

1.1. Affinity Purification of RNAs and RNA–Protein Complexes

A number of RNA aptamers have been used for the purification of RNAs and RNA–protein complexes from cellular extracts, these include aptamers that bind to immobilized streptavidin, tobramycin, streptomycin, or sephadex (5–8). We are most experienced with the streptavidin binding aptamer (S1 aptamer) and focus on this example throughout this chapter. The streptavidin binding aptamer has many desirable properties; it binds to immobilized streptavidin with high affinity ($K_d \sim 70$ nM), the binding is stable to high salt conditions (400 mM NaCl), and the aptamer is quickly and cleanly eluted from the affinity resin by the binding of the small molecule D-biotin to streptavidin. Initial experiments utilized the full SELEX derived sequence to purify RNase P from yeast cells (5) (see Figs. 1 and 2). However, only the minimal streptavidin binding aptamer is needed for full binding activity and the extraneous sequence can be completely omitted (Fig. 1) (9). The streptavidin binding aptamer has been used to purify the multi subunit ribonucleoprotein RNase P from both yeast (5) and human cells (10), mutant ribosomes from bacteria (11), and also telomerase from yeast cells (12).

It is important to note that in each of these published studies the aptamer was placed internally by insertion into a solvent-accessible stem loop within the structured target RNA (see Fig. 2). Early attempts to use the streptavidin binding aptamer as a flanking tag (i.e., placed either 5′ or 3′ to the target RNA) were not successful (unpublished communications). These studies targeted unstructured RNAs (mRNAs and pre-mRNAs) and the tagged constructs were found to be functional in vitro, but failed to perform in vivo, in a manner consistent with the aptamer becoming unwound or otherwise obscured by protein binding and RNP formation in the cellular environment. In contrast, the tobramycin aptamer has been used as a flanking tag to successfully purify pre-mRNA complexes (13). Comparison of the two approaches suggests that the use of a G-C rich stabilizing stem at the base of the tobramycin aptamer was instrumental for success. In this chapter, we will only describe the use of aptamers tags placed internally within a structured target RNA. Although these constructs are more difficult to design and test, we have had greater success when employing the aptamer tag internally.

1.2. Localization of RNAs and RNA–Protein Complexes

Fluorescence *in situ* hybridization (FISH) is an efficient mechanism for localizing RNAs and their associated complexes (14). FISH uses a fluorescently labeled oligonucleotide probe complementary

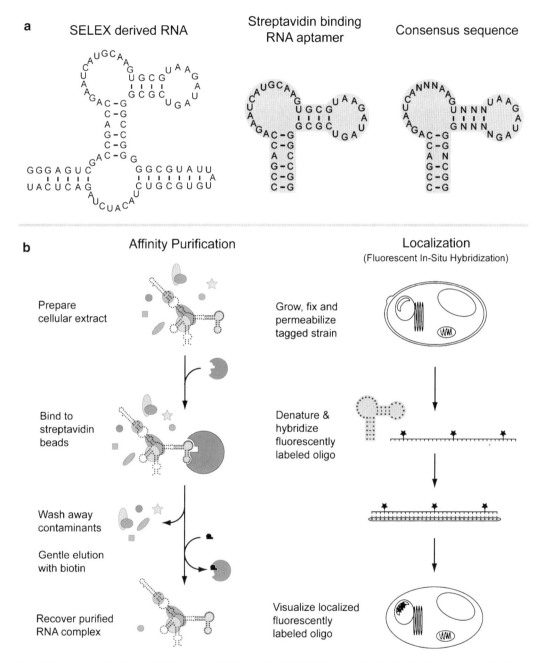

Fig. 1. The streptavidin binding RNA aptamer. (a) The original SELEX derived streptavidin binding sequence is shown along with the minimal streptavidin binding aptamer (*shaded*). The aptamer was derived from a population of species and the consensus sequence from multiple isolated clones is shown. (b) The streptavidin binding aptamer has been used for affinity purifications of ribonucleoprotein complexes (*left*) and the sequence has also been targeted as a unique hybridization region for fluorescent in situ hybridizations to localize ribonucleoprotein complexes (*right*).

to the sequence of interest. Although any RNA can be visualized with an appropriate antisense probe, there can be distinct advantages to exploiting an existing aptamer insertion into a target RNA as a unique target region for FISH. Cells with the aptamer

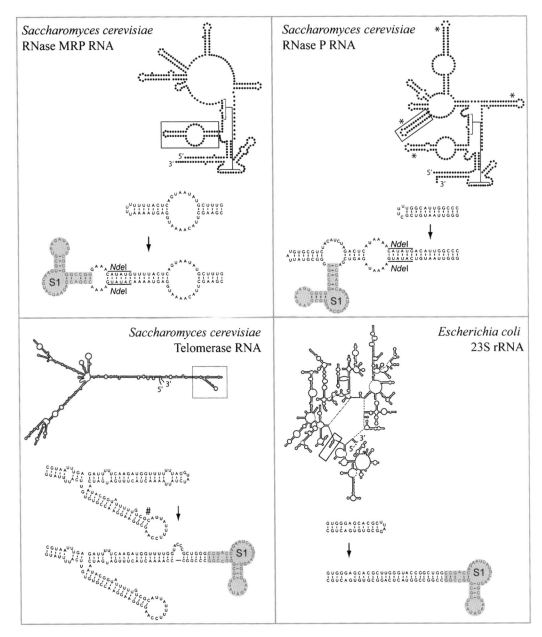

Fig. 2. Insertion of the streptavidin binding aptamer into ribonucleoprotein complexes. (*Yeast RNase P and RNase MRP*) Aptamer insertions into the yeast RNase MRP and RNase P holoenzymes have been used to purify these ribonucleoprotein complexes. The yeast RNase P has been extensively tagged, initially with the full SELEX derived sequence (shown) and subsequently with the minimal streptavidin binding aptamer sequence. The yeast RNase P can be tagged in four different locations that are indicated (*). In each case, these positions are known to be phylogenetically variable and solvent exposed and the aptamer insertion does not alter the growth or pre-tRNA processing significantly. (*Yeast telomerase*) The yeast telomerase was tagged at two positions, the successful aptamer insertion is shown and the unsuccessful position is also indicated (#). The purified telomerase was subsequently shown to be active in a telomerase assay. (*Bacterial ribosomes*) The streptavidin aptamer was used to specifically isolate ribosomal subunits containing a known lethal mutation. Coexpression of the lethal mutation alongside the wildtype sequence was necessary to sustain bacterial growth.

tagged complex are first fixed with formaldehyde and permeabilized on a microscope slide. The probe is then hybridized to the aptamer, and the complex is directly detectable using fluorescent microscopy (see Note 1). The localization can be further extended by the simultaneous hybridization of multiple fluorescent probes. This multicolor FISH permits one to see the location of an aptamer tagged RNA relative to that of other labeled features, such as the nucleus or nucleolus.

The aptamer sequence can serve as a unique site for the identification of only a single population of RNA species. When a functional tagged RNA is coexpressed alongside the wild type RNA it becomes possible to further manipulate the tagged RNA and examine the effects of various mutations on the localization of the tagged RNA. In this situation, the strain is kept viable by the presence of the wild type RNA and the localization of only the mutated RNA population can be detected using the unique tag. Similar approaches have been used to examine nucleo–cytoplasmic shuttling of snRNAs in heterokaryons by detecting the unique sequence region of a functional hybrid U6 snRNA in the presence of the wild type U6 snRNA (15). We have used the streptavidin binding aptamer sequence to successfully localize the RNase P RNA and have also observed mis-localization of a number of RNA mutants (Fig. 3) (16).

There are also other practical reasons for utilizing an existing aptamer insertion as this helps to ensure that the target region is solvent exposed, aiding effective strand invasion and hybridization of the fluorescently tagged oligonucleotide to the target RNA. This approach also allows the same fluorescent oligonucleotide probe to be used for the examination of different aptamer tagged target RNA species. Developing unique antisense probes for each target RNA has drawbacks since each fluorescent probe is costly to produce and must be tested to ensure effective strand invasion and hybridization to the target.

2. Materials

2.1. Design of Tagged RNAs

1. Software allowing secondary structure prediction. Downloadable programs can be found for both the PC (RNAstructure: http://rna.urmc.rochester.edu/rnastructure.html) and Mac (Mulfold: http://iubio.bio.indiana.edu/soft/molbio/mac/). A second program for the PC is (RNAdraw: http://www.rnadraw.com) and links to web-based RNA folding servers can also be found at this address.

428　Walker et al.

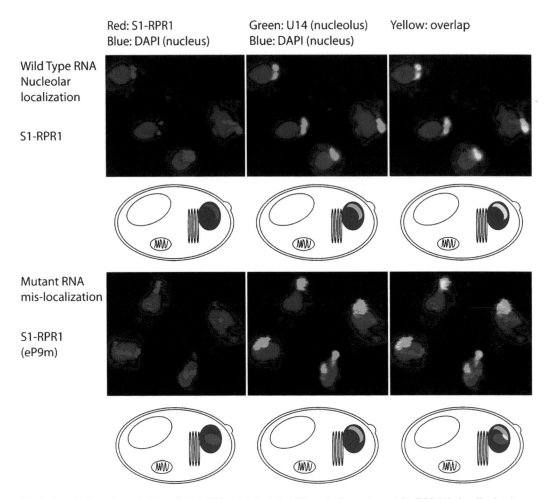

Fig. 3. Localization of yeast RNase P. The RNA subunit of the RNase P ribonucleoprotein (RPR1) is localized using a fluorophore labeled antisense oligo (*red*) targeting the streptavidin binding aptamer insertion into the RNA subunit (S1-RPR1). The U14 RNA is also visualized using an antisense oligo (*green*) and serves as a nucleolar marker. The overlap (*yellow*) shows that the wild type RNA is predominantly nucleolar. The complex becomes mislocalized to the nucleoplasm (*blue*) when an RNA mutation within the eP9 stem loop is introduced into the RNA subunit [S1-RNA (eP9m)]. Adapted from ref. 16 with permission of Cold Spring Harbor Press.

2.2. Construction of the Tagged RNA Sequence

2.2.1. Two Step PCR

1. Genomic DNA or plasmid bearing the target RNA sequence.
2. Oligonucleotides (Integrated DNA Technologies).
3. DNA Polymerase, supplied with commercial buffer (Pfu – Stratagene) (Taq – Roche).
4. Agarose gel electrophoresis equipment.
5. Gel purification mini-prep kit (Promega).
6. Restriction enzymes, supplied with commercial buffer (New England Biolabs).
7. T4 DNA ligase, supplied with commercial buffer (New England Biolabs).

8. Competent *E. coli* cells DH5α or XL1-Blue.
9. Access to an automated DNA sequencing facility.

2.2.2. Site-Directed Mutagenesis

1. Suitable vector bearing the target RNA sequence.
2. QuikChange mutagenesis Kit (Stratagene).
3. Oligonucleotides (Integrated DNA Technologies).
4. Restriction enzymes, supplied with commercial buffer (New England Biolabs).
5. T4 DNA ligase, supplied with commercial buffer (New England Biolabs).
6. Competent *E. coli* cells DH5α or XL1-Blue.
7. Access to an automated DNA sequencing facility.

2.3. Preparation of Yeast Extracts

1. Yeast strain carrying the appropriate tagged RNA.
2. Growth media (YPD) 10 g/l Bacto Yeast extract, 20 g/l Bacto peptone, and 20 g/l Dextrose. Combine and autoclave.
3. 1× Lysis Buffer: 50 mM Hepes at pH 7.4, 10 mM $MgCl_2$, 100 mM NaCl, 1 mM DTT, 0.1% Triton-X100, 10% glycerol, and Complete® protease inhibitors (Roche).
4. Acid washed glass beads 425–600 μm.
5. Protein content assay, Micro Bicinchoninic acid assay (Pierce).

2.4. Affinity Purification

1. Streptavidin agarose (Sigma #S1638).
2. Avidin from egg white (Sigma).
3. 1× Lysis buffer, as above but without Complete®.
4. Ultra-Free MC centrifugal filter device (Millipore).
5. D-biotin (Sigma).

2.5. Localization of Tagged RNAs via Fluorescent In Situ Hybridization

2.5.1. Preparation of Fluorescently Labeled Oligonucleotide Probes

1. Oligonucleotides, FPLC purified bearing appropriate amino modifications for labeling (Integrated DNA Technologies) (see Note 2).
2. Fluorescent Dyes, purchased as an *N*-hydroxysuccinimide (NHS) activated ester (Invitrogen or Amersham).
3. Anhydrous dimethyl sulfoxide (DMSO).
4. Labeling buffer: 0.1 M sodium tetraborate at pH 8.5.
5. Micro Bio-Spin 30 columns (Bio-Rad).

2.5.2. Fixation and Permeabilization of Yeast Cells

1. 40% Paraformaldehyde.
2. Sorbitol buffer: 1.2 M sorbitol and 0.1 M potassium phosphate at pH 7.5.

3. Spheroplast buffer: 1.2 M Sorbitol, 0.1 M Potassium Phosphate at pH 7.5, 1× Vanadyl Ribonucleoside Inhibitor (New England Biolabs), 28 mM β-mercaptoethanol, and 0.06 mg/ml PMSF at pH 7.5, store frozen.

4. Zymolyase 20T (Seikagaku Biobusiness Corp, via amsbio.com).

5. Microscope slides: eight-well not poly-lysine coated (MP Biomedicals LLC).

2.5.3. Hybridization of Fluorescent Oligonucleotide Probe

1. 20× SSC: 300 mM NaCl and 30 mM Citrate at pH 7.0.
2. 10× SSCP: 1.5 M NaCl, 0.15 M NaCitrate, and 0.2 M $NaHPO_4$ at pH 6.0.
3. Denaturating solution: 70% Formamide, 2× SSC. Make fresh for each hybridization.
4. Hybridization mix: 50% Formamide, 2× SSCP, 10% Dextran Sulfate, and 0.4 mg/ml ssDNA. Prewarm an aliquot of 50% (w/v) Dextran Sulfate for the hybridization mix to 70–72°C.

2.5.4. DAPI Staining and Slide Mounting

1. Post-hybridization wash: 50% Formamide, 2× SSC.
2. DAPI: 4′,6-diamidino-2-phenylindole (Roche).
3. Prolong Antifade Kit: Mounting Solution Components A & B (Molecular Probes/Invitrogen).

2.5.5. Visualization of Aptamer Tagged RNA

1. Fluorescent microscope with deconvolution technology or a confocal microscope. We use a Nikon Microscope with deconvolution.
2. CCD (charge couple device) cameras. We use a cooled CCD camera which allows the detection of low signals.
3. Imaging software: We use Esee (ISIS) for microscope imaging to take down background and superimpose images from different channels. We also use a combination of Isee (ISIS) and Photoshop (Adobe) for manipulation of the images.

3. Methods

3.1. Guidelines for Designing Aptamer Tagged RNAs

When introducing the aptamer sequence into a target RNA it is vital that both the aptamer and the target RNA maintain their correct structures. The S1 aptamer has been most successfully applied when used internally, within structured target RNAs (5, 10–12) and examples are shown in Fig. 2. In each of these cases, the target RNA had a predicted secondary structure that was used to aid the design of the successful aptamer tagged RNA.

3.1.1. Identifying Sites for Aptamer Insertion into a Target RNA

It is essential to start with some knowledge of the structure of the target RNA. The S1 aptamer has been used most effectively when it has been placed at the end of a solvent exposed stem structure. It is equally important that the stem-loop does not have a conserved function so that it can be freely manipulated without affecting the function of the target RNA. In most cases, the three dimensional structure is not available, however a secondary structure can be predicted and in many cases this data is available for a particular target RNA across several species (17–19). A phylogenetic analysis of RNA structure can be extremely useful to identify stem structures that are variable between species. Variations in the length of a stem and, most importantly, in the sequence of the stem-loop are often good indicators that the stem structure does not have an essential conserved function within the target RNA. These phylogenetically variable stems are more likely to tolerate manipulations and are therefore good targets for the insertion of the aptamer sequence. Although the identification of variable stems and stem-loops provides a good starting point these stems also need to be solvent exposed in order to provide accessibility of the aptamer tag to the affinity resin and allow purification. A complementary approach is to use chemical or enzymatic footprinting data to identify RNA structures that are accessible to these probes and therefore exposed to solvent. The combination of both phylogenetic and footprinting analyses will allow a solvent accessible and variable stem-loop structure(s) to be identified that have a greater likelihood of being successfully tagged.

In the case of the yeast RNase P RNA, both phylogenetic studies (20) and footprinting data (21) are available and there are five solvent exposed stem-loops that also do not have significant sequence conservation (16). To date, we have successfully introduced the S1 aptamer at each of four different positions within the RNase P RNA (Fig. 2). In each case, the aptamer insertion allows the purification of the enzyme and a thorough analysis of the growth phenotype and pre-tRNA processing profiles establishes that these aptamer insertions have no detectable effect on enzyme function ((5, 16) and unpublished data, S.C. Walker).

When phylogenetic or footprinting data is not available the best approach is to use trial and error in combination with best judgment based on what is known about the particular target RNA under study. In all cases it is still possible that the insertion of the aptamer sequence might have an adverse affect upon the folding of the target RNA. The potential effects of the aptamer sequence on the folding of the target RNA can be tested by folding the proposed sequence *in silico* using secondary structure prediction software. If the target RNA is large (>200 nt) it can often be more reliable to truncate the sequence and predict the folding of only the aptamer tagged stem and the surrounding sequence (approx. ±100 nt).

Ultimately, the only true test of correct aptamer folding and function within a target RNA is to produce and test the tagged target RNA in vivo. We strongly recommend the creation of a minimum of two constructs tagged at different positions. The tagged RNAs should be designed using all available knowledge (secondary structure, phylogenetic analysis, footprinting, or three dimensional structure) to guide the choice of stem structures that will be tagged.

3.1.2. Designing the Hybrid RNA

Once appropriate sites have been identified within the target RNA it is useful to draw out the final designs for the stem structure bearing the aptamer sequence. Different linker designs have been successfully used to fuse the aptamer to the target RNA including both rigid and flexible linkers of various lengths, examples are shown in Fig. 2. A long and flexible linker may provide better access to the affinity resin leading to a successful purification. However, it is also likely that a longer linker will have a greater potential to interfere with the correct folding of the target RNA and large flexible regions can be very sensitive sites for nuclease digestion. We have used both rigid and flexible linker designs and have not observed any significant differences when these have been used at the same position within the yeast RNase P RNA (S. C. Walker, unpublished data). However, since each target RNA will behave differently it is important to recognize that in certain contexts, flexible linkers may be sensitive to nucleases or that rigid linkers may not allow for a productive interaction with the affinity resin. In such cases, the investigator can respond by redesigning the linker or repositioning the aptamer tag.

3.2. Constructing the Hybrid Aptamer Target RNA Sequence

We describe two different approaches to facilitate the introduction of the aptamer sequence into an appropriate position within the target RNA. In each case we assume that the target RNA is being expressed from a plasmid bearing appropriate promoter and terminator signals.

3.2.1. Construction of the Aptamer Tagged RNA Sequence Using Two-Step PCR

After the aptamer insertion has been designed it is useful to split the sequence into upstream and downstream fragments (Fig. 4). Amplification of the downstream fragment will require two oligonucleotides (Primer 1 and Primer 2). The first oligonucleotide (Primer1) must have sufficient complementarity to the 5′ end of the fragment sequence (20–25 nt) and the aptamer sequence, which does not anneal to the template, is added to the 5′ end of this oligonucleotide. The second oligonucleotide (Primer 2) must have sufficient complementarity to the 3′ end of the fragment sequence (20–25 nt) and a suitable restriction site can be added at the 5′ end of the oligonucleotide to facilitate cloning.

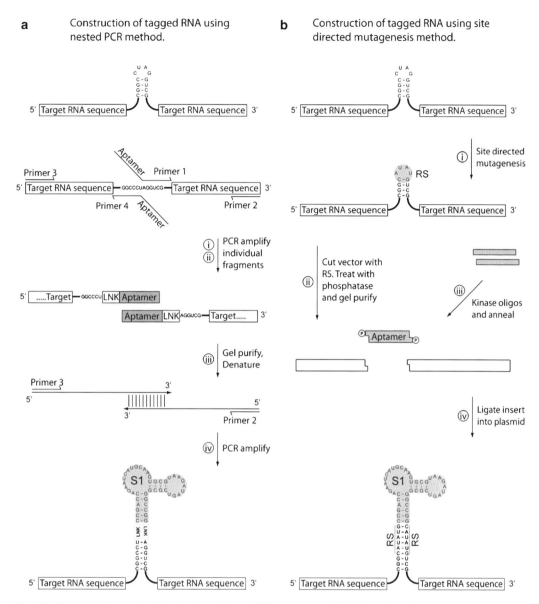

Fig. 4. Strategies for the construction of aptamer-tagged RNA sequences. (**a**) The insertion of the aptamer at the required position using a PCR based strategy. Individual PCR fragments are produced, (i) and (ii), where the aptamer sequence is introduced at the desired location by incorporation into the 5′ regions of the primers. The two aptamer-tagged sequence fragments are isolated and used in trace amounts as template for a further round of PCR. The outlying primers (2 and 3) can be used to amplify the entire region and produce the final product bearing an accurately inserted aptamer within the target RNA sequence. (**b**) The insertion of an aptamer at a required position using a cloning based strategy. Firstly, site-directed mutagenesis is used to introduce a suitable restriction site into the target RNA sequence (plasmid borne). The restriction site is used to facilitate the introduction of a synthetic insert bearing the required aptamer sequence. The presence of a correctly orientated insert can be performed by PCR screening colonies prior to sequencing.

To produce the upstream fragment a similar method is used where the aptamer is added at the 3′ end of the fragment (Primer 4) and the restriction site, or other required sequence (see Note 3), at the 5′ end of the fragment (Primer 3). The two fragments are then used as template in a third PCR where they are joined together via the introduced aptamer sequence and amplified by the outlying primers (Primer 2 and Primer 3) as outlined below.

1. The downstream fragment (Fig. 4a (i)) is produced by PCR using primers 1 and 2 using a suitable template DNA encoding the target RNA sequence.
2. The upstream fragment (Fig. 4a (ii)) is produced by PCR using primers 3 and 4 using a suitable template DNA encoding the target RNA sequence.
3. The PCR inserts are checked on an agarose gel next to size markers and gel-purified.
4. The two purified PCR fragments are mixed in equimolar amounts and used as template (1–5 ng total) in PCR with primers 2 and 3 to amplify the entire fragment bearing the tagged RNA sequence and introduced 5′ and 3′ restriction sites. The first few rounds of the PCR create the full length gene fragment due to the overlapping half gene fragments which will be extended by the DNA polymerase.
5. The correct size of the PCR fragment is checked on an agarose gel next to size markers and the DNA is gel isolated.
6. The fragment is digested with the appropriate restriction enzymes at the termini and ligated into a suitable plasmid at the correct restriction sites.
7. The ligation reaction is transformed in to competent bacterial host cells (e.g., *E. coli* DH5α or XL1-Blue cells) and plated onto selective media.
8. Plasmid is isolated from individual colonies and the PCR insert is verified by appropriate restriction digests and DNA sequencing.

3.2.2. Insertion of the Aptamer Sequence via a User Created Restriction Site

A second cloning strategy can provide some advantages in certain situations. The target stem-loop structure is mutated to a single hexameric restriction site and the aptamer sequence is purchased as two complementary oligonucleotides and inserted into this restriction site. After insertion into the target RNA, the introduced aptamer sequence is flanked by two complementary copies of the initial restriction site. The overall cloning strategy is represented in Fig. 4b. This sequential approach allows the effects of the mutated stem-loop upon target RNA function to be tested in vivo and is a good indicator of the likely success of the aptamer insertion at that position. A second advantage is that the resulting restriction site(s) can be used to facilitate further manipulations of the target RNA at this site. When using a single restriction site the direction of the insert must be screened by PCR or by sequencing. In some cases the sequential introduction of two restrictions sites, to facilitate the insertion of the aptamer sequence, may be necessary.

For the specific introduction of restriction sites into a particular sequence, we routinely use the Quikchange® kit (Stratagene™). This kit uses two user designed oligonucleotides bearing the mutated sequence to generate two mutant strands from a plasmid template. The parental (methylated) DNA is digested, using *DpnI*, and the resulting annealed double stranded nicked cDNA molecules are transformed into *E. coli* where the nicked cDNA is repaired to generate a plasmid bearing the required mutations. For more exact details of the technique, including oligonucleotide design, we refer the reader to the product manual.

1. Restriction site(s) are introduced into the plasmid sequence by site-directed mutagenesis using the Quikchange® kit (Stratagene™).

2. The plasmid is cut with the appropriate restriction enzyme(s). After digestion, the cut vector is treated with alkaline phosphatase to remove terminal phosphates and prevent re-ligation to itself. The DNA is then purified from an agarose gel in preparation for ligation with the insert.

3. The aptamer sequence including any linker is purchased as two individual DNA oligonucleotides bearing the correct overhangs for ligation. In order to facilitate ligation these oligonucleotides are individually phosphorylated with polynucleotide kinase (300 pmol in a 50 µl volume). The enzyme is removed by phenol/chloroform extraction and the two oligonucleotides are annealed in equal concentration, by heating to 65°C and cooling on ice, to create a double stranded insert.

4. The insert is ligated into the vector at the restriction site.

5. The ligation reaction is transformed in to competent bacterial (*E. coli* DH5α or XL1-Blue) cells and plated onto selective media.

6. Plasmid is isolated from individual colonies and the presence and correct orientation of the inserted aptamer is verified by appropriate restriction digests and automated DNA sequencing.

3.3. Preparation of Tagged Yeast Strains

Once a plasmid has been constructed that will allow the expression of the aptamer tagged target RNA, this can be used to prepare a yeast strain for study. In the yeast system it is possible to use a knockout strain such that the only copy of the target RNA in the final yeast strain is the tagged RNA on the introduced plasmid (see Note 4). In the resulting yeast strains, the aptamer tagged RNA is the only source of the RNA. When the target RNA is essential, the viability of these yeast strains is the direct indicator that the particular aptamer insertion is tolerated and does not interfere with the essential function of the target RNA. When the target RNA is not essential or when co-expression is the only option, then efforts to assess the integrity of the isolated complexes through an appropriate activity assay are required. An example is the successful processing of pre-tRNA substrates in vitro that was shown by the human RNase P holoenzyme, isolated using the streptavidin aptamer (10). In contrast, co-transcription of the tagged RNA alongside the chromosomal wild type sequence has been exploited in order to specifically isolate known lethal mutants of the 23S rRNA sequence for further study (11).

3.3.1. Preparation of Yeast Extracts for Binding

The particular method used to generate crude extracts will vary depending upon the system in which the tagged RNA is being expressed. In general, any protocol for the preparation of crude extracts should focus upon firstly maintaining the temperature at 4°C throughout and secondly producing the extract and binding it to beads as quickly as possible to minimize ribonuclease (and protease) degradation of the RNP complexes. For the preparation of extracts from yeast cultures we use the following protocol.

1. A single colony of the appropriate yeast strain is grown to saturation in YPD media at 30°C and used to inoculate 1 l of YPD media.

2. The culture is grown at 30°C until an OD_{600} of around 1–2 and the cells are harvested by centrifugation at 4,000×g for 15 min at 4°C.

3. The pellet is washed and resuspended in 10 pellet volumes of ice-cold sterile water, then re-pelleted by centrifugation at 4,000×g for 5 min at 4°C. This washing procedure is carried out three to four times.

4. The pellet is resuspended in 1 ml of lysis buffer per 2 g of wet cell paste, containing Complete® protease inhibitors (Roche).

5. The cells are lysed by vortexing with of 1/3 volume (~1 ml) of acid washed beads, 425–600 μm (Sigma), for 20–30 min. Efforts to keep the mixture cool during the lysis procedure are strongly recommended. Periodic cooling on crushed dry ice can be used to keep the sample cool, taking care to avoid freezing the sample.

   For larger scale preparations (>20 l culture), we have carried out lysis by passing the resuspended cells (step 4) through a microfluidiser eight to ten times (model #110Y, Microfluidics Corp.). Again, efforts to keep the mixture well cooled throughout the lysis process are strongly recommended.

6. The extract is cleared by centrifugation at 14,000×g for 20 min at 4°C. A second ultracentrifugation step is optional at 142,000×g for 1 h at 4°C.

7. The protein content in the cleared extract is determined using a Micro BCA assay (Pierce).

3.4. Isolation of Aptamer-Tagged RNA Complexes from Extracts Using Streptavidin

In the case of the streptavidin aptamer, a background of free biotin and biotinylated cellular material can be blocked by the addition of egg white avidin (Sigma) to the extract prior to binding to the streptavidin affinity resin (the RNA affinity tag does not bind to egg white avidin, only streptavidin). For optimal yields from different extracts, it can be useful to experimentally determine the correct amount of avidin required to block biotin in the extract. In our hands, extracts produced from yeast cultures can be sufficiently blocked by the addition of 10 µg avidin per milligram of protein in the extract. The streptavidin aptamer can stably bind to the affinity resin at up to 400 mM NaCl allowing purification under different salt conditions to be attempted if desired. A control pull down can be performed by pre-blocking the streptavidin affinity resin with D-biotin prior to incubation with the cell extract.

1. Block the extract with 5–20 µg of avidin per milligram of protein in the extract. Incubate at 4°C for 10 min prior to binding to the affinity resin.

2. Incubate the extract with 10–20 µl of streptavidin beads per milligram of protein in the extract. Binding should be carried out at 4°C for 1 h in 1× lysis buffer at an appropriate volume (five to ten times the bed volume). The beads are separated by centrifugation $4,000 \times g$ for 5 min at 4°C.

3. Wash the beads with 20 bed volumes of 1× lysis buffer five times for 5 min each at 4°C.

4. Transfer the beads to an Ultrafree-MC centrifugal filter unit (0.45 µm pore size Mllipore) and wash twice with 5 bed volumes of 1× lysis buffer at 4°C.

5. Elute the beads by incubating for 0.5–1 h at 4°C with 2 bed volumes of 1× lysis buffer containing 5 mM D-biotin.

3.5. Localization of Tagged RNAs via Fluorescent In Situ Hybridization

3.5.1. Preparation of Fluorescently Labeled Oligonucleotide Probes

The aptamer sequence is localized by hybridizing a fluorescently labeled antisense oligonucleotide and detecting its fluorescence via microscopy. The oligonucleotide is synthesized with reactive amine group modifications that will facilitate chemical coupling of the fluorescent dye. There are a number of reactive amine modifications available and a subset of these are shown in Fig. 5. The positioning of the reactive amine groups within the oligonucleotide is important as the signal from the fluorescent dye can be quenched if the dye molecules are too close to each other (<10 nucleotides) or placed adjacent to a guanine. The following antisense probe is suitable for the detection of the streptavidin binding aptamer sequence and is modified to accept up to three fluorescent dye molecules.

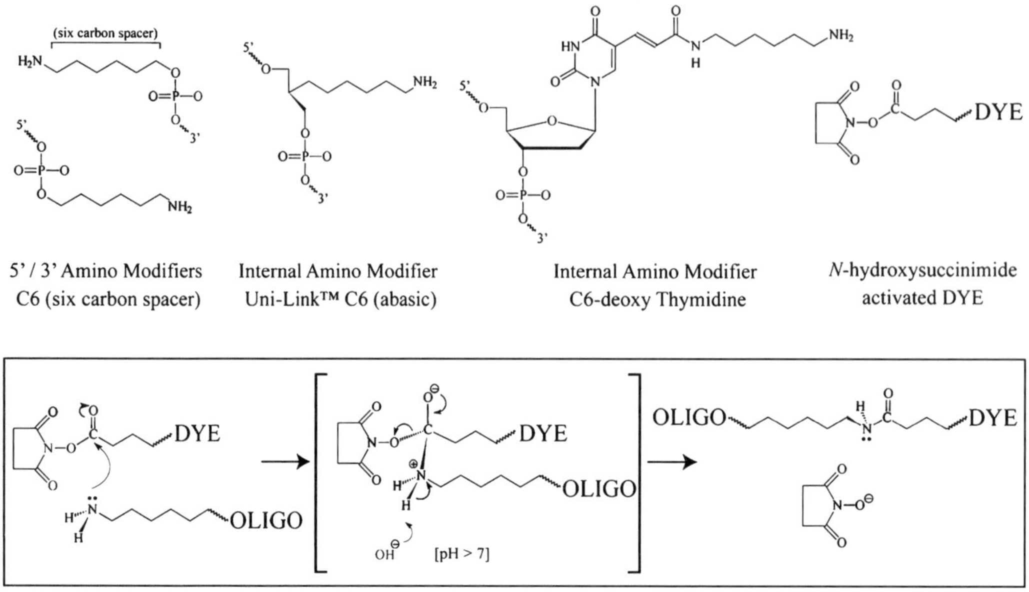

Fig. 5. Labeling amine modified oligonucleotides using activated ester chemistry. A number of amino modifications are available for oligonucleotide synthesis; common amino modifiers are shown including end modifications (5′ and 3′) and internal modifications. The fluorophore (DYE) is purchased as an activated ester form to facilitate coupling to the oligonucleotide via the amine modification. The reactive chemistry is summarized; the amine reacts with the ester group resulting in the formation of a stable carboxamide bond between the oligonucleotide and the fluorescent dye. The reaction is facilitated by the enhanced stability of the *N*-hydroxysuccinimide leaving group.

Anti-S1 probe 5′-GAC(T)ATCTTACGCAC(T)TGCATGATTC(T)GGTCGGT-3′

(T) = internal amino modifier (C6-deoxy Thymidine)

The amine groups can be coupled to a fluorescent dye by using amine reactive fluorescent dyes. The reaction is facilitated by activated ester chemistry where the oligonucleotide amine reacts with an activated ester group linked to the fluorescent dye resulting in the formation of a stable carboxamide bond between the oligonucleotide and the fluorescent dye (see Fig. 5). The chemistry is facilitated by the presence of a stable leaving group on the ester linkage of the fluorescent dye. Fluorescent dyes are available with a variety of leaving groups that react with amines in similar ways, including succinimidyl esters, sulfosuccinimidyl esters, tetrafluorophenyl esters and sulfodichlorophenol esters. Activated ester fluorescent dyes are unstable and have a limited shelf life (6–12 months) when stored cold (< –20°C) in the lyophilized form. The dyes must be dissolved in anhydrous solvent (DMSO) immediately prior to their use.

The reaction is set up with an excess of the amine reactive fluorescent dye and it is essential to remove this prior to purification of the labeled oligonucleotide. Generally, the amine reactive chemistry does not proceed to completion and the unreacted oligonucleotide must be separated from the labeled oligonucleotide species before these can be used for FISH. It is necessary to first remove the excess unreacted dye before attempting a high resolution purification of the labeled species. The final purification can be carried out by either preparative denaturing PAGE or reverse phase HPLC, we describe the use of preparative denaturing PAGE. When using oligonucleotides with multiple reactive amine modifications, the reaction products will be a mixture of mono-, di- and tri-labeled species. Each conjugated dye molecule will retard the migration of the oligonucleotide (by approx. 1–2 nt) and will allow the separation of the labeled species from the unlabeled oligo.

1. Dissolve the amine modified oligonucleotide to a final concentration of 3 mM (approx. 42 µg/µl for the anti-S1 probe) using deionized water (see Note 2). This stock solution can be stored at –20°C.

2. Prepare a fresh solution of 0.1 M sodium tetraborate buffer and adjust to pH 8.5 with HCl.

3. Prepare the amine reactive fluorescent dye by dissolving it in fresh, anhydrous DMSO to a final concentration of 15 mM. Fluorescent dyes are light sensitive and it is best to use a foil wrapped or amber colored Eppendorf tube. For each labeling reaction prepare 15 µl of the dye solution, any remaining unused solution is unstable and it is preferable to prepare fresh solution for each set of labeling reactions.

Cy3 Mono NHS Ester	(766 Da)	15 mM	~11.5 mg/ml
Cy5 Mono NHS Ester	(792 Da)	15 mM	~11.9 mg/ml
Oregon Green 488-X_1 NHS Ester	(623 Da)	15 mM	~ 9.4 mg/ml

4. Combine the following in a foil wrapped or amber colored eppendorf. The final molar ratio of fluorescent dye to reactive amine is (12:1), in this case the concentration of the anti-S1 probe is adjusted from 18 to 6 nmol as it bears three reactive amines.
    - 14 µl of fluorescent dye solution (210 nmol)
    - 2 µl of DNA oligonucleotide (6 nmol oligo/18 nmol reactive amine)
    - 75 µl of labeling buffer
    - 4 µl of deionized water.

Incubate the tube in the dark at room temperature using a rocker/shaker to gently mix the reaction, allow the reaction to proceed overnight.

5. An initial ethanol precipitation step will remove a significant amount of unreacted or degraded fluorescent dye. Precipitate the oligonucleotide by adding 10 µl of sodium acetate (3 M NaOAc pH 5.2) followed by 500 µl of ice cold ethanol. Invert the tube to mix and incubate at −80°C for at least 1 h (preferably overnight). Pellet the DNA by centrifugation at ≥14,000×g for 30 min. Carefully remove the supernatant and wash the pellet twice with 70% ethanol to help remove excess unreacted fluorescent dye. Resuspend the pellet in 25–50 µl of buffer (10 mM Na-Phosphate pH 7.4).

6. Remaining unreacted dye is removed by spin column gel filtration chromatography. Carefully load the resuspended oligo (25–50 µl) onto the center of a prespun Bio-Spin 30 column and spin to recover the labeled probe according to the manufacturer's protocol.

7. Add an equal volume of formamide to the oligonucleotide and load onto a 15% denaturing polyacrylamide gel (approx. 1–1.5 mm thickness). Load additional lanes as tracking markers, with bromophenol blue (BPB) and any leftover unused amine reactive fluorescent dye respectively. Run the gel until the BPB is approximately three-quarters of the way down the gel (the BPB migrates at approx. 15 nt). Transfer the gel onto saran wrap and locate the bands using a hand-held UV lamp and an X-ray film intensifying screen. Unreacted oligonucleotide will run faster than the labeled oligonucleotides and will form a dark UV shadow. The conjugated oligonucleotides will run with the dye color and can also be located by their UV shadow, although this may be less intense due to the fluorescence of the dye. Cut out the labeled oligonucleotides and elute from the gel slices overnight using the crush and soak method.

8. Precipitate the labeled oligonucleotides and resuspend in deionized water. Read the concentration of oligonucleotide in a spectrophotometer at 260 nm and adjust the final concentration to 2 µM.

3.5.2. Fixation and Permeabilization of Yeast Cells

1. Grow 50 ml of cells in synthetic minimal medium to midlog phase (OD = 0.2–0.4 at A_{600}).

2. Crosslink cells by adding 5 ml of 40% paraformaldehyde (3.6% final concentration) to each culture and incubate for 30 min at room temperature, while shaking (see Note 5).

3. Harvest the fixed cells by centrifugation 4,000×g for 5 min. Wash the cells by resuspending in 10 ml of ice-cold sorbitol buffer and repeat centrifugation to pellet. Repeat this wash step.

4. Digest the cell wall by first resuspending the cell pellet in 1 ml of spheroplast buffer and then adding Zymolyase up to 0.2 mg/ml final concentration. Mix gently and incubate at 37°C for 30–45 min.

5. Harvest the spheroplasted cells by centrifugation 4,000×g for 2 min at 4°C. Wash the spheroplasts by resuspending in 1 ml of ice-cold sorbitol buffer and repeat centrifugation to pellet. Resuspend in a final volume of 500–800 µl ice-cold sorbitol buffer.

6. Spot 20 µl of spheroplasted cells onto each well of several slides. Incubate at 4°C for 30–60 min, and then aspirate excess liquid from each well. Wash each well with a droplet of ice cold sorbitol buffer and carefully aspirate. Store in slide jar filled with 70% ethanol at −20°C for at least 16 h before use.

3.5.3. Hybridization of Fluorescent Oligonucleotide Probe

1. Prepare the fluorescent oligonucleotide probe(s) by adding 30 ng (approximately 1 µl of a 2 µM solution) of each probe to 20 µl of hybridization mix; this is enough for one well on a slide (see Notes 6 and 7).

2. Denature slides by incubating for 10 min in prewarmed (72°C) denaturing solution.

3. Wash the slides in a series of cold (−20°C) ethanol washes: 70, 80, 90, and 100% ethanol for 1 min each. Allow slides to air dry at room temperature and immediately add denatured probe (step 4) once dry.

4. Denature probe by incubating at 72°C for 10 min. Immediately load 20 µl probe/hybridization mix to each spot.

5. Incubate at 36°C in dark, in a slide warmer/humid chamber for 24–48 h.

3.5.4. DAPI Staining and Slide Mounting

1. Incubate slides in post-hybridization wash buffer for 20–30 min.

2. Wash slides twice in 1× PBS for 5 min.

3. Dry slides on paper towel in the dark room, take care not to dry completely. To speed up drying, you can aspirate liquid surrounding wells.

4. Spot ~15 µl of 1 mg/ml of DAPI onto each well. Incubate for 3 min.

5. Wash slides twice in a jar of 1× PBS for 2 min.

6. Dry slides on a paper towel in the dark room, take care not to dry completely.

7. Combine mounting solution components A & B when ready to put on cover slip. Prewarming component B to 50°C aids pipetting. Add one drop of mounting solution to each well and gently press on the cover slip. Dry overnight in the dark

3.5.5. Visualization of Aptamer Tagged RNA

on a flat surface. Once dry the slides can be viewed directly or sealed and stored upright in a covered slidebox at −20°C.

High resolution imaging will be required for fine discrimination of molecular compartments. Current methods generally employ either laser confocal or computer assisted deconvolution of Z-stacks acquired on a widefield fluorescence microscope. The user will be limited by the tools available and each approach has its own advantages, we refer the reader to the following review (22).

4. Notes

1. The sensitivity of FISH is such that the threshold level of detection is in the region of 10–20 copies of a localized target RNA per cell. For low copy targets, the sensitivity can be increased by utilizing oligonucleotides with multiple attached fluorophores.

2. It is important that the modified oligonucleotide is free from any contaminating amine compounds that would also react with the amine reactive fluorescent dye (including Tris buffers). We recommend purchasing the oligonucleotide with FPLC purification and resuspending in purified water. Additional chloroform extractions (3×) followed by ethanol precipitation can be performed if contamination is suspected.

3. For the rapid production of templates to be used for in vitro transcription of the target RNA, the T7 RNA polymerase promoter sequence (5′-TAATACGACTCACTATAGG-3′) can be encoded by the outlying primer and thus incorporated into the final PCR product. In vitro RNA transcripts produced from such templates can be used to test the binding of the tagged RNA to the affinity matrix in the absence of cellular components.

4. When the knockout strain involves a target RNA that is essential for yeast growth it is necessary to first supply the gene on a "rescue" plasmid in the absence of the chromosomal copy. When the rescue plasmid is marked with the *URA3* auxotrophic marker this can be counterselected using 5-fluoroorotic acid (5-FOA).

5. The cells must be fixed with formaldehyde under normal growth conditions before centrifugation or any other processing steps are performed. If the cells are manipulated prior to fixation it is possible that fixation will capture a stress response which can significantly impact the final observations and the interpretation of localization data.

6. Multiple labeled oligonucleotide probes can be used with different fluorophores to highlight known locations within the yeast cell. For example, we have routinely employed an antisense probe against the yeast U14 RNA as a marker for the nucleolus.

7. When performing an in situ hybridization experiment, proper controls are necessary to ensure that observable signal is specific. Specificity can be determined with competition studies using both labeled and excess unlabeled probe. Excess unlabeled probe can displace the specific binding of the labeled probe, but will not affect nonspecific binding of labeled probe.

Acknowledgments

This work was supported by a National Institute of Health grant (R01GM082875) to DRE.

References

1. Ellington, A. D., and Szostak, J. W. (1990) In vitro selection of RNA molecules that bind specific ligands. *Nature* 346, 818–22.
2. Wilson, D. S., and Szostak, J. W. (1999) In vitro selection of functional nucleic acids. *Annu Rev Biochem* 68, 611–47.
3. Nimjee, S. M., Rusconi, C. P., and Sullenger, B. A. (2005) Aptamers: an emerging class of therapeutics. *Annu Rev Med* 56, 555–83.
4. Nygren, P. A., Stahl, S., and Uhlen, M. (1994) Engineering proteins to facilitate bioprocessing. *Trends Biotechnol* 12, 184–8.
5. Srisawat, C., and Engelke, D. R. (2001) Streptavidin aptamers: affinity tags for the study of RNAs and ribonucleoproteins. *RNA* 7, 632–41.
6. Srisawat, C., Goldstein, I. J., and Engelke, D. R. (2001) Sephadex-binding RNA ligands: rapid affinity purification of RNA from complex RNA mixtures. *Nucleic Acids Res* 29, E4.
7. Hartmuth, K., Urlaub, H., Vornlocher, H. P., Will, C. L., Gentzel, M., Wilm, M., and Luhrmann, R. (2002) Protein composition of human prespliceosomes isolated by a tobramycin affinity-selection method. *Proc Natl Acad Sci U S A* 99, 16719–24.
8. Bachler, M., Schroeder, R., and von Ahsen, U. (1999) StreptoTag: a novel method for the isolation of RNA-binding proteins. *RNA* 5, 1509–16.
9. Srisawat, C., and Engelke, D. R. (2002) RNA affinity tags for purification of RNAs and ribonucleoprotein complexes. *Methods* 26, 156–61.
10. Li, Y., and Altman, S. (2002) Partial reconstitution of human RNase P in HeLa cells between its RNA subunit with an affinity tag and the intact protein components. *Nucleic Acids Res* 30, 3706–11.
11. Leonov, A. A., Sergiev, P. V., Bogdanov, A. A., Brimacombe, R., and Dontsova, O. A. (2003) Affinity purification of ribosomes with a lethal G2655C mutation in 23 S rRNA that affects the translocation. *J Biol Chem* 278, 25664–70.
12. Shcherbakova, D. M., Sokolov, K. A., Zvereva, M. I., and Dontsova, O. A. (2009) Telomerase from yeast Saccharomyces cerevisiae is active in vitro as a monomer. *Biochemistry (Mosc)* 74, 749–55.
13. Hartmuth, K., Vornlocher, H. P., and Luhrmann, R. (2004) Tobramycin affinity tag purification of spliceosomes. *Methods Mol Biol* 257, 47–64.
14. Long, R. M., Elliott, D. J., Stutz, F., Rosbash, M., and Singer, R. H. (1995) Spatial consequences of defective processing of specific

yeast mRNAs revealed by fluorescent in situ hybridization. *RNA* **1**, 1071–8.
15. Olson, B. L., and Siliciano, P. G. (2003) A diverse set of nuclear RNAs transfer between nuclei of yeast heterokaryons. *Yeast* **20**, 893–903.
16. Xiao, S., Day-Storms, J. J., Srisawat, C., Fierke, C. A., and Engelke, D. R. (2005) Characterization of conserved sequence elements in eukaryotic RNase P RNA reveals roles in holoenzyme assembly and tRNA processing. *RNA* **11**, 885–96.
17. Zuker, M. (2003) Mfold web server for nucleic acid folding and hybridization prediction. *Nucleic Acids Res* **31**, 3406–15.
18. Hofacker, I. L. (2004) RNA secondary structure analysis using the Vienna RNA package. *Curr Protoc Bioinformatics* **Chapter 12,** Unit 12 2.
19. Hofacker, I. L. (2007) RNA consensus structure prediction with RNAalifold. *Methods Mol Biol* **395**, 527–44.
20. Tranguch, A. J., and Engelke, D. R. (1993) Comparative structural analysis of nuclear RNase P RNAs from yeast. *J Biol Chem* **268**, 14045–55.
21. Tranguch, A. J., Kindelberger, D. W., Rohlman, C. E., Lee, J. Y., and Engelke, D. R. (1994) Structure-sensitive RNA footprinting of yeast nuclear ribonuclease P. *Biochemistry* **33**, 1778–87.
22. Lichtman, J. W., and Conchello, J. A. (2005) Fluorescence microscopy. *Nat Methods* **2**, 910–9.

Part VI

Use of Bioinformatics to Identify *Cis*-Acting Motifs and Structures in RNAs

Chapter 27

Identifying and Searching for Conserved RNA Localisation Signals

Russell S. Hamilton and Ilan Davis

Abstract

RNA localisation is an important mode of delivering proteins to their site of function. *Cis*-acting signals within the RNAs, which can be thought of as zip-codes, determine the site of localisation. There are few examples of fully characterised RNA signals, but the signals are thought to be defined through a combination of primary, secondary, and tertiary structures. In this chapter, we describe a selection of computational methods for predicting RNA secondary structure, identifying localisation signals, and searching for similar localisation signals on a genome-wide scale. The chapter is aimed at the biologist rather than presenting the details of each of the individual methods.

Key words: Bioinformatics, mRNA localisation, RNA secondary structure

1. Introduction

In the past few decades, RNA has taken an increasingly central position besides DNA in biology. Indeed, from a catalytic and structural point of view, RNA is a much more interesting molecule than DNA. RNA plays key roles in all aspects of biology, ranging from transcription of genes into nascent RNA and then mRNA that is often localised to particular parts of the cytoplasm. In a screen of 3,370 RNAs in the *Drosophila* embryo using fluorescent in situ hybridization (FISH), it has been shown that 71% show some level of localisation (1). RNA localisation in the nervous system is of particular functional importance (2, 3), as defects in RNA localisation have been implicated in Fragile-X syndrome (4), spinal muscular atrophy (5), and spinocerebellar ataxia (6). Therefore, the bioinformatics of RNA primary, secondary, and tertiary structures is of great importance to the understanding of the wide variety of functions that RNA can perform.

Intracellular mRNA localisation is a very common means of delivering proteins to their site of function, through targeted translation at the site of localisation. This mode of posttranscriptional regulation of gene expression has been identified in most of the model organisms and is known to be very common in *Drosophila*, where it has been tested systematically (7, 8). There are several mechanisms by which mRNAs localise, including diffusion, localised synthesis or degradation, and active transport (9). Here, we discuss mRNAs that localise by the most common of these mechanisms, actively transported mRNAs by molecular motors. Localised transcripts contain discrete *cis*-acting signals that fold and associate with a variety of *trans*-acting proteins that dictate their intracellular destinations. Amongst other proteins, the *trans*-acting factors associate with molecular motors, anchoring complexes, and translational regulatory proteins (Fig. 1a). The molecular motors with their RNA cargoes then travel to their destination along microtubule or actin filaments. In general, the RNA signals to which these proteins associate are best thought of as being identified through a combination of primary sequence and structural features. However, the localisation signals themselves have been defined in detail in only very few examples. In most cases, little is known about the primary sequence or structural features that direct the mRNA to a particular part of the cell. Very few localisation signals have been mapped to the minimally required motif so there is little known about what confers localisation of an mRNA to a particular location in the cell. The *gurken* and *I* factor have been mapped to their minimally required localisation signals; however, the exact sequence and structure requirements determining localisation have yet to be elucidated (10). Other Dynein-mediated localised mRNAs to have their signals mapped include: *bicoid* (11), *oskar* (12, 13), *fs(1)K10* (14), *orb* (15), *nanos* (16), *eve* (17), *hairy* (18), and *wingless* (19).

mRNAs contain primary, secondary, and tertiary structures. The primary sequences are comprised of four nucleotides: the purines adenosine (A) and guanine (G) and the pyrimidines cytosine (C) and uracil (U). RNA secondary structure consists of base pairings between the nucleotides, which can be canonical (e.g. A:U, U:A, G:C, C:G) or non-canonical (e.g. G:U) (20–22). The base pairings then give rise to features such as stems, loops, bulges, internal loops, and pseudoknots (Fig. 1b). Bases important for maintaining mRNA secondary structure are under evolutionary pressure so are usually well conserved between homologous sequences. Base pairings can be maintained through compensatory mutations, where the base pair is in the same position but the sequence differs (AU to GC). Consistent mutations occur where a base pair is maintained, but by a weaker non-canonical interaction (GC to GU) or vice versa. The tertiary structure of RNA contains all the secondary structure features organised in three

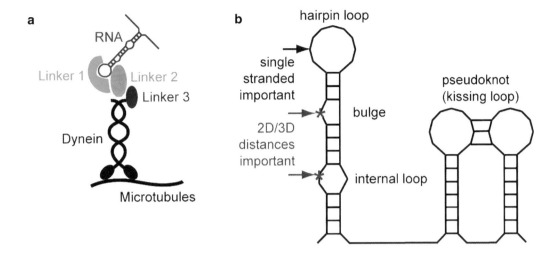

Fig. 1. *RNA secondary structure.* (**a**) Schematic of the Dynein-mediated RNA transport particle. The RNA binds to the motor via one or more transacting factors and linker proteins, such as Staufen, Squid, and Egalitarian. (**b**) RNA secondary structure can contain several features, such as stems, hairpin loops, internal loops, and bulges. The single-stranded regions, and the distances between them, are particularly important for specific RNA:protein interactions. (**c**) Common file formats for RNA secondary structure (FASTA, Stockholm, and CT).

dimensions, such as stem regions that come together to form helices in three dimensions. Full-length RNA structures can be divided into structural motifs, which are important for the function of the mRNA and can be defined at the primary, secondary, and/or tertiary level (21, 23). The most interesting of these motifs are those that are conserved across different species, as they are likely candidates for structural features that provide the specificity for biological function, including determining the site of localisation of the RNA.

In this chapter, we focus on the computational tools that are used for the secondary structure prediction, alignment, and identification of RNA signals involved in intracellular mRNA localisation. However, the same tools are generally applicable to the analysis of RNA signals in any kind of biological context. The aim of this chapter is to survey the array of tools that are currently available for searching, aligning, and characterising RNA localisation signals. Most of the papers describing the methods are aimed at computational biologists, and there is a distinct lack of reviews that are accessible to biologists with a non-computational background. Therefore, this review concentrates on intuitive explanations of the different tools and their application, and where available are illustrated with examples of their usage in the mRNA localisation field.

2. Materials

2.1. Computer Environments

As most command line tools are released as open source, this tends to favour the Linux and Mac OSX operating systems. The command line versions of the programs usually provide the most flexibility for use with a full range of customisable options. However, many of the tools are available as web servers and so can be accessed from any computer with a web browser and internet access. A further issue is the computational complexity of many of the algorithms requiring access to high-performance computers (HPC). For example, a typical desktop would be unable to perform Infernal (24) searches across a genome within a practical time frame, so an HPC cluster would be more appropriate. However, in most universities there are HPC facilities available for use by academic researchers, which also provide help on how to use the resources.

2.2. Software

In addition to a web browser for using the online versions of the tools, it is recommended to use a text editor. Most of the tools described in this chapter require the input RNA sequences to be in plain text rather than the default document formats of the common word processor programs. Sequences can be saved in plain text with most word processing programs; however, there are more suitable programs for editing sequence data. For example, the RNA Alignment Editor in Emacs (RALEE) text editor is designed for editing RNA secondary structure alignments on multiple platforms (Linux, Mac OS, and Windows) (25) (Table 1). The colour highlighting of secondary structure elements on each of the sequences makes alignment editing a much more straightforward task. There are a variety of file formats for RNAs and their secondary structures, the most common of which are FASTA, Stockholm, and connect (CT) (Fig. 1c).

Table 1
A list of all the computational tools described in this chapter, organised by chapter subheadings. Each entry has a brief description, Internet address and citation

Program/resource	Brief description	References
A. Text editors		
RALEE	RNA Alignment Editor in Emacs http://personalpages.manchester.ac.uk/staff/sam.griffiths-jones/software/ralee/	(25)
B. Single sequence RNA secondary structure prediction		
Mfold/UNAFold	Minimum free energy secondary structure prediction for single RNA sequences http://mfold.bioinfo.rpi.edu/	(28, 29)
RNAfold (RNAVienna)	Predicts minimum free energy secondary structures for single RNAs http://www.tbi.univie.ac.at/RNA/	(30)
contraFOLD	RNA secondary structure searching prediction based on experimentally determined parameters and probabilistic models http://contra.stanford.edu/contrafold/	(32)
Sfold	RNA secondary structure prediction using a Boltzmann statistical ensemble http://sfold.wadsworth.org	(33)
Pfold	Prediction of RNA secondary structures using stochastic context free grammars on aligned RNA sequences http://www.daimi.au.dk/~compbio/pfold	(34)
RNAkinetics	Calculated RNA secondary structure based on local sequences, rather than the global sequence http://bioinf.fbb.msu.ru/RNA/kinetics	(35)
Vienna+P	Calculates RNA secondary structure in the presence of RNA binding proteins http://bioserv.mps.ohio-state.edu/Vienna+P	(36)
C. Pseudoknot secondary structure prediction		
Pknots	RNA secondary structure prediction for pseudoknots http://selab.janelia.org/software.html	(44)
ILM	Prediction of RNA secondary structures with pseudoknots http://cic.cs.wustl.edu/RNA/	(46)
HotKnots	Pseudoknot RNA secondary structure prediction http://www.cs.ubc.ca/labs/beta/Software/HotKnots/	(45)
D. Measuring secondary structure similarity		
RNAdistance	Calculates RNA secondary structure global similarity based on tree alignment http://www.tbi.univie.ac.at/RNA/	(30)
RNAforester	Local comparison of RNA structures based on tree alignments http://bibiserv.techfak.uni-bielefeld.de/rnaforester/	(47)

(continued)

Table 1
(continued)

Program/resource	Brief description	References
E. Getting homologous sequences		
Rfam	Database of RNA families http://rfam.sanger.ac.uk	(51)
UTRdb	Database for eukaryotic 3′UTR and 5′UTR sequences http://utrdb.ba.itb.cnr.it	(53)
NDB	Database of RNA three-dimensional structures http://ndbserver.rutgers.edu	(54)
Ensemble	Provides sequences data for a large number of model organisms http://www.ensembl.org	(55)
12× Drosophilids	The genome sequences for the 12 sequenced *Drosophilids* http://flybase.org	(56)
BRALIBASE	Database of RNA structural alignments for the benchmarking of RNA alignment programs http://projects.binf.ku.dk/pgardner/bralibase/	(41)
BLAST	Sequence searching based on local sequence alignments http://blast.ncbi.nlm.nih.gov/Blast.cgi	(50)
Infernal	Models sequence and secondary structure similarity for searching sequence databases for homologous RNA structures http://selab.janelia.org/software.html	(24)
RNApromo	Uses a SCFG approach for searching for RNA motifs using experimentally determined or MFE RNA secondary structures http://genie.weizmann.ac.il/pubs/rnamotifs08/rnamotifs08_exe.html	(52)
F. Search for motifs – sequence and structure		
REPFIND	Finds clustered repeats in nucleotide sequences http://zlab.bu.edu/repfind/	(57)
MEME/MEMERIS	Searching for sequence motifs/searching for sequence motifs in single stranded RNA secondary structures http://meme.sdsc.edu/meme4_3_0/intro.html http://www.bioinf.uni-freiburg.de/~hiller/MEMERIS/	(62, 63)
RNAalifold	Prediction of RNA secondary structure from aligned RNA sequences utilising the covariance between sequences http://www.tbi.univie.ac.at/RNA/	(104)
RNAMotif	Searches sequence databases for user defined sequence and secondary structure motifs http://casegroup.rutgers.edu/casegr-sh-2.5.html	(68)
RNALfold (RNAVienna)	Prediction of locally stable RNA structures from within larger sequences and can be performed on a genome scale http://www.tbi.univie.ac.at/RNA	(69)
ERPIN	Identifies conserved RNA structure motifs in sets of unaligned sequences http://tagc.univ-mrs.fr/erpin/	(71)
DNAMAN	Software package for multiple sequence alignment and analysis http://www.lynnon.com	commercial
RNA2DSearch	A bioinformatics pipeline for identifying similar localisation elements http://www.rna2dsearch.com	(70)
RNABob	Pattern match based searching for RNA secondary structures http://selab.janelia.org/software.html	(13)

(continued)

Table 1 (continued)

Program/resource	Brief description	References
G. Align-then-fold methods		
QRNA	Non coding gene finder, comparing synonymous vs. compensatory mutations between two sequences http://selab.janelia.org/software.html	(77)
RNAz	Calculates a structural conservation index by comparing averaged MFE structures against individual MFE structures for a set of RNAs http://www.tbi.univie.ac.at/~wash/RNAz/	(78)
CMFinder	RNA secondary structure searching based on covariance models http://bio.cs.washington.edu/yzizhen/CMfinder/	(79)
PETfold	RNA secondary structure prediction based on evolutionary and thermodynamic information – the sequences need not be homologues http://genome.ku.dk/resources/petfold/	(80)
H. Simultaneous RNA secondary structure prediction and alignment of RNAs		
Dynalign	Simultaneous alignment and secondary structure prediction for two or more RNA sequences http://rna.urmc.rochester.edu/dynalign.html	(83)
FOLDALIGN	Simultaneously folds and aligns RNA sequences based on local or global alignments http://foldalign.ku.dk/	(82)
CARNAC	Predicts RNA secondary structures by simultaneously aligning and folding the sequences http://bioinfo.lifl.fr/RNA/carnac/	(84)
I. RNA tertiary structure		
AANT	Amino Acid–Nucleotide Interaction database, a resource for searching and visualising experimentally determined interactions http://aant.icmb.utexas.edu	(93)
MC-Fold/MC-Sym	Molecular modelling for RNA tertiary structures from sequence http://www.major.iric.ca/MC-Pipeline/	(89)

3. Methods

Unless an RNA localisation signal is solely defined at the primary sequence level then methods more sophisticated than *BLAST* or *MEME* must be employed (26). If the RNAs are distantly related then it is also necessary to take into account the secondary structure. The accuracy of secondary structure prediction from single sequences can be improved by utilising covariance information from homologous sequences. The sequences must contain enough variation to show the covariance so that selection of sequences for prediction is crucial. This is of particular importance in the single-stranded regions of the RNA secondary structures as this is where specific RNA:protein interactions are likely to occur. The covariance between sequences is calculated from alignments of the homologous RNAs. The alignments can be achieved by several methods. First, structures can be aligned from "single sequence secondary structure prediction" methods (e.g. *MFOLD*). Second, "align-then-fold" methods can be used. Or third, "simultaneous folding and alignment" methods provide another alternative. Each of these methods is outlined in the sections below, followed by methods involving RNA tertiary structure prediction and searching.

3.1. Single Sequence Secondary Structure Prediction

The prediction of secondary structure from sequence alone has been developed around maximising the number of the base pairings in the structure. The structures are most commonly scored by the minimum free energy (MFE) giving the thermodynamic stability of the overall secondary structures (27). Table 1 has a full list of the methods outlined in this section. *MFOLD/UNAFold* (28, 29) and *RNAfold* (30) are derived from the same Zucker–Stiegler algorithm (31) for calculating the MFE structures for single RNA sequences, so therefore produce comparable results. A dynamic programming algorithm utilises experimentally determined energy parameters to calculate secondary structures with the lowest free energies and an RNA secondary structure prediction algorithm based on probabilistic models (32). *contraFOLD* improves on the accuracy of both thermodynamic and stochastic context free grammars (SCFGs) based algorithms (see Note 2). *Sfold* assigns probabilities to representative secondary structure motifs from a Boltzmann probability distribution. The motifs are then clustered by structural similarity into ensembles of similar structures (33). The MFE structure is not guaranteed to be included in the representative set but can be added manually during the analysis. *Sfold* is particularly useful in determining whether protein binding sites are accessible throughout the ensemble of structures. *Pfold* predicts the secondary structure of RNAs using SCFGs from an alignment of related RNA sequences (34). *RNAKinetics* calculates the kinetics of folding for a growing RNA sequence thus taking into account that as RNA molecules are produced they form dynamic rather than static structures (35). The result is an ensemble of structures assigned with probabilities. *Vienna+P* is a modification of the RNAVienna package, and models the secondary structure of RNAs in the presence of single-stranded RNA binding proteins (30, 36). The method applies general models for RNA binding proteins and calculates the probability of a nucleotide binding for each of the nucleotides in the RNA. This method does not make use of biological data, such as known RNA binding motifs or the number of nucleotides typically involved in RNA:protein interactions. However, the binding of proteins is very likely to alter the secondary structure in vivo, so represents very important progress in RNA secondary structure prediction. The method also calculates FRET values for fluorophores bound to end nucleotides in RNAs.

There have been several studies assessing the accuracy of RNA secondary structure approaches using experimentally derived structures. Mathews found *Mfold* to have an accuracy of 73% for a large database of 700 nt or less sequences (37). Doshi et al. found an accuracy of 41% for a reference set of 16S and 23S RNA (38). Dowel and Eddy found 56% accuracy for *Mfold*, 55% for *RNAfold*, 50% for *Pknots*, and 39% for *Pfold* for a reference set of RNase P, SRP, and tmRNAs (39). There is also a database of RNA structure alignments, *BRALIBASE*, for benchmarking RNA structure prediction methods (40, 41) (Table 1). Single sequence RNA secondary structure prediction is always most appropriate for initial assessment of the RNA of interest. Once the secondary structure has been determined, this guides the choice of the next methods to employ (see Fig. 2).

3.2. Pseudoknot Secondary Structure Prediction

Pseudoknots provide a challenge in the field of RNA secondary structure prediction due to the added level of complexity (Fig. 1b). The Fragile-X mental retardation protein (FMRP) is a RNA binding

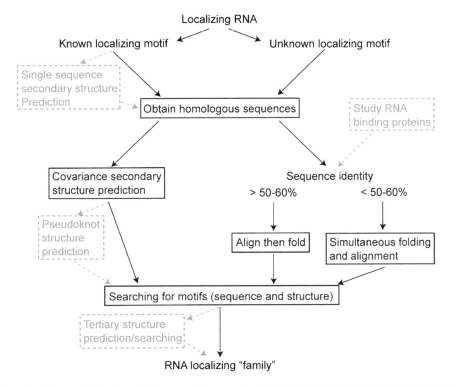

Fig. 2. Flow chart of computational methods representing a generalised approach for the secondary structure prediction and searching for RNA localisation signals. The *blue boxed text* refers to the main subheadings in this chapter. *Green dashed lines* represent methods that are optional as they might not be appropriate in all cases.

protein (KH and RGG domains) and has been proposed as a component of mRNP transport granules (reviewed in (42)). FMRP was recently shown to specifically target mRNA with "kissing-loop" pseudoknots. The loss of FMRP function in Fragile-X syndrome is likely to be due to the inability of the kissing-loop mRNAs to bind FMRP (43). *Pknots* uses a dynamic programming algorithm and thermodynamic energy models for canonical base pairings and ones specific to pseudoknots (44). w*Hotknots* uses a heuristic algorithm to predict non-pseudoknot substructures, which are built up to form a series of secondary structure predictions (45). *Hotknots* utilises an extended version of the standard thermodynamic parameters used in MFE predictions to assess the addition of pseudoknots to the series of predicted structures. The iterative loop matching (*ILM*) approach first predicts stems using a standard non-pseudoknot MFE method and then iteratively predicts further helices (46). The algorithm can either predict the secondary structure of single sequences or use the comparative information in a set of aligned homologous sequences. See Table 1 for a list of the methods outlined in this section.

3.3. Measuring Secondary Structure Similarity

To determine if two RNAs share a common secondary structure or structural motif is possible to calculate a similarity measure between structures (Fig. 2). Although it is trivial to calculate sequence identity to assess the similarity of two sequences, the assessment of the similarity or dissimilarity of RNA secondary structures is more complex. *RNAdistance* encodes RNA secondary structure as tree representations and then performs a global comparison and scores the dissimilarity of the two structures (30). *RNAforester* also encodes structures as trees, but compares structural similarity on a local scale, the advantage being that the algorithm does not get trapped trying to find similarity across the entire length of the sequences being compared (47) (see Note 1 and Table 1). *RNAforester* was recently used in a study of the *internal transcribed spacer 2* (ITS2) gene, which is used as a marker for the classification of new species. ITS2 is a fast evolving gene, but contains conserved secondary structures. The core of the RNA secondary structure was shown to contain a large stem-loop structure, made up from four smaller stems and is conserved across all eukaryotes (48, 49).

3.4. Obtaining Homologous Sequences

In order to improve on the secondary structure prediction of the single sequence methods, it is necessary to make an alignment of homologous sequences (Fig. 2). The alignment can then exploit the covariance of the sequences to determine the nucleotides important for conserving secondary structure. *BLAST* performs pairwise local sequence alignments, where seed alignments are made against a sequence database and are lengthened upon matching (50). BLAST returns a list of sequence hits matching the query sequence ranked in order of their statistical significance. Probabilistic models can be used to search for homologous sequences by finding similarity in both sequence and secondary structures. *Infernal* utilises covariance models (CMs), to describe both the sequence and the secondary structure of the RNAs (24) (see Note 2). *Infernal* is used in the *Rfam* project to annotate the RNAs (51). *Rfam* is a database for RNA families, which are stored as sequence alignments and covariance models (51). Seed alignments annotated with a consensus secondary structure are created for each family, which are used to build covariance models. *RNApromo* utilises SCFGs (see Note 2) to encode an experimentally determined or MFE predicted secondary structure to search for local structure motifs (52). The method was recently applied to the Fly-FISH database of localisation patterns in *Drosophila* embryos and found motifs in 9 of the 94 sets of colocalised mRNAs (1, 52). These covariance models are then used to retrieve further members of the RNA family. *UTRdb* contains the mRNA sequences of the untranslated regions (UTRs) of eukaryotic genomes (53). RNA localisation signals are most commonly, but not exclusively found in the UTRs of the mRNAs as the coding regions are conserved for proteins that they encode (8). Therefore, well conserved secondary structures in UTRs are likely indicators of functional importance. The nucleic acid database (*NDB*) contains the three-dimensional coordinates for 4,406 deposited RNA structures (54). The *Ensembl* project is an integrated resource for the genome annotations and sequence databases for model organisms (55). *Drosophila* is a model organism with a wealth of collated information. One major advantage of using *Drosophila* is the availability of the genome sequences for 12 *Drosophilids*. This allows the covariance information for sequences to be observed by searching across all 12 genomes (56). See Table 1 for a full list of the resources outlined in this section.

3.5. Searching for Motifs: Sequence and Structure

RNA sequences involved in mRNA localisation contain motifs which bind the proteins involved in their translational regulation, transport, and anchoring. The following methods are applicable to search for motifs at the primary and secondary structure levels (see Fig. 2 and Table 1). *REPFIND* finds repeats clustered in nucleotide sequences and has been used to find such repeats in *Vg1* mRNA, which is localised to the vegetal cortex of oocytes in *Xenopus laevis* (57). The *Vg1* localisation signal has been mapped to 340 nt and contains four repeat motifs (E1, E2, E3, E4) which abolish or reduce localisation if deleted (58, 59). The E2 motif, when present as two copies, has been shown to be necessary for localisation and a sequence motif of UUUCAC was found to be crucial (60). The E2 motif was also found to be present in *VegT*, also localised to the vegetal cortex in *Xenopus* oocytes. The E2 motif occurs as five copies in the *VegT* 3 UTR and was also found to be sufficient for localisation (61). *REPFIND* was developed to search for other RNAs containing the E2 motif, specifically the CAC repeat. The same searches were not possible using BLAST (50) due to the short repeat sequences. The CAC motif was found to be present in all the RNAs localising to the vegetal cortex and is conserved across all chordates (57). *MEMERIS* searches for sequence motifs in the context of single-stranded regions and is an extension of the *MEME* algorithm for sequence only motif searching (62, 63). The rational for the searches in single-stranded regions is that some RNA-binding proteins are known to target these single-stranded regions. For example, RNA-recognition motifs (RRM) and K-homology (KH) domains specifically bind RNA via approximately four nucleotides in single-stranded regions (64). *RNAalifold* predicts consensus secondary structures consistent with a set of aligned homologous RNA sequences (65). *RNAalifold* was recently improved with the implementation of the *RIBOSUM* (66) substitution matrices and improved gap scoring in the alignments. *RNAalifold* was used to identify possible binding sites in the GRIK4 3'UTR, which has been implemented in bipolar disorder (67). *RNAMotif* is an RNA motif search algorithm able to search at the primary, secondary, and tertiary structure levels (68). The algorithm allows any base interaction, including non-canonical base pairings to be defined. A recursive algorithm then searches the motif descriptor against a set of sequences to find matches. *RNAMotif* also provides the option of defining a scoring system for the matches, which can also be used to assess the similarity between sequences. *RNALfold* predicts locally stable RNA secondary structures from within long sequences such as whole genomes (69). The predicted stable structures are calculated for a user defined length by folding the entire sequence in overlapping windows of the defined length (see Note 1). *RNALfold* was utilised as part of a pipeline, *RNA2DSearch*, for the identification of localisation signals in transposable elements similar to *gurken* and the *I* factor in *Drosophila*. The transposons, *G2* and *Jockey* were identified in the search and were found to localise in a similar manner (70). *ERPIN* identifies conserved RNA motifs in sets of unaligned sequences annotated with an RNA secondary structure (71). Each single-stranded or double-stranded region is given a profile score, and a dynamic programming algorithm searches for the occurrence of these regions in databases of sequences. The advantage of this method is the ability to search for less rigorously defined regions and pseudoknots. The myelin basic protein (MBP) mRNA is localised to the myelin compartment of oligodendrocytes via an 11 nucleotide hnRNP A2 response element (A2RE) (72). The A2RE is necessary and sufficient for localisation and is present in other mRNAs localising to the myelin compartment (αCaMKII, neurogranin, ARC, Gag, Vpr, and PKMζ) (73). However, a simple sequence search for the A2RE reveals many instances for the A2RE that do not localise. In a recent study, it was found that the A2RE motif contains a tertiary structure kink-turn motif where a bulge in a stem presents the single-stranded bulge nucleotides for protein binding (74). The study used a modified version of *DNAMAN*, however *RNAMotif* would also be able to search for kink-turn motifs with both sequence and secondary structure definitions. An investigation into the localisation signal common to the *Drosophila* genes, *K10* and *Orb* used *RNABob* to show that the signal is not found in other *Drosophila* genes when defined at the secondary structure level (75).

3.6. Align-then-Fold Methods

The align-then-fold methods rely on accurate multiple sequence alignments and a recent evaluation of various alignment methods by Gardner et al. revealed that the *Clustal* multiple sequence alignment program consistently performs well for sets of sequences above 55% sequence identity (41, 76). Below the 50–60% sequence identity "twilight zone" then algorithms using the simultaneous folding and alignment method, incorporating structural information, are required to achieve accurate alignments (see Fig. 2). The *QRNA* non-coding gene finding algorithm uses a comparison of regions of two aligned sequences containing synonymous substitutions, indicating a coding region, against compensatory mutations that test for the presence of stems in a conserved RNA secondary structure (77). *QRNA* utilises hidden Markov models where the input is two aligned sequences (see Note 2). *RNAz* finds thermodynamically stable and conserved RNA secondary structure motifs in sets of aligned sequences (78). A structure conservation index is calculated through the use of a support vector machine (SVM) and the method is suitable for scanning on a genome-wide scale. *CMFinder* finds motifs in sets of unaligned RNA sequences using covariance models (79). One key advantage of *CMFinder* is that the method is applicable to unrelated sequences. Most align-then-fold algorithms assume that the input sequences are evolutionarily related. *PETfold* predicts the structure of aligned RNA sequences by combining evolutionary and thermodynamic information (80). *PETfold* is an extension of the conserved base pair identification of *Pfold* (34) and adds thermodynamic energy parameters (see Table 1).

3.7. Simultaneous Folding and Alignment Methods

These methods utilise the Sankoff algorithm (81) and are applicable when the sequence identity of the set of sequences is below the "twilight zone" of 55–60% identity (see Fig. 2). *FOLDALIGN* simultaneously aligns and predicts the secondary structure of two RNA sequences, then scans the sequences for common local structural motifs (82). The algorithm can also be configured to align the sequences locally or globally across the entire length of the sequences. There is also a multiple sequence alignment version called *FOLDALIGNM*. *Dynalign* aligns and predicts the MFE secondary structure of two unaligned sequences and the predictions can be constrained to experimentally characterised base pairings (83). *CARNAC* predicts secondary structures common to a set of homologous sequences, calculating the structures of all the sequences as stable stems which are then assessed to whether they are common to all the sequences (84). The computational demands of the simultaneous folding and alignment algorithms mean that they should only be realistically considered with alignments below the "twilight zone," as the speed and accuracy of *Clustal* are sufficient for more similar sequences. See Table 1 for a full list of the methods outlined in this section.

3.8. 3D Structure Searching and Prediction

Despite the advances in sequence searching and secondary structure prediction algorithms they fail to take the three-dimensional structure of the RNAs into account. Methods predicting and searching for three-dimensional RNA localisation signals incorporate primary, secondary, and tertiary structure information and will provide the most insight into the parts of the motif determining localisation (see Fig. 2). The searching for structural motifs has been successfully applied in *Saccharomyces cerevisiae* where *ASH1* mRNA is localised to the bud tip of daughter cells and binds the localisation machinery via the RNA-binding protein She2p. There are four known localisation signals, three from the coding region and one from the 3′UTR (85–87), each forming a stem loop. Olivier et al. identified a consensus of two cytosine bases within the stem-loops separated by six other nucleotides corresponding to approximately 28Å (see Fig. 1b). Furthermore, they used *MC-SEARCH* and *RNAMotif* (68), and were able to search through a database of RNA three-dimensional structures to find further examples containing the motif. Two of which were found to localise the bud tip in the same manner as *ASH1* mRNA (88). The computational prediction of RNA tertiary structure is a key goal in the study of RNA localisation signals. The experimental determination of RNA tertiary structure by NMR and X-ray crystallography can be an expensive and time-consuming affair. The *MC-FOLD/MC-SYM* pipeline predicts RNA tertiary structure from sequence (89) predicting all possible "nucleotide cyclic motifs" describing the basic building blocks of the secondary structure. These building blocks are then matched to a database of experimentally determined RNA three-dimensional structures, also processed into these building blocks. A predicted three-dimensional structure is then built up by using the *MC-SYM* algorithm. A recent study comparing the NMR structure and molecular dynamics (MD) simulation of an

RNA tetraloop (UUUU) concluded that the MD simulation and force-fields used in the MD were in good agreement with the NMR structure (90). Although the tetraloop is a fairly simple example of RNA tertiary structure prediction, the force-fields utilised in the MD are likely to be applicable to larger RNA structures such as stem loops involved in RNA localisation signals (see Table 1).

3.9. Study of the RNA-Binding Proteins

Most of the methods described in this chapter do not take into account the proteins that bind the RNA signals. By studying known binding partners of the RNAs, it is possible to deduce a vast wealth of information about the likely RNA signals. For example, whether proteins bind single- or double-stranded RNA and whether the interactions are specific or non-specific. This is particularly relevant to the development of biochemical pull-down assays to identify proteins binding the RNA signals (91). In a recent study of the prediction of protein–RNA interactions by Shulman-Peleg et al., it was observed that the nucleotides interacting with the protein are formed by at least two consecutive bases (92). In addition, over 30% of the interacting bases are π-stacked, allowing the bases in hairpin loops to be available for aromatic interactions while also increasing the stability of the loop. The *amino acid–nucleotide interaction* (*AANT*) database is a resource for searching and visualising experimentally determined interactions (93). The MS2 system for visualising native mRNAs in vivo, was originally developed by Bertrand et al. in *S. cerevisiae* to label *ASH1* mRNA (94) and has since been used to study many other transcripts in a variety of models, including, *nanos*, *bicoid*, *oskar*, and *gurken* mRNA localisation in *Drosophila* (95, 96). mRNAs are engineered to contain multiple copies of the MS2 loop, which then bind an MS2-coat protein fused to green fluorescent protein (GFP). The MS2-coat protein specifically binds the MS2 loops in the RNA, and is able to discriminate against the similarly structured Qβ loops. The Qβ-coat protein shares a low sequence identity to the MS2-coat protein (~20%); however the tertiary structures of the binding domains are very similar, both are alpha/beta two-layered sandwiches (97). A study by Horn et al. investigated the structural basis for the specific binding of MS2 and Qβ coat proteins to their respective RNA loops (98). It was shown that the MS2-coat protein can be engineered to specifically bind the Qβ loops by mutating just two key residues (Asn87 and Asp91). It was also shown that the Qβ-coat protein could specifically bind the MS2 loops by mutating the equivalent residues. This study also gave some insight into which parts of the RNA secondary and tertiary structures are important for specific binding. In the Qβ loops, mutational analysis shows that the last A in the hairpin loop and the bulged A are the only bases that are important for binding affinity when mutated.

3.10. Closing Remarks

There are now a bewildering and comprehensive range of routines developed and used to search and align RNA structure at all levels. The choice of routine for a particular application is not always straight forward. To some extent the applications of some of the routines described here are overlapping. Nevertheless, it is important to enter into a detailed and continuous dialogue between the biologists and bioinformaticians. The most important issues to be included in this dialogue are to consider examples of previous applications of the rivalling method and the available bioinformatics resources, including staff time and computer infrastructure.

4. Notes

Within cells, RNAs are always extensively decorated with proteins, so that global secondary structure prediction is unlikely to represent the full picture for predicting protein binding. It is therefore important to study the locally stable secondary structure predictions for the RNA of interest. Long et al. proposed a two-step binding of mRNAs to their targets where a short region of RNA sequence with a key secondary structure is first bound, followed by a second step of secondary structure relaxation and binding elongation (99). Furthermore, Heale et al. showed that the target substructures of target RNAs are also important for the interaction with siRNA in mammalian cells (100).

A covariance model is a probabilistic model simultaneously describing both the primary and secondary structures of an RNA. The models are analogous to hidden Markov models (HMMs) used for sequence only alignment and searching (101). Covariance models are a form of SCFGs (102, 103), where the nucleotides and base pairings are described in the form of probabilities. By aligning homologous sequences to the original RNA sequence it is then possible to determine the probabilities of insertions, deletions, and nucleotide changes. These probabilities, for sequence and secondary structure, are encoded in a model. Covariance models are used in secondary structure prediction, multiple sequence and structure alignment, and similarity searching against sequence databases.

Acknowledgement

RSH and ID are supported through a Wellcome Trust Senior Research Fellowship (081858) to ID.

References

1. Lecuyer, E., Yoshida, H., Parthasarathy, N., et al. (2007) Global analysis of mRNA localization reveals a prominent role in organizing cellular architecture and function, *Cell* **131**, 174–187.
2. Dahm, R., Kiebler, M., and Macchi, P. (2007) RNA localisation in the nervous system, *Semin Cell Dev Biol* **18**, 216–223.
3. Hengst, U., and Jaffrey, S. R. (2007) Function and translational regulation of mRNA in developing axons, *Semin Cell Dev Biol* **18**, 209–215.
4. Jin, P., and Warren, S. T. (2003) New insights into fragile X syndrome: from molecules to neurobehaviors, *Trends in Biochemical Sciences* **28**, 152–158.
5. Paushkin, S., Gubitz, A. K., Massenet, S., and Dreyfuss, G. (2002) The SMN complex, an assemblyosome of ribonucleoproteins, *Current Opinion in Cell Biology* **14**, 305–312.
6. Mutsuddi, M., Marshall, C. M., Benzow, K. A., Koob, M. D., and Rebay, I. (2004) The Spinocerebellar Ataxia 8 Noncoding RNA Causes Neurodegeneration and Associates with Staufen in Drosophila, *Current Biology* **14**, 302–308.
7. St Johnston, D. (2005) Moving messages: the intracellular localization of mRNAs, *Nat Rev Mol Cell Biol* **6**, 363–375.
8. Palacios, I. M., and St Johnston, D. (2001) Getting the message across: the intracellular localization of mRNAs in higher eukaryotes, *Annu Rev Cell Dev Biol* **17**, 569–614.
9. Palacios, I. M. (2007) How does an mRNA find its way? Intracellular localisation of transcripts, *Semin Cell Dev Biol* **18**, 163–170.
10. Van De Bor, V., Hartswood, E., Jones, C., Finnegan, D., and Davis, I. (2005) gurken and the I factor retrotransposon RNAs share common localization signals and machinery, *Dev Cell* **9**, 51–62.
11. Gavis, E. R., Lunsford, L., Bergsten, S. E., and Lehmann, R. (1996) A conserved 90 nucleotide element mediates translational repression of nanos RNA, *Development* **122**, 2791–2800.
12. Kim-Ha, J., Webster, P. J., Smith, J. L., and Macdonald, P. M. (1993) Multiple RNA regulatory elements mediate distinct steps in localization of oskar mRNA, *Development* **119**, 169–178.
13. Munro, T. P., Kwon, S., Schnapp, B. J., and St Johnston, D. (2006) A repeated IMP-binding motif controls oskar mRNA translation and anchoring independently of Drosophila melanogaster IMP, *J Cell Biol* **172**, 577–588.
14. Macdonald, P. M., Kerr, K., Smith, J. L., and Leask, A. (1993) RNA regulatory element BLE1 directs the early steps of bicoid mRNA localization, *Development* **118**, 1233–1243.
15. Macdonald, P. M., and Struhl, G. (1988) cis-acting sequences responsible for anterior localization of bicoid mRNA in Drosophila embryos, *Nature* **336**, 595–598.
16. Serano, T. L., and Cohen, R. S. (1995) A small predicted stem-loop structure mediates oocyte localization of Drosophila K10 mRNA, *Development* **121**, 3809–3818.
17. Davis, I., and Ish-Horowicz, D. (1991) Apical localization of pair-rule transcripts requires 3' sequences and limits protein diffusion in the Drosophila blastoderm embryo, *Cell* **67**, 927–940.
18. Bullock, S. L., Zicha, D., and Ish-Horowicz, D. (2003) The Drosophila hairy RNA localization signal modulates the kinetics of cytoplasmic mRNA transport, *Embo J* **22**, 2484–2494.
19. dos Santos, G., Simmonds, A. J., and Krause, H. M. (2008) A stem-loop structure in the wingless transcript defines a consensus motif for apical RNA transport, *Development* **135**, 133–143.
20. Stombaugh, J., Zirbel, C. L., Westhof, E., and Leontis, N. B. (2009) Frequency and isostericity of RNA base pairs, *Nucleic Acids Res* **37**, 2294–2312.
21. Leontis, N. B., Lescoute, A., and Westhof, E. (2006) The building blocks and motifs of RNA architecture, *Curr Opin Struct Biol* **16**, 279–287.
22. Lemieux, S., and Major, F. (2002) RNA canonical and non-canonical base pairing types: a recognition method and complete repertoire, *Nucleic Acids Res* **30**, 4250–4263.
23. Holbrook, S. R. (2005) RNA structure: the long and the short of it, *Curr Opin Struct Biol* **15**, 302–308.
24. Nawrocki, E. P., Kolbe, D. L., and Eddy, S. R. (2009) Infernal 1.0: inference of RNA alignments, *Bioinformatics* **25**, 1335–1337.
25. Griffiths-Jones, S. (2005) RALEE--RNA ALignment editor in Emacs, *Bioinformatics* **21**, 257–259.
26. Hamilton, R. S., and Davis, I. (2007) RNA localization signals: deciphering the message with bioinformatics, *Semin Cell Dev Biol* **18**, 178–185.

27. Eddy, S. R. (2004) How do RNA folding algorithms work?, *Nat Biotechnol* **22**, 1457–1458.
28. Zuker, M. (2000) Calculating nucleic acid secondary structure, *Curr Opin Struct Biol* **10**, 303–310.
29. Markham, N. R., and Zuker, M. (2008) UNAFold: software for nucleic acid folding and hybridization, *Methods Mol Biol* **453**, 3–31.
30. Hofacker, I. L. (2003) Vienna RNA secondary structure server, *Nucleic Acids Res* **31**, 3429–3431.
31. Zuker, M., and Stiegler, P. (1981) Optimal computer folding of large RNA sequences using thermodynamics and auxiliary information, *Nucleic Acids Res* **9**, 133–148.
32. Do, C. B., Woods, D. A., and Batzoglou, S. (2006) CONTRAfold: RNA secondary structure prediction without physics-based models, *Bioinformatics* **22**, e90–98.
33. Ding, Y., and Lawrence, C. E. (2003) A statistical sampling algorithm for RNA secondary structure prediction, *Nucleic Acids Res* **31**, 7280–7301.
34. Knudsen, B., and Hein, J. (2003) Pfold: RNA secondary structure prediction using stochastic context-free grammars, *Nucleic Acids Res* **31**, 3423–3428.
35. Danilova, L. V., Pervouchine, D. D., Favorov, A. V., and Mironov, A. A. (2006) RNAKinetics: a web server that models secondary structure kinetics of an elongating RNA, *J Bioinform Comput Biol* **4**, 589–596.
36. Forties, R. A., and Bundschuh, R. (2009) Modeling the interplay of single-stranded binding proteins and nucleic acid secondary structure, *Bioinformatics*.
37. Mathews, D. H., Disney, M. D., Childs, J. L., Schroeder, S. J., Zuker, M., and Turner, D. H. (2004) Incorporating chemical modification constraints into a dynamic programming algorithm for prediction of RNA secondary structure, *Proc Natl Acad Sci U S A* **101**, 7287–7292.
38. Doshi, K. J., Cannone, J. J., Cobaugh, C. W., and Gutell, R. R. (2004) Evaluation of the suitability of free-energy minimization using nearest-neighbor energy parameters for RNA secondary structure prediction, *BMC Bioinformatics* **5**, 105.
39. Dowell, R. D., and Eddy, S. R. (2004) Evaluation of several lightweight stochastic context-free grammars for RNA secondary structure prediction, *BMC Bioinformatics* **5**, 71.
40. Gardner, P. P., and Giegerich, R. (2004) A comprehensive comparison of comparative RNA structure prediction approaches, *BMC Bioinformatics* **5**, 140.
41. Gardner, P. P., Wilm, A., and Washietl, S. (2005) A benchmark of multiple sequence alignment programs upon structural RNAs, *Nucleic Acids Res* **33**, 2433–2439.
42. Bassell, G. J., and Warren, S. T. (2008) Fragile X syndrome: loss of local mRNA regulation alters synaptic development and function, *Neuron* **60**, 201–214.
43. Darnell, J. C., Fraser, C. E., Mostovetsky, O., et al. (2005) Kissing complex RNAs mediate interaction between the Fragile-X mental retardation protein KH2 domain and brain polyribosomes, *Genes Dev* **19**, 903–918.
44. Rivas, E., and Eddy, S. R. (1999) A dynamic programming algorithm for RNA structure prediction including pseudoknots, *J Mol Biol* **285**, 2053–2068.
45. Ren, J., Rastegari, B., Condon, A., and Hoos, H. H. (2005) HotKnots: heuristic prediction of RNA secondary structures including pseudoknots, *Rna* **11**, 1494–1504.
46. Ruan, J., Stormo, G. D., and Zhang, W. (2004) An iterated loop matching approach to the prediction of RNA secondary structures with pseudoknots, *Bioinformatics* **20**, 58–66.
47. Hochsmann, M., Toller, T., Giegerich, R., and Kurtz, S. (2003) Local similarity in RNA secondary structures, *Proc IEEE Comput Soc Bioinform Conf* **2**, 159–168.
48. Schultz, J., Maisel, S., Gerlach, D., Muller, T., and Wolf, M. (2005) A common core of secondary structure of the internal transcribed spacer 2 (ITS2) throughout the Eukaryota, *Rna* **11**, 361–364.
49. Schultz, J., Muller, T., Achtziger, M., Seibel, P. N., Dandekar, T., and Wolf, M. (2006) The internal transcribed spacer 2 database–a web server for (not only) low level phylogenetic analyses, *Nucleic Acids Res* **34**, W704–707.
50. Altschul, S. F., Gish, W., Miller, W., Myers, E. W., and Lipman, D. J. (1990) Basic local alignment search tool, *J Mol Biol* **215**, 403–410.
51. Gardner, P. P., Daub, J., Tate, J. G., et al. (2009) Rfam: updates to the RNA families database, *Nucleic Acids Res* **37**, D136–140.
52. Rabani, M., Kertesz, M., and Segal, E. (2008) Computational prediction of RNA structural motifs involved in posttranscriptional regulatory processes, *Proc Natl Acad Sci U S A* **105**, 14885–14890.
53. Mignone, F., Grillo, G., Licciulli, F., et al. (2005) UTRdb and UTRsite: a collection of sequences and regulatory motifs of the untranslated regions of eukaryotic mRNAs, *Nucleic Acids Res* **33**, D141–146.
54. Berman, H. M., Olson, W. K., Beveridge, D. L., et al. (1992) The nucleic acid database.

A comprehensive relational database of three-dimensional structures of nucleic acids, *Biophys J* **63**, 751–759.

55. Hubbard, T. J. P., Aken, B. L., Ayling, S., et al. (2009) Ensembl 2009, *Nucl. Acids Res.* **37**, D690–697.

56. Stark, A., Lin, M. F., Kheradpour, P., et al. (2007) Discovery of functional elements in 12 Drosophila genomes using evolutionary signatures, *Nature* **450**, 219–232.

57. Betley, J. N., Frith, M. C., Graber, J. H., Choo, S., and Deshler, J. O. (2002) A ubiquitous and conserved signal for RNA localization in chordates, *Curr Biol* **12**, 1756–1761.

58. Mowry, K. L., and Melton, D. A. (1992) Vegetal messenger RNA localization directed by a 340-nt RNA sequence element in Xenopus oocytes, *Science* **255**, 991–994.

59. Deshler, J. O., Highett, M. I., and Schnapp, B. J. (1997) Localization of Xenopus Vg1 mRNA by Vera protein and the endoplasmic reticulum, *Science* **276**, 1128–1131.

60. Gautreau, D., Cote, C. A., and Mowry, K. L. (1997) Two copies of a subelement from the Vg1 RNA localization sequence are sufficient to direct vegetal localization in Xenopus oocytes, *Development* **124**, 5013–5020.

61. Kwon, S., Abramson, T., Munro, T. P., John, C. M., Kohrmann, M., and Schnapp, B. J. (2002) UUCAC- and vera-dependent localization of VegT RNA in Xenopus oocytes, *Curr Biol* **12**, 558–564.

62. Bailey, T. L., and Elkan, C. (1994) Fitting a mixture model by expectation maximization to discover motifs in biopolymers, *Proc Int Conf Intell Syst Mol Biol* **2**, 28–36.

63. Hiller, M., Pudimat, R., Busch, A., and Backofen, R. (2006) Using RNA secondary structures to guide sequence motif finding towards single-stranded regions, *Nucleic Acids Res* **34**, e117.

64. Lunde, B. M., Moore, C., and Varani, G. (2007) RNA-binding proteins: modular design for efficient function, *Nat Rev Mol Cell Biol* **8**, 479–490.

65. Bernhart, S. H., Hofacker, I. L., Will, S., Gruber, A. R., and Stadler, P. F. (2008) RNAalifold: improved consensus structure prediction for RNA alignments, *BMC Bioinformatics* **9**, 474.

66. Klein, R. J., and Eddy, S. R. (2003) RSEARCH: finding homologs of single structured RNA sequences, *BMC Bioinformatics* **4**, 44.

67. Pickard, B. S., Knight, H. M., Hamilton, R. S., et al. (2008) A common variant in the 3′UTR of the GRIK4 glutamate receptor gene affects transcript abundance and protects against bipolar disorder, *Proc Natl Acad Sci U S A* **105**, 14940–14945.

68. Macke, T. J., Ecker, D. J., Gutell, R. R., Gautheret, D., Case, D. A., and Sampath, R. (2001) RNAMotif, an RNA secondary structure definition and search algorithm, *Nucleic Acids Res* **29**, 4724–4735.

69. Hofacker, I. L., Priwitzer, B., and Stadler, P. F. (2004) Prediction of locally stable RNA secondary structures for genome-wide surveys, *Bioinformatics* **20**, 186–190.

70. Hamilton, R. S., Hartswood, E., Vendra, G., et al. (2009) A bioinformatics search pipeline, RNA2DSearch, identifies RNA localization elements in Drosophila retrotransposons, *Rna* **15**, 200–207.

71. Lambert, A., Fontaine, J. F., Legendre, M., et al. (2004) The ERPIN server: an interface to profile-based RNA motif identification, *Nucleic Acids Res* **32**, W160-165.

72. Ainger, K., Avossa, D., Diana, A. S., Barry, C., Barbarese, E., and Carson, J. H. (1997) Transport and localization elements in myelin basic protein mRNA, *J Cell Biol* **138**, 1077–1087.

73. Carson, J. H., Gao, Y., Tatavarty, V., et al. (2008) Multiplexed RNA trafficking in oligodendrocytes and neurons, *Biochim Biophys Acta* **1779**, 453–458.

74. Tiedge, H. (2006) K-turn motifs in spatial RNA coding, *RNA Biol* **3**, 133-139.

75. Cohen, R. S., Zhang, S., and Dollar, G. L. (2005) The positional, structural, and sequence requirements of the Drosophila TLS RNA localization element, *Rna* **11**, 1017–1029.

76. Larkin, M. A., Blackshields, G., Brown, N. P., et al. (2007) Clustal W and Clustal X version 2.0, *Bioinformatics* **23**, 2947–2948.

77. Rivas, E., and Eddy, S. R. (2001) Noncoding RNA gene detection using comparative sequence analysis, *BMC Bioinformatics* **2**, 8.

78. Washietl, S., Hofacker, I. L., and Stadler, P. F. (2005) Fast and reliable prediction of noncoding RNAs, *Proc Natl Acad Sci U S A* **102**, 2454–2459.

79. Yao, Z., Weinberg, Z., and Ruzzo, W. L. (2006) CMfinder–a covariance model based RNA motif finding algorithm, *Bioinformatics* **22**, 445–452.

80. Seemann, S. E., Gorodkin, J., and Backofen, R. (2008) Unifying evolutionary and thermodynamic information for RNA folding of multiple alignments, *Nucleic Acids Res* **36**, 6355–6362.

81. Sankoff, D. (1985) Simultaneous Solution of the RNA Folding, Alignment and Protosequence Problems, *SIAM Journal on Applied Mathematics* **45**, 810–825.

82. Havgaard, J. H., Lyngso, R. B., and Gorodkin, J. (2005) The FOLDALIGN web server for pairwise structural RNA alignment and mutual motif search, *Nucleic Acids Res* **33**, W650–653.
83. Mathews, D. H. (2005) Predicting a set of minimal free energy RNA secondary structures common to two sequences, *Bioinformatics* **21**, 2246–2253.
84. Touzet, H., and Perriquet, O. (2004) CARNAC: folding families of related RNAs, *Nucleic Acids Res* **32**, W142–145.
85. Chartrand, P., Meng, X. H., Singer, R. H., and Long, R. M. (1999) Structural elements required for the localization of ASH1 mRNA and of a green fluorescent protein reporter particle in vivo, *Curr Biol* **9**, 333–336.
86. Chartrand, P., Meng, X. H., Huttelmaier, S., Donato, D., and Singer, R. H. (2002) Asymmetric sorting of ash1p in yeast results from inhibition of translation by localization elements in the mRNA, *Mol Cell* **10**, 1319–1330.
87. Gonzalez, I., Buonomo, S. B., Nasmyth, K., and von Ahsen, U. (1999) ASH1 mRNA localization in yeast involves multiple secondary structural elements and Ash1 protein translation, *Curr Biol* **9**, 337–340.
88. Olivier, C., Poirier, G., Gendron, P., Boisgontier, A., Major, F., and Chartrand, P. (2005) Identification of a conserved RNA motif essential for She2p recognition and mRNA localization to the yeast bud, *Mol Cell Biol* **25**, 4752–4766.
89. Parisien, M., and Major, F. (2008) The MC-Fold and MC-Sym pipeline infers RNA structure from sequence data, *Nature* **452**, 51–55.
90. Koplin, J., Mu, Y., Richter, C., Schwalbe, H., and Stock, G. (2005) Structure and dynamics of an RNA tetraloop: a joint molecular dynamics and NMR study, *Structure* **13**, 1255–1267.
91. Czaplinski, K., Kocher, T., Schelder, M., Segref, A., Wilm, M., and Mattaj, I. W. (2005) Identification of 40LoVe, a Xenopus hnRNP D family protein involved in localizing a TGF-beta-related mRNA during oogenesis, *Dev Cell* **8**, 505–515.
92. Shulman-Peleg, A., Shatsky, M., Nussinov, R., and Wolfson, H. J. (2008) Prediction of interacting single-stranded RNA bases by protein-binding patterns, *J Mol Biol* **379**, 299–316.
93. Hoffman, M. M., Khrapov, M. A., Cox, J. C., Yao, J., Tong, L., and Ellington, A. D. (2004) AANT: the Amino Acid-Nucleotide Interaction Database, *Nucl. Acids Res.* **32**, D174–181.
94. Bertrand, E., Chartrand, P., Schaefer, M., Shenoy, S. M., Singer, R. H., and Long, R. M. (1998) Localization of ASH1 mRNA particles in living yeast, *Mol Cell* **2**, 437–445.
95. Jaramillo, A. M., Weil, T. T., Goodhouse, J., Gavis, E. R., and Schupbach, T. (2008) The dynamics of fluorescently labeled endogenous gurken mRNA in Drosophila, *J Cell Sci* **121**, 887–894.
96. Weil, T. T., Forrest, K. M., and Gavis, E. R. (2006) Localization of bicoid mRNA in late oocytes is maintained by continual active transport, *Dev Cell* **11**, 251–262.
97. Orengo, C. A., Pearl, F. M., and Thornton, J. M. (2003) The CATH domain structure database, *Methods Biochem Anal* **44**, 249–271.
98. Horn, W. T., Tars, K., Grahn, E., et al. (2006) Structural basis of RNA binding discrimination between bacteriophages Qbeta and MS2, *Structure* **14**, 487–495.
99. Long, D., Lee, R., Williams, P., Chan, C. Y., Ambros, V., and Ding, Y. (2007) Potent effect of target structure on microRNA function, *Nat Struct Mol Biol* **14**, 287–294.
100. Heale, B. S., Soifer, H. S., Bowers, C., and Rossi, J. J. (2005) siRNA target site secondary structure predictions using local stable substructures, *Nucleic Acids Res* **33**, e30.
101. Eddy, S. R. (2004) What is a hidden Markov model?, *Nat Biotechnol* **22**, 1315–1316.
102. Eddy, S. R., and Durbin, R. (1994) RNA sequence analysis using covariance models, *Nucleic Acids Res* **22**, 2079–2088.
103. Durbin, R., Eddy, S. R., Krogh, A., and Mitchison, G. J. (1998) *Biological Sequence Analysis: Probabilistic Models of Proteins and Nucleic Acids*, Cambridge University Press.
104. Hofacker, I. L., Fekete, M., and Stadler, P. F. (2002) Secondary structure prediction for aligned RNA sequences, *J Mol Biol* **319**, 1059–1066.

Chapter 28

Computational Prediction of RNA Structural Motifs Involved in Post-Transcriptional Regulatory Processes

Michal Rabani, Michael Kertesz, and Eran Segal

Abstract

mRNA molecules are tightly regulated, mostly through interactions with proteins and other RNAs, but the mechanisms that confer the specificity of such interactions are poorly understood. It is clear, however, that this specificity is determined by both the nucleotide sequence and secondary structure of the mRNA. We developed RNApromo, an efficient computational tool for identifying structural elements within mRNAs that are involved in specifying post-transcriptional regulations. Using RNApromo, we predicted putative motifs in sets of mRNAs with substantial experimental evidence for common post-transcriptional regulation, including mRNAs with similar decay rates, mRNAs that are bound by the same RNA binding protein, and mRNAs with a common cellular localization. Our new RNA motif discovery tool reveals unexplored layers of post-transcriptional regulations in groups of RNAs, and is therefore an important step toward a better understanding of the regulatory information conveyed within RNA molecules.

Key words: Bioinformatics, Motif prediction, Post-transcriptional regulation, RNA secondary structure, SCFGs

1. Introduction

RNA molecules undergo diverse post-transcriptional regulation of gene expression, including regulation of RNA transport and localization, mRNA translation, and RNA decay (1–3). In many cases, such post-transcriptional regulation occurs through elements on the mRNA molecule that interact with the hundreds of RNA binding proteins (RBPs) that exist in the cell (4). A well-known example is the iron responsive element (IRE), a secondary structure RNA motif located on UTRs of members of the iron metabolism and transport pathway (5). The binding of the RBPs Irp1 and Irp2 to IRE elements affects the translation rate of the mRNA, and by that coordinates the response to changing levels of iron in the environment. Other examples include a 118-nucleotide stem-loop structure through which mRNAs are transported to the yeast bud tip by the RBP She2 (6) and the RBP Sbp2 that is involved in mediating UGA redefinition from a stop codon to selenocysteine by binding specific stem-loop structures, termed SECIS elements, in the 3'UTR of selenoproteins (7). In other cases, elements on the mRNA molecule interact with other RNAs that direct the regulatory effect. For example, the recognition and binding affinity of a microRNA to its mRNA target is determined by both the sequence and structure of the target mRNA (8–11).

The examples above suggest that the post-transcriptional regulation of mRNAs is determined not only by its linear nucleotide sequence, but also by its secondary structure. Thus, a key goal is to understand the involvement of mRNA secondary structures in such regulation. One approach is to identify recurring patterns, termed motifs. However, linear sequence motifs, which are commonly found in DNA sequences, are not suitable in this case. Instead, we wish to identify motifs that combine primary and secondary structural elements, and are therefore better suited to describe functional elements in RNA molecules.

We developed RNApromo (RNA prediction of motifs), a new computational method to identify short structural motifs in sets of long unaligned RNAs (12). RNApromo predicts motifs that include both primary sequence and secondary structure elements. We successfully applied RNApromo to analyze several specific sets of experimental data. First, we identified common structural elements in fast-decaying and slow-decaying mRNAs in yeast, and linked them with binding preferences of several RBPs. We also predicted structural elements in sets of mRNAs with common subcellular localization in mouse neurons and fly embryos. Finally, by analyzing pre-microRNA stem-loops, we identified structural differences between pre-microRNAs of animals and plants, which provide insights into the mechanism of microRNA biogenesis.

RNApromo is therefore an important tool in the analysis of RNA sequences and in understanding the regulatory and mechanistic information conveyed within them. Improving our understanding of this information reveals unexplored layers of post-transcriptional regulations, and promotes our overall understanding of cellular regulatory processes.

Here, we shortly describe the algorithmic principles of RNApromo, and provide a detailed guide to how to apply RNApromo for specific sets of RNAs in question.

2. Materials

2.1. RNApromo Online Resources

1. An online version of RNApromo (12) (with limited options) is available at http://genie.weizmann.ac.il/pubs/rnamotifs08/rnamotifs08_predict.html.

2. The full version of RNApromo (12) (an executable for Linux machines) can be downloaded from: http://genie.weizmann.ac.il/pubs/rnamotifs08/rnamotifs08_exe.html. This version also includes the ViennaRNA (13, 14) and RNAcontrafold (15) executables. Specific instructions and system requirements are provided online. The main RNApromo algorithm is easy to use with the script rnamotifs08_motif_finder.pl provided for download.

2.2. Additional Online Resources

1. The emboss software package (16) is available for download from: http://emboss.sourceforge.net/index.html. We use the implementation for the Needleman–Wunsch global alignment algorithm available with this package (see http://emboss.sourceforge.net/apps/cvs/emboss/apps/needle.html) to calculate the sequence similarity in our sets.

2. The ViennaRNA (13, 14) package is available from: http://www.tbi.univie.ac.at/~ivo/RNA. We use the RNAfold and RNAsubopt algorithms for thermodynamic secondary structure prediction available with this package. We use RNAplot to draw RNA secondary structures, and as a basis for the motif drawing. ViennaRNA is also provided with the RNApromo download.

3. RNAcontrafold (15), an additional structure prediction algorithm is available from: http://contra.stanford.edu/contrafold/. RNAcontrafold is also provided with the RNApromo download.

3. Methods

We present RNApromo, an efficient motif discovery scheme for identifying short RNA motifs from a set of unaligned input RNAs. RNApromo takes as input a set of unaligned RNAs, and predicts common motifs in these RNAs, describing both sequence-based and structure-based motif features.

3.1. The Algorithmic Principles of the Motif Discovery Scheme

RNApromo is a new probabilistic motif prediction method, based on stochastic context free grammar (SCFG) representation of RNA motifs, for finding local motifs in a set of unaligned RNAs.

Some of the most successful approaches for motif discovery are based on advanced probabilistic models, as these are highly suitable to capture the observed variation in input sets. SCFGs are a class of probabilistic models proposed for modeling common sequence and structure in a set of input RNAs (17, 18), which replaced the thermodynamic considerations in several existing RNA motif discovery tools (19, 20). However, currently available SCFG applications optimize the model's parameters by essentially considering all possible secondary structures of the input sequences. This approach is successful mainly when it is possible to exploit co-variation between organisms to infer the secondary structure at the motif position. However, when one wishes to identify short motifs, using data from a single organism, co-variation data cannot be used. Moreover, enumerating all possible secondary structures results in a rather high time complexity of the algorithm, making it unfeasible to scan large RNA sets or long RNA sequences.

Unlike currently available SCFG applications, RNApromo restricts the search space to a predefined and limited number of structures for each input RNA, by incorporating prior knowledge about the secondary structure. The motif prediction algorithm is composed of two main parts. In the first part, several heuristics are employed to identify good motif candidates. Specifically, we search for short structural elements that are more common than expected in the input structures, and use them to build an initial SCFG model of the motif. In the second part, the initial SCFG model is refined using an EM algorithm. The algorithm iterates between a step of searching for the best alignment of the current SCFG model to the set of sequences, and using the aligned positions to learn new parameters for the model. This can be done efficiently using a dynamic programming approach specifically fitted for this problem (18).

As we showed (12), RNApromo is able to identify short RNA motifs from the full length sequence in which they embedded. It performs well even on sequences from a single organism, and in the presence of noise in the input set or the structural data.

3.2. Input #1: A Set of Unaligned RNA Sequences

The first, and most important, kind of input to the algorithm is *a set of unaligned RNA sequences* that are expected to share a common motif.

1. Using the full version of the algorithm, there is no limitation on the number of sequences in each set, or their length. However, it is not recommended to have less than eight sequences in a set. Note that as the set becomes bigger (either by sequence length or number of sequences) running time is longer.

2. Limiting the sequence similarity of an input set of RNA sequences. High sequence similarity between different RNAs

within a gene set can skew the results and produce high scores even when a functional motif is not present. Therefore, we recommend filtering out sequences with high sequence similarity.

(a) Using the emboss (16), implementation of the Needleman–Wunsch global alignment algorithm, align each pair of sequences to each other and retrieve the sequence similarity of the alignment (sequence similarity is defined as the percentage of matches between the two sequences over the reported aligned region, including any gaps in the length).

(b) The sequence similarity for an input set is the maximal sequence similarity between any two sequences in the set.

(c) Make sure the maximal sequence similarity within the set is not higher than 90%, by removing one sequence from each pair of sequences with 90% or more sequence similarity.

3.3. Input #2: A Set of Suggested Secondary Structure

The second type of information that RNApromo algorithm needs is, for each input RNA sequence, *a set of suggested secondary structures*. Secondary structures are described using the standard "bracket notation."

1. RNApromo uses prior information about the input sequences' secondary structure to limit the search space. Therefore, it will search motifs only on the suggested structures, and will ignore any other options.

2. There is no limit as to the maximal number of possible structures to provide per sequence, but at least one structure must be provided. Some of the structures can be of only subsegments of the sequence, but at least one structure must be provided for each position in the sequence. Note that as the number of structures per sequence increases, the running time will be longer.

3. What kind of input RNAs structures can the algorithm use?

(a) Ideally, *experimental information* about RNA structure is used, and only the few structures that are consistent with the structural data are given as input.

(b) Alternatively, utilizing existing *thermodynamic-based secondary structure prediction tools* (13, 14) (with 50–90% prediction accuracy (21)) for predicting a small set of thermodynamically stable folds. The main advantage is in integrating thermodynamic considerations into the algorithm (which are not fully embedded into standard covariance model applications). We use the ViennaRNA (13) package and the RNAsubopt (14) thermodynamic folding program to predict all possible folds up to free energy of MFE+0.1.

(c) Certainly, *other structure prediction tools*, such as those based on probabilistic models (15), can also be used to derive a set of highly probable folds as an input to RNApromo. We use the probabilistic folding program CONTRAfold (15) to predict a single structure to each sequence. We did not see any significant difference from using the RNAsubpot predictions.

4. Limiting sequence length for structure prediction. Secondary structure prediction algorithms have relatively long running times, especially for long sequences. Moreover, their accuracy in predicting long range interactions is relatively limited. Therefore, we predict input RNAs secondary structure separately for segments of 200 bp length, with 100 bp overlap.

3.4. Statistical Evaluation of the Biological Signal in the Input Sequences

We use a standard cross-validation test to assess the quality of the biological signal in a given input set. Note that in this analysis we do not attempt to characterize a motif in the input, but to estimate the statistical significance of the hypothesis that such a motif exists.

1. *Negative examples.* For each set of input RNAs, we produce a set of random sequences of equal size, and with the same dinucleotide distribution.

2. *Cross-validation.* In this standard setting, the input set is randomly partitioned into k sets (e.g., $k=5$), and we learn an RNA motif from each of the k possible combinations of $k-1$ sets. We then use that motif to assign a likelihood value to the set of RNAs that were held out when learning it and to the negative examples. Eventually, there is one likelihood value for each real (positive) example and k likelihood values for each negative example. Assigning higher likelihood values to the held out sequences compared with the negative examples suggests that there is a biological signal that is specific to the input RNAs.

3. *AUC score* (area under the curve score). We use the standard receiver operating characteristic (ROC) curve and its associated AUC measure to evaluate the significance of the difference in likelihood; a predicted motif assigns to the input set and to the negative examples. The AUC is between 0 and 1, where values higher than 0.5 indicate that the likelihoods assigned to the real held out motifs are greater than those assigned to the permuted sequences, while values lower than 0.5 indicate the opposite situation. We use the AUC score as an indicator for the ability of a learned motif to separate the real (positive) examples from the permuted (negative) ones.

4. *Background score distribution.* We calculate a background distribution for the AUC scores by repeating the above analysis for many different sets of randomly selected UTR sequences with size similar to that of the tested set and taken from the same organism. Using the background distribution, we can assign a *p*-value to each AUC score to assess its statistical significance as a true biological finding.

3.5. Statistical Models for Motifs in the Input Sequences

After validating that the hypothesis that the input sequences share a motif is statistically significant, we attempt to find the specific description of such motifs.

1. *Learning a motif for entire input set.* We use ten different initialization points, each time learning a new motif. Each of these motifs will be assigned a motif-AUC score, measuring its ability to separate positive from negative examples. Eventually, we chose the model with maximal motif-AUC score as a statistical representation of the predicted motif.

2. *Motif positions and sequence likelihood scores.* Each motif is also assigned to a specific position in each of the input sequences, which is the position of that sequence predicted to express that motif. The algorithm attempts to find the position that best match the motif. Matching a segment of a sequence to a motif is measured by calculating a likelihood score for that segment. The algorithm tries all possible segments, and chooses the one with the maximal score. Some sequences will have higher likelihood scores, suggesting that the motif instance in these sequences match the predicted motif better than other sequences.

3. *Graphical representation of a motif.* The result of the learning process is a probabilistic model. In order to allow a more schematic view of the large number of parameters learned by the model, we provide a simplified graphical representation of a model. This representation is based on the familiar two-dimensional drawing of RNA secondary structures, yet also includes some information about probabilities. The basic shape is that of the most probable structure according to the model probabilities. The color of each structural element (single nucleotides, or pairs) is darker as the probability assigned to it by the model is higher. Sequence elements with high probability are also indicated in this representation, with a color scale that represents their probability (green for low probability and red for high). Looking at this representation, one must bear in mind that although the structure shown is the most probable one, other structures, sometimes quite different, can also have a relatively high probability.

3.6. Predicting an Existing Motif in New Sequences

1. After predicting a motif on a specific input set, it is sometimes beneficial to find the motif in other sequences, which are not part of the original input set. To do that, you must provide RNApromo with the new sequences, and the parameter file of the existing model.

2. A separate script (rnamotifs08_motif_match.pl) allows to perform this analysis easily. The script is provided with the main RNApromo download.

3.7. Example #1: Predicting Known Motifs from the Human Genome

A simple example is using RNApromo to detect a well known human motif – the IRE motif (see Fig. 1).

1. *Input sequences.* We start by estimating the sequence similarity in the input set of sequences. Maximal sequence similarity is less than 50%, so there is no need to filter out sequences.

2. *Predicting secondary structure.* We use the RNAsubopt option, and predict all possible folds up to free energy of MFE+0.1, in 200 bp segments (with 100 bp overlap).

3. *Evaluating input set biological signal.* We use RNApromo to first calculate an AUC score for the input set (AUC = 0.8). We can use a background distribution to assign that score a p-value ($p < 6.5 \times 10^{-4}$). Therefore, the biological signal is statistically significant. Since a background distribution must be calculated separately for each organism, and adjusted to the size of the input sequences, RNApromo does not provide the p-value directly, and it needs to be calculated separately.

4. *Predicting a statistical model.* We use RNApromo to build a statistical model of the motif, based on the input sequences and structures. The output includes the top five predicted motifs, and their likelihood values. The algorithm's output includes two files per suggested model: tab file (containing the model parameters) and an excel (xls) file (containing the information about the specific position of the motif in each input sequence and a per-sequence likelihood score). The motif with the highest motif-AUC score (here number 3) is considered to be the best prediction, and indeed it matches the known IRE motif.

3.8. Example #2: Predicting Pri-microRNA Motifs

microRNAs are a class of small (21–24 nucleotides) noncoding RNAs that play a significant role in regulating gene expression and mRNA stability (22). During their processing, mature microRNAs are cleaved from a pre-microRNA precursor with a stem-loop structure. We used RNApromo to see whether these pre-microRNA precursors have some additional structural characteristics (see Fig. 2).

1. *Input sequences.* We constructed two types of input sets. First, we used real pre-microRNA sequences of all organisms for

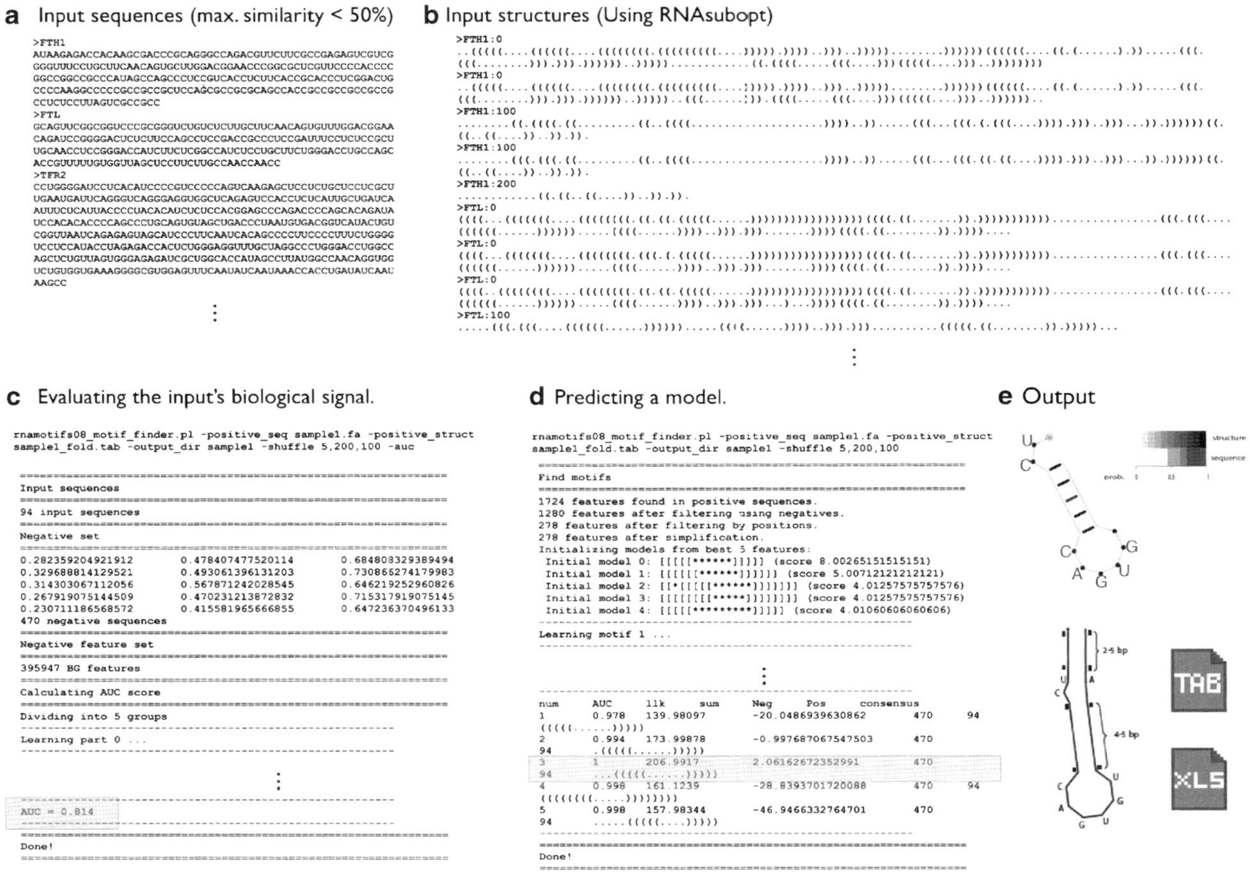

Fig. 1. Example of identifying the IRE motif. (a) Example of input sequences (three out of ten sequences). Similarity is measured by needle (see text) to be <50%. (b) Example of input structures folded by RNAsubopt (see text). (c) Evaluating the input sequences biological signal by calculating AUC score. Here AUC=0.8. Using a separate background distribution can assign a p-value to this measurement (here $p<6.5\times10^{-4}$). The specific command line used is also showed. (d) Predicting a statistical model of the motif in the input sequences. The program outputs five options (default), and we chose model #3, the model with the highest model-AUC (model-AUC=1). The specific command line used is also showed. (e) A graphical representation of the selected model (top) and a description of the known IRE motif (bottom). The algorithm's output includes two files per suggested model: tab file (containing the model parameters) and an xls file (containing the information about the specific position of the motif in each input sequence and a per-sequence likelihood score). The graphical representation is produced from the tab file using the RNAmodel_plot.pl script provided with RNApromo.

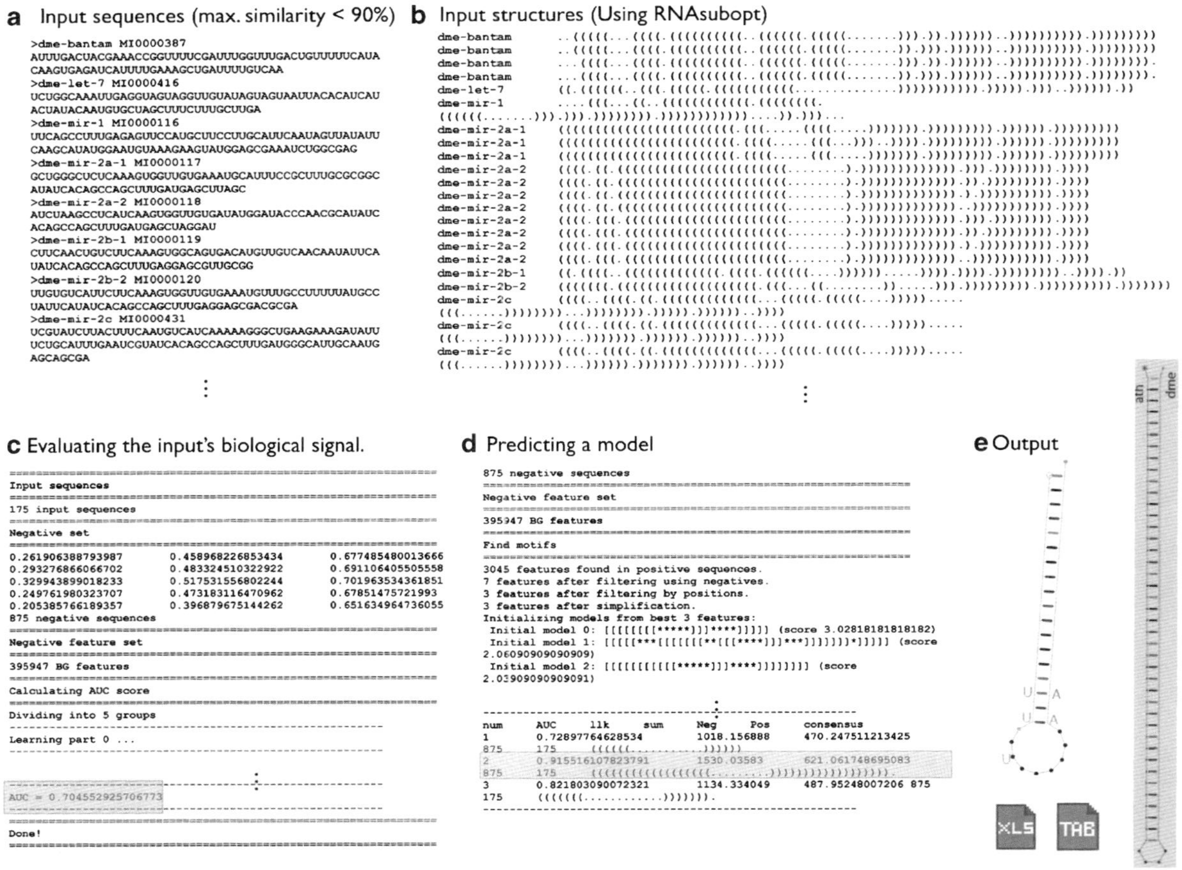

Fig. 2. Example of identifying the stem-loop motif within D. melanogaster pre-microRNA sequences. (a) Example of input sequences (8 out of 157 sequences). Similarity is measured by needle (see text) to be <90%. (b) Example of input structures folded by RNAsubopt (see text). (c) Evaluating the input sequences' biological signal by calculating AUC score. Here AUC = 0.7. (d) Predicting a statistical model of the motif in the input sequences. The program outputs three options, and we chose model #2, the model with the highest model-AUC (model-AUC = 0.92). (e) A graphical representation of the selected model (*left*) compared with a model calculated from random transcribed stem-loops in D. melanogaster genome. Note the longer stem and shorter loop in that model. The algorithm's output includes two files per suggested model: tab file (containing the model parameters) and an xls file (containing the information about the specific position of the motif in each input sequence and a per-sequence likelihood score). The graphical representation is produced from the tab file using the RNAmodel_plot.pl script provided with RNApromo.

which at least ten pre-microRNA sequences appear in the miRBase database (23, 24). Second, we extracted a set of random stem structures from genomic transcripts of each organism, selecting stems with at least 20 bp and an overall size of 50–200 bases. Both types of sets are expected to have stem-loop structures, but we are more interested in seeing the differences between the two types of stem-loops.

2. *Predicting secondary structure.* To predict secondary structure, we use the RNAfold option, predicting the MFE structure, in 200 bp segments (with 100 bp overlap).

3. *Evaluating input set biological signal.* We use RNApromo to first calculate an AUC score for all the input sets, and assign that score a *p*-value. As expected, in all input sets the biological signal is statistically significant.

4. *Predicting a statistical model.* We use RNApromo to build a statistical model of the motif, based on the input sequences and structures. The motif with the highest likelihood value is considered to be the best prediction, and indeed it matches the expected stem-loop structure.

5. *Further analysis of the motifs.* We identified several differences between plant and animal pre-microRNAs. Plant pre-microRNAs have a small loop (5.6 bases on average), a long stem (average 38.1 bp), and an overall length of 160 bases on average. However, animal pre-microRNAs usually have a longer loop (average 10.9 bases), shorter stem (average 32.8 bp), and a much shorter overall length (88 bases on average). On the other hand, the motifs for arbitrary stem-loop transcripts are highly similar between animals and plants. Moreover, in plants, there is no evident difference from real pre-microRNA motifs, whereas in animals, stem-loops of arbitrary transcripts and stem-loops of animal pri-microRNAs have markedly different lengths. Thus, our results suggest that in animals, the nuclear RNase Drosha recognizes and cleaves its pri-microRNA targets, perhaps by measuring the length of their loop, among other features. Longer loops, which are characteristic of pri-microRNAs, enable the enzyme to specifically recognize these transcripts. These results further suggest that in plants, Drosha recognition is not part of the microRNA biogenesis process, and thus a different mechanism may exist. Plant microRNAs might be processed, similarly to other siRNAs, in a mechanism that requires only the Dicer enzyme. Overall, the predicted motifs represent novel and experimentally testable findings, and demonstrate the potential of RNApromo for uncovering post-transcriptional regulatory mechanisms.

4. Notes

1. The version of RNApromo available as an online tool is very limited, and therefore we recommend to download and install the full version. Specific limitations include: cannot compute an AUC score for the input set, cannot work on input sets larger than 5,000 bp (total size).

2. RNApromo does not provide a p-value for the AUC score. This is because calculating a p-value is specifically matched to the organism, and the size of the input set. Therefore, users who wish to calculate such a statistical estimate for the AUC scores must produce the background distributions independently, and use them to calculate p-values.

3. Structure prediction does not have to be done separately. The main RNApromo script provides options to fold the sequences as part of the main run. However, if you need to run RNApromo several times on an input, it is better to calculate the structures separately to save running time.

4. For additional comments and frequently asked questions see our website: http://genie.weizmann.ac.il/pubs/rnamotifs08/rnamotifs08_faq.html.

References

1. Arava, Y., et al., Genome-wide analysis of mRNA translation profiles in Saccharomyces cerevisiae. Proc Natl Acad Sci U S A, 2003. **100**(7): p. 3889–94.

2. Shepard, K.A., et al., Widespread cytoplasmic mRNA transport in yeast: identification of 22 bud–localized transcripts using DNA microarray analysis. Proc Natl Acad Sci U S A, 2003. **100**(20): p. 11429–34.

3. Wang, Y., et al., Precision and functional specificity in mRNA decay. Proc Natl Acad Sci U S A, 2002. **99**(9): p. 5860–5.

4. Anantharaman, V., E.V. Koonin, and L. Aravind, Comparative genomics and evolution of proteins involved in RNA metabolism. Nucleic Acids Res, 2002. **30**(7): p. 1427–64.

5. Hentze, M.W., M.U. Muckenthaler, and N.C. Andrews, Balancing acts: molecular control of mammalian iron metabolism. Cell, 2004. **117**(3): p. 285–97.

6. Olivier, C., et al., Identification of a conserved RNA motif essential for She2p recognition and mRNA localization to the yeast bud. Mol Cell Biol, 2005. **25**(11): p. 4752–66.

7. Krol, A., Evolutionarily different RNA motifs and RNA-protein complexes to achieve selenoprotein synthesis. Biochimie, 2002. **84**(8): p. 765–74.

8. Kertesz, M., et al., The role of site accessibility in microRNA target recognition. Nat Genet, 2007. **39**(10): p. 1278–84.

9. Robins, H., Y. Li, and R.W. Padgett, Incorporating structure to predict microRNA targets. Proc Natl Acad Sci U S A, 2005. **102**(11): p. 4006–9.

10. Long, D., et al., Potent effect of target structure on microRNA function. Nat Struct Mol Biol, 2007. **14**(4): p. 287–94.

11. Zhao, Y., E. Samal, and D. Srivastava, Serum response factor regulates a muscle-specific microRNA that targets Hand2 during cardiogenesis. Nature, 2005. **436**(7048): p. 214–20.

12. Rabani, M., M. Kertesz, and E. Segal, Computational prediction of RNA structural motifs involved in posttranscriptional regulatory processes. Proc Natl Acad Sci U S A, 2008. **105**(39): p. 14885–90.

13. Hofacker L.I., F.W. Stadler P.F., Bonhoeffer L.S., Tacker M., Schuster P., Fast Folding and Comparison of RNA Secondary Structures. Monatshefte fur Chemie, 1994. **125**: p. 167–88.

14. Wuchty, S., et al., Complete suboptimal folding of RNA and the stability of secondary structures. Biopolymers, 1999. **49**(2): p. 145–65.
15. Do, C.B., D.A. Woods, and S. Batzoglou, CONTRAfold: RNA secondary structure prediction without physics-based models. Bioinformatics, 2006. **22**(14): p. e90–8.
16. Bleasby A. Rice P., Longden I. *EMBOSS: The european molecular biology open software suite.* Trends in Genetics, 16(6):276–277, 2000.
17. Eddy, S.R. and R. Durbin, RNA sequence analysis using covariance models. Nucleic Acids Res, 1994. **22**(11): p. 2079–88.
18. Sakakibara, Y., et al., Stochastic context-free grammars for tRNA modeling. Nucleic Acids Res, 1994. **22**(23): p. 5112–20.
19. Holmes, I., Accelerated probabilistic inference of RNA structure evolution. BMC Bioinformatics, 2005. **6**: p. 73.
20. Yao, Z., Z. Weinberg, and W.L. Ruzzo, CMfinder--a covariance model based RNA motif finding algorithm. Bioinformatics, 2006. **22**(4): p. 445–52.
21. Wiese, K.C. and A. Hendriks, Comparison of P-RnaPredict and mfold–algorithms for RNA secondary structure prediction. Bioinformatics, 2006. **22**(8): p. 934–42.
22. He, L. and G.J. Hannon, MicroRNAs: small RNAs with a big role in gene regulation. Nat Rev Genet, 2004. **5**(7): p. 522–31.
23. van Dongen S. Bateman A. Enright A.J. Griffiths-Jones S., Grocock R.J. *miRBase: microRNA sequences, targets and gene nomenclature.* nuc. acid res., 34:D140–4.
24. Griffiths-Jones S. *The microRNA registry.* nuc. acid res., 32:D109–11.

INDEX

A

ABC method 94–95
Acetonitrile 5, 144
Acrylamide/bisacrylamide solution, 20, 24. *See also*
    Polyacrylamide gel electrophoresis
Actin 189, 258, 268, 270, 277, 308,
    309, 311, 353–356, 360, 448
Actin-rich structure 127
Adenine 217, 226, 229, 233, 240, 326
Adherent cell line 251, 260
ADH1 terminator 223, 232
Affinity purification 275, 323–332, 370,
    371, 387–404, 423–443
Affymetrix GeneChip ENCODE 2.0R array 416
Affymetrix gene microarray 414–415
Agarose 36, 37, 60, 78, 85, 89, 97–99, 121, 128, 130,
    191, 195–197, 242–244, 306–307, 310, 311,
    316–317, 326, 327, 331, 357, 360–361, 365,
    376–378, 402, 414, 428, 429, 434, 435
    gel electrophoresis 97, 306–307, 310, 316–317,
        402, 428
    pads 195–197
Alexa Fluor 20, 27, 72–74, 88, 91, 98, 108, 113,
    358, 360
Algae 15
Align-then-fold methods 453, 454, 459
Alignment 273, 449–460, 462, 470
    global 453, 469, 471
    model 462, 470
    Needleman–Wunsch algorithm 469, 471
Alkaline phosphatase 32, 56, 61, 62, 84, 91, 104,
    112, 126, 228, 435
Amino-allyl dUTP 289
Amino-allyl modified Thymine 208, 293
Aminoallyl UTP 378–379
Amino reactive dyes 5
Ammonium sulfate 240, 326
Ampicillin 191, 194
Analysis of variance (ANOVA) 415
Anhydrotetracycline (ATc) 191–198
Antibiotics 36, 73, 79, 163, 167, 168, 172, 194,
    251, 254, 255, 261, 348, 373, 396, 403

Antibody
    alkaline phosphatase-conjugated antibody 84, 104
    anti-digoxigenin antibody 109, 114, 129
    anti-HA agarose-conjugated 327
    enzyme-conjugated antibody 85
    fluorophore-conjugated antibody 104, 111, 114
    immobilization 411–412
    peroxidase-conjugated antibody 84, 410
    T7 antibody 411, 413
Anti-fading agent 218
Aprotinin 179, 207, 210, 227, 327, 391
Arginine 232, 240, 326
Argonaute 371
Ascidian 49–68
AsHeading protein 221, 268
Aspartate 57, 127–128, 134, 240, 326
Aspergillus niger 6
    catalase 6
    TypeVII glucose oxidase 6
ATc. *See* Anhydrotetracycline
AUC score (area under the curve score) 472
Autofluorescence. *See* Background
Avidin 95, 350, 390, 393, 398, 400, 429, 437
Axons
    endoaxoplasmic ribosomal
        plaques (EARPs) 127
    invertebrate unmyelinated 126
    isolated axoplasm 131, 132
    isolation of 337, 338, 341–342, 345
    Mauthner axon dissection 130
    Mauthner axons 130–132, 135
    Mauthner neurons 127
    motor neurons 127, 135, 340
    myelinated axons 126, 127
    pulling 130–132
    pulling solution 127–128, 130, 132
    RNA isolation 345
    RNAs 338, 341, 343, 345, 346, 349
    vertebrate myelinated axons 126
Axoplasm 127, 131, 132
Axoplasmic sprays 132
Axoplasmic whole mount 131, 132, 135, 136

B

Background
    autofluorescence 22, 27, 100
    distribution 473–475, 478
    staining 87, 96, 99–100, 121, 122
Base pairs 120, 146, 448, 454, 456, 458–460, 462
BCIP. *See* 5-bromo–4-chloro–3-indolyl phosphate
Beacon Designer 171
Benjamini–Hochberg method 415
Binding sites 7, 91, 118, 238, 268, 271, 273–275, 454, 458
Bioinformatics 414–417, 447, 452, 462
Biotin
    elution 393, 399
    moieties 390, 398
Blastomere isolation 54
β-Mercaptoethanol (β-ME) 207, 210, 307, 357, 362, 393, 430
Bony fishes 127
Bovine serum albumin (BSA) 6, 21, 26, 46, 55, 57, 63, 65, 73, 76, 87, 95, 108–110, 129, 133, 179, 207, 212, 252, 289, 305, 339, 343, 346, 358, 362, 374, 375, 393, 411
Boyden chamber
    modified 337, 338, 340, 344
BPB. *See* Bromophenol blue
Bracket notation 471
Bradford 373, 375
    reagent 357, 361
Brain
    cryostat sections 109, 115
    freezing 373, 398, 418, 436
    perfusion 64, 65, 67, 120
    stem 130, 131
Breast cancer 360
Bromoacetamide 144, 147–150, 154
5-bromo–4-chloro–3-indolyl phosphate (BCIP) 52, 56, 59, 61, 62, 67, 84–85, 87, 91, 93, 95
Bromophenol blue (BPB) 20, 25, 307, 328, 357, 393, 440
BSA. *See* Bovine serum albumin
Bud. *See* Yeast
Budding yeast. *See* Yeast
Buffers
    blocking 26–27, 56, 61, 64, 109, 114, 129, 289, 294, 358–360, 364, 409, 420
    blocking reagent 56, 87, 111, 129
    buffer B 5, 8, 144, 148, 179, 183, 206, 207, 210, 211
    calmodulin elution 227, 231
    elution 20, 25, 179, 182, 227, 231, 234, 331, 374, 377
    extraction 408, 417
    immunoprecipitation 408–409
    M2 lysis 357
    MWG 289
    PBS-CM 358–360
    phosphate buffered saline (PBS) 33, 86, 108, 128, 162, 163, 178, 192, 304, 339, 356, 373, 393, 418
    polysome lysis buffer (PLB) 373, 375, 380, 408–410, 417–418
    proteinase K 382, 412
    running 85, 307, 317, 357, 363
    saline–sodium citrate buffer (SSC) 86, 87, 108, 113, 120, 252, 289
    sample loading 357, 363
    semi-dry transfer 357–358, 363, 365
    separating 357, 362–363
    sodium carbonate 108, 112, 206, 208
    stacking 357, 363
    TE 144–145, 391
    TNB blocking 129
    TNT 111, 119, 133
    transcription 13, 33, 37, 72, 85, 88, 128

C

Calmodulin affinity resin 227, 231
Campenot chamber 338
Cancer 164, 354, 356, 360, 371, 413
Carasius auratus 130
Carbon source 206, 216, 244, 296, 326
    galactose 206, 244, 296
    glucose 206, 326
Carrier DNA 251, 258
CCD camera 5, 12, 115, 155, 163, 180, 184, 185, 187, 203, 251, 254, 271, 279, 430
cDNA
    microarray 315, 370, 374, 379, 414
    synthesis 183, 293, 317, 339, 342, 346
Cell 3, 15, 43, 50, 71, 83, 103, 125, 141, 159, 175, 189, 203, 221, 237, 249, 265, 287, 301, 323, 335, 353, 369, 388, 408, 424, 448, 467
    body 104–106, 125–126, 336, 338, 341–344, 347–349, 355, 359–361
    body fraction 355, 359, 361
    culture 5, 9, 22–23, 126, 161, 178, 194, 196–198, 210, 225–226, 304, 326–328, 337–339, 356–357, 359–360, 373, 402
    mammalian cells (*see* Mammalian cells)
    yeast (*see* Yeast)
    cycle 189, 218
    homogenizer 305, 311, 331
    lines (*see* Mammalian cells)
    lysate preparation 408, 410–411, 418
    lysis 297, 391–392, 395–396, 417, 418
        animal cells 390, 403
        yeast cells 391, 395–396

migration and invasion 176, 354
polarity.................... 16, 71, 221, 222, 237, 323
Cellular fractionation...............................301–320, 323–332
differential centrifugation 303, 305, 308, 311–313
sequential detergent extraction 303–305, 308
Central nervous system (CNS).....................126, 127, 130, 336–337, 343
Centrifuge... .... 33, 37, 72, 74, 79, 112, 115, 116, 148, 150, 164, 166, 182, 183, 210, 217, 230, 233, 255, 290–292, 304, 306, 311, 313–315, 326, 328, 329, 332, 341–343, 345, 361–363, 375, 376, 379, 382, 393–399, 401, 404, 412, 418, 419
Cerebellum 104, 130
Chambered coverglass 163, 168
Chemical stripping 209
Chemotactic gradients... 354
Chlamydomonas reinhardtii 15
Chloramphenicol (Chp) 191, 194
Chlorophyll autofluorescence 15, 16, 22, 27
Chloroplast..15–17, 20, 22, 26–28
Chordate... 50, 458
Chromatography 19, 22, 163, 169, 358, 440
chromatogram 8, 148, 149
paper..358
Chromosomal integration.. 215, 225
CHX. *See* Cycloheximide
Ciona intestinalis..50, 54, 60
CLIP. *See* Cross-linking-immunoprecipitation
Cloning cylinders 251, 255
CMfinder... 453, 459
CNS. *See* Central nervous system
Coating solution 9
Collagenase 73, 75, 79, 339–341, 348
Colony PCR... 228, 246
Colorimetric ISH 91–93, 95, 126
Comparative threshold (C_t) method................346–348, 350
Confocal ISH. *See* Fluorescence microscopy, in situ hybridization
Confocal microscopy. *See* Fluorescence microscopy
Conjugation..................................... 111, 148, 149, 154, 155
CONTRAfold...451, 454, 469, 472
Controlled-pore glass (CPG) 144, 147
Coplin jar........................21, 23, 26, 207, 212, 213, 217
Cortex.........................50, 53, 57, 59, 63–65, 67–68, 77, 104, 107, 154, 457–458
cortical areas 127
cortical axoplasm 132
cortical layer.. 127
Co-translational translocation 302
Cotton-tipped applicator................................341, 345, 348
Counting..4
algorithm 4
cells...4, 168
spots..4

Covariance......................................452–454, 456–457, 459, 462, 471
Cover slips.................5, 25, 33, 59, 111, 128, 145, 192, 207, 245, 252, 270, 294, 305, 358, 374, 441
CPG. *See* Controlled-pore glass
Cre recombination.. 241, 244
Cre/loxp..215, 238, 239
Cre-mediated excision 215
Cre recombinase 238, 239, 241, 244
C18 reverse-phase column.........................5, 144, 148, 149
Cross-linking.. 372, 400
agent..210, 381
reaction 395, 397
reversal 399
Cross-linking-immunoprecipitation (CLIP)................. 372
Cryosection 126, 127
Cryotomy... 135
ΔC_t 346, 347
$\Delta\Delta C_t$ 346
Cyber-T...383
Cycloheximide (CHX) 289, 297, 304
Cysteine..240, 326, 408
Cytosine arabinoside 339, 341
Cytoskeleton... 268, 308, 311, 336, 388
Cytosol....................16, 17, 160, 164, 290–292, 294, 297, 302–303, 307–312, 318, 324, 326

D

DAB. *See* 3,3'-Diaminobenzydine
Danio rerio. *See* Zebrafish
DAPI. *See* 4',6-Diamidino–2-phenylindole
DDW. *See* Double-distilled water
Decay rate.. 468
Deconvolution software.......... 252, 254, 258, 271, 279, 280
Deep-sequencing.. 414
Deionized formamide...6, 86, 108, 306
DeltaVision 245, 252, 270–271, 277–280
Denaturation 20, 24–25, 75, 98, 130, 133, 183, 226, 229, 241, 243, 246, 293, 310, 311, 324, 377, 379, 391, 439–441
denaturing formaldehyde agarose gel electrophoresis 306–307, 316–317
denaturing profile 150–152
denaturing solution.................. 127, 130, 132, 430, 441
Dendrite.............................. 103–107, 111, 115, 125–126, 335–338, 340, 353, 354
Dendritic spine.. 104–106
Deoxyribonuclease (DNase)....................45, 54–55, 60, 74, 85, 89, 116, 128, 162, 319, 330, 372–374, 394, 401, 402, 418
Deoxyribonucleic acid (DNA).................. 4, 18, 34, 53, 72, 85, 108, 128, 143, 162, 176, 189, 203, 226, 238, 249, 271, 289, 303, 324, 339, 362, 370, 391, 414, 423, 447, 468

Deoxyribonucleic acid (DNA) (*Cont.*)
    linearized DNA template 60, 74, 78
        polymerase chain reaction (PCR).................. 37, 97, 227, 242, 434
    polymerase............................ 96–97, 191, 242, 428, 434
    precipitation... 238, 242
Deoxyribonucleotide triphosphate (dNTP)
        solution 191, 242–244, 330, 394, 402
DEPC. *See* Diethylpyrocarbonate
Desiccant... 155
Dessicator... 207
Dextran sulfate 6, 108, 110, 119, 120, 129, 133, 430
DGC. *See* Drosophila Gene Collection
4′,6-Diamidino–2-phenylindole (DAPI)..............16, 18, 35, 41–44, 109, 114, 160, 179, 184, 186, 212, 213, 256, 305, 309, 314, 430, 441–442
3,3′-Diaminobenzydine (DAB).....................67, 84–85, 88, 93–95, 98
Dibasic potassium phosphate 240, 326
DIC. *See* Differential interference contrast
Dicer... 477
5,6-Dichlorobenzimidazole riboside (DRB) 339, 341, 345
Diethylpyrocarbonate (DEPC).................... 5, 6, 66, 72, 74, 75, 79, 85, 86, 96, 97, 106, 128, 134, 206–208, 210, 211, 217, 303–304, 315, 316, 318, 319, 326–327, 402
Differential interference contrast (DIC)..............11, 16–18, 22, 105, 106, 152, 197, 257
Differentiation..49–50, 58, 221, 323
DIG. *See* Digoxigenin
Digitonin..303–305, 308, 309, 318
Digoxigenin (DIG) 32–37, 41–45, 84, 85, 87, 91–95, 98, 108, 112, 133
    nucleotides.. 134
    RNA..110, 116, 128
        (*see also* RNA probes)
    RNA labeling mix...................................54–55, 60, 130
DiI..53, 57, 64, 67–68
Dimethyl sulfoxide (DMSO)........................20, 24, 95, 96, 112, 128, 129, 132, 226, 229, 297, 304, 393, 429, 438, 439
Dinucleotide distribution 472
Dissection..............................34, 38–39, 106, 130, 270, 338
Dissociation constant 185
Dithiothreitol (DTT).................... 54–55, 60, 74, 110, 116, 128, 144, 148, 179, 227, 241, 289, 290, 296, 304–305, 327, 328, 331, 357, 364, 373, 374, 391–393, 408, 409, 429
DMF. *See* N,N-dimethylformamide
DMSO. *See* Dimethyl sulfoxide
Dodecylmaltoside 303
Dorsal root ganglion (DRG) 337–341, 344, 347–349

Double-distilled water (DDW).....................85–88, 96, 97, 238–244, 252, 326–328, 330, 391, 393, 400, 402
Dounce homogenizer 291, 326
DRB. *See* 5,6-Dichlorobenzimidazole riboside
3D representation 11
DRG. *See* Dorsal root ganglion
Drosha.. 477
Drosophila...................................... 4, 9, 31–47, 49–50, 65, 152, 237, 265–281, 303, 413, 417, 447, 448, 457–459, 461
    collection 33–34
    dissection 34, 38–39, 270
    double FISH...35, 42, 44
    egg chamber... 152, 275
    embryos and tissues 33–36, 39–44
    fixation...33–34, 38–39
    mounting 35, 36, 41–42
    nurse cell 152
    oocyte..269–271, 276, 279–280
    ovariole 154
    single FISH 34–35, 39–42
    storage..35, 39, 41–42
Drosophila Gene Collection (DGC) 36
Drosophila melanogaster. *See Drosophila*
DTT. *See* Dithiothreitol
Dual wavelength detector.. 5, 144
Dynabeads...381, 409, 418, 419
Dynein... 448, 449

E

ECL. *See* Enhanced chemiluminescence
EDTA. *See* Ethylenediaminetetraacetic acid
Egalitarian...449
EGFP. *See* Enhanced green fluorescent protein
EGTA. *See* Ethylene glycol tetraacetic acid
eIF4A protein... 190, 198
Electrophoretic mobility
        shift assay..............................179, 180, 182–183
Electroporator 251
Elution stream 148, 149
EM algorithm.. 470
Emboss software package 469
Embryo..............................4, 32, 49, 83, 106, 160, 266, 303, 323, 336, 413, 447, 468
Emission wavelength... 151, 152
EMT. *See* Epithelial-mesenchymal transition
ENCODE... 416
End-labeled oligonucleotides 208, 316–317
Endogenous mRNA.................. 15–16, 175–187, 237–246, 266, 268, 276, 324
Endoplasmic reticulum (ER)..........................16, 52, 53, 57, 63–65, 204, 287, 290, 292, 301–320, 324, 326, 387–388

Endosome 160–161
Enhanced chemiluminescence (ECL) 307, 318, 358, 364
Enhanced green fluorescent protein (EGFP) 176–178, 185, 186, 190
Enzyme-conjugated anti-tag antibody. *See* Antibody
Epifluorescence 18, 22, 128, 132, 197, 203, 206, 216
Epithelial-mesenchymal transition (EMT) 354–355
Equilibration buffer 6, 10
ER. *See* Endoplasmic reticulum
Erlenmeyer flask 19, 23, 394
Escherichia coli 182, 190, 193–197, 435
 competent 179, 191, 251, 429, 435
 HB101 218
 tRNA .... 6, 21, 25, 108, 110, 207, 211 (*see also* RNA, tRNA)
Ethanol 5–7, 9, 21, 24, 33, 36–37, 44, 57, 61, 63, 67, 74, 78, 97, 107, 110, 112, 116, 117, 128, 130, 144, 148, 150, 191, 207, 211, 238, 242, 252, 256, 292, 306, 315, 342, 357, 362, 376, 378, 382, 401, 409, 412–413, 440–442
Ethylenediaminetetraacetic acid (EDTA) 19, 20, 23, 57, 72–74, 85, 89, 144, 161, 163, 167, 179, 191, 226, 227, 240, 293, 304, 306, 327, 357, 360, 370, 372–374, 377, 378, 380, 382, 391–393, 408, 409
Ethylene glycol tetraacetic acid (EGTA) 57, 73, 127, 227, 304–305, 357
Excitation wavelength 20, 21, 152
Eyebrow tool 128, 132

F

False discovery rate (FDR) 380
FASTA 449, 450
Fast axonal transport 126
Fast penetration 127
Fast Red 56, 84, 87, 88, 91, 93, 94
FBS. *See* Fetal bovine serum
F_{buffer} 152
F_{closed} 152
FDR. *See* False discovery rate
Fertilization 50, 53, 54, 89
Fetal bovine serum (FBS) 109, 161, 163, 178, 183, 184, 251, 255, 304, 356, 373, 392
Fibroblasts 161, 162, 164, 337, 338, 347, 353, 354
Fibrosarcoma 356
Filters (polycarbonate; 0.4, 1, 3 and 8 μm pores) 354
Fire-polish 99, 339, 341, 347
FISH. *See* Fluorescence in situ hybridization
FITC. *See* Fluorescein isocyanate
Fixation 6, 9, 19, 23–24, 27, 33–40, 45, 53, 55, 57, 59, 61, 66, 67, 73, 75–76, 78, 80, 84, 86, 89, 94, 95, 97, 98, 113, 121, 135, 204, 210–211, 217, 305, 429–430, 440–442

Flagella 15–17
Flow cytometry 191–196
Fluorescein isocyanate (FITC) 21
Fluorescence 4, 15, 35, 53, 85, 104, 128, 142, 190, 203, 239, 293, 324
 intensity 81, 150, 165, 170, 171, 185, 187, 218, 222, 273–275, 279, 280
Fluorescence in situ hybridization (FISH) 4, 11, 15–28, 32, 34–45, 103–122, 203, 204, 208, 211, 212, 216–218, 251, 253, 256–257, 260, 261, 296, 324, 347, 424–425, 427, 439, 442, 447, 457
 combined with protein immunostaining 65
 high-throughput FISH in 96-well plates 32
 multiple transcripts (multi-color) 85, 93–94, 100, 427
  sequential 85
  simultaneous 85, 92–93, 427
 single molecule 4, 6, 11
 single transcript (one color) 343
Fluorescence microscopy
 artifacts 389
 autofluorescence 15, 16, 22
 background 324
 confocal microscope 12, 22, 28, 35, 53, 62, 63, 65, 77, 128, 132, 155, 203, 216, 278, 430
 image acquisition 5, 277
 image deconvolution 104, 277
 immunofluorescence 308, 309, 311
 in situ hybridization 20–21, 50, 104, 106–109, 111–119, 128–129, 132–133, 203–204, 208–213, 231, 324, 347, 424–425, 429–430, 437–442
Fluorescent dyes 3, 4, 20, 27, 59, 100, 135, 203, 290, 293, 370, 372, 429, 437–440, 442
 Alexa Fluor 20, 27
 Alexa Fluor 594, 7, 11, 96, 143
 Cy3 .... 100, 293, 370, 439
 Cy5 .... 100, 293, 370, 439
 Tetramethylrhodamine (TMR) 7, 11, 12, 143, 152
 Tetramethylrhodamine–5-iodoacetamide 144, 147, 148, 150
 Texas Red C5 bromoacetamide 144, 147, 148
 tyramide 32, 35, 41–43, 45, 46, 53, 56–57, 59, 66, 85, 88, 98, 104, 107, 111, 115, 119, 122, 129, 133
Fluorescent granules 245
Fluorescent-labeled DNA/RNA probes .... 324. *See also* Fluorescence in situ hybridization (FISH); Fluorescent dyes
Fluorophore-conjugated antibody. *See* Antibody
F_{open} 152
Formaldehyde 6, 9, 33, 34, 38, 45, 55, 57, 61, 62, 66, 67, 73, 121, 217, 230, 305–307, 311, 316–317, 319, 365, 377, 381, 390–392, 395, 397, 400, 426–427, 442

Formamide................5, 6, 9, 12, 20, 21, 25, 26, 34, 55–57, 66, 67, 86, 87, 90–91, 98, 106, 108, 110, 113–114, 117, 118, 120, 122, 129, 135, 144, 207, 211–213, 217–218, 252, 256, 306, 316, 430, 440
Fragile-X syndrome 447, 456

G

ΔG... 146, 147, 154, 155
Galactose........................ 206, 215, 226, 229, 230, 239–241, 244, 246, 289, 290, 296
  promoter 223, 224
GAPDH. *See* Glyceraldehyde 3-phosphate dehydrogenase
G+C content 6–7, 12, 111, 120, 121, 155, 208, 217–218
Gene expression..........................50, 59, 71, 83, 84, 100, 104, 159, 165, 175, 249, 260, 348, 369, 372, 407, 448, 467, 474
  analysis..84, 165, 407
  arrays.. 370, 372
  spatial pattern 31
  temporal pattern 31
Gene Pulser Cuvette... 251
GeneSpring GX10 software 415
Gene tagging 237, 324
Genome...............15, 20, 54, 204, 238, 241, 246, 250, 261, 274, 287–297, 324, 350, 370, 387, 389, 396, 397, 407, 416, 417, 450, 452, 453, 457–459, 474, 476
Genotype..39, 270, 276–277, 280
Germ cell..50, 51, 53, 265–266
Germ granule 53
GFP. *See* Green fluorescent protein
Giant axon.. 127
Glass bottom tissue culture dishes 12, 163, 166, 168, 184, 252
Glass bottom tissue culture plates....................10, 252, 258
Glass cover slides 195
Glioma.. 356
Glucose............................6, 12, 57, 127, 206, 225, 226, 233, 240, 241, 326–328
Glutamic acid 232, 240, 326
Glutamine..5, 9, 161, 356
Glyceraldehyde 3-phosphate dehydrogenase (GAPDH) 308, 310, 312, 320, 413, 414, 420
Glycerol phosphate dehydrogenase promoter (pGPD) 223–225, 228
Glycine solution 391, 392, 395
Glycogen............................74, 110, 117, 130, 382, 393, 401, 409, 412
Goldfish..127–128, 130, 135
Golgi..16, 326, 331
G10 protein 415
Green fluorescent protein (GFP) 15, 16, 85, 95, 176, 204–206, 215, 217, 218, 222–226, 228, 229, 231, 233, 238, 239, 241, 245–246, 250–254, 258–262, 267, 269, 274, 275, 277–278, 281, 324, 347, 371, 389–392, 394–398, 400, 403, 461
Growth cone..104, 105, 126, 336, 347, 353
Guanidinium-HCl 292

H

Hairpin.............................6, 142, 143, 147, 154, 155, 198, 449, 461
Halocarbon Oil...700, 145, 154
Hemaglutinin (HA) epitope............................. 325
Heparin........................34, 86, 97, 109, 115, 227, 289, 292, 297, 373, 374, 380, 382
HEPES. *See* 4-(2-hydroxyethyl)–1-piperazineethanesulfonic acid
High numerical aperture 5, 12
High pressure liquid chromatography (HPLC)........... 5, 8, 19, 22, 106, 108, 111, 129, 144, 148–150, 439
  reverse phase C–18 column 5, 8, 144
High-throughput FISH in 96-well plates. *See* Fluorescence in situ hybridization (FISH)
Hippocampal neurons ....5, 11, 105, 106, 337, 340. *See also* Mammalian cells
  culture.. 111, 113
  fixation... 113
Hippocampus 5, 9, 104
Histidine..226, 240, 243–245, 326
Hoechst dye..53, 62, 355
HO endonuclease 221–222
Homologous recombination 176, 215, 238, 239, 324
Human Gene 1.0 ST array 414–415
HuR...413, 414, 417
Hybridization 3, 15, 31, 49, 84, 104, 128, 142, 160, 203, 222, 289, 324, 424, 447
  buffer.........................6, 21, 25, 26, 34, 55, 57, 66, 108, 110, 117, 118, 120, 144, 152, 289, 294
  chamber........................... 213, 256, 294, 374, 379, 383
  kinetics.. 151, 164
  oven... 20, 26, 307
  solution 10, 25, 26, 66, 68, 90, 93, 99, 113, 114, 118, 129, 133, 135, 256, 307, 316, 374, 379, 383
  temperature... 26, 64, 121, 320
  time 58, 135
Hydrophobicity 8, 148
4-(2-hydroxyethyl)–1-piperazineethanesulfonic acid (HEPES)................ 57, 73, 86, 127, 179, 227, 293, 304, 305, 356, 373, 379, 408, 429
Hypothalamous-hypophyseal tract................... 126

I

iCycler iQ5 145
I factor 448, 458
IGB. *See* Integrated Genome Browser
IgG sepharose 227, 230
IKA/Vibrax shaker 395–396
Imaging 3–13, 54, 59, 62, 72, 73, 75–77, 81,
    141, 142, 150–155, 159–173, 175–187, 190,
    195, 203–205, 208, 216–218, 222, 245,
    249–262, 268, 270–271, 275–281, 337, 338,
    358, 430, 442
    2D 452, 458
    3D 11, 115, 141, 257, 277, 460–461
    image acquisition computer and software 5, 145,
        163, 271, 279
    image analysis software 165, 252, 258, 271
        image J 135, 136, 145, 154, 156, 163,
            165, 168, 170, 172, 196, 197, 252
        MetaMorph 165, 252, 271, 280, 281
    image processing 5, 11
    live cell 142, 150, 159–173, 204, 205, 208,
        216, 218, 245, 251–252, 257, 258, 270
    MATLAB image processing 5
    multiplex mRNA imaging 4, 7, 32
    real time 268
    TIFF images 11, 280
Immunoblot 179, 308, 311, 349, 371, 376, 377
Immunofluorescence 15–28, 46, 73, 75–77, 80,
    109, 114, 136, 260, 308, 309, 311, 358, 365
Immuno-fluorescence in situ hybridization
    (FISH) 15–28
Immuno-labeling 58–59, 61–62, 64
Immunoprecipitation (IP) 179–181, 183, 349,
    370, 372, 375, 377–380, 383, 407–410,
    412–414, 418–420
INT. *See* 2-(4-iodophenyl)–3-(4-nitrophenyl)–5-
    phenyltetrazolium chloride
Integrated Genome Browser (IGB) 416, 417
Integrated strain 245
Internal transcribed spacer 2 (ITS2) 456
Iodoacetamide 144, 147, 148,
    150, 154
2-(4-iodophenyl)–3-(4-nitrophenyl)–5-phenyltetrazolium
    chloride (INT) 93
IPTG. *See* Isopropyl-β-thiogalactoside
IRE. *See* Iron responsive element
Iron metabolism 467
Iron responsive element (IRE) 467–468,
    474, 475
Isolated cortex 53, 59, 63–64, 67
Isomer 16, 20, 150
Isopropyl-β-thiogalactoside (IPTG) 179, 182,
    191–195, 197
ITS2. *See* Internal transcribed spacer 2

K

K-homology (KH) domains 455–456, 458
Kimwipes 26, 27, 213
Kinetics 151, 164, 191, 196, 197, 249, 250,
    451, 454

L

LacZ reporter 213, 215, 271
Laminin 5, 9, 337, 339, 340, 348
Laplacian of Gaussian filter 11
Leucine 226, 240, 326
Leupeptin 179, 207, 210, 227, 327, 357,
    364, 391
L-glutamine 5, 9, 161, 356
Ligase-T4 DNA ligase 251, 428, 429
Likelihood score 473–476
Linear elution gradient 148, 149
Liquid nitrogen 115, 230, 233, 375, 395, 398
Lithium acetate (LiOAc) 240–243, 391, 394
Locked nucleic acid (LNA) 142–144, 147–150,
    152, 154, 155
Loxp. *See* Cre recombination
Lumbar ventral roots 126
Lysine 9, 232, 240, 326
Lysosome 160

M

Magnesium chloride (MgCl$_2$) 20, 26, 56, 57, 72,
    86, 87, 108, 109, 113, 114, 119, 120, 128, 144,
    145, 227, 241–244, 293, 304, 305, 313, 330,
    358, 373, 374, 380, 382, 391, 392, 406, 427, 429
Magnetic beads 408, 410–412
Maleimide 154
Mammalian cell culture 4, 9, 161, 304, 392–393,
    397–398, 402
    media 19, 339, 340, 343, 345, 373
        complete culture 339–341, 343, 345
        DMEB/F12 339–341
        Dulbecco's Modified Eagle's Medium
            (DMEM) 178, 183, 184, 251, 304,
            339, 340, 373, 392
    neuronal cell culture 104, 106–109, 111–115, 126
Mammalian cells
    cancer cells 164, 354, 356, 360, 371, 413
    Chinese hamster Ovary (CHO) cells 4
    DU145 prostate cancer cells 354, 356
    embryonal carcinoma cells 160
    E18 Primary Hippocampal neurons 5, 9, 106
    fibroblasts 161, 162, 164, 337, 338, 347
    HEK293T 308, 318, 319, 397
    HeLa cells 178, 181, 183, 184, 186, 251, 255,
        261, 302, 373, 380–382, 413
    hippocampal neurons 5, 337, 340

Mammalian cells (*Cont.*)
  HT1080 fibrosarcoma cells 354, 356
  lysis 297, 392–393, 398
  MCF-7 161
  MDA-231 breast cancer cells 354–356, 360
  MDA-435 breast cancer cells 354, 356
  MEF 160, 170
  MSV-MDCK-INV cells 354
  P19 160
  stem cells 164, 410
  transfection 178, 185–187, 254, 255, 258, 261, 262, 392, 396–398, 402
  U87 glioma cells 354, 356
  U251 glioma cells 354, 356
Marker extraction 239, 241, 244
Mass spectrometry analysis 225, 226, 371, 500
Maternal mRNA 49–68, 265
Mauthner neurons. *See* Axons
Melting temperature 68, 120, 147, 150
MEME algorithm 296, 458
Metastatic tumor cell lines 356
Methionine 232, 240, 245, 246, 326, 394
  starvation 395
MET25 promoter 215, 246, 390
Mfold 142, 144, 145, 154, 156, 171, 267, 451, 454, 455
Mice 115, 120, 127–128, 132
Microarrays 290, 293, 315, 355, 413
  analysis 294–296, 324, 354, 365, 369–383, 407, 414–417
  hybridization
    best-fit trend line 295
    fluorescent dye labeling 370, 372
    fold-change 295
    labeled cDNA 294
    laser scanner 294
    reference sample 294
    spotted microarray 293, 294, 379
    standard deviation 295
Microfilaments 59, 189
Microfluidics 337, 436
Microinjection
  apparatus 73, 145, 163
  needles 72
Microparticles 339, 343–345, 349, 350
Micropinocytosis 160
microRNA. *See* RNA
Microscopy
  CCD camera 5, 12, 115, 155, 163, 180, 184, 185, 187, 203, 251, 271, 279, 430
  confocal microscope 12, 22, 35, 53, 62, 63, 65, 77, 128, 132, 155–156, 203, 216, 278, 430
  cover slips 41, 59, 132, 134, 192, 245, 252, 277, 305, 313, 358

  slides 23, 25, 35, 109, 115, 119, 192, 207, 430
  wide-field fluorescence microscope 5, 115
Microtubules 59, 104, 105, 268, 270, 277, 318, 354, 448
Minimum free energy (MFE) 146, 451–454, 456, 457, 460, 471, 474, 477
miRBase database 477
Mitochondria
  isolation of 287–297
Mitochondrial DNA (mtDNA) 186
Mitochondrial Hsp70 .... 355, 358, 364
Mitochondrial RNA (mtRNA) 176
Mitotic inhibitors 337
Molecular beacons
  Black Hole Quencher 143, 161, 162
  Cy5 .... 143, 152, 155, 161–163, 166–168, 170, 171, 173
  2'-O-methyl 142–144, 147–152, 154, 155, 161, 162
Molecular Extinction Coefficient (MEC) 209
Molony murine leukemia virus reverse transcriptase 328
Mosaic development 50
Motif
  prediction 467–478
  sequence 452, 457–459, 468
  structure 452, 454, 457, 459
Motor proteins 221, 371
Mounting medium 6, 7, 12, 13, 58, 120, 206–207, 213, 305, 314, 358, 365
Mouse 21, 34, 35, 58, 59, 73, 96, 107, 109, 114, 121, 127, 161, 179, 339, 358, 364, 375, 410, 413, 418, 468
mRNA
  β-actin 11, 126, 134, 135, 254, 260, 343, 347, 349, 350, 353, 354, 413, 414
  γ-actin 347, 349
  anchoring 49–68, 154, 324, 388, 448, 457
  Arc/Arg3.1 .... 104, 107
  ASH1 .... 212, 221, 226, 268, 389, 460, 461
  bicoid (bcd) 266, 448, 461
  αCaMKII 459
  CaMKIIα 103, 104, 106
  c-myc 162
  compartmentalization 301, 336
  detection 27, 62, 109, 111, 141–156, 381
  eve 448
  expression 54, 83–100, 103, 216, 246, 254, 260, 388, 395
  fs(1)K10 448
  G2 458
  Gag 459
  GFAP 349
  Gurken (grk) 266, 272, 448, 458, 461
  hairy 448

isolation
    from adult worms 4
    from bacteria... 4, 9
    from cultured mammalian cells.......................... 4, 9
    from drosophila embryos................4, 9, 33, 49, 265, 268–271, 274–276, 279, 303, 457
    from fruit fly 4, 9
    from tissue sections...................................... 49, 115
    from yeast 4, 9, 49, 203–218

Jockey..458
K-ras.. 161
LacZ..................................... 197, 198, 213, 215, 271
localization.................. 18, 32, 50, 52, 58, 59, 67, 71–81, 103–105, 126, 134, 176, 185, 223, 238, 288, 296, 301–320, 324, 347, 353, 354, 371, 389, 407–420, 467
localization elements..............................213, 221, 222
MAP2..11, 104, 107, 349
MAP1B... 105, 107
myelin basic protein (MBP) 458
nanos (nos) 266, 448, 461
neurogranin 459
4-Oct... 160, 162
Orb 448, 459
oskar (osk).................. 152–153, 265, 266, 281, 448, 461
oxytocin 126
PKMζ..459
postplasmic/PEM 50–54, 59
sorting... 303
TAF 2
trafficking 213, 324, 388
transcript........................13, 83–85, 88, 90–94, 98, 100, 125, 128, 130, 175, 181, 211, 213, 215, 221, 246, 249, 250, 256, 257, 261, 262, 267, 273, 276, 294, 303, 324, 345, 349, 387, 388, 407
vasopressin 126
VegT..458
Vg1 457
visualization....................................239, 241, 245
Vpr...459
wingless..448
mRNP complex...................................141, 381, 388
MS2, 172, 176, 204–206, 213, 215, 218, 222, 238, 239, 246, 249–257, 259, 260, 265–281, 389, 391, 392, 461. *See also* RNA aptamer
MS2 aptamer-tagged mRNA.........389–392, 394, 397, 402
MS2 coat protein (MS2-CP).......... 204, 205, 222, 238, 250, 253, 257, 260–262, 268, 274, 324, 389, 461
MS2-CP-GFP....................... 205, 215, 218, 246, 250–254, 258–262, 395
m-TAG...237–246, 324
Multiple cloning site (MCS)............................... 193, 453
Multiple sequence alignment...........................452, 459, 460
Myo4..222, 225, 226, 371

N

NADH dehydrogenase (ND)................................ 176, 184
Nail polish............................. 6, 11, 13, 20, 26, 35, 41, 63, 65, 213, 216, 218, 305, 314
NanoDrop spectrophotometer.............................. 379, 414
NanoOrange...339, 343, 346, 350
Nascent polypeptide 302
NBT. *See* Nitroblue tetrazolium
NBT/BCIP....................52, 56, 59, 61, 62, 67, 84–85, 87, 93, 95
ND. *See* NADH dehydrogenase
Nerve growth factor (NGF) 339, 343, 346, 349, 350
Neurobasal medium..5, 9, 111
Neurofilament 126
Neurons............................. 9, 11, 53, 103–122, 125–127, 335–338, 340–344, 348, 349, 353, 354, 468. *See also* Axons
    cell bodies................................ 105, 106, 125, 126, 135, 336–338, 341–343
    cytoplasm.. 125
    projections 126, 127
Neurotrophins 342, 343, 345, 349
    coupling to microparticles 343
    local application 343
NeutrAvidin 163, 165, 169–171
NGF. *See* Nerve growth factor
NHS-monoester.. 379
Nitroblue tetrazolium (NBT) 84–85
Nitrocellulose membranes 290, 307, 310, 317, 358, 363, 364, 419
2-(N-morpholino)ethanesulfonic acid (Mes) 327
N,N-dimethylformamide (DMF)....................5, 7, 56, 106, 112, 144, 148, 155
Non-coding RNA. *See* RNA
Nonidet P–40 (NP40)............................227, 303–305, 308, 311, 357, 391–393, 408
Northern blotting 197, 306–308, 310–312, 314–317
Nuclear export......................................203–204, 387–388
Nuclear localization 165, 215, 272
Nuclear localization signal (NLS)..................215, 224, 225, 228, 267
Nuclease-free water 54, 60, 106, 144–145, 293, 306, 315, 319
Nucleic acid 83, 97, 98, 108, 112, 126, 134, 142, 150, 151, 159–160, 204, 315–316, 335, 378, 423
    probe.. 176, 203
Nucleotide........................3, 33, 42, 78, 142, 146, 147, 152, 154, 204, 208, 224, 225, 252, 268, 271, 293, 454, 458, 460, 462
    sequence..........................180, 222, 238, 267, 268, 452, 457, 468
Nycodenz...325–327, 329, 332

O

OD. *See* Optical density
Oleate/oleic acid 326–328, 331
Olfactory receptors 126
Olfactory tract 126
Oligo dT. *See* Poly(dT) oligonucleotides
Oligonucleotide 3, 4, 6–8, 12, 15, 19, 20, 24,
    25, 27, 54, 104–109, 111–115, 120, 142, 143,
    145–149, 151, 159–161, 182, 189–191, 208,
    209, 225, 227, 266–267, 306, 310, 316,
    427–430, 432, 434, 437–443
    amino-modification 111, 208, 429, 437–438
    design 7, 19, 111, 112, 120, 143, 204,
        293, 435
    digoxigenin (DIG) labeling 27, 32, 33, 91,
        104–109, 111–115
    fluorophore labeling 7, 24, 104, 106, 108, 109,
        111–113, 115, 147–149, 158, 267, 324, 442
    in situ hybridization (ISH) 91, 204
    labeling 4, 19, 24, 106–108, 111–113, 438
    purification 4, 8, 24, 112, 148, 149
2'-*O*-methyl RNA 142–144, 147–152, 154–155,
    161, 162
Oocyte 50, 54, 58–61, 63, 71–81, 125,
    152, 154, 265–266, 268–271, 273, 275–281,
    353, 457, 458
Open reading frame (ORF) 121, 213, 215, 224,
    238, 239, 241, 243–244, 246, 330
Optical density (OD) 150, 194, 209, 210, 216,
    217, 229, 230, 233, 242–246, 290, 291, 328,
    362, 436, 440
Optical slices 7, 11, 136
ORF. *See* Open reading frame
Oxalyticase 210, 216
Oxidation 147, 326
Oxidative stress 18

P

PABP. *See* Poly-A binding protein
PAGE. *See* Polyacrylamide gel electrophoresis
Parafilm 10, 25, 26, 66, 67, 113–114, 120, 208,
    211–213, 256, 314, 319, 357, 360, 361
Paraformaldehyde (PFA) 19, 23, 33, 55, 86, 89, 90,
    92, 95, 97, 108, 109, 113–115, 117, 119, 120,
    128, 132, 207, 210, 241, 245, 252, 256, 313,
    358, 359, 365, 429, 440
    for fixation 19, 97, 429, 440
PARPs. *See* Periaxoplasmic ribosomal plaques
Pasteur pipettes, glass 99, 341, 347
PCR. *See* Polymerase chain reaction
PEG. *See* Polyethylene glycol
Penicillin-Streptomycin solution 73, 340, 356, 373
Pepstatin 179, 207, 210, 227, 327, 391
Peptone 206, 226, 240, 289, 327, 429

Periaxoplasmic ribosomal domains 127
Periaxoplasmic ribosomal plaques (PARPs) 135
    markers 134
Peripheral nervous system (PNS) 126, 336–338
Peroxisomes 323–332
Pex30 .... 325, 326
PFA. *See* Paraformaldehyde
Phalloidin, Alexa fluor 568, 358, 360
Phallusia mammillata 50, 54
Phase contrast 134, 328–329
Phenol:chloroform extraction 97, 129, 378,
    401, 412
Phenotype 276, 431
Phenylalanine 232, 240, 326
Phenylmethanesulphonylfluoride (PMSF) 179, 207,
    289, 327, 357, 364, 391, 430
1-Phenyl–2-thiourea (PTU) 89, 98
Phosphoramidite 144, 147
Phosphoribosyl-aminoimidazole 217
Photostability 7, 274
Phototoxicity 216, 218, 253
Pigmentation 89, 98, 99, 217
Pipette tip 7, 9, 25, 61, 148, 230, 381
piRNA 143
Pixel 11, 136, 279, 294
Plant 22, 84, 177, 323, 353, 371, 468, 477
Plasma membrane 59, 64, 160, 167, 289, 290,
    303, 308, 309, 326, 331
Plasmid 36, 44, 97, 109, 116, 178, 182–184, 189,
    191–195, 197, 198, 216, 218, 224, 225,
    227–229, 232, 238, 239, 242–245, 251, 253,
    255, 391, 397, 403, 428, 432–435, 442
    linearization 36, 60, 74, 78, 88, 96, 97, 116,
        121, 130, 228, 232
    MS2-CP-GFP 239, 241, 245, 251, 253, 258,
        395, 396
    MS2-CP-GFP-SBP 391, 392, 394–396
    pACYCDuet1 193
    pETDuet 1 193
    pG14-MS2-CP-GFP 215
    pSL24MS2 .... 251, 254
    pSL-MS2–6 plasmid 215
    purification 225
    YCP111-MS2-CP-GFP-ΔNLS 215
    YCP111-MS2-CP-GFP 215
PLIER16 algorithm 415
PMSF. *See* Phenylmethanesulphonylfluoride
PNS. *See* Peripheral nervous system; Post-nuclear
    supernatant
P-num 146, 147
Poly-A binding protein (PABP) 371, 411, 413, 414
Polyacrylamide gel electrophoresis (PAGE) 20, 24–25,
    234, 238, 290, 291, 307, 317–319, 328, 331,
    357–358, 362–363, 377, 393, 399–400,
    409–410, 412, 419, 439

acrylamide/bis-acrylamide mix 20, 24, 357, 362–363
   primer purification 238, 434
PolyA+ transcripts 203–204
Polycarbonate filters. *See* Filters
Poly-d-lysine 5, 9, 318
Polyethylene glycol (PEG) 169, 170, 226, 229, 241, 243, 327, 329, 391
Polyethylene glycol (PEG) 1500, 327
Polyethylene terephthalate (PET)
   inserts 340
      coating for cell culture 318, 337, 339
      size determination 339
      membrane filters 357, 358
Poly-l-lysine 20, 23, 111, 207, 209–211, 339, 340
Polymerase chain reaction (PCR) 4, 34, 36–37, 39, 40, 45–46, 97, 130, 150, 183, 215, 225, 227, 228, 232, 237–246, 324, 326, 330–332, 339–340, 342, 345, 346, 401, 402, 428–429, 432–434, 442
Poly(dT) oligonucleotides 203
Polysomes 302, 303, 306–312, 373, 388, 408
Polysomes, Polysome gradient 310, 314
POPO 128, 132, 134
Post-nuclear supernatant (PNS) 326, 329–332
Post-synaptic density 126, 336
Post-synthesis protocol 147–148
Post-transcriptional 369
   regulation 370, 407, 448, 467–478
      mRNA translation 103, 273
      RNA decay 467
      RNA localization 369–370, 372, 467
      RNA stability 273
      RNA transport 273, 467
      sub-cellular localization 42, 336
Potter-Elvehjem small clearance teflon pestle 329
P-Phenylenediamine 206
Precipitation 37, 45, 52, 53, 59, 61–62, 66, 67, 74, 78, 91, 97, 130, 134, 179–181, 183, 238, 239, 242, 297, 319, 349, 370, 377, 378, 382, 393, 399, 401, 409, 418, 440, 442
Pre-hybridization 21, 39, 40, 86, 90, 113, 114, 117, 118, 256, 374, 379
Primary antibody 21, 22, 26, 28, 36, 44, 46, 47, 57, 63, 65, 76, 80, 94–96, 114, 305, 307, 310, 314, 317, 358, 364, 410, 420
Primers 36, 97, 179, 183–185, 191, 225, 227, 232, 238, 243–244, 246, 253, 256, 330, 339, 346, 347, 378, 402, 413, 414, 419, 432–434
   design 238, 241
   random hexamers 394, 402
Probabilistic models 454, 457, 462, 470, 472, 473
Probabilistic motif prediction 469

Probes .... 3, 15, 31, 52, 71, 83, 103, 126, 141, 159, 176, 203, 252, 266, 291, 310, 324, 349, 358, 370, 414, 424–425. *See also* RNA probes
   accessibility 46, 99, 100, 115, 126
   coupled probes 8, 32, 45
   delivery
      microinjection 163, 164
      microporation 163, 165
      toxin-based cell permeabilization 161, 162
      transfection 160
   intensity value 414
   preparation 32–33, 36–37, 128–130, 208–209, 211, 217
   RNA probes 3, 31, 32, 36, 37, 54–55, 59, 60, 66, 84, 99, 107, 116, 176, 177, 324, 370, 379
   single label probes 85, 208
Proline 240, 326
ProLong Antifade 129, 133, 430
Promoters
   ADH1 .... 223, 224
   MET25 .... 215, 246, 390
   T7/lacO 192, 193
   T7/tetO 260
Pronase E 86, 99
Prostate cancer 356, 414
Protease inhibitor 233, 304, 305, 327, 329, 392, 408, 409, 417, 429, 436
Protein A 373–376, 408–411, 418, 419
Protein A/G beads 373–376, 409–411, 418, 419
Proteinase K 34
   buffer 382, 408, 412
   digestion 46, 408, 419
Protein synthesis 126, 193–194, 196, 287, 288, 302, 336, 338, 354
   machinery 126
Protein targeting 18, 98, 288, 381, 403, 413, 420
Proteome 301, 407
Proteomics 336, 355
Protocol
   hybridization 90–91
   mounting 92
   preparation of embryos
      dechorionization 89
      dehydration 89
      fixation 89
      permeabilization 90
      prehybridization 90
      rehydration 89
      washing 89–90
   RNA-probe hybrid detection
      antibody-binding 91
      blocking 91
      washing 91
Proton sponge 160–161

Pseudopod (PS) fraction 355, 359–361, 363–365
Pseudopodia 337, 338, 347, 353, 354, 356, 357, 359–362, 365
PS fraction. *See* Pseudopod fraction
PUF proteins 288, 372
PUM-HD. *See* Pumilio homology domain
Pumilio 185
PUMILIO1 176
Pumilio homology domain (PUM-HD). 176–182, 184, 185
p-value 415–417, 473–475, 477, 478
Pyrenoid 16, 17, 22

Q

Qβ 461
QIAshredder homogenizer 357
qRT-PCR. *See* Quantitative reverse-transcription PCR
QuantaMaster 145
Quantigene viewRNA 27
Quantile normalization 416
Quantitative reverse-transcription PCR (qRT-PCR) 296, 349
real-time PCR instrument 150, 348
Quencher........................ 142, 143, 147, 155, 159–161, 173, 266, 267
black hole quencher 1 (BHQ1) 143
black hole quencher 2 (BHQ2) 143, 152, 155
dabcyl 143, 152, 155

R

Rabbit 21, 58, 127–128, 131, 132, 135, 375, 410, 418
Raffinose 226, 229, 230, 233
Random hexamer primers 394
RaPID. *See* RNA-binding protein purification and identification
Rat...9, 11, 43, 106, 127, 132, 135, 136, 348, 349
Ratiometric imaging........................ 161, 163–165, 169–171
RBP. *See* RNA-binding protein
Receiver operating characteristic (ROC) curve 472
Recombination 54–55, 60, 161, 176, 215, 216, 218, 238, 239, 241, 244–246, 274, 276, 304, 324, 371, 408
Reconstitution 56–58, 97, 129, 176, 177, 186
RedStar fluorescent protein 215
Replica plating... 239
Restriction enzymes 60, 96, 110, 116, 121, 128, 129, 191, 228, 251, 428, 429, 434, 435
Retention time 8, 148, 150
Reverse transcription 180–181, 184, 289, 293, 324, 328, 339–340, 342, 345, 346, 350, 401, 402

Reverse transcription polymerase chain reaction (RT-PCR) 180–181, 183, 197, 253, 256, 296, 315, 324–326, 328, 330–331, 337, 338, 342, 349, 402, 403, 413–415, 419
Rho/ROCK signaling 354
RiboGreen 339, 342, 346
Ribonuclease H 142
Ribonucleic acid (RNA) 3, 31, 53, 71, 84, 104, 126, 141, 159, 175, 189, 203, 221, 237, 250, 266, 287, 302, 324, 335, 353, 369, 387, 407, 423, 447, 467
buffer 34, 64, 206–207, 345, 357
co-purification 408
decay 320, 369–370, 388, 467, 468
dynamics 16, 141–142, 189, 206, 218, 249–251, 266, 275, 277, 280, 288, 347, 454, 460–461
expression 52, 54, 58, 83–100, 103, 141, 152, 175, 192, 197, 211–213, 216, 223, 224, 231, 233, 246, 249, 254, 260, 273, 388, 467
fold................... 144, 145, 427, 451, 454, 455, 469, 477
granules 52, 104, 105, 107, 238, 239, 245, 246, 276, 354, 395
isolation 315, 323–332, 339, 342, 345, 355, 357, 362, 371, 374, 375, 378–379, 388, 389, 396, 399, 401, 402, 410, 437
kinetics 151, 191, 196, 197, 250, 451, 454
labeling mix 37, 54–55, 60, 85, 88, 128, 130
microRNA biogenesis 468, 477
non-coding RNA 42, 43, 197, 370, 388, 416, 474
normalization 339, 346, 350
plot 150–151, 469, 475, 476
polymerase 32, 33, 37, 54–55, 60, 74, 88, 89, 96–97, 109, 116, 128, 130, 134, 193, 378–379, 442
polymerase II 250
polymerase priming site 85
precipitation 45, 59, 74, 78, 378, 382, 409
pre-microRNA 468, 474–477
pri-microRNA 474, 476–477
processing 4, 268, 431
purification 89, 328, 330, 353–365, 389, 393, 412–414
purity 335, 345, 347, 349
quantification 339, 342, 346
recovery 402
5S rRNA 198
tagging 176, 190, 191, 193–194, 215, 225, 259, 389
transcription 72, 94, 246, 249–250, 416, 442
translation 467
unaligned 459, 468–471
visualization 189, 192
zipcodes (*see* RNA localization)

Ribonucleoprotein (RNP) 126, 194, 221, 222, 224–231, 233, 234, 270, 274, 275, 278, 370–372, 375, 377, 380–382, 387–404, 408, 424–426, 428, 436
    particles 126, 270, 274, 278
Ribonucleoprotein (RNP) complex purification............ 225
    calmodulin elution buffer 227, 231
    CBB... 227, 231
    cell growth and harvest 229–230
    cell homogenization... 230
    elution buffer 227, 234, 382
    extract buffer 226, 227, 230, 233
    HCB..227, 230, 231
    TAP purification.. 224–231
    TCB 227, 231
Ribonucleotide 142, 150, 151
Riboprobe. See RNA probes
Ribosomal RNA (rRNAs)...................... 126, 197, 310, 350, 377, 378
Ribosomes............................. 53, 65, 67, 134, 136, 288, 297, 302, 310–312, 335, 371, 380, 424, 426
Ribotrap.. 389
RIP. See RNA-binding protein immunoprecipitation
RIP-Chip. See RNA-binding protein immunoprecipitation-microarray profiling
RNA. See Ribonucleic acid
RNA aptamer................................. 189–190, 193–194, 197, 198, 224, 237–246, 387–404, 423–443
    D8.. 389
    MS2................................. 176, 204, 213–215, 239, 246, 250–251, 254–257, 259–262, 265–281, 389, 461
        aptamer-tagged mRNA 389, 391, 392, 402
        bacteriophage..................... 204, 222, 249, 268, 274, 324, 389
    MS2 coat protein (MS2-CP)....................204–205, 222, 238, 274, 324, 389, 461
    MS2-CP fused streptavidin binding protein (MS2-CP-GFP-SBP) 389–390, 396
    MS2-CP-GFP plasmid..............................389–392, 394
    MS2 stem-loop repeats
        2xMS2 213
        6xMS2 ....213, 215, 259
        12xMS2 ....213, 239, 259
        24xMS2 .... 215, 249, 251, 253–255, 261
        sequence...205, 250–251, 271
    S1 RNA aptamer 389, 424, 428, 430, 431
    U1A aptamer 222–224, 227, 231
        location and number on RNA construct........... 224, 227, 228
        PCR-based preparation 228, 232
        RNA-binding construct (U1Ap-GFP)...... 224–225
        RNA construct (U1A$_{tag}$-RNA)......................... 224
        RNA-protein interaction 222
        RNA visualization 222, 224

RNP complex purification (see Ribonucleoprotein complex purification)
    sequence..............................222–224, 227–228, 231
    subcloning-based preparation 225–226
    U1A pre-mRNA 222, 224, 231
    U1A protein...222, 244–245
RNA-binding protein (RBP)................................ 134, 190, 204–206, 221–223, 267, 288, 296, 369–383, 389, 407–408, 410, 411, 415–417, 419, 458, 460, 461, 467–468
    Irp1 467
    Irp2 467
    PUF... 288, 372
    Sbp2.. 468
    She2 ....222, 225, 226, 371, 389, 460, 468
    She3............................... 222, 225, 226, 371, 389
    ZBP-1..134, 136
RNA-binding protein immunoprecipitation (RIP).. 407–420
RNA-binding protein immunoprecipitation-microarray profiling (RIP-Chip) 369–383, 407–408, 414
RNA-binding protein purification and identification (RaPID).. 387–404
RNA contraFOLD................................451, 454, 469, 472
RNA-folding software... 142
RNAi. See RNA interference
RNA interference (RNAi)................................ 21–22, 302
RNA localization................................... 43, 71, 72, 77, 212, 223–225, 230, 237, 301, 335, 369–370, 372, 408, 423–443
    elements... 213, 221
    motifs...347, 449, 457, 460
    zipcodes 213, 221
RNA Oligowalk 146, 147
RNA probes........................... 3, 31–33, 36–37, 45, 54–55, 58–60, 64–66, 84, 85, 88, 90, 99, 104, 107, 115, 116, 133, 152, 176, 177, 185, 324, 370, 379. See also Synthesis of labeled RNA probes; Tagged riboprobes
    alkaline size reduction... 116
    antisense probes 43, 52, 53, 60, 96, 97, 135, 154, 425, 427, 437, 443
    biotin labeling..42, 44, 66
    control sense probes...52, 96, 97
    cross-hybridization 90
    digoxigenin labeling.........32, 60, 91, 104, 105, 115, 116
    fluorescein labelling 91–94
    in vitro transcription 58, 66, 74–75
    preparation for FISH.. 36
    probe design 208
    radioactive labeling 83
    RNA probe-antibody complex
        fixation.. 89
        pre-staining...87, 88, 91

RNA probes (*Cont.*)
    stop reaction 89, 90, 92, 99
    visualization 189, 192, 239, 241, 245
    washing...45, 381, 382
    specificity...........................19–20, 68, 121, 152, 176, 181,
        185, 204, 208, 355–356, 371, 443, 449
RNApromo 452, 457, 468–472, 474–478
RNA-protein complexes......... 250, 393, 398–399, 423–443
RNA-protein interactions 141, 171, 222, 380,
    390, 396, 449, 454–455, 461
RNaqueous-micro kit 339, 342
RNA-recognition motifs (RRM) 458
RNase........5, 6, 32, 33, 36, 37, 44, 55, 57, 64, 66, 68, 72, 74,
    79, 85, 87–89, 96, 106, 110, 118, 128, 130, 162,
    206, 207, 210, 212, 233, 292, 303–305, 316, 319,
    328, 331, 342, 346, 349, 350, 362, 372–374, 378,
    380, 381, 389, 391, 392, 394, 402, 408, 409, 413,
    418, 424, 426–428, 431, 432, 436, 455, 477
    inhibitors 33, 37, 66, 85, 88, 207, 210,
        212, 380, 402
    RNase-free DNase................ 74, 89, 128, 162, 319, 394
RNAse-free BSA...6, 207, 212
RNAsin...................54–55, 60, 74, 128, 327, 329, 391, 392
RNA structure.........................115, 144, 146, 198, 427, 431,
    449, 451, 452, 455, 457, 461, 462, 467, 471
    bulges.. 198, 449
    consensus...176, 181–182, 458
    double-stranded regions 458
    internal loops..222, 448, 449
    locally stable secondary structure...............452, 458, 462
    minimum free energy....................... 146, 155, 451, 454,
        471, 474
    prediction...................427, 451, 452, 455, 461, 467–478
    primary structure 447, 448, 457–458, 460, 462, 468
    pseudoknots.............................. 448, 451, 455–456, 458
    secondary structure 293, 447–456, 458,
        459, 467–469, 472, 473
    similarity.. 415, 451, 452, 454,
        456–458, 470–471
    single sequence 451, 453–455
    single-stranded regions......................146, 449, 454, 458
    stem loops................................... 205–206, 213, 431, 448,
        460, 461, 477
    stems..........................147, 204, 431, 448, 449, 456, 459
    structural properties... 150
    structure alignment.....................................450, 455, 462
    suboptimal structure 145–146
    tertiary structure.............................145, 274, 447–449,
        453, 454, 458–461
    tetraloop 460–461
RNAsubopt 469, 471, 474–476
RNA, tRNA 6, 21, 25, 74, 86, 97, 108, 110,
    113, 119, 129, 133, 176, 198, 207, 211, 252,
    319, 374, 379, 393, 398, 426, 431, 436

RNP. *See* Ribonucleoprotein
ROC curve. *See* Receiver operating characteristic (ROC) curve
RPMI 1640 ....356, 360–361
RRM. *See* RNA-recognition motifs
rRNAs. *See* Ribosomal RNA
RT-PCR. *See* Reverse transcription polymerase chain reaction

S

Saccharomyces cerevisiae.................... 204, 240, 288, 371, 460
Saccharomyces genome database (SGD).......................... 296
    gene ontology term finder 296
    slim mapper tools... 296
Salmon sperm DNA......................... 21, 25, 108, 113, 129,
    133, 229, 251, 258, 391
    sonicated..34, 207, 211, 226
SAM. *See* Significance analysis of microarrays
SCFG. *See* Stochastic context free grammar
Sciatic nerve 126
SDS. *See* Sodium dodecyl sulfate
SDS-PAGE. *See* Polyacrylamide gel electrophoresis
Sealing wax 208
Secondary antibody 21, 22, 27, 36, 46, 47, 63,
    65, 76, 95, 119, 208, 305, 314, 318, 320, 358,
    364, 410, 420
Selection marker. *See* Yeast, selection marker
Selenocysteine 468
    selenocysteine insertion site (SECIS)
        elements... 468
    selenoprotein 468
Sephadex G25/G50 column..................... 72, 107, 112, 121,
    179, 182, 207, 208, 307, 317
Sequence
    alignment............273, 450–454, 456–460, 462, 470, 471
    identity 456, 459, 461
    motifs........................ 452, 458–460, 468–470, 472, 473
    similarity.................................. 452, 456–458, 460, 462,
        469–471, 473–476
Sequencing................................ 204, 371, 414, 429, 433–435
Sequential detergent extraction 303–305, 308–311
Serine..232, 240, 326, 408
Signal
    hypothesis 302
    sequence 176, 302, 453, 475, 476
Signal P server... 296
Signal recognition particle (SRP).............43, 301–303, 455
    receptor 302
Signal-to-background ratio 26, 150–152, 198
Significance analysis of microarrays (SAM) 380, 383
Silver nitrate 144, 148
Silver staining 226, 233, 400
Simultaneous folding and alignment methods 454,
    459–460

SMD. *See* Stanford microarray database
Sodium acetate (NaOAc) 5, 7, 33, 37, 45, 108, 110, 112, 116, 144, 148, 150, 292, 306, 440
Sodium azide 21, 320, 374, 420
Sodium bicarbonate............................. 5, 7, 20, 24, 27, 144, 148, 161, 206–207, 293
Sodium chloride (NaCl)... 86, 304
Sodium deoxycholate (DOC).................303–305, 308, 309
Sodium dodecyl sulfate (SDS).................................. 20, 55, 56, 72, 179, 227, 289, 294, 304, 306, 307, 315–317, 319, 320, 327, 328, 331, 332, 357, 358, 363, 364, 374, 375, 379, 381, 382, 393, 408, 412
Sodium hydroxide (NaOH)..............................86, 119, 306
Sodium orthovanadate... 179
Somatic cell 50, 51, 53
Sonicated salmon sperm DNA. *See* Salmon sperm DNA
Sorbitol.................................. 206, 241, 245, 289–291, 296, 327, 429, 430, 440, 441
Spectral unmixing... 173
Spectrofluorometer.. 145, 151
Spectrophotometer.......................8, 98, 113, 164, 378, 379, 413–414, 440
Speed vacuum (SpeedVac).............. 112, 113, 208, 211, 399
Spermidine 72, 128, 134
Spheroplasts. *See* yeast
Spinal cord...................................... 127, 130–132, 134, 135
Spinal muscular atrophy 447
Spinning disc confocal microscopy......... 152, 154, 155, 278
Spinocerebellar ataxia 447
Splicing..................... 43, 204, 266, 268, 271, 272, 387, 417
Squid.. 449
SRP. *See* Signal recognition particle
SRP54... 302
Stanford microarray database (SMD)............................ 380
Staufen... 269, 449
Stem-loop. *See* RNA structure
Stochastic context free grammar (SCFG) 451, 452, 454, 457, 462, 469, 470
Streptactin-HRP conjugate 410
Streptavidin 34–35, 41–45, 66, 424–429, 436, 437
Streptavidin-binding protein 389
Streptavidin-conjugated beads.......................390, 393, 398
StreptoTag... 389
Stringency............................12, 68, 90, 120, 217, 307, 317, 320, 380–382, 389
Subcellular fractionation................................324–325, 347
Succinimidyl ester...7, 20, 154, 438
Sucrose density gradient... 290
Sulfonamides 150
Support vector machines (SVMs) 459
SVMs. *See* Support vector machines
Synaptosome(s) 336
Synthesis..................................3, 7, 32, 54–55, 60, 106, 126, 147–150, 159, 175, 190–196, 198, 250, 280, 287–288, 301, 323–325, 335–336, 338, 354, 369, 378–379, 388, 438, 448
  automated.. 147, 148
  column... 144, 147
  cycle... 147
  DNA...4, 143, 183, 293, 317, 339, 342, 346, 429, 435
  standard 143, 147–148
Synthesis of labeled RNA probes........................37, 54–55, 58–60. *See also* RNA probes
  in vitro synthesis 84, 88
  purification 60
  quality assessment..37, 89, 98
  storage... 35, 41
Synthesizer 394
DNA/RNA 144
System Gold.. 144

T

Tadpole..50, 52, 53
Tagged riboprobes .... 98, 115. *See also* RNA probes
  biotin......................................33, 35, 37, 42, 44, 66, 84, 85
  digoxigenin (DIG)............................32–35, 37, 42–45, 54, 58, 60, 64, 66, 84, 91–95, 98, 104, 105, 110, 111, 115, 116, 121, 128, 130
  fluorescein.......................................66–67, 84, 85, 91–94
  radioactive.. 83
Tandem affinity purification (TAP)......................224, 226, 227, 230–231
TAP. *See* Tandem affinity purification
Taq polymerase 242–244, 330
TEA. *See* Triethylammonium acetate
Template plasmid 238, 242–244, 435
5' Terminal modifier 147
Tetracycline 191, 392, 395–397, 403
Tetracycline repressor (TR) protein.............................. 396
Tetramethyl rhodamine iso-thiocyanate (TRITC).......... 21
TEV protease 231
Thermal cycler...145, 152, 382
Thermal denaturation profiles....................................... 152
Thermodynamic characteristics................................... 150
Thermodynamic energy models. *See* Minimum free energy
Thermodynamic structure prediction 146, 456, 459, 469, 471
Threonine...232, 240, 326
Tiling array...414, 416–417
Tiling analysis software (TAS) 416, 417
Time-lapse analysis 195–197
Tiny molecular beacon 141–156
Tissue culture plates. *See* 6-well plates
Total RNA isolation kit.. 357

Index

Trans-acting factors 222–224, 266, 268, 269, 448, 449
Transcription 31–33, 37, 60, 72, 85, 91, 116, 128, 175, 191, 193, 204, 246, 249, 250, 289, 293, 349, 369–370, 387, 389, 436, 442, 447
   site 13, 74, 213, 215, 253, 256, 257, 261, 262
Transcriptional fragments 414, 416–417
Transcriptome 125, 297, 303, 407
Transcripts 43, 83, 91, 92, 100, 116, 125, 134, 203, 204, 215, 221, 224, 231, 249–250, 267, 294, 295, 297, 342, 343, 346, 347, 372, 380, 388–390, 395, 416, 442, 448, 461, 477
Transfection 105, 160, 172, 176, 178, 183–187, 253–255, 258, 261, 262, 392, 395–398, 402, 403
Translation 50, 94, 103, 131, 175, 213, 237, 249, 259, 287, 288, 302, 336, 347, 355–356, 388, 407, 448, 457, 467–468
Translational machinery 127, 354
Transparent finger nail polish 6, 35, 41
T-Rex™ 390
Triethylammonium acetate (TEA) 5, 144
N,N,N,N-tetramethyl-ethylenediamine (TEMED) 357
Tris-HCl 56, 57, 72, 73, 87, 108–110, 127–129, 144, 145, 179, 226, 227, 240, 241, 305, 307, 327, 328, 357, 374, 382, 391–393, 409, 419
TRITC. *See* Tetramethyl rhodamine iso-thiocyanate
Triton X-100 .... 24, 27, 38, 73, 95, 96, 179, 207, 213, 256, 305, 313, 358
Trituration 339, 341, 347
Trityl-protected sulfhydryl 147, 148
Trizol 306, 310, 311, 313–316, 318, 319, 378
Trypsin 167, 168, 184, 251, 255, 258, 304, 360, 391
   solution 161, 163, 178, 356, 392, 396
Tryptophan 226, 240, 326
TSA. *See* Tyramide signal amplification
T-test 383
Tubulin 43, 44, 189, 273, 308–311, 318, 420
Tween 80, 327
Tweezer 38, 131, 134, 145, 208, 270, 277, 356–357, 360
Twilight zone 459, 460
Two-sample analysis 416
Tyramide signal amplification (TSA) 32, 35, 84–85, 98, 111, 115, 119
Tyramide working solution 119, 129, 133
Tyrosine 232, 240, 326

U

U1–70k 411
Ultra-pure water 19, 22, 327, 372–373, 379, 393, 394, 401, 402
Ultrasonic cell disruptor 393
Ultraviolet light (UV) 5, 20, 25, 112, 306, 310, 314, 315, 349, 371–372, 378, 379, 440
Untranslated regions (UTR) 452, 457, 467, 473
   3'UTR 96, 121, 213, 215, 222, 224, 237–239, 241, 246, 249, 254, 255, 257, 259, 261, 271, 273, 288, 302–303, 390, 417, 452, 458, 460, 468
   5'UTR 121, 213, 215, 223, 224, 259, 273, 452
Uracil 204, 226, 240, 244, 326, 448
U1 RNA 222
Urochordata 50
UTR. *See* Untranslated regions
UV. *See* Ultraviolet light
U3 (E3) zipcode 224

V

Valine 232, 240, 326
Vanadyl ribonucleoside complex (VRC) 6, 21, 26, 27, 108, 207, 210, 212, 227
Ventricle 130, 131
Venus 176, 178, 185, 274
ViennaRNA package 451, 452, 454, 469, 471
Velocity 145, 154–156
VRC. *See* Vanadyl ribonucleoside complex

W

Wavelength 5, 8, 20–22, 78, 113, 134, 144, 148–152, 171, 278, 279
6-Well plates 312, 337, 340, 343–345, 357–360
Western blot 312, 355, 359, 413, 418
Western blotting 256, 310, 314, 327, 363–365, 396–397, 399–400, 419–420
   blocking buffer (TBST) 358, 364, 409, 420
   one-hour 410, 413, 420
   ponceau red 409, 420
   PVDF 409, 419
   SDS-PAGE (*see* Polyacrylamide gel electrophoresis)
   tris-HCl gel 419
Whatman discs 23, 32, 36–37, 88, 306, 316, 317, 358, 360, 363, 392, 396
Whole mount axonal preparation. *See* Axons
Wilcoxon signed rank test 416
Wild-type strain 225, 240, 243–245, 427

X

Xenopus laevis 71, 75, 457

Y

Yeast 4, 33, 49, 74, 86, 129, 154, 191, 203, 221, 237, 259, 265, 288, 302, 324, 353, 371, 389, 424, 468

bud 215, 221, 222, 224, 225, 389
    localization 215, 303, 371
    tip 204, 212, 215, 221–223, 324, 460, 468
extract 55, 206, 226, 239–241, 246, 289,
    327, 389, 403, 429, 436, 437
mating-type 221–222
media 210, 217, 239, 240, 244, 296
    rich medium (YPD) 206, 226, 229,
    230, 233, 429, 436
    synthetic medium 243
plasmids 225, 229, 239, 391, 435, 442
    high-copy plasmids (2m) 224, 225, 227
    integrating plasmids 232
    low-copy plasmids (CEN/ARS) 198,
    224–225, 227
selection marker 238, 239, 244
    *ade*2 .... 217, 225, 233
    *his*3 .... 225, 227, 239, 240, 245
    *met*15 .... 240, 245–246
    SpHis5/HIS3 245–246
    *ura*3 .... 225, 240, 245–246
spheroplasts 209–213, 216, 217, 243, 244,
    291, 296, 325–329, 430, 441

transformation 225–226, 229, 232,
    238–243, 246, 391, 394
tRNA 74, 86, 129, 133, 319, 374, 379,
    393, 398, 426, 435–436
Yeast rich media (YEP) 206, 213
Yellow fluorescent protein (YFP) 176, 250,
    257, 260–262, 274, 279
YEP. *See* Yeast rich media
YFP. *See* Yellow fluorescent protein
YOYO 128, 132, 134–136

Z

ZBP-1. *See* RNA-binding protein
Zebrafish 84, 95, 98
    histological sections 83
    transgenic lines 54
    whole mount embryos 83–100
Zeocin 392, 396, 403
Zipcodes. *See* RNA localization
Z section 11, 355, 360
Zygotic expression 50, 53, 58
Zymolase 241, 243, 244, 326, 327, 331
Zymolyase 100T 207, 210, 216